Deep-Sea Biology

DEEP-SEA BIOLOGY
Developments and Perspectives

N. B. MARSHALL

Garland STPM Press
New York and London

15 14 13 12 11 10 9 8 7 6 5 4 3 2 1

Library of Congress Cataloging in Publication Data

Marshall, Norman Bertram.
  Deep-sea biology.
  Bibliography: p.
  Includes index.
  1. Marine biology. 2. Benthos. I. Title.
QH91.M29   574.92   79-25526
ISBN 0-8240-7228-6

First published in the UK in 1979
Blandford Press Ltd
Link House, West Street
Poole, Dorset BH15 1LL

First U.S. edition published by
Garland STPM Press 1980
136 Madison Avenue
New York, New York 10016

Printed in England

Contents

Preface and Acknowledgements   ix

Chapter 1   Introduction   1

Chapter 2   Deep-sea environments   10

    The deep-sea floor   11
    Oceanic waters: the pelagic environment   20
    Life zones and life forms   40

SECTION I   DEEP-SEA LIFE

Chapter 3   Oceanic plants   49

    Plant groups   50
    General features of distribution   56
    The euphotic zone   57
    Patterns of primary productivity   61

    Animal life in the deep sea   67

Chapter 4   Pelagic animals   68

    Micro-zooplankton   68
    The zooplankton   74
    Nekton (and micronekton)   105
    Benthopelagic animals   134

Chapter 5   Life on the deep-sea floor   143

    The infauna   143
    The epifauna   158
    The errant epifauna   189
    Benthic fishes   212

SECTION II *DEEP-SEA FOOD WEBS*

Introduction 219

Chapter 6 *Pelagic food webs* 221

Epipelagic food webs 221
Mesopelagic food webs 230
Bathypelagic food webs 253
Organization of midwater life 259

Chapter 7 *Food webs near the deep-sea floor* 270

The benthopelagic fauna 272
Food resources of benthopelagic fishes 274
Energetics and locomotion 278
Standing stocks 280

Chapter 8 *Food webs of the deep-sea floor* 283

Organic food sources 283
The fall of organic material 285
The size spectrum of organisms in benthic food webs 288
Hydrothermal 'oases' 306
Trench faunas 307
The rate and mode of life 310

SECTION III *LIFE FROM DAY TO DAY: PHYSIOLOGICAL ASPECTS*

Chapter 9 *Against gravity* 315

Oceanic plants 316
The buoyancy of planktonic animals 316
Buoyancy of the micronekton 323
Buoyancy of the nekton 324
Buoyancy of midwater fishes 324
Vertical migration and buoyancy 335
Buoyancy of benthopelagic animals 338
Buoyancy relations of benthic animals 339
The biological significance of neutral buoyancy 340

Chapter 10 *Bioluminescence: light from life* 342

Luminescence and living space 343
Light and life at midwater levels 344
Bioluminescence near and on the deep-sea floor 371

Chapter 11   *The sensory systems of deep-sea*
            animals   379

            Visual systems   380
            The chemical senses: smell and taste   402
            Lateral-line (distance-touch) senses
               in fishes and invertebrates   413
            Ears and sounds in fishes   422

Chapter 12   *Early life histories*   428

            The early life of midwater animals   429
            The early life of benthopelagic animals   455
            Early life histories in the benthos   458
            Early life and ecology   463

SECTION IV   *ASPECTS OF BIOGEOGRAPHY*

Chapter 13   *Ecological biogeography*   471

            Pelagic distributions   473
            Benthopelagic distributions   490
            Benthic distributions   492
            Variation, diversity and zoogeography   499

            Bibliography

            Index

Preface and acknowledgements

Though existing in the largest living space on earth, the forms, activities and distribution of deep-sea organisms are essentially similar over wide belts of the ocean. Such inherent simplicity and the intriguing nature of deep-sea life have encouraged me to write for all students of the oceans, both academic and otherwise. I have tried to be intelligible but not to write down. Moreover, I have not written for those awe-inspiring creatures the educated layman and the general reader.

Since the intent and extent of this book are covered in the introductory chapter, I am left with grateful remembrance of things past. My liking for marine biology grew during my research under Professor A. C. (now Sir Alister) Hardy, on the distribution and ecology of plankton in the north sea and adjacent oceanic waters. My interest in the systematics and biology of deep-sea fishes expanded during twenty-five years in the British Museum of Natural History. There was also stimulating collaboration both ashore and at sea, with colleagues in Denmark and the USA. More recently I have joined two biological cruises on RRS *Discovery*, thanks to the courtesy of The Institute of Oceanographic Sciences, Godalming, where I am continuing my investigations.

This book has been written during and after the time I was Professor and Head of the Department of Zoology and Comparative Physiology Queen Mary College, University of London. One gets more time to write and research as an Emeritus Professor and I have recently been granted facilities for doing so by the Department of Zoology of Cambridge University. My thanks are due to both institutions and also to the Plymouth Marine Laboratory, where I am enjoying my tenure as Ray Lankester Investigator. This book was completed during my visit as Visiting Scientist at the Bigelow Laboratory for Ocean Sciences, Boothbay Harbor, Maine. I much appreciated the liveliness and hospitality afforded me by this young laboratory.

Lastly, the loneliness of the long-distance writer was made readily bearable by the ever-present help and encouragement of my wife. Her outstanding gifts as an illustrator illuminate many pages; indeed the originals of the two colour plates in this book have hung in the Royal Academy Summer Exhibition in London, a tribute to the care and quality of her work.

N. B. M. 1979

Maps and diagrams by Helen Downton and D.W. Graphics.

Introduction 1

This book emerged from my attempts to revise *Aspects of Deep-Sea Biology*, published in 1954. Apart from a reluctance to use scissors and paste, deep-sea biology has grown remarkably over the past twenty-five years. Such developments are the concern of the present book, which is hardly large enough to cover all aspects of the subject. Thus, the design of new deep-sea gear is appreciated but not described. For those interested, there is an admirable review by Clarke (1977a). Concerning the treatment of deep-sea biogeography, the reader is referred to the beginning of Chapter 13. Even so, my aim has been to produce an integrated survey of deep-sea life. Ideas, including some of my own, are freely discussed.

The main features of the earth beneath the ocean and the physical nature of deep-sea waters are considered in the opening chapter, which ends with an introduction to deep-sea life zones and life forms. Recent discoveries that the floor of the deep ocean is very different from the continents are bound to interest deep-sea biologists. 'The mountains of the ocean are nothing like the Alps or the Rockies, which are largely built from folded sediments. There is a world encircling mountain range— the mid-ocean ridge—on the sea bottom, but it is built entirely of igneous rocks, of basalts that have emerged from the interior of the earth. Although the undersea mountains have a covering of sediments in many places, they are not made of sediments, they are not folded and they have not been compressed' (Bullard, 1969). On the undersea mountains live myriads of animals, particularly attached forms, which of all deep-sea organisms are least accessible to the biologist.

The deep-sea floor is geologically young and grows outwards from the rift in the mid-ocean ridge. Near parts of the continents the spreading tectonic plates have formed deep-sea trenches, which although the deepest parts of the ocean, are by no means necessarily the most devoid of life. Moreover, along such oceanic spreading centres may emerge hydrothermal plumes that form deep-sea 'oases', inhabited, inter alia, by giant beard-worms and clams. And what is the marine biogeographer to make of the vast unfolding of the Atlantic Ocean since early Mesozoic times? The controversies over continental drift are recently fossilized bones of contention.

The organismic nature of oceanic waters—the steady-state balance of water, salts and heat—should appeal to the biologist. Knowledge of the formation and circulation of the water masses, including their dissolved content of oxygen, is necessary in biogeography. The near-surface circulation of oceanic waters is one of the factors behind the pattern of oceanic productivity. Investigations on the penetration of light and the formation of light fields in the ocean are essential in studies of primary production, vertical migration and bioluminescent camouflage.

In the chapter on oceanic plants (the phytoplankton) the emphasis is on patterns of oceanic productivity in space and time, which, apart from their reflection in the standing stocks of deep-sea animals, are related to the broad features of animal biogeography, particularly at deep midwater and benthic levels. Moreover, though there is continual and often rapid change in the communities of phytoplankton, the dominant species, at least, fall into certain distributional patterns, some of which are like those of pelagic animals. There is also a relationship between the kinds of plants and the oceanic distribution of nutrient concentrations—from the diatoms of richer waters to the micro-flagellates and blue-green algae of impoverished surroundings.

The section on pelagic and bottom-dwelling animals is not simply an introduction to the main kinds of deep-sea forms. The two chapters lead to the rest of the book. As I say later, '... the intention is not only to review the main groups of deep-sea animals but to see how they are organized, both structurally and functionally, for life in the main divisions of the deep sea'. The stress is on functional morphology in relation to ecology. In the deep sea, perhaps more than in any other major part of the biosphere, one appreciates the aptness of both Cuvieran and Darwinian points of view. There are the relationships of structure to function and of both to living space, Cuvier's 'conditions of existence'. Concepts of natural selection, adaptation, adaptability, fitness, survival, competetive exclusion and so forth, are never far from mind. The two viewpoints overlap, but both are needed.

Survey of deep-sea animals reminds us that the inventory of living organisms is far from complete. There are thousands of unknown species in the deep sea, particularly among the very small animals (meiofauna) that live just below the surface of the sediments. Close study of the deepsea meiofauna began so recently that one is not surprised that there are a great many nematode worms and harpacticoid copepods—to take but two dominant groups—to be named and discovered. Even among deep-sea fishes, known by some 2,000 species, there are likely to be a thousand unknown kinds.

The discovery of new groups of deep-sea animals also enlarges our knowledge of the forms of life that are possible in the biosphere. Systematic and other studies of the beard-worms (Pogonophora), which live by absorbing dissolved organic substances, are confined largely to the past twenty-five years. Much more recently, two major groups of protozoans have been discovered on the deep-sea floor. The Xenophyophoria, which include the largest living protozoans, some with radiating pseudopodia up to 12 cm long, live at all depths from littoral to abyssal reaches. The Komokiacea, a new superfamily of foraminiferans, are widely distributed and abundant over the deep-sea floor. Photographs taken at depths of more than 8,000 metres in trenches in the southwest Pacific have revealed hemichordate worms that may well represent a new group and there are also unusual kinds of solitary sea-squirts (Fig. 1). Of the 'living fossils', the most exciting event was the discovery of living monoplacophoran molluscs during the Danish Deep-Sea *Galathea* Expedition (1950–52).

There need be no apology for devoting a considerable section of this book to the animals (in the terms outlined above). The closer one gets to the animals the better one's understanding of deep-sea biology. The point is well made by Hutchinson (1975), who writes: 'Modern biological education ... may let us down as ecologists if it does not insist, and it still shows too few signs of insistence, that a wide and quite deep understanding of organisms, past and present, is as basic a requirement as anything else in ecological education. It may best be self-taught, but how often is this difficult process made harder by a misplaced emphasis on a quite specious modernity?'

The integration of deep-sea life in food webs is considered before the section on everyday physiological activities. Through food webs organisms form ecosystems. In the pelagic sphere complete ecosystems (producers, consumers and decomposers) exist only in the epipelagic zone, which is also the night-time feeding ground of many mesopelagic animals. Close study of daily vertical migrations is essential to understanding the position in the water column and feeding strategies of mesopelagic animals from copepods to fishes. Through such concepts as ecological niche and competitive exclusion one begins to appreciate how migrators and non-migrators fit together in the water column (species-packing). At the same time, of course, one needs to know the range of food organisms taken (are the consumers opportunists or selective feeders?) and when they feed. But there are no migrators in the sunless bathypelagic zone, which is the largest and most deserted of all major living spaces in the ocean. After considering trophic strategies from copepods to fishes and the closer carnivorous nature of food webs at deep midwater levels,

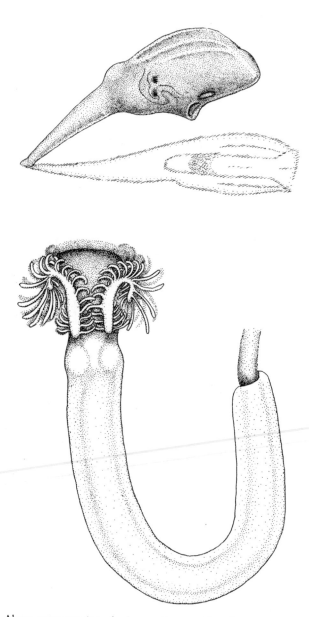

Fig. 1   Above, reconstruction of a large solitary ascidian (length 25 cm) as seen in bottom photographs at depths of 7,057 to 7,075 m in the New Britain Trench.

Below, reconstruction of a hemichordate representing a hypothetical new group of lophenteropneusts as seen in bottom photographs at depths beyond 6,750 m in trenches of the Southwest Pacific. (Redrawn from Lemche et al., 1976)

my concern is with the simple organization of the fishes in relation to their food-poor surroundings. The conclusion is that entry into the largest and most deserted life zone on earth has demanded a drastic loss of complexity.

Consideration of food webs near the deep-sea floor focusses attention on the existence of a benthopelagic fauna. My attention was first drawn through study of the systematics and biology of the rat-tailed fishes (Macrouridae), a dominant group in this fauna. It is now clear that the fauna ranges in size from copepods to large cephalopods and fishes. Of the fishes, the most versatile feeders, such as the rat-tails, are the most successful. But weight for weight the food requirements of a cod are likely to be about twenty times those of a rat-tail.

On the deep-sea floor recent *in situ* measurements of benthic respiration suggest that life proceeds very slowly by comparison with inshore communities of benthic organisms. Indeed, one wonders how the deeper dwelling forms, especially under the poorly productive central waters of subtropical regions, manage to make any sort of living. But recent investigations indicate that it is the relatively swift fall of faecal material (particularly from copepods) to the deep-sea floor, rather than the slow 'perpetual snowfall' of organic flakes, that provides most of the energy of the deep-sea benthos. Moreover, carnivores such as sea-stars and sea-anemones are restricted to the richer (eutrophic) parts of the ocean. Small forms of suspension-feeders and deposit-feeders live also under the food-poor (oligotrophic) regions. The chapter ends with consideration of two special benthic environments, hydrothermal 'oases' and trenches, which are not only interesting in themselves but also through comparison and contrast throw light on 'conditions of existence' over the main expanses of the deep-sea floor.

The section on the physiological aspects of life from day to day begins with a chapter on the buoyancy of oceanic organisms. The density of living substance is little more than that of seawater, which makes the lack of special buoyancy devices understandable in such active muscular animals as arrow-worms, copepods, euphausiid shrimps and sergestid prawns. The same is true of diverse squids, while many midwater teleost fishes lack a gas-filled swimbladder. All such animals, being somewhat heavier than sea water, must swim upwards from time to time to keep their level in the water column. Even so, the fishes, especially those of bathypelagic levels, have such lightly ossified skeletons and reduced watery muscles that their positively buoyant components lift them close to neutral buoyancy.

At mesopelagic levels about three-quarters of the fishes (in numbers of individuals) have some kind of gas-filled swimbladder, which is lacking in all bathypelagic species. Near the bottom, from slope to abyssal levels, nearly all of the benthopelagic teleost fishes have a gas-filled swimbladder containing gas-retaining and gas-tension generating structures (retia mirabilia) that become longer with increasing depth ranges. The sharks and chimaeras have a large, buoyant liver, charged with squalene. The implications of these facts are considered.

The pelagic invertebrates with special buoyancy devices include gelatinous forms of zooplankton, cnidarian, ctenophoran and tunicate. Like certain 'pill-box' kinds of diatoms, these animals derive ionic uplift and neutral buoyancy through the partial exclusion of heavy sulphate ions from their body fluids. In diverse midwater squids and certain crustaceans neutral buoyancy is attained through the storage of ammonium ions in place of other, and collectively heavier, cations. The chapter ends with discussion of the biological significance of neutral buoyancy in the ocean.

'Perhaps the most striking feature of oceanic life is the capacity in so many different kinds of organisms to produce light of a phosphorescent quality. Bioluminescence, as this is called, occurs in many surface-living midwater and bottom-dwelling organisms, and is particularly highly developed in deep-sea fishes, squids and crustaceans. The simplest forms of light-producing life are found among the bacteria' (Marshall, 1954). The most exciting recent discovery is of the precise use of complex light organs in mesopelagic fishes, decapod crustaceans and squids to camouflage the user. By matching the blue wavelengths, intensity and light field of mesopelagic twilight, these animals become invisible to their predators. Camouflage through body coloration is also highly developed in many midwater invertebrates, some of which are not luminescent. All such forms have an overall or partly red coloration, a pigment cloak that reflects virtually no blue light, whether bioluminescent or solar. But certain predatory fishes bear large cheek photophores that produce red light. Bioluminescent camouflage may be unmasked by certain fishes and squids that see through yellow lenses. Light lures and sexual differences in light organ patterns are also considered. Less is known of the biological significance of luminescence near and on the deep-sea floor, but recent investigations show that many benthic invertebrates, particularly echinoderms, are light producers.

After bioluminescence it is natural to turn to visual organs in the chapter on the sensory systems of deep-sea animals. Fishes are considered first. Their optical adaptations to dim light are through large eyes with wide-open pupils and matching lenses: their retinal adaptations include long closely packed receptor units (rods), which may be grouped (as macro-receptors) or disposed in several banks, special reflective backing (tapeta) and, above all, a high density of golden light-absorbing pigments in the rods. Adaptations to enlarge the binocular field, and thus obtain a better 'fix' on nearby objects, involve the development of lensless (aphakic) spaces in the pupil, which may be associated with temporal foveae in the retina, and tubular eyes. In the latter the concentration of the eyes in one main binocular field is correlated with special retinal devices (diverticula), lens-pads or wave guides to cover light or lights outside the main field of vision. Below the reach of the sun both midwater and bottom-dwelling fishes, with the notable exception of dwarf free-living males of the bathypelagic angler-fishes, tend to have small, deficient eyes or very regressed eyes.

The eyes of cephalopods, though of individual design, parallel those of fishes in several striking respects. The same is true of the eyes of alciopid polychaete worms. Concerning the eyes of oceanic crustaceans, perhaps the most interesting are the structural, physiological and ecological contrasts between euphausiids with spherical eyes and those with bilobed eyes.

Turning to the chemical senses, the main emphasis in consideration of the midwater fishes is on sexually dimorphic olfactory systems (large in males and regressed in females), which are most highly developed in the bathypelagic fauna. Near the deep-sea floor many of the fishes, such as the rat-tails, use olfactory means to find their food, particularly the occasional carcasses that fall to the bottom. The taste-bud system of rat-tails is also considered. Among the invertebrates the most exciting recent discovery concerns the fine structure and functioning of the long antennae in sergestid and other midwater prawns. Scent trails, such as may be left by falling particles of food, are precisely followed.

Lateral line ('distance-touch') sensory systems are highly developed in many deep-sea fishes. The simplest, consisting only of free-ending organs, are found in most of the bathypelagic species. Many of the fishes from mesopelagic and bottom-dwelling levels have a mixed system of free-ending and canal-enclosed neuromasts. Special attention is given to the lateral line systems of rat-tails and certain benthic fishes (chlorophthalmids). But fishes are not the only oceanic animals with hair-bearing sensors to detect nearby vibrations in the medium. Such receptors are found also in arrow-worms, ctenophores and pelagic prawns.

Apart from the hagfishes, the inner ears of deep-sea (and other) fishes are well developed. The chapter closes with consideration of sound-producing mechanisms in benthopelagic fishes, in which the hearing (saccular) parts of the ears are large. Sound signals produced by the males may well play a leading part in sexual congress near the deep-sea floor.

The chapter on early life histories is concerned largely with reproductive strategies and larval ecology. In the section on midwater animals the early life history patterns in the fishes, which produce buoyant eggs, are contrasted with those in crustaceans, whose eggs sink after they are shed. The crustaceans produce either many small eggs, which hatch into simple larval stages, such as nauplii, or relatively few large eggs, which yield advanced kinds of larvae. In the latter species the start of trophic life is postponed until the young are large enough not to be restricted, as are nauplius larvae, to minute kinds of food organisms. Arrow-worms and cephalopods also produce eggs that hatch into advanced young. Nearly all the midwater fishes produce numerous buoyant eggs that presumably float upwards as they develop. Certainly the larval stages are passed in the productive surface waters, where the young feed and grow until the approach of metamorphosis, when they descend towards the adult habitat. Like many of the larger midwater crustaceans, the smaller species of fishes are evidently annual forms. At bathypelagic levels the life history may span two or more years, as in female ceratioid anglerfishes. But the dwarf (progenetic) males have a free-living existence of less than a year.

Of the benthopelagic fishes that swim near the deep-sea floor, some (eels, notacanths and halosaurs) produce eggs that give rise to pelagic leptocephalus larvae. Considering their diversity, wide distribution and abundance, little is known of the early life history of rat-tailed fishes (Macrouridae). It is possible that they spawn and spend their early life near the deep-sea floor, as do the benthic species, some of which are synchronous hermaphrodites.

Recent investigations suggest that many of the larger invertebrates of the deep-sea floor have no particular spawning season, which seems understandable below a surface layer with no marked seasonal changes in primary production and in physically seasonless surroundings. Some are continuous (trickle) reproducers: others reproduce asynchronously. But there are exceptions.

Concerning their early life history, it seems that most of the larger invertebrates of the deep-sea benthos lay relatively large eggs that hatch into larvae that live on their yolk (lecithotrophy) during a short pelagic (presumably near-bottom) existence. Fewer produce still larger eggs that develop directly into advanced young. Small, long-lived larvae that depend on minute planktonic food (planktotrophic larvae) are known in some species, but they are in a minority and mostly confined to the upper continental slope. The chapter closes with a discussion of larval types and their dispersal by currents.

The concluding chapter, entitled 'Aspects of Biogeography', is devoted largely to the ecological aspects of distribution in the deep sea. Investigations on the distribution of mesopelagic fishes in the Atlantic show that subarctic, temperate, subtropical and tropical regions may be subdivided into certain provinces, each characterized by an individual assemblage of species. These and other investigations show that the distributions of some midwater fishes cross the boundaries of water masses: at the other extreme there are species with headquarters within a particular water mass. Concerning the latter, there are dwarf species that live within the poorly productive central water masses, whereas their larger relatives are centred in richer peripheral waters. Few species are able to live in oxygen minimum layers, which apparently restrict the ranges of the majority, as in the Arabian Sea and the Bay of Bengal.

While distribution patterns in the mesopelagic zone are much like those in epipelagic waters, many of the larger bathypelagic animals tend to be centred under the eutrophic regions of the ocean. This, as we saw, is true also of the larger carnivorous invertebrates of the deep-sea floor. The larger kinds of deposit-feeders are also eutrophically disposed. Ecological factors in the zonation of benthic invertebrates are also discussed.

Lastly, in a section on variation, diversity and zoogeography, the main emphasis is on recent investigations of genetic variability in deep-sea animals. Genetic variability, like species diversity, increases from cold to warm regions and tends to be high in certain benthic animals of the deep sea. It seems that where biological interactions are complex, particularly in food webs, genetic variability is greatest. The section ends with discussion on genetic variability within the range of widely distributed species.

Introductions should be brief, which means that readers must be left to discover other recent developments in deep-sea biology in the chapters to come. At the end of the book I hope that the search will have been absorbing and stimulating. The deep sea is now closer to us than it used to be.

Deep-sea environments

'Contrary to what is often imagined, biology is not a unified science.' Marine biologists will surely agree with Jacob (1974), but whatever their interests, whether physiological, biochemical, ethological, ecological or biogeographical, they know they must endeavour to be more than biologists. Proper pursuance of their studies demands at least some familiarity with correlated knowledge about the physical nature of marine environments. As Pantin (1968) says, biological sciences are unrestricted'... and their investigator must be prepared to follow their problems into any other science whatsoever.'

'The cradle of life on earth is certainly the oceans' (Bernal, 1967), and throughout geological time sea water has been a benign medium for very diverse forms of plant and animal life. Since the beginning of Cambrian times, about 600 million years ago, Durham (1967) estimates, there have been at least four million and probably as many as ten million species of (preservable) marine organisms. In the modern ocean there are at least 160,000 species of animals, making them nearly twice as numerous as the non-insect forms of animal life on land. Moreover, certain forms, such as radiolarians, comb-jellyfishes, lamp-shells, arrow-worms, barnacles, cephalopods, beard-worms, echinoderms and tunicates live only in the sea. Foraminifera, sponges, cnidarians, bryozoans, sharks and rays are among those that are almost entirely marine (Chapters 4 and 5).

Like Wordsworth's rocks, stones and trees, the ocean is 'rolled round in earth's diurnal course'. Even more than rocks and stones the ocean endures. 'The obvious things that no one comments on are often the most remarkable; one of them is the constancy of the total volume of water through the ages. The level of the sea, of course, has varied from time to time. During the ice ages, when much water was locked up in ice sheets on the continent, the level of the sea was lower than it is at present and the continental shelves of Europe and North America were laid bare. Often the sea has advanced over the coastal plains, but never has it covered all the land or even most of it. The mechanism of this equilibrium is unknown; it might have been expected that water would be expelled gradually from the interior of the earth and that the seas would grow steadily, or that water would be dissociated into hydrogen and oxygen in the upper atmosphere and that the hydrogen would escape, leading to a gradual drying up of the seas. These things either do not happen or they balance each other' (Bullard, 1969).

Bullard writes above of the fully formed ocean that can be seen through the study of rocks and fossils. To trace the origin of all this water, or to appreciate why our planet might better be called Ocean rather than Earth, we must first realize that water, air and life (hydrosphere, atmosphere and biosphere) are made of elements that are not prevalent in the earth's crust or even in volcanic rocks that have come from deep within the earth (Turekian, 1968). We are led to study the elements in the entire Solar System and driven to conclude that water was not derived by condensation during the formation of the earth but came from subsequent 'degassing' inside the newly formed earth. Relics of this process are apparent today in the explosions of volcanoes, for water seems to be part and parcel of molten rock (magma). The expansion of water vapour in the fierce heat below a plugged volcanic vent may eventually lead to an awesome explosion of steam, rocks and lava. It seems clear that all or most of the ocean grew from the 'volcanic breath' of the earth's mantle (Dineley, 1973). Indeed, the volcanic bed of the ocean extends over some seven-tenths of the earth's surface. We turn first to the bed, then to the waters.

The deep-sea floor

The configuration of the deep-sea floor is revealed by scrutiny of soundings, and nearly all have been made in the past fifty years. Since the nineteen-twenties, when echo-sounders came into use, a vast and varied, yet orderly topography has been brought into view. Survey of this submarine landscape with echo-sounders (and other sound sources) has also helped geologists to reveal the volcanic nature of the crust that bears the ocean.

There are three fundamental parts of the ocean floor: the continental margins, the ocean basin floors and the mid-oceanic ridge system (Figs. 2–4). Over most of the earth the continental margin is capped by a continental shelf, which is the submerged perimeter of the continent, ever subject to marine erosion and sedimentation. The shelf cants gently seaward (gradient about 1 : 1,000) and at depths ranging from 20 to 550 metres (average 133 m) gives way to the steeper gradients (> 1 : 40) of the grandest escarpments on earth, the continental slopes.

The continental shelves, which are generally supported by sedimentary strata or by igneous and metamorphic rocks, have a mean width of 78 kilometres and underlie 7·5 per cent of the total expanse of the ocean. Excluding the littoral or intertidal waters, the seas over the shelves form the neritic province of the ocean: the rest is the oceanic province and also the province of this book. But before leaving the shallow seas, which include the coral seas, it must be said that these waters hold a greater diversity of life and are much more productive than the deep ocean. For instance, we know some 12,000 species of marine fishes and about 10,000 live in the shallow seas: the others are deep-sea species. Over nine-tenths of the world's annual catch of fishes (about 50 million tonnes) consists of species that depend on the most productive waters over the continental shelves. By comparison with the richest shelf waters, the deep ocean is a deserted living space.

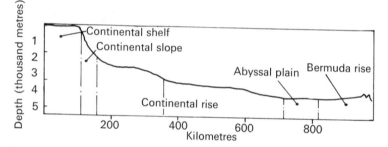

Fig. 2 Profile of the deep-sea floor off the northeastern United States. (Adapted from Heezen, Tharp and Ewing, 1959)

The continental slope descends to depths near 2,000 metres or more before merging with the gentler gradients (1 : 100–1 : 700) of the continental rise. Off the northeastern United States, for instance, the transition between continental shelf and slope is about 100 km off shore. At the foot of the continental slope, which is nearly 160 km off shore, the depth is nearly 2,000 metres. Towards the end of the continental rise, which is nearly 650 km from the shore, the depth is about 5,000 metress (see Fig. 2). This lowest reach of the continental margin averages about 4,000 metres in depth over the ocean. But over fringing areas of the western and northern Pacific Ocean, and off Chile and Peru, the continental margin descends into trenches with depths exceeding 6,000 metres.

Over most of the ocean the continental slopes and rises are covered with terrigenous or hemipelagic muds, mixtures of pelagic ooze and debris weathered from the continents and described as 'a sediment comprised of more than 30 per cent of continentally derived sand and silt grains; it is often green, black or slightly red, due to varying degrees of oxidation of the abundant organic matter'. More than 50 per cent of the particles forming this sediment are coarser than 30 microns in diameter, and the rate of sedimentation, which is appreciably higher than that of any other deep-sea sediment, varies between depths of 5 and 100 centimetres per thousand years (Heezen and Hollister, 1971).

Nearness to the land and hence the relatively rich shelf waters has much to do with the high content of organic matter in terrigenous muds, which in turn goes far to being the basis for the relatively great productivity of bottom-dwelling life over and on the continental margins. In particular, the upper reaches of the continental slope are not only near to the land and shelf, but also comparatively near to sources of organic material from dead pelagic organisms. It is not surprising, then, that slopes and upper rises support more abundant populations than most of the deeper parts of the ocean.

Abyssal plains, typically the most level stretches of the deep-sea floor, are also covered and indeed formed by land-derived sediments. They lie at the foot of continental rises and are widespread in the Atlantic and Indian Oceans and very extensive off the continental margin of Antarctica. (Gradients range between 1 : 1,000 and 1 : 10,000.) The sediments over these plains resemble the sands, silts and gravels recovered from submarine canyons. Submarine canyons, such as the Hudson Canyon, the Congo Canyon and the Ganges Canyon, fissure the continental slopes throughout the world. Most are only a few hundred metres deep but some cut 2,000 metres into the slope. All abyssal plains are connected by canyons or other channels to landward sources of sediment and down these fissures run fast turbidity currents charged with a great mass of sediment. This sediment, which contains considerable amounts of organic matter, spreads over the plain. Photographs taken over the Sohm Abyssal Plain, south of the Great Banks of Newfoundland '... revealed very muddy bottom water, and it seems likely that some of the finest grained clays have not settled out from the most recent turbidity current' (Heezen and Hollister, 1971). It is to be expected that some abyssal plains should be productive places for benthic life.

The ocean basin floors include the parts of the deep-sea floor between the continental margins and the major oceanic ridge systems. Apart from the deep-sea trenches the ocean basins form the deepest reaches of the ocean, mostly with depths of 4,000 metres or more. In the Atlantic Ocean, the main basins are the North American Basin (western North Atlantic), Canaries and Cape Verde Basins (eastern North Atlantic), Brazil and Argentine Basins (western South Atlantic), Angola, Cape and Agulhas Basins (eastern South Atlantic) and the Atlantic–Antarctic Basin (southern South Atlantic) (Fig. 4). More than half the area of these basins extends to depths of 5,000 metres or more, and much the same is true of the major basins of the Indian and Pacific Oceans. In the Northern

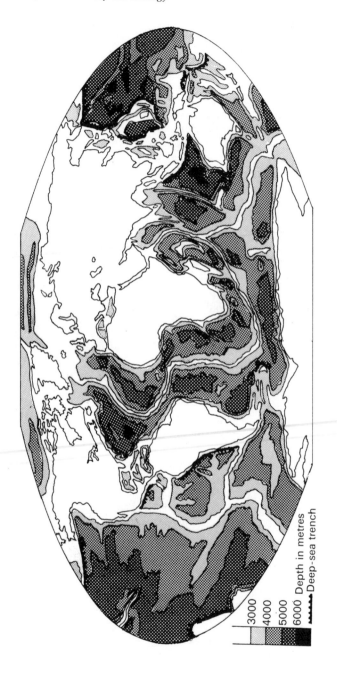

Fig. 3 General bathymetric chart of the ocean. (Adapted from Heezen and Hollister, 1971)

3000
4000
5000
6000 Depth in metres
Deep-sea trench

Pacific Ocean, which is without a mid-oceanic ridge system, there is one vast North Pacific Basin, extending from the deep-sea trenches in the west to the continental margin of North America. Of course, in culinary (and relative) terms, the ocean basins look more like saucers than basins (see Fig. 4), but their vast expanses are relieved here and there by islands and sea mounts, all the result of volcanic activity. 'Thousands of bountiful volcanic vents have poured out sufficient lava from the earth's mantle to build the massive underpinnings of oceanic islands, coral atolls and sea mounts. However, milllions of abyssal hills, knolls and peaks have been produced by other less ample vents which failed to produce enough to build to the surface of the sea before their activity was snuffed out' (Heezen and Hollister, 1971).

At depths of 5,000 metres or more the ocean basins are covered with the sediment known as red clay and at shallower reaches with calcareous oozes (Fig. 5). Red clay, which covers half the deep-sea floor of the Pacific and a quarter of each of the Atlantic and Indian Oceans, is: 'A chocolate-brown, extremely fine-grained sediment ... It is composed of clay minerals, residue from dissolved plankton shells, wind-borne continentally derived silt, volcanic particles and in high latitudes red clay contains an admixture of ice-rafted pebbles, rocks and sand. Accumulation rates are the lowest of any deep-sea sediment, ranging from one-tenth to one millimetre per thousand years' (Heezen and Hollister, 1971). Indeed, red clays, which contain very little organic material, are the deserts of the deep-sea floor. Life on red clay consists largely of small invertebrates and they are thinly spread.

The mid-ocean ridge system, which is a complex mountainous chain that winds round the world for a distance of 50,000 kilometres and forms a third of the deep-sea floor, is surely the most impressive major feature of our planet's physiography (Fig. 4). Here is the volcanic birthplace of the deep-sea floor, which is formed of vast tectonic plates that grow slowly outward from the mid-ocean ridge. The crest of the ridge system '... is cleft by a deep median rift valley. Deep gashes cut into the sea floor nearly at right angles to the crest and flanks of the ridge continue across the ocean basin floor until they are obscured beneath the sediments accumulated in the continental margins. Sea mounts and islands lie in straight rows along these fracture zones and together with the linear scarps and ridges, which are parallel to the crest of the Mid-Ocean Ridge, they impose a vivid fabric-like texture on the ocean floor' (Heezen and Hollister, 1971).

The ridge is between 100 and 4,000 kilometres wide and from its foothills to the crest depths range from about 5,000 to 2,000 metres. The lower flanks are covered with red clay, the higher mountainous regions with calcareous oozes containing the remains of foraminiferans and coccoliths. Actually, the valleys of the system are filled with ponds of

Fig. 4   Main features of the bottom topography of the ocean. (From Heezen and
Hollister, 1971) Mid-ocean ridges are cross-hatched about the broken line of
the rift-valley system.

16

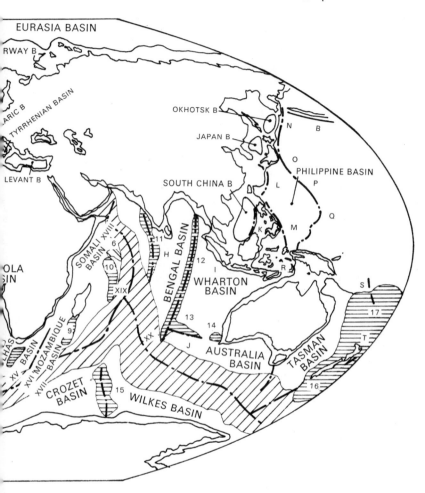

Plateaus:

1. Rockall Plateau
2. Azores Plateau
3. Sierra Leone Plateau
4. Rio Grande Plateau
5. Walfisch Ridge
6. Falkland Ridge
7. Agulhas Plateau
8. Mozambique Ridge
9. Madagascar Ridge
10. Mascarene Ridge

11. Chagos-Laccadive Plateau
12. Ninetyeast Ridge
13. Broken Ridge
14. Naturaliste Plateau
15. Kerguelen Plateau
16. New Zealand Plateau
17. Melanesia Plateau
18. Galapagos Plateau
19. Jamaica Plateau

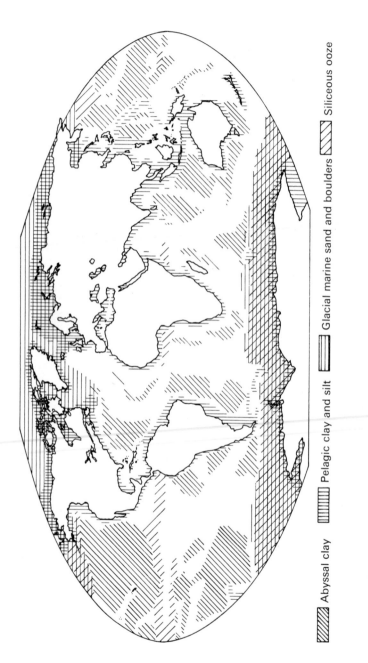

Fig. 5 Distribution of deep-sea sediments (the white areas represent calcareous oozes). (Adapted from Heezen and Hollister, 1971)

Abyssal clay   Pelagic clay and silt   Glacial marine sand and boulders   Siliceous ooze

sediment, which show clearly on the records of powerful, low-frequency echo-sounders. The peaks are just dusted with sediment. From ridge crest to ocean basin the covering of sediment gradually thickens.

Over its upper reaches, at least, there is presumably some variation in the rate of sedimentation, for the ridge system lies under waters of differing primary productivity (usually measured by the amount of carbon assimilated by the plant plankton under a square metre of the sea surface per day). Starting off east Greenland in the Jan Mayen area and proceeding through Iceland and along the mid-Atlantic to the latitude of Gibraltar, the ridge passes under waters of very high to high productivity.* Proceeding under the deep blue Sargasso Sea, where productivity is low, the ridge ranges under highly productive waters up to equatorial regions, but through most of the South Atlantic productivity falls to low values and the ocean is again deep blue. Level with Cape Horn the ridge turns east and then northeast into the Indian Ocean to about latitude 30°S. Over this stretch the ridge is under highly productive seas, but as the ridge turns north to the east of Malagasay there is a belt of low productivity. The ridge continues northwards to the equator then northwestward into the Gulf of Aden and the Red Sea. Over this stretch the waters increase in productivity from moderate to very high values.

Retracing our giant steps along this part of the Indian mid-oceanic ridge, the system continues southeastward and then eastward across the ocean until it arrives roughly midway between Australia and Antarctica. This expanse passes under moderately productive waters, then enters high regions between New Zealand and Antarctica. South of New Zealand the ridge continues eastward but soon turns east-northeast then northward into the South Pacific, where productivity is low. In the equatorial region the ridge is once more under more productive waters and over its northward continuation towards the Gulf of California the seas become even richer (see Fig. 16). The significance of this varying path of productivity in relation to animal life on the ridge has yet to be unravelled.

Even so, deep-sea cameras have shown biologists aspects of life on the ridge system that could never have been revealed by dredge or trawl. The volcanic heart of the ridge is seen in broken pillars of basalt along the rift valley, where hot lava is chilled by the sea. Here life is rare. The peaks of the ridge are covered with attached organisms, such as sponges, sea-fans, sea-pens and bryozoans. On the ponds of sediment in the valleys there are sea-cucumbers, brittle-stars, sea-urchins, bristle-worms,

---

*Very high productivity is an assimilation of more than 250 mg of $C/m^2/day$. The comparable figures for high productivity, moderate productivity and low productivity are 150–250, 100–150 and < 100, respectively.

acorn-worms, sea snails and crabs. Fishes, such as rat-tails, cruise over the ponds.

The deepest parts of the ocean are in trenches at depths below 6,000 metres. Most trenches have V-shaped sides and a level floor. Thirty-one have been listed, and most lie along the margins of the Pacific Ocean: others indent the borders of the northern Indian Ocean and in the Atlantic follow the outer loops of the Caribbean and Scotia Arcs (Figs. 3 and 4). Deep-sea trenches are elongated and narrow depressions with lengths ranging from a few hundred to over 2,000 kilometres and mean widths from 25 to 120 kilometres.

Starting from New Zealand and moving clockwise round the Pacific, the main trenches and their maximum depths (in metres in parentheses) are: the Kermadec Trench (10,047), Tonga Trench (10,800), New Hebrides Trench (9,165), Philippine Trench (10,497), Mariana Trench (11,524), Ryukyu Trench (7,507), Idzu-Bonin Trench (9,810), Japan Trench (8,412), Kurile–Kamchatka Trench (10,542), Aleutian Trench (7,679), Middle American Trench (6,662) and the Peru–Chile Trench (8,055). The main trenches of the Indian Ocean are the Chagos Trench (5,408), Java Trench (7,450), Amirante Trench (9,074), Mauritius Trench (5,564) and the Diamantina Trench (8,230). In the Atlantic lie the Puerto Rico Trench (8,385), Cayman Trench (7,930), Dominican Trench (ca. 6,200), Romanche Trench (7,856) and South Sandwich Trench (8,428). Most trenches, then, are parallel to island arcs or parts of the continents with coastal cordillera (Fig. 4). It is now evident that such trenches are formed along collision lines between tectonic plates.

Though the deepest parts of the ocean, trenches are not necessarily the most barren parts. The deepest levels are filled with land-derived sediments, which may be rich in organic material and several trenches underlie waters of high productivity. Invertebrate life may thus be relatively abundant, and indeed, diverse. Bruun called these animals the hadal fauna, for present exploration indicates that each trench contains a proportion of (endemic) species that have yet to be found elsewhere (see also pp. 307–9).

Oceanic waters: the pelagic environment
Water, salt and heat balance

Physical oceanographers study water, salt and heat balance in the ocean, and so do physiologists in living organisms. Both ocean and organisms tend to maintain steady states. One that is often taken for granted, the principle of conservation of volume, follows from the fact that water is no more than slightly compressible. Thus, the fact that water flows out of basins such as the Mediterranean and Red Sea with no change in sea level means that water must be flowing in at the same rate.

Elsewhere the conservation of volume has profound effects on the productivity of the ocean. For instance, in the Antarctic Ocean and off parts of the western sea board of the African and American continents, surface waters are displaced and water rich in nutrient salts moves up to fill their place. We shall see later that the upwelling leads to prolific crops of plant plankton (p. 64–5).

The sea gains water from the land, rain and melting ice and loses it through evaporation and freezing. Most of the earth's rain water comes from the very surface of the ocean (carrying much particulate matter) and much finds its way back to its source. Clearly, salinity is increased by evaporation and reduced by precipitation. Indeed, if the mean values of surface salinity and the difference between evaporation and precipitation (in cm) are plotted against latitude, there is a close parallel between the two curves. Salinity, which is the total amount of dissolved material in parts per thousand (‰) of sea water, is at a minimum (ca. 34·5‰) near the Equator, rises to a maximum (35·8) in about latitudes 20°N and 20°S and then decreases to 34·0‰ or less towards high latitudes.

Though the salinity varies, the proportion of the main ions in sea water is almost constant, which C. R. Dittmar discovered by making careful analyses of 77 water samples collected all over the world by HMS *Challenger*. Thus, percentage values are: chloride 55·26; sulphate 7·68; magnesium 3·76; calcium and strontium 1·20; potassium 1·12; and sodium 30·58. This uniformity of relative composition is due to constant mixing and circulation and it is vital to marine plants and animals which all have near-oceanic cell sap or body fluids. If any one of the main ions changed in proportion from place to place, marine organisms would need to be ever using energy reserves to maintain a steady inner state. Even so, some animals do contain concentrations of certain ions that differ considerably from those in sea water. For instance, relative to sea water, magnesium ions are lower in concentration in the body fluids of many crustaceans and chordates but higher in some bivalves, gastropods and cephalopods. Potassium ions, which are particularly vital for nervous and muscular systems, are concentrated by molluscs and crustaceans. In the body fluid of the squid, *Loligo*, the concentration of potassium is more than twice that of sea water.

There is also the principle of conservation of salt, which '... asserts that the total amount of dissolved salts in the ocean is constant. When one first learns that the rivers of the world contribute to the sea a total of about 3×10^{12} kg of dissolved solids per year, the conservation of salt seems to be contradicted. In principle it is, but in practice it is contradicted only to a negligible extent. The amount of salt in the oceans is about 5×10^{19} kg and therefore the amount brought in each year by the rivers increases the ocean salinity by about one part in three million per year. But we can only measure the salinity of sea water to an accuracy of about ± 0·003‰, or about 300 parts in three million taking the mean ocean salinity as 35‰. In other words, the oceans are increasing in salinity each year by an amount which is three hundred times smaller than our best accuracy of measurement. So for all practical purposes we can assume that the average salinity of the oceans is constant at least over periods of tens or even hundreds of years' (Pickard, 1963).

The surface of the sea covers 71 per cent of the earth and thus intercepts most of the sun's heat that falls on our planet. Even so, most of the ocean is cold, for though tropical latitudes gain much heat, cold waters, especially in polar regions, sink and fill the ocean basins.

Radiant energy from the sun reaches the earth directly or as radiation scattered by the atmosphere, particularly by clouds (sky radiation). At the sea surface, some of this energy is scattered: the rest enters the sea, is absorbed and raises the temperature of the water. As already intimated, the intensity of solar energy falling on a unit area of surface from a beam of given cross-section decreases from equator to poles (as the cosine of the latitude). At a given latitude the diurnal and seasonal range in the sun's elevation will have a differential radiant effect like that of latitudinal variation. Again, when the sun is directly overhead (90° elevation) a unit area of sea surface will receive more radiant heat from a given beam than at slanting elevations (and the amount decreases as the size of the angle of elevation). Concerning the reflection of direct radiant energy at the sea surface, the lower the elevation of the sun, and the more disturbed the water surface, the greater the amount of reflection. For sky radiation, which comes from all directions and varies mainly with cloud cover, about a tenth is reflected.

The sea surface loses heat through long-wave radiation to the atmosphere, through heat lost by evaporation and through heat conducted away from the surface, either into the water or out of the water. If these three loss quantities are Q_b, Q_e and Q_h respectively and Q_s is the rate of flow of solar energy through the sea surface, then the heat balance is $Q_s = Q_b + Q_e + Q_h$.[*]

The outstanding thermal feature of the ocean, as already hinted, is that there is an upper warm layer of water, mainly in tropical and subtropical latitudes, floating, as it were, on a cold ocean. Anton Bruun, who believed the isotherm for 10°C to have wide significance in marine biogeography, called the warm water above this isotherm the thermosphere and the cold underlying waters the psychrosphere. In the Atlantic

---

[*] These quantities are expressed in langleys—one langley (ly) is one gram calorie per square centimetre—and Q_s is the rate of flow of solar energy through one square centimetre of sea surface.

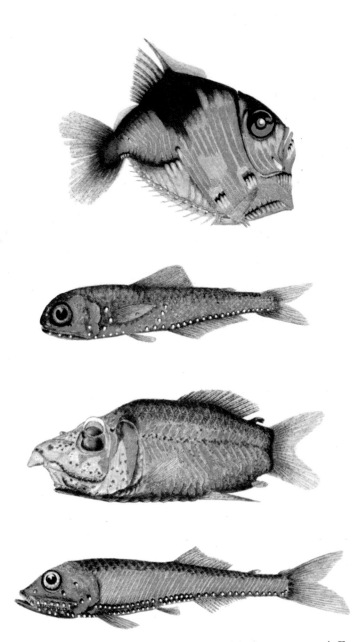

Silvery fishes from midwater levels of the deep sea (size range 4 to 10 cm). From top to bottom; a hatchet-fish, *Sternoptyx diaphana*; a lantern-fish, *Myctophum punctatum*; an argentinoid fish, *Opisthoproctus grimaldii*, with tubular eyes, and a gonostomatid, *Vinciguerria attenuata*. Though not visible in the third fish, each species has a light organ system.

and Pacific Oceans the thermosphere extends between latitudes 50°N and 50°S and to depths between 300 and 1,000 metres (see Fig. 8). If the warm and cold water spheres are so defined, the former would seem to occupy no more than one-tenth of the entire volume of the ocean. Indeed, three-quarters of this volume has temperatures between 0° and 6°C (and salinities between 34·6‰ and 34·8‰). The mean temperature of the entire ocean is 3·5°C (and the mean salinity 34·7‰). (See also Pickard, 1963.)

The temperature structure of the ocean has, of course, a more complex pattern. A profile of temperature through the warm and cold water

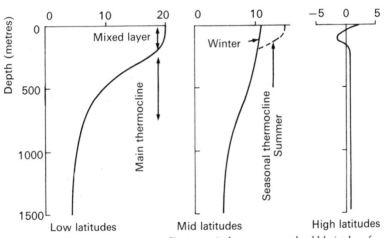

Fig. 6   Typical mean temperature profiles in tropical, temperate and cold latitudes of the open ocean. (From Pickard, 1963)

spheres traces a sigmoid curve (Fig. 6). The uppermost, vertical part of the curve shows an isothermal surface layer ranging from about 10 to 200 metres in thickness. Temperatures in this layer range between 30° and 15°C, depending on season and latitude. In a zone below this, extending between depths of 300 and 1,000 metres, temperatures fall rapidly to values near 5°C, then in the rest of the water column the temperature decreases slowly to levels between 1° and 2°C at 4,000 metres, the mean depth of the ocean.

The surface layer is mixed to isothermal consistency by the wind, and is thus called the 'mixed layer'. It becomes thicker in winter, especially in middle latitudes. In the western North Atlantic, for instance, the mean depth of the mixed layer is between 40 and 120 metres in winter and between 10 and 60 metres in summer. The layer is separated from cooler underlying waters by a thermocline, a layer in which there is a rapid decrease of temperature with depth. The depth of maximum decrease is the axis of the thermocline and is marked by the upper bend of the sigma. This discontinuity layer, for reasons to be given, is called the 'seasonal' thermocline. The epithet is not entirely apt, for over much of the subtropical belts the 'seasonal' thermocline persists throughout the year, and, as will be seen more fully in Chapter 3, tends to bar the upward movement of nutrient salts into the surface waters. Hence productivity of the plant plankton is so limited that subtropical regions are the largest and most deserted parts of the ocean.

The zone between 300 and 1,000 metres, where there is a second rapid decrease in temperature, is known as the permanent thermocline. Below are the cold, deep and bottom waters.

Poleward from the subtropical belts there are marked changes in temperature profiles. In temperate waters the thermocline below the mixed layer develops only in the summer: hence the term 'seasonal'. The permanent thermocline is also much less marked, and on reaching polar waters, thermoclines have vanished. Indeed, there is little change in temperature from top to bottom (see Fig. 6). In the coldest polar waters there is an increase from negative values ($-1.9°C$ to $0°C$) to slightly positive values at depths below 200 metres (see Fig. 6).

The ocean is immense and water absorbs great quantities of heat with relatively little change in temperature. Except for liquid ammonia, the heat capacity of water is the highest of all solids and liquids. Thus, after leaving the surface layer, great quantities of heat are moved by the circulation from one part of the ocean to another. (The transfer of water properties through currents is termed 'advection', to distinguish it from transfer through diffusion.) Moreover, there are no extreme ranges or sudden changes in temperature in the ocean, which in this respect (and others) is a more benign environment than the atmosphere. On land the most abundant and ubiquitous vertebrates are birds and mammals, which are warm-blooded and able to regulate their body temperature. In the sea the warm-blooded homeothermal whales seals and penguins, which had land-dwelling ancestors, range from polar to tropical regions. The only inveterate warm-blooded dwellers of the ocean among the vertebrates are members of the tuna family (Thunnidae) and mako-shark family (Isuridae). The bluefin tuna, which has some power of temperature regulation, ranges from warm waters off the Bahamas to cool seas off Norway.

Even so, many marine organisms are eurythermal, able to exist over a wide range of temperature. The most remarkable are the diverse invertebrates and fishes that migrate up and down each day between cool deep waters (near $10°C$) to surface waters in the tropics with temperatures in excess of $20°C$ (see Chapter 6). There are planktonic organisms that live in the surface waters of the ocean from polar to tropical regions.

But any one of these species may consist of temperature races, each adapted to a relatively narrow range of temperature, the entire complex covering the span of environmental temperatures. Temperature tolerances of marine organisms range from these eurythermal (generalist) kinds to specialists that exist within narrow limits of temperature and are described as stenothermal.

Density

Density in the ocean is fundamental to marine science. For physical oceanographers, a synoptic survey of density helps them to discover the origins, equilibrium levels and spread of water masses (pp. 26–33). The density of marine organisms matches or is a little more than that of the sea, and biologists are thus led, inter alia, to investigate the buoyancy relations of diverse plants and animals (Chapter 9). For animals, a crucial metabolic advantage of weightlessness in winter is that they can stay at a preferred level with the minimum expenditure of energy.

Density, which is expressed in grams per cubic centimetre, depends on pressure as well as on salinity and temperature. Pressure in water increases by one atmosphere* for each increase in depth of 10 metres, and water is slightly compressible. Thus at the surface, sea water of salinity 35 ‰ and temperature 5·0°C has a density of 1·02813, which at a depth of 4,000 metres and hydrostatic pressure of 400 atmospheres is increased by compression to a value of 1·04849.

Since physical oceanographers usually compare water masses at the same depth or over the same interval of depth, densities are reduced to atmospheric (i.e. hydrostatic pressure, $p = 0$), but salinities and temperatures are taken as measured at the depths involved. In the open ocean values of density range from about 1·024 to 1·030 g/cm³. Naturally, the saltier the water, the more its density, while warm water is less dense than cold. More precisely, the rate of increase of density with salinity at atmospheric pressure is about 8×10^{-4} gm/cm³ per 1‰. The rate of decrease of density with increasing temperature is more involved and ranges from 5×10^{-5} per 1°C change at 0°C to 34×10^{-5} per 1°C change at 30°C.

The ocean is more than 800 times denser than the atmosphere, and is thus more buoyant by the same factor. Sea water sustains, in more than one sense, a marvellous diversity of organisms from bacteria to blue whales. It is hardly fortuitous that 'protoplasm', which arose in the sea,

---

*1 standard atmosphere = 14·696 lb per square inch = 1·033227 kg per square centimetre = 101·325 kilonewtons (kN) per square metre.

is only a little denser than its birth fluid, or that the cell sap of plants and the body fluids of animals have a near-oceanic salt balance. Marine animals almost entirely depend, directly or indirectly, on the microscopic plants in the plankton, which have buoyant and motile means to stay in surface waters, where light is strong enough for photosynthesis (p. 58). Beside supporting whales, seals and fishes, all with firm muscular bodies and heavy skeletons, the buoyant properties of the sea have led to its exploitation by forms that consist largely of a fragile, gelatinous skeleton. Medusae, jellyfishes, siphonophores and pelagic tunicates are among such forms.

Since marine animals have much the same compressibility as the sea, their change of density with depth will match that of their environment. Gases, of course, are highly compressible, and animals that depend on a gas-filled float for neutral buoyancy have evolved means of regulating their density against changes of pressure (see Chapter 9).

Changes of pressure have less obvious effects on organisms. Certain fishes and planktonic animals of coastal waters respond in various ways to quite small changes in pressure, and it will be interesting to discover if the same is true of their relatives in the deep sea. Changes of pressure have effects on the activities and enzyme systems of marine animals. Certain deep-sea species tolerate wide ranges of pressure: others seem to be less tolerant.

Water masses and circulation

Through tracing the origin and distribution of each oceanic water mass, physical oceanographers have helped marine biologists to understand better the distribution of deep-sea organisms. Water masses keep their identity and support certain characteristic communities of invertebrates and fishes (pp. 278–83).

Water masses have individual ranges of temperature, salinity and density, and in 1918 the Norwegian oceanographer, Bjorn Helland-Hansen, realized how these properties could be used effectively to define a water mass. Besides plotting the values of temperature and salinity against the depths of sampling at any particular station, Helland-Hansen found that characteristic thermohaline curves are produced when one property is plotted against the other. Such a curve is known as a TS diagram, and there is an advantage in using temperature, salinity (and density) as identification marks of water masses. Near the surface, as we saw, these properties may be changed by variations in solar radiation, evaporation, precipitation and so forth (pp. 22–3), but away from the surface they are unchanged, except through mixing processes. They are 'conservative properties'. Even so, other features, such as oxygen content or concentration of nutrients (e.g. phosphates), can help in identifying water masses. When using these features one has to remember that they are 'non-conservative' properties: their concentrations are changed by living organisms (see p. 63).

If there are enough stations in depth and area for synoptic survey, one can see where a water mass begins and ends. That each water mass has a characteristic TS diagram may be appreciated when the TS curves for an ocean are plotted on one graph. Wider understanding comes from comparing the oceans, which have certain water masses in common. Thus, the Atlantic Ocean, Indian Ocean and South Pacific Oceans contain water masses that are given these names: Antarctic Bottom Water, Antarctic Intermediate Water, Subantarctic Water and Central Water (Figs. 8 and 166). Apart from the last kind, these water masses, as implied by their names, are formed in antarctic and subantarctic regions of the Southern Ocean. Their origin is near the surface and, together with water masses formed near the surface in high latitudes of the North Atlantic and North Pacific, they fill about three-quarters of the ocean, which as we saw, is predominantly cold.

We begin then, in the Southern Ocean, which merges widely with the three other oceans. After moving into the Atlantic Ocean, which has been well explored, we will then compare and contrast the Atlantic with the Pacific and Indian Oceans.

During the winter the surface waters around Antarctica become colder and saltier, and by summer, when the days are longer and ice melts, become warmer and less saline. In winter surface temperatures are between $-1.9\,^\circ$C and $1\,^\circ$C, rising to between $-1\,^\circ$C and $4\,^\circ$C in summer. Winter conditions are particularly severe over the shelf of the Weddell Sea, where temperatures reach $-1.9\,^\circ$C and the formation of sea ice leads to water of relatively high salinity ($34.6\permil$). Such water has a high density (1.0279), and after mixing with Circumpolar Water, it sinks down the continental slope and flows away from Antarctica over deeper reaches of the ocean floor. This is Antarctic Bottom Water and it moves northward and eastward into the basins of the Southern Atlantic, Indian and Pacific Oceans* (Fig. 7). In the Atlantic this water type has been traced as far north as $40\,^\circ$N, about the latitude of Spain. There is evidence of Subpolar Bottom Water off eastern Greenland and in the Norwegian Sea, but this water has clearly nothing like the volume of Antarctic Bottom Water.

North of a narrow east wind drift near the land, surface waters off Antarctica are driven clockwise round the continent as a broad west wind drift. Due to the force inherent in the earth's rotation (Coriolis Force),

---

* Water of even higher density is formed in the Ross Sea. Indeed, there is evidence that Antarctic Bottom Water is formed all round the continent but the main source is in the Weddell Sea.

there is a northerly component in this drift. Over southerly reaches, the properties of Antarctic Surface Water, which has a thickness of 100 to 250 metres, vary with the seasons (winter temperatures reach − 1·9°C and salinities nearly 34·5‰), but during their oblique drift to the north these waters absorb summer heat and at latitudes between 60°S and 50°S temperatures are close to 2°C and salinities below 34‰. Between these latitudes and along a narrow circumpolar front called the Antarctic Convergence, Antarctic Surface Water meets Subantarctic Water of higher

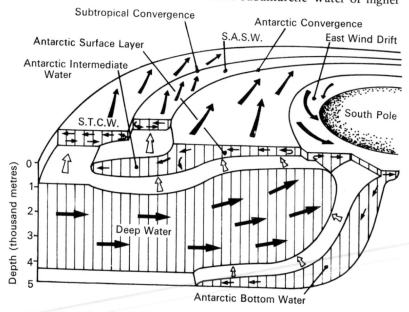

Fig. 7   Three-dimensional representation of the water masses and circulation of the Southern Ocean.
S.A.S.W.—Subantarctic Surface Water.
S.T.C.W.—Subtropical Central Water. (From R. I. Currie, after David, 1965a) See also Figs. 9–11.

temperature (4°C or more) and begins to sink. On the way down it mixes strongly with deeper and warmer waters that are rising above Antarctic Bottom Water and, after finding its level at about 1,000 metres, continues north as Antarctic Intermediate Water (temperature 2°C, salinity 33·8‰). In the Atlantic this water mass crosses the Equator in the west. North of the Equator in the east a smaller intermediate water mass is formed by the sinking of Mediterranean water to 1,000 metres or more as it flows into the Atlantic. The warm deep water is not formed in the Southern Ocean but far to the north in the surface waters of the North Atlantic.

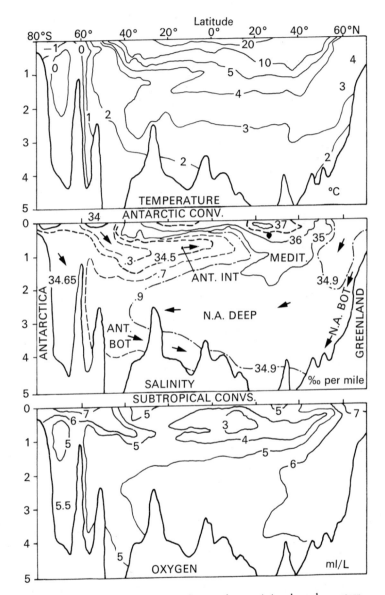

Fig. 8   South to north longitudinal section of water characteristics along the western
         trough of the Atlantic Ocean. (From Pickard, 1963, after Wüst)
         ANT. INT, Antarctic Intermediate Water; ANT. BOT, Antarctic Bottom
         Water; N.A. BOT, North Atlantic, Bottom Water, N.A. DEEP, North Atlantic
         Deep Water; MEDIT, Mediterranean Water.

South of Greenland at about latitude 60°N relatively saline waters that are chilled in winter, sink and then flow south between depths of 1,500 and 4,000 metres as the very voluminous North Atlantic Deep Water Mass (Fig. 8).

North of the Antarctic Convergence at about latitude 40°S there is a second circumpolar front, the Subtropical Convergence, where cool Subantarctic Water meets a warmer mass of water. Here, over a much broader front than met at the Antarctic Convergence, water sinks to levels between Antarctic Intermediate Water and the surface mixed layer to form South Atlantic Central Water. North Atlantic Central Water is formed along another broad front at about 40°N in the northwestern Atlantic. These Atlantic Central Waters, which extend to depths of 300 metres on either side of the Equator and deepen to 600–900 metres at mid-latitude, have temperatures ranging from about 17° to 7°C and salinities from 36·5 to 34·5‰. It will be seen that the Central Water Masses form most of the thermosphere (see Figs. 8 and 166).

The Southern Subtropical Convergence may be taken also as the southern boundary of the Atlantic, Indian and Pacific Oceans. The rest of the ocean is formed of 'seas'. One group, such as the Arctic Sea, the Bering Sea, the Sea of Okhotsk, the Gulf of Mexico, the Caribbean Sea, the Mediterranean Sea, the Red Sea, the Sea of Japan and the South China Sea, are partly enclosed by land or island arcs, and thus have restricted openings into the main ocean. The term 'sea' is given also to certain unconfined parts of the ocean that have individual oceanographic features. The Barents Sea, the Norwegian Sea, the Arabian Sea and the Tasman Sea are among such seas.

Conditions in the Pacific Ocean and Indian Ocean may now be seen in comparison and contrast to the above outline of water masses in the Southern Ocean and Atlantic Ocean. Antarctic Bottom Water and Antarctic Intermediate Water move into southerly parts of the Pacific and Indian Oceans. Intermediate Water in both oceans flows north to meet Equatorial Water Masses, soon to be described. Intermediate Water is formed also in the northwestern Pacific, evidently by vertical mixing in the subantarctic region, not by the sinking of surface waters. In the northwestern Indian Ocean an intermediate water mass between depths of 1,000 and 1,500 metres may be traced to the sinking of warm and saline waters (salinity 36‰, temperature 15°C) that flow out of the Red Sea.

There is little or no formation of deep water in the Pacific Ocean, and from a level of 2,500 metres downwards the Deep and Bottom Water (temperature 1° to 3°C, salinity 34·6 to 34·8‰) must come entirely from the Antarctic Circumpolar Current. This is also the source of some of the deep water in the Indian Ocean; as is North Atlantic Deep Water, which enters at depths below 2,000 metres in the southwest. Two more deep-water masses are formed within the ocean itself. North Indian Deep

Water forms in the Arabian Sea from Red Sea water and a much smaller outflow from the Gulf of Oman. The Red Sea Water moves towards Africa and may be traced as far south as 5° to 6°S. Further south, between 10° and 16°S, North Indian Deep Water mixes with Antarctic Intermediate Water and Antarctic Bottom Water and is transformed into South Indian Deep Water (temperatures 3·0 to 1·0°C, salinities 34·78 to 34·60‰).

The Central Water Masses in the South Pacific and South Indian Ocean arise, of course, at the Southern Subtropical Convergence. In the North Pacific Central Water is formed along a front at about latitude 30°N. North of this is a broad ocean-spanning mass of Subarctic Water of low salinity (33·5 to 34·5‰) and rather low temperatures (2° to 4°C). In both oceans the Central Water Masses lie between depths of about 100 and, 1,000 metres and have temperatures ranging from 17° to 4°C, salinities from 36·3‰ to 34·2‰.

Between the Central Water Masses of the Pacific and north of the Central Water in the Indian Ocean is a vast Equatorial Water Mass. There is no counterpart in the Atlantic Ocean. Pacific Equatorial Water extends right across the ocean, decreasing in width towards the west and reaching a depth of about 800 metres. It has temperatures of about 15° to 5°C, salinities of about 35·3 to 34·7‰). Indian Equatorial Water, which has a small range of salinity (35·2 to 34·9‰), extends across the ocean to the north of latitude 10°S and reaches a depth of about 1,000 metres. Indian Central Water merges with Antarctic Intermediate Water at a depth of some 1,000 metres. In the Pacific and Indian Ocean, then, Equatorial and Central Waters form most of the thermosphere (see p. 30). The uppermost part consists of the surface mixed layer (see p. 23).

This ends a brief survey of the oceanic water masses seen in terms of temperature, salinity and density (based largely on Pickard, 1963). Water masses maintain their thermohaline (and other) features, which was well seen when observations made in the Atlantic during 1925 to 1927 by the German *Meteor* Expedition were compared with those made during the International Geophysical Year (IGY, 1957–58) on ships of the Woods Hole Oceanographic Institution USA and the National Institute of Oceanography (UK). Relevant comparisons of the two sets of observations showed that the distributions of temperature, salinity and dissolved oxygen were almost identical (see Fig. 8). Marine biologists who know a particular area through repeated observations find also that communities of deep-sea animals may maintain their identity. Diverse species seem to be bound to the physical and biological conditions of water masses: others transcend these conditions (see also Chapter 13).

After thirty years, it seems astonishing that values of dissolved oxygen were still the same in certain parts of the Atlantic. Through a given 'volume' of the ocean, the circulation moves at a certain rate and carries a certain quantity of dissolved oxygen, which is used by resident organisms in the processes of life and decay. Does the correspondence between the Meteor and IGY measurements imply that in some parts of the ocean there is a steady state in the quantity of dissolved oxygen? Is the supply of oxygen so great compared to the oxygen requirements of the organisms that methods of oxygen analysis are unable to detect the withdrawal of oxygen by life? At mid-depths in some parts of the ocean the oxygen content is minute but life abundant. Life in these oxygen minimum layers will be considered later (pp. 484–6).

Convergences, where surface waters rich in oxygen sink into the depths, have been called the lungs of the ocean. The atmosphere is the main source of oxygen dissolved in sea water. The plants in the plankton also produce oxygen during photosynthesis, and at times the surface layer may be supersaturated with this gas. The oxygen content of water is expressed as the number of millilitres (ml) of oxygen gas dissolved in one litre (l) of seawater at NTP. Values range from 0–8 ml/l.

Down from the convergences the circulation of the water masses carries oxygen to nearly all parts of the ocean below the surface layer. The exceptions are the Black Sea and Cariaco Trench (in the Caribbean Sea) where subsurface layers are devoid of oxygen and carry hydrogen sulphide, sure signs of stagnant conditions. As a water mass sinks and finds its level in the ocean, the contained oxygen is used by living organisms and in the oxidation of organic detritus. Once it has left the surface a water mass can only gain oxygen by mixing with another water mass of higher oxygen content. If the circulation is sluggish a water mass will lose much oxygen to its life after travelling several thousand miles. But how fast is sluggish?

Natural radiocarbon (C^{14}) is the most useful of the radionuclide tracers of oceanic circulation. The sea surface gains radiocarbon from the atmosphere. Thus, the concentration of C^{14} in part of a water mass will depend on how long this part has been below the surface* (the residence time) and on the rate of gain or loss of C^{14} to local water masses. To use this method, then, one must make certain assumptions about the exchange of C^{14} between the atmosphere and ocean and about mixing between water masses. In a survey of the distributions of C^{14} in the Pacific and Indian Oceans much emerged. Deep waters were less active than those near the surface, showing, as one would expect, their greater age. In the Pacific, for instance, the activity of the deep water decreased from south to north, which reflects the ageing of the water as it moves north. The northerly component in the deep circulation was calculated to have

---

*Radiocarbon decays to C^{12} with a half life of 5,570 years. If there is no outside source of C^{14}, then from radioactive measurements that give the C^{14}/C^{12} ratio in part of a water mass, one can calculate the time elapsed since this part left the surface.

an average rate of 0·05 cm/sec (43 metres per day). Since the deep water moves so slowly *and* originated in antarctic regions, it is hardly surprising that values of oxygen content (at levels of 2,000 metres or more) have fallen from about 4·5 ml/l in the South Pacific to between 0·5 and 3 ml/l in the North Pacific. Indeed, in the eastern Pacific between 10° and 20°N there is a pronounced oxygen minimum layer with values down to less than 0·1 ml/l between 200 and 1,000 metres. Here productivity is high and it seems that the requirements of life, combined with slow circulation, are the major causes for the oxygen minimum layer.

As already implied, the distribution of radiocarbon can be used to calculate mean residence times for the various water masses. For deep Pacific water the residence time is about 1,000 to 1,600 years, which is approximately twice that of deep water in the Atlantic and Indian Oceans.* In Atlantic deep-water values of dissolved oxygen (3 to 6·5 ml/l) are higher than those in Pacific deep water (see above), which seems reasonable in view of the less sluggish circulation of the former. But deep currents are not invariably slow, as will be seen later (p. 38). At all events, the slow pace of the deep waters may be appreciated in contrast to surface waters, which have a residence time of 10 to 20 years and a lively circulation.

Patterns of circulation near the surface

Circulation in the upper waters is turned by part of the sun's energy that falls on the sea. Evaporation at the surface, which may be facilitated by the stirring motions of microscopic plants and animals in the surface film, yields potential energy in the form of water vapour. Minute particles of salt and organic material also leave the surface film, ejected it seems by breaking bubbles (MacIntyre, 1974). Water vapour and particles are swirled into the atmosphere, where the particles form the nuclei for the condensation of water vapour which releases latent heat, the main source of energy for atmospheric circulation. Winds that blow over the sea raise waves and drive ocean currents. The wind-driven circulation of the ocean thus differs from the thermohaline circulation of the water masses, which as we saw, is maintained by the pull of gravity on dense water types that sink and then flow along stratified levels of equilibrium. The wind-driven circulation consists essentially of currents that flow horizontally in the upper few hundred metres of the ocean. Due to the force inherent in the earth's rotation (Coriolis Force), these currents move along lines to the right of the wind direction in the northern hemisphere and to the left in the southern.

---

*North Pacific Deep Water with a recently estimated age of 3,400 years must be the oldest of all the deep-water masses.

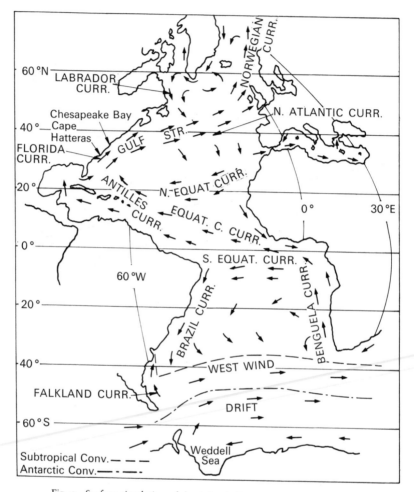

Fig. 9 Surface circulation of the Atlantic Ocean. (From Pickard, 1963)

The patterns of wind-driven circulation in the three main oceans are broadly similar, as one might expect (Figs. 9–11). In equatorial regions the Trade Winds drive a current system that A. Defant called 'the backbone of the circulation'. Near or at the equator there is a westerly flowing South Equatorial Current that it kept in motion by the southeast Trades. To the north of the equator a parallel, westerly flowing North Equatorial Current is driven by the northeast Trades. Both currents cross the ocean and pile up water that returns between them as an Equatorial Counter Current. This current spans the entire width of the Pacific Ocean

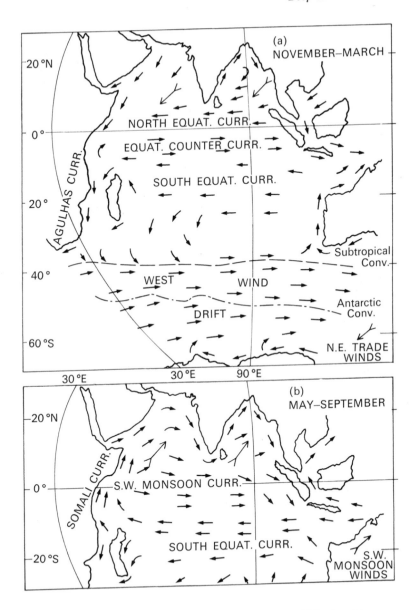

Fig. 10 Surface circulation of the Indian Ocean. (From Pickard, 1963)

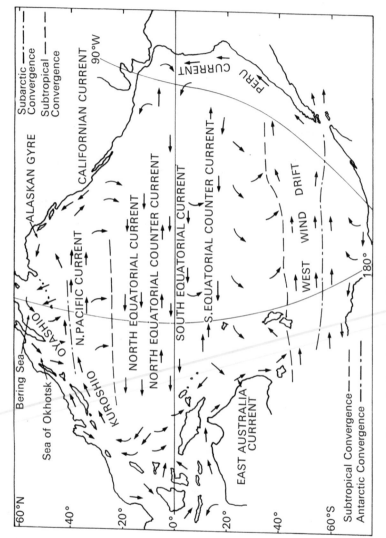

Fig. 11   Surface circulation of the Pacific Ocean. (From Pickard, 1963)

but is prominent only over the eastern half of the Atlantic. There is a parallel, triple current system in the equatorial Indian Ocean from November to March, when northeast Trade Winds (northeast monsoon) maintain the North Equatorial Current, which is close to the equator (with the change to the southwest monsoon from April to September, the North Equatorial Current reverses direction and the Equatorial Counter Current is no longer distinguishable) (Fig. 10).

The Trade Winds not only drive the equatorial currents but also their continuation as vast gyres that encircle the subtropical regions: hence the evocative phrase 'the backbone of the circulation', seems justified. These oceanic gyres thus turn clockwise in the northern parts of the Atlantic and Pacific (the Asiatic land mass precludes the formation of a gyre in the northern Indian Ocean) and anticlockwise in the southern parts of all three oceans (see Figs. 9–11). The gyres overlie the Central Water Masses and, indeed, may extend into these waters.

The North Atlantic gyre includes the Sargasso Sea, the westerly edge of which is the fast-flowing Gulf Stream and largely the continuation of the Florida Current that springs from the Gulf of Mexico. Speeds in the Stream, which reach 250 cm/sec (5 knots), are among the highest recorded in oceanic currents. Detailed surveys have shown that the Gulf Stream consists of narrow streams that trace a sinuous northeasterly path, and occasionally great eddies are detached to turn in the Sargasso Sea for months at a time. The Stream breaks away from the American coast at Cape Hatteras and near the Grand Banks of Newfoundland, where it broadens into the North Atlantic current, it impinges on cold water from the Labrador Current. Part of the North Atlantic current moves eastward across the ocean and then turns south past Spain to complete the gyre in the North Equatorial Current: the other part flows northeasterly and eventually becomes part of the circulation of the Greenland, Norwegian and Arctic Seas (Fig. 9).

The western boundary current of the South Atlantic gyre is the south-flowing Brazil Current, but it is not so fast as the Gulf Stream. The eastern flow of the gyre moves across the ocean with the west wind drift in sub-antarctic waters, and on heading north is edged by the Benguela Current off Southeast Africa.

In the North Pacific the counterpart of the Gulf Stream is the Kuroshio, which flows past Japan and then turns east as the Kuroshio Extension. Like the Gulf Stream, the Kuroshio changes its position from time to time and reaches maximum speeds of 250 cm/sec (5 knots). Again, the Kuroshio Extension, like its counterpart, the North Atlantic current, meets cold waters (in the south-flowing Oyashio). In both regions the mixing of warm and cold waters leads to high productivity and profitable fisheries. The Kuroshio Extension flows east as the North Pacific Current, which edges the cool subarctic waters and on nearing America, part turns south, as the California Current, part northeast to feed the gyre in the Gulf of Alaska. In the South Pacific the easterly flow of the gyre moves with the west wind drift in subantarctic waters and after turning north, forms part of the Peru Current (Fig. 10).

There are undercurrents below parts of the wind-driven circulation. Below the Gulf Stream there is a counter-flowing undercurrent with speeds of 9–18 cm/sec but it is not always present. This was discovered by plotting the movements of Swallow neutral-buoyancy floats, which also revealed that deep currents in the North Atlantic are not invariably sluggish and steady. In the Bermuda area, where slow deep currents were expected, the flow at a depth of 2,000 metres averaged 6 cm/sec (maximum 20 cm/sec) and at 4,000 metres the average was 12 cm (maximum 42 cm/sec). Off Portugal measurements at 1,500 to 4,000 metres indicated currents of 1 to 2 cm/sec (maximum 5 cm/sec), which seems more in keeping with earlier estimations of deep water movement in the Atlantic.

Undercurrents are prominent below the equatorial systems and the first was discovered in the Pacific, where, 'The Equatorial Undercurrent is almost as large in volume transport as the North or South Equatorial Currents and it lies only 100 metres or less below the sea surface along the equator. Yet it was not discovered until 1952.' The outstanding discovery from current surveys, which was announced by Cromwell, Montgomery and Stroup, '. . . was that near the equator and embedded in the well-known westward flowing South Equatorial Current, there was a previously undiscovered eastward flowing current between 70 and 200 metres depth'. An intensive study by Knauss of Scripps Institution of Oceanography, revealed many details of the current. It is a thin ribbon, only 0.2 km thick, but 300 km wide, extending from about 2°N to 2°S. This ribbon certainly stretches over the eastern half of the Pacific and may well be 12,000 km in length. The speed in the eye of the current is up to 150 cm/sec and the volume of water transported approaches that of the Gulf Stream. 'In other words, it is a major ocean current which was unknown before 1952 and had not even been predicted by theory' (Pickard, 1963). There are also Equatorial Undercurrents in the eastern half of the Atlantic Ocean and the Indian Ocean (from about 60°E to 90°E). The speed of the Indian Equatorial Undercurrent is only about half that of its counterpart in the other two oceans.

The wind-driven circulation has profound effects on the lives of deep-sea animals. The upward transport of nutrient salts from subsurface levels to the productive, plant-bearing surface layer is quickened along the edges of swift currents (e.g. the Gulf Stream), or along lines of current divergence, or slowed in the great subtropical gyres that turn above a more or less permanent thermocline. As we shall see, the outstanding patterns of plant productivity are reflected down to all levels in the ocean, certainly affecting the abundance and the community structures of deep-sea faunas (pp. 486–98). Each night, the productive surface waters are the feeding grounds of diverse animals that move up from the upper mid-waters (pp. 234–52). These waters also form the nursery grounds for the young stages of many pelagic and some bottom-dwelling deep-sea animals (Chapter 12). During their planktonic existence the young may be widely dispersed by the lively wind-driven circulation and eventually some will grow into recruits for their parental populations. The quickest parts of the circulation may well carry the young away from the parental grounds but some, as they metamorphose and descend, may return in the undercurrents (p. 252). Gyres and swirls will, of course, tend to keep planktonic forms within the confines of their circulation. If the ocean may be likened to a marine organism, the living units must above all evolve the right kind of organization and staying power to survive in the great organism's circulation.

Light in the ocean

Short-wave energy streaming from the sun heats and illuminates the ocean. Light penetrating the sea is both scattered and absorbed. Both processes are involved in the submarine attenuation of light which increases exponentially with depth and varies with the wave length. Virtually all of the energy in waves shorter than the visible (ultra-violet) or longer than the visible (infra-red) is absorbed in the top metre of the sea. At a depth of 10 metres most of the energy is in the blue-green part of the spectrum, though there is still a trace of visible red light. Thus, reddish organisms appear dark or black at depths below 10 metres, for at these levels there is little or no red light to be reflected by these organisms. It is thus not altogether surprising that diverse midwater invertebrates have adopted vanishing coats of red pigment (pp. 362–4).

Blue-green rays are still detectable near a level of 1,000 metres in the clearest oceanic waters, but no more than 5 per cent of the light energy penetrates to 50 metres. Even so, the light is strong enough for plant photosynthesis at depths down to 150 metres in the most transparent regions, where between 50 and 100 metres there is also enough light to allow divers to work.

The twilight zone of the ocean thus falls between the productive (euphotic) zone and the greatest reach of blue-green sunlight, which is probably close to 1,200 metres (Clarke, 1970). The diversity of animal life in this upper midwater (mesopelagic) zone, greatly exceeds that in the vaster, deeper midwater (bathypelagic) zone, where there is no trace of sunlight. The twilight zone is full of animals that migrate daily to the more productive waters that move overhead. The most complex animals, notably decapod crustaceans, squid and fishes, have very sensitive eyes, able, most probably, to detect daily changes in light level and so regulate their vertical migrations. The twilight zone is also full of luminescent animals and with few exceptions, the light produced is blue-green and close in wavelength to the ambient solar light. Furthermore, eyes are most sensitive to blue-green rays, which seems almost inevitable in this marvellously integrated twilight world (pp. 380–402).

Just below the surface, where the sun's rays are reflected and refracted, the field of light is diffused and complex. At deeper levels most of the radiation is downward and approaches the zenith with increasing depth. At depths between 300 and 400 metres in clear oceanic waters the field of light is likely to be symmetrical around the zenith, and the polar diagram looks rather like the cross-section of a fish. Indeed, by matching this ambient field of light with the light of their own photophores, certain fishes of the twilight zone vanish from the view of predators (pp. 351–362).

Life zones and life forms

It must now be evident that water masses and pelagic forms of life are disposed in horizontal strata in the ocean (Fig. 12). Pelagic organisms drift and move in submarine space: they are adapted for life at all levels between the air/sea interface and the interface between sea and substrate. Benthic organisms live on the ocean floor. For many millions of years, certainly from Cambrian times (600 million years ago), the vast pelagic and benthic living spaces have been the scenes for a succession of evolutionary plays known as adaptive radiations. Today, according to Thorson (1971), there are about 160,000 species of marine animals, of which 98 per cent are benthic and 2 per cent pelagic. Perhaps about 5 per cent of the total number of marine animals live in the deep sea, where the proportions of pelagic forms is perhaps greater. At all events this proportion certainly differs in large groups with both pelagic and benthic representatives. There are considerably more benthic, deep-sea cnidarians, polychaete worms, crustaceans and molluscs than of related pelagic forms. But pelagic deep-sea fishes are about as diverse as bottom-dwelling kinds.

Pelagic organisms, which range in size from bacteria to the largest whales, are broadly classed as planktonic or nektonic forms. Plankton (from the Greek *plankton*, neuter of *planktos*, wandering) consists of diverse organisms that are non-motile or have limited powers of movement: they thus drift, or tend to drift, with the sea. Pelagic bacteria are classed as bacterioplankton, pelagic plants as phytoplankton and pelagic animals as zooplankton. Bacteria and the plants are microscopic forms, and so are some of the animals, such as the Protozoa and the early larval stages of invertebrates. Most adult members of the zooplankton range

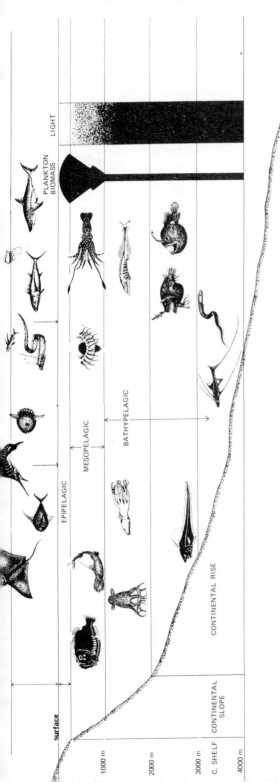

Deep-sea environments

Fig. 12 A section through the ocean. Scale on left in metres.
Epipelagic animals (top row) from left to right. Giant devil-ray (*Manta*), ocean bream (*Tarachtichthys*), spearfish (*Tetrapturus*), ocean sunfish (*Mola*), ribbon-fish (*Regalecus*), and above flying-fish, long-fin tuna, Potuguese-man-of-war (*Physalia*), on surface and great white shark (*Carcharodon*).
Mesopelagic animals, left to right. Hatchet-fish (*Agyropelecus*), viper-fish (*Chauliodus*), coronate jellyfish (*Atolla*), squid (*Histioteuthis*). Bathypelagic animals, left to right. *Vampyroteuthis*, midwater octopod (*Amphitretus*) both of which live also at mesopelagic levels, *Cyclothone*, and immediately below, two female ceratioid angler-fishes (*Linophryne* and *Melanocetus*).
Bottom-dwelling fishes, left to right. Rat-tail (*Nezumia*), a benthopelagic form, tripod-fish (*Bathypterois*), a benthic form, and a deep-sea eel (*Synaphobranchus*), a benthopelagic form.
On the far right is seen the amount of plankton (biomass) showing that by far the greatest amount is in the epipelagic and mesopelagic zones. The light field shows that there is virtually no sunlight below a depth of 1000 metres. (From Marshall, 1971.)

41

in size from millimetre-long copepods to large jellyfishes (e.g. *Cyanea*) over a metre in diameter. The most ubiquitous forms are copepod crustaceans: other important groups are cnidarians, ostracod crustaceans, pteropod molluscs, arrow-worms, and the pelagic tunicates.

Members of the nekton (from the Greek *nekton*, neuter of *nektos*, swimming) have sufficient powers of locomotion to make their way against currents. Cetaceans, seals, turtles, fishes and squid are classic forms of nektonic animals. But where does the plankton end and the nekton begin? The commonest deep-sea fishes belong to the genus *Cyclothone* and the smallest species range from 25 mm to 50 mm in length. Several species of euphausiid shrimps fall within this size range, or even exceed it, and the group as a whole is often regarded as planktonic. Large euphausiids must have just as good powers of movement as small *Cyclothone*. Indeed, certain of the deep-sea prawns, which are larger than euphausiids, prey on *Cyclothone*. It seems reasonable then, to have a category called the micronekton (Marshall, 1954), which would consist largely of decapod crustaceans and the smallest cephalopods and fishes. At all events, nektonic animals are larger than, say, 15 mm and they belong to groups with plenty of muscle. Such animals, which need to maintain large, energy-consuming muscles, are mostly omnivores or carnivores. Many planktonic animals, notably copepods and the pelagic tunicates, are entirely or largely herbivorous.

Animals that move or float on the sea surface have the collective name pleuston. The movers are water-striders (*Halobates* spp.) of the insect order Hemiptera: the floaters have some kind of gas-filled buoyancy chamber and include various cnidarians (e.g. *Physalia*, *Velella*, *Porpita* and certain sea-anemones), stalked barnacles, purple snails (*Ianthina* spp.) and nudibranch molluscs (e.g. *Glaucus*). Their headquarters are on the warmer parts of the ocean, which seem very remote from the levels of deep-sea life. Each day, though, many animals of the upper midwaters move into the surface layer (pp. 234–52) and some, as we shall see, feed close to the surface, where they may well fall prey to the cnidarians, especially to *Physalia*, the Portuguese man-of-war.

Organisms that live in the uppermost few centimetres of the sea form the neuston. There are bacteria, tiny species of phytoplankton (nanoplankton) and diverse planktonic animals. Relatively few species of neustonic animals stay near the surface throughout the entire day, but from time to time they are joined by others, especially during the night. In subtropical and tropical regions many members of the neuston (e.g. pontellid copepods, arrow-worms, pelagic tunicates and larval stages of crustaceans and fishes) may be suffused with blue or purple pigments (David, 1965; Herring, 1967). Again, life close below the surface seems remote from that in the depths, but young stages of upper midwater (mesopelagic) animals, such as larval sergestid prawns and lantern-fishes, may form part of the neuston. Furthermore, neuston nets catch adult lantern-fishes (e.g. *Gonichthys coccoi*, *Symbolophorus veranyi*, *Myctophum punctatum*), which evidently take some neustonic food during their nightly stay in the surface waters. Indeed, land-dwelling insects that have been blown out to sea and eventually come to rest on the surface film may make up much of the food of certain lantern-fishes. Concerning the neuston, one must remember, though, that it is concentrated in a very thin layer and must thus be no more than a subsidiary source of surface-dwelling food for nektonic animals, particularly in tropical and subtropical waters. In these regions the overall impression one gains of the near surface (o–10 cm) zone (compared to the layer from the surface to 30 metres) is that of a poor feeding ground with a rather low density of invertebrates and fishes.

The uppermost waters of the open ocean are known as the epipelagic zone. Here, in the subtropical and tropical belts, is the swim of flying squid, flying-fishes, skippers, sauries, grey sharks, thresher-sharks, mako sharks, tuna-fishes and billfishes (Fig. 12). The sharks, tunas and billfishes are Gullivers compared to the Lilliputian fishes in the waters of the deep-sea world. Many thousands of the largest lantern-fishes and stomiatoid fishes are needed to equal the weight of a large bluefin tuna or a blue marlin.

What is the lower limit of the epipelagic zone? Tuna fishes live in the upper 200 metres and are sometimes concentrated near the 'seasonal' thermocline (p. 24), but this varies in level between about 10 to 250 metres. Planktonic plants live in the surface layer and sunlight is strong enough for photosynthesis down to 150 metres in the clearest oceanic waters (p. 58). Not far below this depth small luminous fishes with silvery sides (e.g. the stomiatoids such as *Bonapartia*, *Vinciguerria* and *Valenciennellus* begin to appear in nets towed at successively deeper levels below the surface. We shall accordingly, and somewhat arbitrarily, take the lower limit of the epipelagic zone to be about 150 metres, which is also close to the mean depth of the outer edge of the continental shelf (p. 11), where shallow seas give way to deep seas.

If the mesopelagic zone is regarded as a twilight environment, then the transition between this zone and the dark underlying bathypelagic zone is the threshold of light (see Fig. 12). Measurements of underwater light by Clarke (1970) and his colleagues have shown that in very clear oceanic waters sensitive instruments will detect sunlight down to about 1,200 metres (p. 39). There are not enough measurements to give a reliable estimate of the mean depth of light penetration in the entire oceanic province, but this is likely to be close to 1,000 metres. In subtropical and tropical waters the mesopelagic zone contains the main thermocline (p. 24), but everywhere below this zone (except in basin seas) temperatures at a given level are virtually uniform and range between 5°C and negative values. The bathypelagic zone is thus uniformly cold and dark (apart from flashes of bioluminescence). Considering that the headquarters of deep-sea animals are below the warm oceanic belts, what are the biological features of the twilight–thermocline–mesopelagic and the dark–cold–bathypelagic zones? Full treatment of the mesopelagic and bathypelagic faunas comes later in this book: here we look briefly at copepods and fishes, the two ends of oceanic food chains. At depths below 1,000 metres there is a distinct and rather diverse bathypelagic fauna of calanoid copepods, distinguishable from mesopelagic and epipelagic kinds, some of which are related forms (Grice and Hulsemann, 1965). Concerning the fishes, the transition between the mesopelagic and bathypelagic zones is marked by certain dark-coloured species of lantern-fishes of the genera *Lampadena* and *Taaningichthys*. Lantern-fishes, most of which migrate daily between their daytime levels and the upper waters, are classic meso-pelagic animals. Ceratioid angler-fishes, dark-coloured species of bristle-mouth fishes (*Cyclothone*), whale-fishes and gulper-eels, which are non-migrators, are typical dwellers at bathypelagic levels. This is not to say that they are entirely confined to the bathypelagic zone. The vertical ranges of certain black species of *Cyclothone* and angler-fishes (e.g. *Melanocetus* and *Cryptopsaras*) may extend into lower mesopelagic levels. The transition between the two midwater zones, whether physical or biological, is by no means abrupt. (See also Marshall, 1971.)

Besides housing many kinds of vertical migrators and the reflections of some as sound-scattering layers (which do not occur at bathypelagic levels), the mesopelagic zone is also, as already stated, full of luminous animals. At depths below 1,000 metres luminous flashes become much less frequent (Clarke, 1970). Indeed, the specific diversity of life and the amount of life (biomass) are much reduced at bathypelagic compared to mesopelagic levels. For instance, for every species of the bathypelagic fish fauna (about 150 spp.) there are six mesopelagic species. The biomass of zooplankton decreases markedly between the surface layer and 1,000 metres but at deeper levels the decline is less rapid. Beyond 2,000 metres there is a gradual and steady exponential fall in biomass (Vinogradov, 1968).

In the deeper parts of the bathypelagic zone below 4,000 metres there seems to be a distinct abyssopelagic fauna, which contains, inter alia, members of the copepod family Spinocalanidae. Species of this family live also near the deep-sea floor as part of a benthopelagic fauna, which ranges from these millimetre-sized copepods to certain fishes (macrourids) and cirrate octopods that reach lengths of about a metre. Part-time members of the benthopelagic fauna include pelagic forms (e.g. medusae and jellyfishes) and also some benthic species (e.g. amphipods).

Animals of the deep-sea floor, like their benthic relatives on the shelf, live partly or entirely above the interface between sea and land (epifauna) or within the upper layer of sediment (infauna). The infauna consists largely of minute (meiobenthic) forms (e.g. foraminifera and nematode worms) and small species of burrowing bivalves, crustaceans and polychaete worms. The epifauna is not only represented by members of the taxa that flourish on the continental shelves, but there are others that are predominantly deep-sea forms, notably the glass-sponges, beardworms (Pogonophora) and elasipod sea-cucumbers.

Down the continental slope and rise and moving deeper into the ocean basins there is a decline in the biomass and species diversity of the benthos. The decline is most marked in most of the epifauna but is much more gradual in the relatively small forms of the infauna (p. 494). Some specialists differentiate a bathyal fauna, extending down the slope to about 2,000 metres, from an abyssal fauna at deeper levels. Critical survey of these faunas is beyond the scope of this book, but see pp. 492–5 for discussion on the ecological aspects of benthic zonation.

SECTION I
DEEP-SEA LIFE

Oceanic plants 3

During the long and arduous antarctic voyages of HMS *Erebus* and HMS *Terror* between 1839 and 1843, Sir James Clark Ross had Dr Joseph Hooker with him as naturalist to the expedition.

Dr Hooker was no doubt well aware of the Royal Society's instructions for the expedition, but he must have been particularly impressed by a letter from Alexander von Humboldt to the First Lord of the Admiralty. The letter drew attention to the far-reaching significance of the work of the German microscopist, Dr Ehrenberg, on the fine structure of certain rocks. Dr Ehrenberg had shown that the beds of 'rotten stone' found in various parts of the world were made of immense deposits of diatom frustules and the skeletons of Radiolaria and sponges; and from other observations on living diatoms and Radiolaria had concluded that similar rocks ought to be forming in the present age—at the bottom of the sea.

During the summer the antarctic seas teem with diatoms. Hooker saw signs of their presence almost everywhere; as brown stains on the icebergs and pack ice, as a wash of pale ochreous colour in the brash and pancake ice and as a mucous scum floating in the ice-free waters. He saw with Ehrenberg that the diatoms with their indestructible siliceous coverings must sink to form sediments on the sea floor after they die, but he also became aware of the unique importance of the living diatoms in the economy of the seas. They could be found in the stomachs of small Crustacea and other planktonic creatures. They were the invisible plant life of the ocean '... a microscopic herbage, ... probably maintaining in the South Polar Ocean that balance between the vegetable and animal kingdom which prevails over the surface of our globe'. Hooker's remarkable observations and predictions were soon followed by the independent discoveries of the Danish botanist Anders Sandøe Ørsted. During a voyage from Denmark to the West Indies in 1847, Ørsted found that microscopic algae were the cause of discoloured water in aquaria and was led to collect sea water samples for microscopic examination. Once again he saw microscopic algae, and like Hooker, discerned that these minute plants must be the main primary source of food for marine animals.

Under the microscope the sculptured patterns in the siliceous cell walls of diatoms are both beautiful and distinctive. From the middle of the nineteenth century onwards many naturalists delighted in studying these patterns; and they described more and more new species. In 1883 von Stein monographed another group of microscopic plants, the peridinians or Dinoflagellata. They too were evidently of importance as food, for much of the material studied came from the stomachs of the transparent, barrel-like salps of the plankton. During the *Challenger* Expedition, Sir John Murray was able to show that the calcareous coccospheres and rhabdoliths of deep-sea deposits are actually parts of the skeletons of minute plants (Coccolithophores) that live in the surface waters of the ocean.

Soon after 1880 Victor Hensen began to study the biology of the drifting life of the ocean, the plankton as he called it. Hensen particularly wanted to get some measure of the production of nutrient material by the plants (the phytoplankton). He reasoned that a conical silk net of a particular size drawn vertically through the upper plant-bearing layer of the sea should catch all the plants in the cylindrical column of water through which it would pass. Having counted the plants it would then be possible to calculate how many were living below a unit of sea surface. But later on Hans Lohmann pointed out that very small and important members of the phytoplankton, such as the coccolithophores, passed through the meshes of the finest silk nets. This fine nanoplankton fraction could only be obtained by means of special filters or by centrifuging samples of sea water so as to throw the tiny cells out of suspension.

Plant groups

Lohmann convincingly revealed the plant-like nature of the Coccolithophoridae: within the cells are one or two brown bodies containing chlorophyll. Like the dinoflagellates, the coccolithophorids are propelled by two protoplasmic flagella and both have a tendency to move towards the light (phototaxis). Reproduction, as in all groups of the phytoplankton, is by binary fission, though there is evidence of sexual means.

Coccolithophorids, otherwise known as calcareous flagellates, are so-called from the minute calcareous plates (coccoliths) that form around them and which vary in size from 1 to 35 microns (1 micron, $1 \mu =$ 0.001 mm). The numbers and ornamentation of the plates differ from species to species (Fig. 13), but may also vary within a species. Coccoliths may be lacking, and if formed the cells may be without flagella. Indeed,

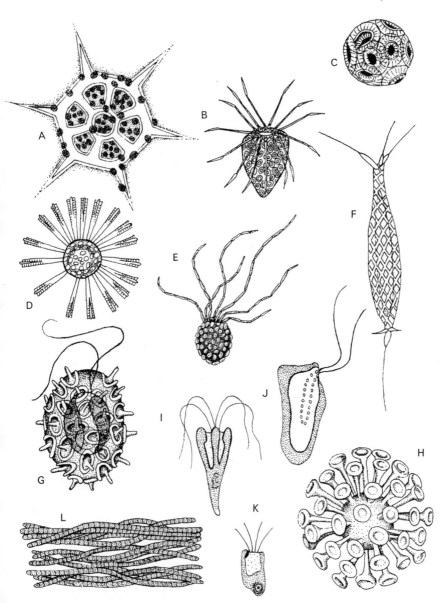

Fig. 13 Flagellates and a blue-green alga of the phytoplankton. A, *Distephanus speculum*, a silicoflagellate; B–H, Calcareous flagellates (coccolithophores); B, *Michaelsarsia splendens*; C, *Coccolithus pelagicus*; D. *Thorosphaera elegans*; E, *Ophiaster formosus*; F, *Calciosolenia sinuosa*; G, *Syracosphaera subsalsa*; H, *Discosphaera thomsoni*; I–K, Microflagellates; L, the common oceanic blue-green alga, *Trichodesmium*. (Redrawn from Smayda, 1971)

the form of coccolith and the nutritional mode of the motile individuals may differ considerably from those of non-motile individuals of the same species. Much has been learned of coccolith structure and taxonomy through the use of electron-microscopes. But 'essential details of the morphology, life history, physiology and ecology of the coccolithophorids are still unknown' (Smayda, 1971). There is much to be gained, both in pure and applied biology, from better knowledge of these minute plants which, after all, are known from Jurassic times and are the main constituent of chalk deposits of the Cretaceous.

Most of the calcareous flagellates range in size from 5 to 25 microns. Micro-flagellates, which belong to several classes of algae, are even smaller. They have one or more flagella and many species contain green or yellow-brown photosynthetic pigments (Fig. 13). Some are so small (1–2 microns) and delicate as to be almost beyond detection. Mary Parke, who has established pure cultures of several species in the Plymouth Marine Laboratory, has done much to elucidate their structure, life cycles and taxonomy. When this diverse and elusive assemblage of micro-organisms is better known, we should, above all, hope to gain closer appreciation of their significance in oceanic productivity and their importance as food for minute members of the zooplankton and for pelagic tunicates.

Dinoflagellates, known as micro-fossils from Silurian times, are considerably larger than calcareous flagellates and micro-flagellates. Most species fall in the size range from 10 to 1,000 microns: the largest are species of *Noctiluca*. There are two main groups, armoured and unarmoured. Armoured species form a transparent cell wall of cellulose that soon divides into platelets pierced by minute holes. The pattern and armature of the platelets vary from species to species (Fig. 14). The cell wall of unarmoured species is thin and undivided. There are two flagella and in many forms one flagellum is housed in a groove around the cell: the other is attached so that it trails backwards (Fig. 14). The grooved flagellum rotates the cell, while the other drives it forwards. In some species both flagella arise at one end of the organism.

The many plant-like dinoflagellates contain yellow, brown or greenish photosynthetic bodies. Forms of photosynthetic species (zooxanthellae) are symbiotic in diverse invertebrates and in the open ocean occur in various radiolarians (p. 73). There are also heterotrophic forms, which are able to absorb and process dissolved organic substances as a source of energy. Some species even ingest minute particles of food.

Many dinoflagellates are luminescent. The light is produced in cellular organelles known as scintillons, which flash more frequently by night than by day. Whenever luminescence in the sea looks uniform from a distance and is without large blobs of light, the display is most likely to come from dinoflagellates. At times 'brilliant, short-lived bursts of

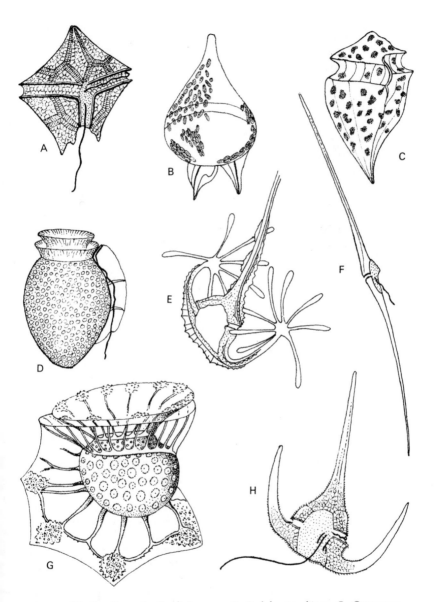

Fig. 14 Dinoflagellates. A, *Peridinium* sp.; B, *Podolampas bipes*; C, *Oxytoxum tessalutum*; D, *Dinophysis acerta*; E, *Ceratium ranipes*; F, *Ceratium fusum*; G, *Ornithocercus steinii*; H, *Ceratium tripos*.

light may dart across the sea surface and the crests of rolling waves glow as the dinoflagellates activated into emission through water disturbances create an aquatic version of *aurora borealis*' (Smayda, 1971).

Diatoms have been falling as deposits on the earth since Triassic times. Under the modern ocean there are great belts of diatom ooze around Antarctica and across the North Pacific.

Diatoms form transparent but rather thick cell walls of hydrated, amorphous silica. Just inside the cell wall is a thin layer of cytoplasm that holds the greenish or brownish photosynthetic bodies (chloroplasts). Actually, the cell wall is formed in two halves, part of one fitting closely inside a corresponding part of the other. When the cell divides in two, each new cell is enclosed in one half of the old cell wall. A new cell wall is then formed inside the old half. Thus, as one vegetative division follows another, diatoms gradually decrease in volume. The shrinkage is not indefinite but stops at a fixed cell size that varies from species to species. There is now an auxospore stage when the vegetative cells form male and female cells. In certain species it seems that the male cells, which have flagella, swim into a cell containing eggs and fertilize them. The fertilized egg cells, which are next extruded through the opening that admitted the male cells, enlarge to full size and form the normal siliceous cell wall. The cell divisions of the vegetative stage now begin once more. During unfavourable times, the diatoms of coastal waters also form resting spores, which settle on the bottom and germinate when conditions improve. In the open ocean this kind of renewal is clearly impossible, but if buoyant resting spores are produced, as some suggest, they have yet to be discovered.

There are two main groups of diatoms. The pennate forms (Pennales) are bilaterally symmetrical; the centric forms (Centrales) are radially symmetrical. Pennate species, some of which are mobile, live on the sea floor or attached to seaweeds, grasses and ice. The planktonic species are centric forms of diverse appearance: hair-like, needle-like, pill-box-like, lamellate, stellate, bracelet-like, chain-like and so forth (Fig. 15). Sizes range from 1 to 2,000 microns.

Sources of silicon may be in short supply in the sea, but there is another group of planktonic plants with siliceous cell walls, the Silicoflagellata (Fig. 13). These forms, which bear two flagella and vary in size from 10 to 150 microns, play a minor role in marine productivity.

All the foregoing members of the phytoplankton are eucaryotes, organisms with their genetic material concentrated in a definite nucleus. Procaryotes which have their genetic material dispersed through the cell and not in a nucleus are represented in the phytoplankton by the blue-green algae (Cyanophyta) and bacteria. In blue-green algae the photosynthetic pigments are held by elaborate internal membranes. The well-known tropical species, *Trichodesmium erythraeum*, is a filamentous form that

A stomiatoid fish of the genus *Cyclothone*, the most widely and deeply represented of all deep-sea fish genera. Note the small eyes and light organs. Length about 6 cm.

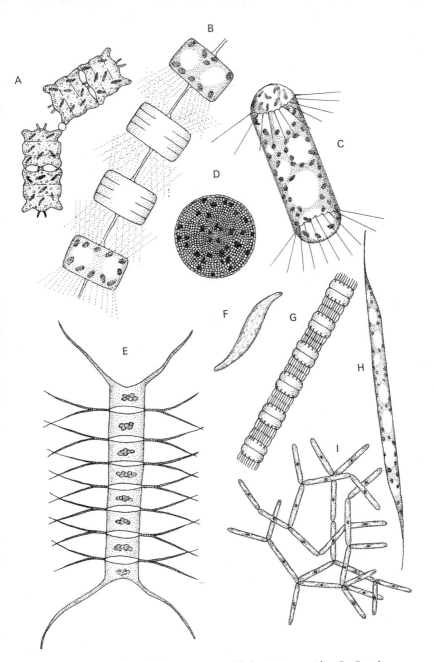

Fig. 15 Diatoms. A, *Biddulphia aurita*; B, *Thalassiosira gravida*; C, *Corethron criophilum*; D, *Coscinodiscus radiatus*; E, *Chaetoceros decipiens*; F, *Gyrosigma* sp.; G, *Skeletonema costatum*; H, *Rhizosolenia hebetata* forma semispina; I, *Nitzschia frigida*.

is often associated in loose bundles (Fig. 13). Blooms of this species, which give the sea a rusty appearance and, indeed, gave the Red Sea its name, may at times cover great stretches of the tropical ocean. It seems that *Trichodesmium* and certain other blue-green algae, fix molecular nitrogen and are thus able to flourish in waters deficient in nitrates and other nutrient salts.

Bacteria abound in the euphotic zone, including the neustonic layer dealt with earlier (p. 42), but decline markedly in the underlying waters, especially at levels below 1,000 metres. Many are attached to organic and inorganic particles and to organisms, both living and dead. Though there are photosynthetic species, planktonic bacteria seem to play a minor role in oceanic productivity '... but in association with detritus they may form an appreciable organic reserve during times of low phytoplankton density' (Parsons and Takahashi, 1973). More significantly, bacteria and other heterotrophic organisms, such as yeasts and moulds, are the decomposers in ecosystems. Recent work by Azam and Hodson (1977) strongly suggests that most of the microbial heterotrophic activity in the ocean is due to bacteria that are *unattached* to larger organic particles. By transforming dead life into nutrient salts and accessory food substances, the decomposers provide the means of new life to the plant producers of the epipelagic ecosystem.

General features of distribution

Phytoplankton communities flourish in the surface waters of the entire ocean. In any given area there is continual and sometimes rapid change in the composition of a community. In a litre of sea water the number of species may vary from less than 10 to greater than 250. Each species waxes and wanes in relative abundance and the cycle is presumably shaped by a complex of physical and biological factors. As we shall see, supplies of nutrient salts are often limited, and as fast as a species grows and reproduces it is consumed by countless planktonic animals. Compared to the zooplankton, the plants of the plankton come and go rather quickly. Even so, numerous phytoplankton species, like their animal counterparts, have spread over most of the ocean. In a detailed study of certain species from ocean-wide samples, Hasler (1976) found that she could divide them into cosmopolitan species (e.g. *Leptocylindrus mediterraneus*, *Nitzschia bicapitator* and *Thalassiosira eccentrica*); warmwater species (e.g. *Coscinodiscus nodulifer*, *Lauderia annulata*, *Nitzschia marina* and *Thalassiosira diporocyclus*) and cold-water species. The latter could be subdivided into bipolar species (e.g. *Nitzschia cylindrus*, *Thalassiosira antarctica*), northern hemisphere species (e.g. *Nitzschia grunowii*, *N. seriata* and *Thalassiosira constricta*) and southern hemisphere species (e.g. *Nitzschia curta*, *N. kerguelensis* and *Thalassiosira gracilis*) (see also pp. 473–4). But our present concern is largely with certain contrasting features in the distribution of the main groups of phytoplankton.

Diatoms are dominant in cool or cold waters. They flourish in polar seas, not only as planktonic species, but also as attached (pennate) forms on the underside of sea ice. Diatoms bloom in profusion during the long summer days in antarctic waters. South of the Antarctic Convergence, where primary productivity is high, diatom ooze (and mud) accumulates on the deep-sea floor. Productivity is lower in the waters north of the convergence and here carbonate oozes are formed. Diatomaceous muds are also formed below the diatom-rich waters of the subarctic water mass in the North Pacific. In middle latitudes diatoms flourish in the cool, nutrient-rich waters that rise to the surface in upwelling regions off the coasts of California, Peru and southwest Africa. Species of diatoms are also prominent in temperate regions but in the subtropical and tropical waters they are generally much less evident and the individuals of most species tend to have thin siliceous walls. An outstanding exception is in the eastern equatorial Pacific, where the remains of diatoms are common in the oozes under the divergence between the equatorial current and countercurrent. As will be seen (p. 64), the divergence draws up nutrient salts to the surface where the diatoms can flourish (see also p. 222).

Calcareous flagellates, dinoflagellates and blue-green algae are often the most prevalent forms in warm oceanic waters. But all three groups are represented in temperate and cold waters. Unlike many diatoms, members of these groups are able to subsist in waters of very low nutrient concentrations. On moving from coastal to oceanic waters there is usually a marked decrease in the size range of species in phytoplankton communities and also a decrease in the minimum requirements for nutrient. There may well be parallel changes in oceanic waters on moving from temperate to tropical regions.

The euphotic zone

Like their land-based relatives, oceanic plants synthesize energy-rich compounds, such as glucose and adenosine triphosphate (ATP), from the low-energy compounds, water and carbon dioxide. The energy for this process comes from sunlight absorbed by plant pigments (photosynthesis). In the ocean the main producers are the microscopic algae of the phytoplankton, which contain light-absorbing chlorophyll and accessory pigments in chromatophores.

The overall photosynthetic process may be represented as

$$nCO_2 + 2nH_2O - \text{visible light} \rightarrow n(CH_2O) + nO_2 + nH_2O$$

Wavelengths absorbed range from 400 to 700 nm (blue to red) and the relative light absorption over this spectrum varies somewhat among the groups of phytoplankton. There is also the differential absorption of light rays by the sea: thus, though chlorophyll can absorb red light, there is virtually none below a depth of 10 metres, where the blue-green part of the spectrum (400 to 600 nm) provides most of the energy for photosynthesis.

The euphotic zone of the ocean, where sunlight is strong enough for photosynthesis, extends from the surface to a depth of about 150 metres. Photosynthesis must be confined to the hours of daylight but respiratory processes, which consume the energy-rich substances formed during photosynthesis, go on all the time. In culture media the respiration of phytoplankton seems to be about 10 per cent of the maximum gross photosynthesis but under certain conditions the respiration rate of some species increases by 2 to 3 times in the light. If dinoflagellates are abundant in samples, respiration rates may be 35 to 60 per cent of the maximum gross photosynthesis and such high rates are thought to be related to the mobility of the dinoflagellates. Whatever the relative levels of photosynthesis and respiration, the difference between the two processes over the life cycle of a species must be sufficient overall to provide for growth and reproduction.

At any given time, the rate of photosynthesis, which can be expressed as the carbon assimilated in milligrams per milligram of chlorophyll per hour, depends on the level of light intensity. Experiments show that there is a light saturation level above which the rate of photosynthesis is not increased, or may even be decreased, by further increments of light intensity. Below this level the rate of photosynthesis falls with decreasing light intensity. Thus, in the euphotic zone the amount of carbon assimilated must fall with increasing depth until a level is reached (the compensation depth) where the amount gained during photosynthesis balances the amount lost during respiration.

If the respiration of phytoplankton is taken as 10 per cent of the maximum photosynthesis (see above), graphs of photosynthetic rate against light intensity show that light values at the compensation depth are about 1 per cent of the surface radiation. The more transparent the sea, the greater is the compensation depth, which in the open ocean reaches a maximum level of about 150 metres. Clearly, a phytoplankton community will only maintain its identity if the populations of the member species are adequately represented at levels above the compensation depth.

As planktonic plants drift in wind-driven circulation (pp. 33–9) they follow the turbulent motions of their medium. Turbulence, which increases with the wind and may be lively for long periods, will mix the phytoplankton below the euphotic zone. Some cells will be mixed back into this zone but many may never return, or stay so long in the twilight that most of their food reserves are consumed by respiration. Furthermore, poorly nourished cells may become less buoyant and sink into the dark (see also below). Turbulent waters are not productive. In the northern North Atlantic, for instance, the spring outburst of phytoplankton follows closely the warming of a stable water column, when the 'seasonal' thermocline (pp. 23–4) returns (Glover, 1967). Turbulent mixing is unable to penetrate the marked density gradient of a thermocline. In tropical and subtropical waters, where the 'seasonal' thermocline is more or less permament, phytoplankton production continues throughout the year. On the other hand, the thermocline bars the upward transfer of nutrient salts and so limits primary production (p. 24).

As already implied, staying in the euphotic zone depends on the buoyancy of the phytoplankton,. Planktonic diatoms, which have cell walls of dense silica and no powers of movement, would seem to be ill equipped for flotation. Yet when diatoms actively growing in sea water are examined through a horizontal microscope they can be seen poised, quite motionless, in midwater, or they may move slowly up and down, carried by small convection currents. Diatoms thus appear to be perfectly buoyant and the reason seems clear. From time to time the protoplasm of the diatom flows together to form a round resting spore: at the same time, the cell loses the watery sap that forms at least three-quarters of its volume. The resting spore enclosed within the siliceous valves of the mother cell sinks in sea water, and so does the spore by itself. Evidently the buoyancy of the diatom must reside in the cell sap, which would have a lighter salt complex than sea water if the cell tended to exclude the heavy divalent ions (Gross and Zeuthen, 1948). An increase in the cell's lipid content would also increase buoyancy. The ionic mechanism is considered more important, especially in cells with large vacuoles (Smayda, 1970). For instance, in the large diatom, *Ethmodiscus rex*, the total ionic content of living cells is about two-thirds of the content in dead cells and sea water.

But there are other considerations. 'It has been argued that in very nutrient-starved conditions, sinking is an active process for phytoplankton, enabling them to enrich their metabolism. Consider a plant cell motionless with respect to the water. Under these conditions there is the possibility of the cell soaking up the nutrients in its immediate vicinity and thus building up a shadow of low nutrient water around the cell. The shadow effect would be overcome by sinking, since the medium surrounding the plant is being constantly changed.

'Effects of phytoplankton sinking can be observed in the vertical distributions of chlorophyll. Where nutrients are low in the surface water and the density of the water is also low, maximum concentrations of chlorophyll are found in the lower limit of the euphotic zone. It has also been noted that these maxima occur where there is an increase in nutrients. Sinking experiments conducted by various workers have indicated that the buoyancy of the phytoplankton is somehow directly related to the physiological conditions of the cell. Hence a healthy cell, which is one enriched with nutrients, is more buoyant than a cell that is nutrient-starved. In contrast, when the stability of the water column is low and vertical mixing is active, maximum growth and concentration of chlorophyll is near the surface' (Yentsch, 1971).

Experiments on living phytoplankton show that the sinking rate (in metres per day) ranges from zero to 30 (Smayda, 1970). Larger and denser organisms sink faster than smaller and less dense ones of the same form (the smaller the organism the greater its relative surface area and frictional resistance to sinking). Relatively large 'bladder-like' diatoms, such as *Ethmodiscus rex*, evidently depend on their very low density to maintain low sinking rates. Besides the relatively high surface area, inherent in their smallness, diatoms have also increased such area through the evolution of spines, setae, gelatinous threads, wing-like projections and so forth (see Fig. 15).

Apart from bacteria, the smallest members of the phytoplankton are the micro-flagellates, which are motile and bounded by fragile cell walls. These forms should easily maintain their level, but the calcareous flagellates and armoured dinoflagellates carry substantial ballast. If both forms tend to swim towards the light, they will rise in the water column, but their calcareous or cellulose armour may well cause them to sink when they are inactive. Certainly, the coccoliths or platelets of numerous species bear projections of diverse form (Figs. 13 and 14) which, as in diatoms, will increase their surface area and retard their rate of sinking. Large dinoflagellates swim at speeds from about 2 to 20 metres per day, and certain species, at least, change their position in the water column during the day. For instance, *Gonyaulax polyedra* 'displays a diel migration in which it absorbs nutrients from 10 to 15 m during the night and swims to the surface during the day, where light is available for photosynthesis and growth, but where nutrients are much lower than at 10 to 15 m' (Parsons and Takahashi, 1973). In some tropical areas dinoflagellates make daily migrations between near surface waters and those of greater nutrient content below the euphotic zone (Eppley et al., 1968). If this kind of adaptive response is common in the flagellated members of the phytoplankton, it may well go far to account for their success in the nutrient-poor waters of the subtropical and tropical zones. In such surroundings, the immobile centric diatoms, which are much more tied to the 'shadow zone' of nutrient depletion around them (p. 59), would seem to be at a disadvantage.

Patterns of primary productivity

Primary production is the rate at which high-energy organic compounds are formed by photosynthesis from the common inorganic substances, carbon dioxide and water (or some other hydrogen donor). The alternative is needed because photosynthetic bacteria use other hydrogen donors. Since photosynthetic organisms make their own foods, they are called autotrophic, a term that also fits chemosynthetic bacteria. But the main primary producers of the ocean are the algae of the phytoplankton.

Primary production may be estimated by measuring the amount of carbon assimilated or oxygen produced over a given period. The first method, which was introduced in 1952 by E. Steeman Nielsen, involves adding a known amount of radioactive carbon (C^{14}) in the form of sodium bicarbonate ($NaHC^{14}O_3$) to bottles of sea water with the natural content of phytoplankton and other organisms. The bottles may be left in the light or suspended in the euphotic zone. After suitable preparation, the amount of C^{14} assimilated by the phytoplankton can be estimated through counts with a Geiger–Muller tube. Oxygen evolved may be measured by using an oxygen electrode or the Winkler titration technique.

The C^{14} technique, which is considerably more sensitive than the oxygen method in regions of low productivity, is the one generally used in the open ocean. Primary productivity is usually expressed in milligrams of carbon assimilated under a square metre of sea surface per day (see Fig. 16).

It will already be evident that primary production depends on a number of factors: the intensity and quality of light, the stability of the water column and the supply of nutrients. Productivity also falls with temperature. Thus the photosynthetic rate of arctic and antarctic phytoplankton per unit of chlorophyll is less than that of phytoplankton in temperate or tropical waters. Salinity and hydrogen ion concentration (pH) may also be involved. Under nutrients are not only included the essential phosphates and nitrates but also accessory growth substances such as vitamins, notably vitamin B_{12}. Vigorous phytoplankton growth also requires quantities of iron and manganese, needed for the formation of chlorophyll. Diatoms require silicates enough to make their cell walls.

Measurements of primary production have now revealed the main patterns of ocean productivity (Fig. 16). Differences in productivity between one region and another are presumably related to productivity controlling factors, but how are these to be unravelled? The way was

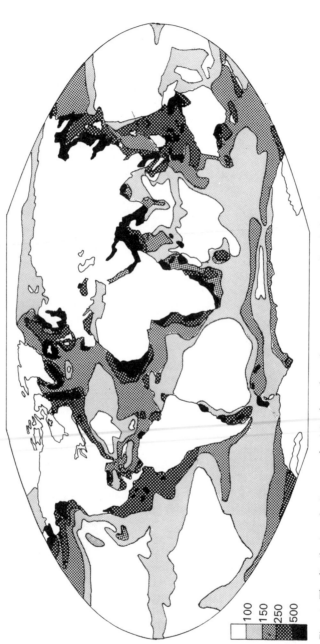

Fig. 16   The distribution of primary productivity throughout the ocean, as estimated with the radiocarbon method in mg carbon per metre² per day. (After Cushing, 1975)

100
150
250
500

shown in 1840 by the German chemist Liebig in his book on the application of chemistry to agriculture and physiology. His statement that the 'growth of a plant is dependent on the minimum amount of food stuff presented' became known as Liebig's law of the minimum. In 1899 another German chemist, Karl Brandt, thought that Liebig's law might also apply to the ocean; that plants might from time to time be limited in their growth and reproduction by lack of nutrient salts, such as phosphates and nitrates, which occur in small quantities in the sea. Later research has amply proved the usefulness of Brandt's suggestion. By measuring the distribution of nitrates and phosphates in the ocean and the cycles in the rate of supply of these salts to the euphotic zone, one may go far towards understanding the productivity of the ocean in space and time.

The subtropical and tropical belts of the ocean, where deep-sea animals are most diverse, will be considered first. In these regions the intensity of light and the stability of the water column are adequate for the growth and reproduction of phytoplankton in all seasons. Indeed, as already stated, primary production is more or less continuous throughout the year. But supplies of phosphates and nitrates in the euphotic zone are meagre, which may be seen in contrast to conditions in the entire water column. In general, the vertical distribution of phosphate and nitrate concentrations (μg atoms per litre) is layered in a characteristic way. The surface layer with its very low and rather uniform concentrations overlies a layer marked in depth by increasing concentrations, which in turn leads to a layer of maximum concentration, usually between depths of 500 and 1,500 metres. Below this is a deep layer in which nutrient salt concentrations show relatively little change with depth.

This stratification of nutrients is bound to life and death in the euphotic zone. The proportions of carbon, nitrogen and phosphorus in the protoplasm of planktonic plants is about $100C : 15N : 1P$ (by atoms), and during growth these elements are drawn from the sea in approximately these ratios. If the uptake of nutrients continues to the point where the most limiting nutrient is reduced below its minimal level, growth of the phytoplankton will cease. And nutrients continually leave the euphotic zone. The plants tend to sink more quickly after death, when processes of decomposition eventually renew the supply of nutrient salts. At a depth of about 200 metres all plant material has been oxidized back into mineral form. Animals of the euphotic zone also sink and decay after death. Moreover, potential nutrient-yielding material leaves the euphotic zone each day as food in vertically migrating animals (see pp. 234–5). In one way or another, nutrient salts accumulate below the euphotic zone. If there is no updraft of nutrient salts into the euphotic zone, the phytoplankton must depend on the local recycling of nutrients by decomposition and the excretion of nutrient-yielding materials by the local animals.

The euphotic zone in subtropical regions overlies the central water masses, which are circumscribed by the great oceanic gyres (p. 37). In these waters a virtually permanent 'seasonal' thermocline underlies a stable, well-lit water column. Evidently the upward turbulent transfer of nutrient salts is so restricted that subtropical waters are the least productive of the major oceanic regions. The waters are so clear and contain so little particulate matter in the form of phytoplankton and other microscopic organisms that there is little to interfere with the blue light scattered by the water molecules. Subtropical waters are a deep blue.

Deep blue waters have been called the deserts of the ocean but they are not uniformly deserted. Surveys in the eastern Pacific have shown that productivity may be enhanced where the thermocline is closest to the surface. In such regions the thermocline may be eroded if wind-induced turbulence is sufficient to plough the sea and bring nutrient salts to the surface waters. A thermal front across the Sargasso Sea separates a cooler and more productive northern region from a warmer and less productive southern region. North and south of the front there is a difference in stability of the water column. The upper part of the water column in the southern Sargasso Seas is well stratified during all months of the year, while that of the north is only well stratified from about May to October (Backus et al., 1969). These contrasting conditions are related to parallel contrasts in the abundance and distribution of mesopelagic fishes (pp. 474–7).

There is a wider range of primary production in equatorial waters, regions of which are several times more productive than are subtropical waters (Fig. 16). North and south of the equator the trade winds drive the west-flowing equatorial currents which, under the influence inherent in the earth's rotation, tend to diverge from the equatorial countercurrent (see also pp. 34–7). Upwelling occurs along the divergences, and is particularly well marked over the eastern half of the equatorial Pacific. Here, the refreshing of the surface waters with upwelled nutrients leads to high productivity (150–250 mgC/m²/day) (Fig. 16) and good standing stocks of zooplankton, which support, inter alia, profitable fisheries for tuna and billfishes.

In the Indian Ocean the North Equatorial Current and the Equatorial Countercurrent are in full flow and divergence from November to March, the time of the Northeast Monsoon. Shcherbinin's (1971) studies of the geostrophic circulation even indicate that the direction of the North Equatorial Current undergoes no radical seasonal changes. Russian surveys also show that the waters about the equatorial divergence (0–10°S) are rich in nutrient salts and high in primary productivity (Maximova, 1971). But the northwestern Indian Ocean, particularly the Arabian Sea, is richer still in nutrients and plankton (Fig. 16). Evidently there is extensive upwelling of nutrient-rich water northwards across the Equator to replace surface water flowing southwards and away from the Arabian Sea (Ryther et al., 1966).

In the Atlantic Ocean, the Equatorial Countercurrent is developed only over the eastern half, and it eventually flows into the Gulf of Guinea. Upwelling and zones of enhanced productivity are evident in unstable and sluggish waters between the above current system and the South Equatorial Current (Bernikov et al., 1966).

In temperate and subpolar waters the winter-cooled surface waters sink and form convection currents, the rising parts of which renew the supplies of nutrient salts in the euphotic zone. When spring returns the timing of the outburst of phytoplankton depends on the development of a stable water column (see also p. 59) and adequate levels of light. Growth and reproduction are mainly in the spring but continue at lesser rates during the summer. Populations and communities decline during autumn as the thermocline breaks down and the days shorten.

The surface layer in the antarctic is well supplied with nutrient salts by the waters that move up to replace the surface waters that sink at the Convergence (p. 28). During the long well-lit days of December and January, when the retreat of the ice has opened great expanses of the sea, and melt-water stabilizes the euphotic zone, production of the phytoplankton is at its height. Productivity declines during autumn as the ice reappears and the days shorten. Both north and south of the Antarctic Convergence parts of the Southern Ocean are among the most productive in the world (Fig. 16). During the productive season teeming stocks of plant and animal plankton support, inter alia, the great whales and millions of seals, penguins and fishes.

Primary productivity is high over great stretches of the continental shelves throughout the world, where nutrient salts and organic material from the land may contribute greatly to plant growth. But some of the most productive near-shore waters are related to desert lands. Off northwest and southwest Africa in the eastern Atlantic, Somaliland and Arabia in the northwest Indian Ocean, and Peru and California in the eastern Pacific, seasonal or variable winds blow the surface waters away from the shore, and to take their place nutrient-rich waters well up from deeper levels. The biology of these coastal upwelling regions is not our present concern, though there is evidence that the upwellings may enrich neighbouring oceanic waters. Off southwest Africa and Peru there is a cyclonic gyre such that the Benguela and Peru Currents leave the coast and flow into the open ocean in a northwesterly direction (Bulatov and Stepanov, 1971). Presumably the offshore transport of nutrient-rich water contributes to the high productivity of both gyres. During the monsoons, particularly the southwest, upwelled water off the Arabian and

Somali coasts may make some contribution to the high productivity of the open Arabian sea.

Lastly, enhanced patches of productivity in the broad oceanic pattern may arise near islands. Oceanic currents rounding islands may generate large eddies and upwelling on the leeward side of the flow sufficient enough to lead to increased productivity. Prominent changes in under-water topography may also interrupt the flow of surface currents enough to cause turbulence and upward transport of nutrient-rich water. On a smaller scale sea mounts may also cause upwelling.

This survey of the patterns of oceanic productivity is all too brief, but the main object is to provide a background for later considerations. There we shall endeavour to see how far differences in productivity in space and time are related to the diversity, distribution and the life-forms of deep-sea animals.

Animal life in the deep sea

The evolutionary radiations of animals have penetrated to all depths of the ocean. In the deep sea, as in neritic waters, bottom-dwelling species (the benthos) outnumber the (pelagic) species of oceanic space (p. 000). The deep and shallow seas are also alike in that animal life is most diverse in the subtropical and tropical belts. There is also a broad taxonomic similarity and continuity between neritic and deep-sea faunas. Thus, while selection pressures in the deep sea have evoked remarkable adaptations in the biota, even the most modified forms do not seem excessively changed.

Pelagic deep-sea animals are represented from surface to near-bottom levels of the ocean. Beside the permanent residents in the depths, diverse species rise up each day to feed in the productive surface waters and/ or use these waters as nursery grounds for their young (Chapter 12). As we shall see, if not altogether clearly (p. 230), the fruits of primary production are adequately partitioned, though not evenly, among the many animal species of the pelagic realm. It is certainly easier to see how enough life-sustaining oxygen is circulated in the water masses to all parts of the ocean (pp. 31–2). And, like the water masses, pelagic life remains recognizably stratified throughout the water column.

Benthic deep-sea animals live at all depths from the threshold of the continental slope to the bottom of the deepest trenches. As on the continental shelves, an epifauna living on the sea floor may be distinguished from an infauna living in the sediments.

Again, in continuity with shallow seas, there is a meiobenthic fauna of small invertebrates (pp. 143–5), which form part of the infauna. Though benthic life is very sparse in the ocean basins below regions of low primary production (pp. 294–300), conditions are such that life is sustainable over virtually all of the deep-sea floor.

The present intention is not only to review the main groups of deep-sea animals but to see how they are organized, both structurally and functionally, for life in the main divisions of the deep sea. Such consideration will serve to introduce later parts of this book.

Pelagic animals

Even if one needs a micronekton to merge zooplankton and nekton (pp. 105–6), the division of pelagic life into drifters and swimmers is still a useful concept. Size, activity and level of organization are the main criteria in subdividing, even if somewhat cursorily, the great range of pelagic animals. The smallest and simplest animals, the protozoans, together with the earliest young stages of various metazoans, are now called micro-zooplankton. Nearly all of the protozoans are active ciliated forms (tintinnids and oligotrichs) and their more passive relatives, the Foraminifera and Radiolaria. Naupliar stages of copepods make up most of the metazoan fraction. Beers and Stewart (1969) define micro-zooplankton as forms that will pass through $200\,\mu$ (0·2 mm) filter cloth, though Strickland extends the upper limit to $500\,\mu$ (0·5 mm). During investigations off Lower California, Beers and Stewart (1969) found that average numbers of micro-zooplankton in the euphotic zone were higher in neritic than in oceanic waters. Protozoa outnumbered the metazoans at all stations and made up nearly all of the animals with sizes up to $100\,\mu$ (0·1 mm). Metazoans, consisting largely of nauplius and postnauplius stages of copepods, were relatively the most abundant in the size fraction between 100 and $200\,\mu$.

Micro-zooplankton

In oceanic waters ciliated protozoans of the micro-zooplankton consist largely of tintinnids (subphylum Ciliophora, order Tintinnoida). Each individual is housed in a case of tectin. These cases, which are secreted by the animals and vary in form from urn-shaped to tubular, may be plain or more often strengthened by agglutinated material, such as sand grains, diatom valves or coccoliths (Fig. 17). Tintinnids range in size from 20 to 640 microns and most species live in warm temperate or tropical regions. At times they become very abundant in coastal waters and the surface layer of the open ocean. Their main food, which consists of organic particles, bacteria and the smallest species of the phytoplankton, is wafted to them in currents created by their apical complex of cilia.

'The remarkable evolutionary versatility of the foraminifera has enabled them to colonize virtually every kind of marine environment with the result that fossil foraminifera are more nearly universally present in the marine sediments of the last 100 million years than any other group of fossils' (Ericson and Wollin, 1967). Living foraminifera are known from more than 1,200 species, of which thirty or so form part of the plankton of oceanic waters: the rest live at all depths on the ocean floor

Fig. 17   A tintinnid protozoan (Ciliophora), *Tintinnopsis ventricosa*, with a test formed of foreign particles. (Redrawn from Meglitsch, 1967)

and will be considered under benthic animals (pp. 145–6). All species secrete a chitinoid test around the main body of protoplasm. The test is pierced by one or more holes through which the protoplasm flows out to form the filamentous pseudopodia. In the planktonic species and others the test is strengthened by calcareous material. Many of the benthic species have agglutinated tests (p. 146). Though foraminifera range widely in size from 20 microns to 190 millimetres, the mean size of the common species is about 500 microns.

The planktonic species live in tests formed of a cluster of interconnecting chambers, which in many species are spirally coiled and look like miniature snail shells. Beside the main test aperture, the walls of the chambers are punctured by many small holes through which emerge the pseudopodia. When fully extended these cytoplasmic filaments not only radiate in all directions, but flow into one another to form a network.

It seems that the main food of the surface-dwelling forms consists of phytoplankton and small ciliates, which are trapped by the pseudopodia and digested on the spot (Figs. 18 and 19).

Planktonic foraminifera are most abundant in the epipelagic zone, where most, if not all, species are thought to spend their early life. The epipelagic species generally have spiny tests and the spines not only support an outer, frothy cytoplasm (Fig. 18) but presumably help to keep the rate of sinking to a low value (see also pp. 316–17). The species without a spiny test have a greater depth range and some (e.g. *Globorotalia crassiformis, G. scitula* and *Hastigerinella digitata*) are evidently mesopelagic or bathypelagic forms (Bé, 1966). Most species live in tropical and

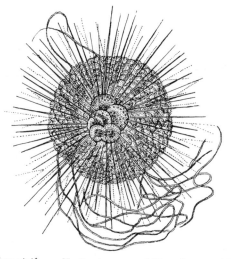

Fig. 18   A living foraminiferan, *Hastigerina murrayi.* Note the central chambers (and projecting spines) surrounded by vacuolated cytoplasm. The pseudopodia are dotted.

subtropical waters, which overlie most of the vast expanses of calcareous sediments that are rich in the shells of planktonic foraminifera (globigerina ooze) and cover about 45 per cent of the deep-sea floor (Fig. 5).

Living Radiolaria, which are known from about 4,400 species, are exclusively marine and with a few exceptions planktonic. The very numerous fossil species go back to Precambrian times. In nearly all living species the endoplasm is housed in a central capsule: the ectoplasm contains, or is contained by, a gelatinous layer. As in foraminiferans, long and fine pseudopodia radiate outwards and finally anastomose to form an elaborate food trapping net. Some species are luminescent (Fig. 19).

All but a few species secrete some kind of skeleton formed of spicules

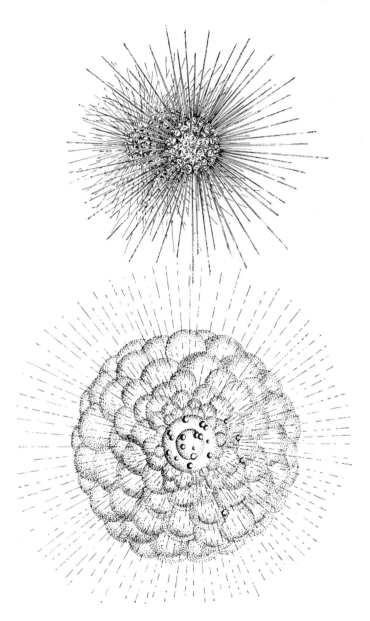

Fig. 19 A spiny-shelled foraminiferan *Globigerina bulloides* (above) and a radiolarian, *Thalassicola pelagica*.

that may be isolated but are much more frequently joined in diverse and beautiful forms, many of concentric and radiating patterns. D'Arcy Thompson (1942) was reminded of snowflakes. 'The little skeletons remind us of snow crystals (themselves almost endless in their diversity), rather than a collection of animals constructed in accordance with functional needs and distributed in accordance with their fitness for particular situations.' The unity in diversity of radiolarian skeletons that so intrigued Sir D'Arcy was rediscovered by Buckminster Fuller (1961), who was charmed to find geodesic structuring in their skeletons, and in those of diatoms. Indeed, the skeletons of numerous radiolarians look very like the geodesic domes designed by Buckminster Fuller and others. As he wrote: 'Employing the fundamental principles of nature's structuring, the average of all the geodesic domes to date enclose space at approximately one per cent of the weight of materials required by conventional architectural and engineering structures of equivalent volume designed for equivalent load stresses and clear span environment controls.'

Upheld by the sea, radiolarians face no such stresses, but they presumably need to be supported by skeletal designs that make the best possible use of rare supplies of building material. Curiously enough, the skeletons of the Acantharia (sometimes classified as a separate order), which use rare strontium in the form of sulphate, look more like snow-crystals than geodesic domes. But the radiating configuration of spicules is integral to their organization: the spicules support a well-formed gelatinous layer that surrounds the central protoplasmic body. In most acantharians contractile threads (myonemes) are attached between main spicules and the surface of the gelatinous layer, the entire system forming an 'appareil hydrostatique' (see pp. 317–18 and Figs. 20 and 112).

The skeleton of the three other main groups of radiolarians is made of silica. The spicules are isolated in certain species but in most are united, often in geodesic array, to form concentric lattices or shells of extraordinary delicacy and diversity. A central capsule is always present. The ectoplasm, part of which may form a frothy layer that may help in flotation (p. 317), usually contains a well-developed gelatinous layer.

The species of two orders (Spumellaria and Nassellaria), like the acantharians, are predominantly epipelagic. The former are very diverse and include many species with concentric, latticed shells. The deep-sea species belong to a third order, the Phaeodaria, notably to the families Challengeridae and Tuscaroridae. Species of the first family, which have an oval or lenticular shell pitted with minute holes, live at mesopelagic and bathypelagic levels. In species of *Challengeria* the size increases with their depth range (from 0·11 to 0·16 mm at about 500 metres to 0·35 to 0·58 mm from 1,500 to 5,000 metres). The tuscarorids, with a spherical, elliptical, fusiform or pyramidal shell (Fig. 21), are also small forms with a wide depth range.

Like the foraminifera, epipelagic radiolarians trap and digest microscopic plants and animals in their pseudopodial complex. They are said to take flagellates, diatoms, tintinnids and even young stages of copepods. In species other than phaeodarians the protoplasm contains the resting

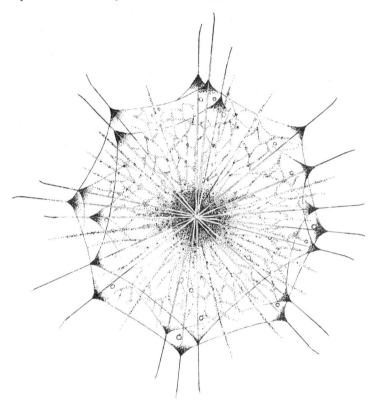

Fig. 20   An acantharian, *Acanthoplegma krohni.*

stages of dinoflagellates known as zooxanthellae. The latter are photosynthetic and there is presumably some form of nutrient exchange between plant cells and their hosts.

Radiolarians are also most diverse in subtropical and tropical regions. At times they are very abundant in productive areas but little is known of their overall success in oceanic waters. During a crossing of the western North Atlantic catches of radiolarians were generally less than those of foraminiferans (see also p. 74). The formation of the siliceous oozes rich in radiolarian skeletons reminds one most of Rachel Carson's (1951) evocation of the 'long snowfall' that forms the deep-sea sediments. 'When

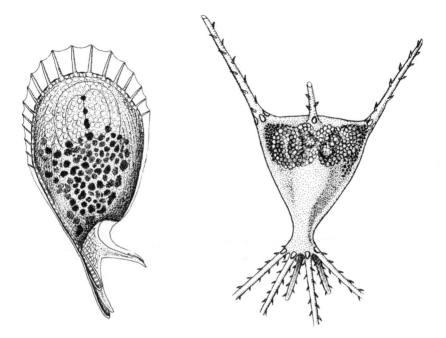

Fig. 21   Two deep-sea radiolarians; left, *Challengeron armatum*; right, *Tuscarora nationalis.*

I think of the floor of the deep sea, the single, overwhelming fact that possesses my imagination is the accumulation of sediments. I see always the steady, unremitting, downward drift of materials from above, flake upon flake, layer upon layer—a drift that has continued for hundreds of millions of years, that will go on as long as there are seas and continents.' It seems odd that radiolarian ooze is almost entirely restricted to the deep-sea floor below the productive waters of the North Pacific Equatorial Divergence. There are also substantial deposits of radiolarian ooze on deep stretches of the ocean floor below the rich waters of the Peruvian Current. Why is so little radiolarian ooze produced in the other oceans?

The zooplankton

Some biological categories are more arbitrary than others, but if the micro-zooplankton is taken to include organisms that will pass through a net with a mesh size of 500 microns, the net, if large enough, would retain all the adult forms of zooplankton, which range in size from 300 micron copepods to the jellyfish *Cyanea capillata*, with an umbrella that may reach a span of about 2 metres. If we remember that the tentacles of a *Cyanea* may extend for 50 metres or more, the size spectrum of zooplankton covers more orders of magnitude than that of the nekton, with a range from a squid or fish of a few centimetres to a 35-metre blue whale.

In taxonomic terms the zooplankton virtually consists of the pelagic forms of Cnidaria (jellyfishes, siphonophores, etc.), Ctenophora (comb-jellyfishes), Polychaeta (bristle-worms), Mollusca (pteropods and heteropods), small crustaceans (mainly copepods and ostracods), Chaetognatha (arrow-worms) and tunicates. Young stages in the life history of these animals also form part of the zooplankton, but are considered elsewhere under early life histories.

The concept of plankton, as already stated, centres on organisms that drift in the hydrosphere. In the ocean the speed of water movements ranges widely from quick-moving surface currents (velocities up to 250 cm/sec in the Gulf Stream and Kuroshio), to the slow circulation of the depths, where velocities may be no more than a fraction of a centimetres per second (pp. 33–9). Velocities of 10 to 50 cm/sec are very common (1 knot is about 55 cm/sec), which are well above sustained swimming speeds of most forms of zooplankton. Though the zooplankton moves with the sea, and indeed, certain species make reliable indicators of water movements, this passive tendency is transcended by strategic activities whereby any species tends to maintain its population and centres of distribution in a moving ocean.

In his comprehensive review of propulsion in aquatic animals Lighthill (1969) shows clearly that the predominant and most successful form of locomotion is by undulatory movement. The smallest forms, such as the flagellates and ciliates, propel themselves by undulation of flagella or cilia. Fishes and whales undulate body and fins to make their way. Annelid and nematode worms throw their bodies into waves. Though propelled by jet reaction, 'the second main alternative to undulatory propulsion in water', squids and certain octopods have fins that are fashioned for undulatory propulsion.

In moving through water, small forms, whose relative motion is very slow, are faced almost entirely by the viscous resistance of their medium, which varies as the first power of their velocity: larger, faster-moving forms have mainly to contend with the inertial resistance, the resistance of the medium to displacements caused by a moving body, which is proportional to the square of the velocity. The relative effect of these two forms of resistance is expressed by the (dimensionless) Reynolds' Number R.

$$R = \frac{LVP}{\mu}$$

where L is some dimension (usually length) defining the size of the body, V is the speed relative to the water (cm/sec) and P/μ (viscosity divided by the density of the medium) is the kinematic viscosity, which is about 0·1 in water. Typical values of Reynolds' Number are: flagellates, about 10^{-3}; small nematode worms (ca. 1 mm), 1; medium-sized copepods (ca. 2 mm), 10; annelid worms (ca. 100 mm), 10^3, and medium-sized fishes (ca. 30 cm), about 10^4. At low values of R (1 or less) viscous resistance is predominant.

The undulatory mode of propulsion is thus effective over a wide range of R (or body size), which is one reason, as Lighthill says, why it is the principal mechanism in the motions of aquatic animals. Jet propulsion is appropriate only in animals large enough to yield relatively high values of Reynolds' Number. In this context it is interesting to survey the locomotion of planktonic organisms. The smallest forms of phytoplankton (nanoplankton) are predominantly flagellates, while the rest are also flagellates or non-motile (most diatoms). The main groups of micro-zooplankton are either ciliates or without means of active movement (foraminiferans and radiolarians). In the zooplankton, however, the small crustaceans and many of the gelatinous forms, which, as we shall see, are dominant groups, have other than undulatory means of locomotion. The copepods, meaning oar-footed, are well named, while the much larger cnidarians and the largest forms of tunicates (Thaliacea) have evolved gelatinous bodies that are shaped to muscular means of jet propulsion (p. 98). Even so, the cnidarians usually start their free-swimming life as small ciliated larvae (planulae) and some of the tunicates have a larval stage with an undulatory tail. The two dominant groups of undulators in the zooplankton are the smallest tunicates (Larvacea) and the arrow-worms.

If biological success is judged by a combination of biomass (the weight of organisms in a given unit of living space), ubiquity and diversity, which in a full sense involves not only the number of species but also the relative abundance of species, the copepod crustaceans are unrivalled in the zooplankton. After the small crustaceans, we shall consider the gelatinous forms of zooplankton (cnidarians, ctenophores and tunicates), and the arrow-worms, which in varying respects meet some of the above criteria. The polychaete worms and molluscs are much more successful as benthic forms.

Small crustaceans (Copepoda and Ostracoda)

Like their arthropod relatives, the insects, spiders and so forth, the Crustacea have segmented bodies and paired, jointed appendages that are covered by a skeleton made largely of chitin and protein. The body is divisible into head, thorax and abdomen. In common with most arthropods, the crustacean appendage is essentially a jointed system of hollow levers that are moved by internal striated muscles. Crustaceans are readily distinguished from other arthropods in developing two pairs of

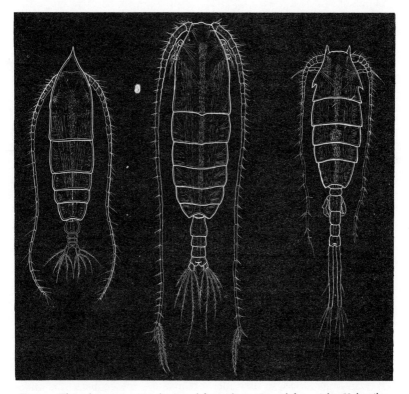

Fig. 22 Three deep-sea copepods viewed from above. From left to right, *Haloptilus acutifrons, Megacalanus princeps* and *Lucicutia bicornuta*. (From Marshall, 1954, after Sars)

antennae on the head. Close behind the mouth is a pair of mandibles followed by two pairs of maxillae.

Copepods have a dorsal carapace that extends over the head and one or two thoracic segments. The first pair of appendages on the thorax (maxillipeds) are succeeded by five pairs of biramous swimming legs in the free-living forms. The abdomen is without appendages. There are about 6,000 species, most of which are marine (Figs. 22 and 23).

Individuals of the pelagic species (about 750) range from about 0·5 to 17 mm in length and live in virtually every part of the ocean. Diversity is greatest in the warmer regions. Much more often than not, copepods dominate the zooplankton. For instance in standard net samples taken during a cruise between New York and Bermuda the volume of epipelagic copepods (measured by displacement) averaged 51 per cent of the total volume of zooplankton (Grice and Hart, 1962). Copepods are also likely to be prominent in catches at mesopelagic and bathypelagic levels. Most species are virtually transparent, but when their tissues become cloudy after preservation, they look like small grains of rice (Figs. 22 and 23). Some of the deep-sea species are tinged in varying degrees with red pigments: a few are black. In some species there are luminescent glands in the skin such that the extrusion of their products causes flashes of light.

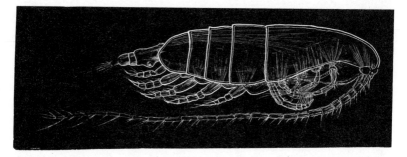

Fig. 23   A deep-sea copepod, *Eugaptilus laticeps*, seen from the side. (From Marshall, 1954, after Sars)

When copepods are cruising, the five pairs of biramous thoracic limbs, which are flattened, paddle steadily backward in sequential strokes. If the animals are disturbed, these limbs beat strongly backward in rapid succession: at the same time the antennae are folded backward along the body. The result is a relatively long dart, and for human eyes at least, such movements in transparent copepods are over in a virtually undetectable flash. The long leap of *Gaussia princeps*, a black, deep-sea copepod that reaches a length of just over 10 millimetres, is quite startling. Copepods of the neuston may even jump out of the sea. Wherever they are, these sudden bursts of speed must often enable copepods to evade their many predators in the zooplankton.

Most of the pelagic copepods (calanoids) have long and well-developed first antennae, which may be suddenly flexed to jerk the animals forwards. When copepods are feeding these antennae are extended outward, and no doubt help them to find their food. Species that feed mainly on phytoplankton are generally the most abundant. The classic instance is

Calanus finmarchicus, which in north temperate and subarctic waters forms much of the food of large whale-bone whales and herring. Herbivorous copepods filter their food through the long fine bristles of their maxillipeds but if the phytoplankton cells are large enough, they are seized one by one. There are also omnivorous and carnivorous species (pp. 232–3). Members of all three trophic groups exist together at surface and deep-sea levels of the ocean. In the tropical ocean the number of copepods (calanoids) in a cubic metre of water falls from about 15 in the surface waters to about 2 at lower mesopelagic levels (500–1,000 m). Between 1,000 and 8,000 metres the number ranges from 0.42 to 0.05 (Vinogradov, 1968). The considerable success of some mesopelagic species is no doubt related to their daily vertical migration in search of better food supplies above the daytime levels of their populations (pp. 234–5).

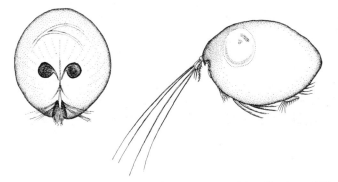

Fig. 24   Ostracods. Left, a front view of *Gigantocypris*; right, side view of *Macrocypridina castanea*.

Ostracods, which have much the same size range as copepods, are very differently made (Fig. 24). The indistinctly segmented body is housed in a laterally compressed, bivalved carapace that may be closed by an adductor muscle. There are two pairs of thoracic appendages but the antennae provide the main propulsive power. Most of the more than 2,000 species are marine and benthic in habit.

Ostracods rival the copepods in ranging adaptability, but the planktonic species are not so abundant. At depths below 1,000 metres the numbers of copepods in a cubic metre are generally much more than ten times those of ostracods. Some of the deep-sea species are luminescent and red or brown in colour. The largest form, *Gigantocypris*, with an almost globular carapace that attains a length of 1 cm, is a bright red (Fig. 24). *Gigantocypris* normally propels itself with a pair of fan-like antennae protruding through a slot in the carapace at the anterior of the animal. Balanced strokes propel the animal forward and movement of one or other antenna individually serves to rotate the animal round its vertical axis' (Macdonald, 1975). Most planktonic forms are eyeless, but the eyes of *Gigantocypris* look like '... two huge forward-facing, headlamp-like reflectors, which perhaps assist in its capture of the small fish, chaetognaths, and copepods found in its stomach' (Clarke and Herring, 1971). If pelagic ostracods are generally carnivores or scavengers it is natural that they are less numerous than copepods. Compared to the herbivorous or omnivorous copepods their place in the food chain is less close to the primary source of food, the phytoplankton (see also pp. 234–5). Like the copepods, some of the ostracods (e.g. *Conchoecia*) undertake daily vertical migrations (Angel and Fasham, 1974). At one station in the eastern North Atlantic, species of *Conchoecia*, especially *C. curta*, formed most of the food of the hatchet-fish, *Argyropelecus aculeatus* (Merrett and Roe, 1974).

Gelatinous forms

The gelatinous forms of zooplankton are cunningly fitted for a marine existence. Though usually jelly-like and fragile, the bodies of cnidarians, ctenophores and tunicates are supported and kept in shape by the sea. In cnidarians the outer and inner tissues of the body wall are separated by a gelatinous layer (mesogloea), which is greatly developed and shaped as a matrix for the parachutes, bells and other parts of planktonic medusae, jellyfishes and siphonophores. Such organization is found also in the comb-jellyfishes (Ctenophora). In pelagic tunicates, gelatinous tissue is a matrix for the 'house' of larvaceans, the tunic of salps and doliolids and the tubular setting of pyrosomids.

The water content of gelatinous tissue in cnidarians is 95 per cent or more. The matrix contains collagenous fibres that not only support it but give the jelly elasticity as well as a relatively high viscosity. In tunicates the test, whatever its consistency, is made largely of a cellulose called tunicin. More generally, gelatinous tissue forms a hydrostatic skeleton that is very well fitted for propulsion by jet reaction. After each pulsation of the bells of jellyfishes and siphonophores, which forces out a jet of water, the elasticity of the jelly and the return flow of water restore the bell to its resting shape in readiness for the next pulse. Much the same is true of salps and doliolids in motion.

When the gelatinous bracts of a siphonophore are detached from the colony they float upwards (Jacobs, 1937). They are positively buoyant, but how is this possible in tissue that seems to consist almost entirely of water and sea salts? At Villefranche in 1957 E. J. Denton and T. I. Shaw found that gelatinous animals (like the large dinoflagellate *Noctiluca*) achieve their buoyancy by excluding some of the heavy ions of sea water, notably sulphate, from their body fluids and replacing these ions by lighter ions such as chloride. Denton (1974) writes: 'The results of the Villefranche experiments were quite clear. For five kinds of gelatinous marine animals, medusae, ctenophores, tunicates and heteropod and pteropod molluscs, the whole animals were usually either close to neutral buoyancy or positively buoyant. This buoyancy could be quantitatively explained in terms of the salt composition of the animals' body fluids. These were always isomotic with seawater but contained less sulphate and more chloride, and also generally less magnesium and calcium but more sodium than seawater. In all of the animals studied the buoyancy in seawater mainly depended on excluding some sulphate from the body fluids and replacing this anion by chloride.' Presumably gelatinous tissue contains cells that exchange heavy for lighter ions. Whatever the mechanism, the energy needed for such differential ionic transport must be easily saved by the energy not needed, as it were, by buoyant gelatinous animals to keep their level in the ocean.

Indeed, the gelatinous nature of these planktonic animals must be seen as integral to their whole organization, which is relatively simple and easy to run. Most of the cellular elaboration of a cnidarian resides not so much in their means of locomotion but in their means of procuring and digesting food and in reproducing themselves. Peter David's splendid photograph in *Deep Oceans* (ed. Herring and Clark, 1971) of a deep-sea jellyfish (*Periphylla*) with a lantern-fish ensnared in its tentacles is a striking illustration of a simply organized animal availing itself of rich and long-lasting nourishment in a highly organized one. The nervous system and muscles of the prey alone, which was just starting to be digested and is about as long (5 cm) as the bell diameter of the predator, are many times the biomass of the simple nerve net and muscle cells involved in the movement of the jellyfish and its tentacles. One must not, of course, forget the outlay of materials needed to build the millions of stinging cells in the tentacles (though nematocysts once formed need no maintenance, for they are dead motile units) (Fig. 25). Even more than their bottom-dwelling relatives, the cnidarians of the plankton are the simplest, deadliest and most effective predators in the ocean.

Tunicates are just as cunningly contrived. They are surely the simplest and most effective herbivores in the plankton. As in cnidarians, the feeding, digestive and reproductive systems rather than the nervous and locomotory systems, subsume most of the cellular organization of a tunicate. By means of a house with a system of fine filters (Larvacea) or by means of a mucus filter net secreted by the endostyle (Thaliacea), tunicates are able to trap the smallest and often very abundant members of the phytoplankton. Such plant food is beyond the filtering capacity of most forms of crustacean herbivores, and the larger species are also available to the tunicates.

Living organisms range from simple to very complex forms, but how is one to express their degree of organization? Some measure may be got by estimating cellular quantities of deoxyribonucleic acid (DNA), which contains the genetic blueprint of all levels of organization in living forms. The more complex the organism the greater is the amount of DNA. At extreme ends of the scale, a human cell contains over two million times as much DNA as that in a simple virus. Sponges have about

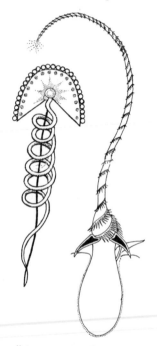

Fig. 25   Right, a stinging-cell (nematocyst) of a cnidarian. Left, a sticky-cell (colloblast) of a comb-jelly (ctenophore).

ten times as much DNA as bacteria. Cnidarians have about six times as much as sponges and in the present context, about twice as much as tunicates, which is remarkable considering that tunicates are related to the chordate ancestors of vertebrate animals. Evidently, DNA content may be a measure of the degree of organization, which is usually but not always related to position in the evolutionary scale.

Simple organisms are easier to run than more complex forms of about the same size, and one way of expressing this difference is by measuring their rate of oxygen consumption (O_2 consumed in c.c. per gram per hour). In the jellyfish *Aurelia* one such measurement is 0·0034 and in a tunicate (*Salpa*) 0·008, which are among the lowest figures for marine animals. This is not altogether surprising. The jellyfish, for instance, essentially consists of simple layers of living tissue dispersed over a massive gelatinous skeleton that is dead apart from its content of amoeboid cells. Moreover, the tissues of a jellyfish or a salp are so thin that they can obtain all their oxygen by simple diffusion: they have no need of respiratory organs. As Schmidt-Nielsen (1975) writes, 'Organisms that are supplied with oxygen by diffusion only, protozoans, flatworms etc. are mostly quite small, less than a mm or so, or have very low metabolic rates, as jellyfish do. Although some jellyfish can be very large, they may contain less than 1% organic matter, the rest is water and salts. They have a very low average rate of oxygen consumption, and the actively metabolizing cells are located along the surfaces where diffusion distances are relatively short.' One can begin to see why planktonic tunicates and cnidarians are sometimes able to flourish in the deep blue deserts of the subtropical ocean.

Thus, by a consideration of all the above attributes, the gelatinous forms of zooplankton are well fitted to their (predominant) place in the economy of oceanic life. But any attempt at a full appraisal must include some consideration of their size. In one respect, as implied earlier, the size range of swimming bells conforms to the nature of the medium in that jet propulsion is effective only when principally resisted by the inertia of water. In the oceanic zooplankton the species with the widest size ranges are gelatinous and jet-propelled, which would still be true if one judged such range only on jet dimensions. Even so, the aspect to be considered here is total size. The greater the size, and the cnidarians also have batteries of noxious deterrents, the less available they are to predators. Predatory neighbours in the plankton can only attack a jellyfish or a salp in piecemeal fashion. To feast on jellyfish the predator must have a mouth as large as that of the ocean sunfish (*Mola*). Salps and medusae are sought by various kinds of large turtles. When the gelatinous forms of zooplankton evolved the means of making and shaping large amounts of their matrix, a wide range of size became possible, yet another biological advantage of their kind of organization.

CNIDARIA

Cnidaria, from the Greek *cnides*, 'nettle', is an apt group name for hydrazoans, jellyfishes, sea-anemones, corals and so forth. There are about ten thousand living species, nearly all marine, and the key to their success and carnivorous way of life resides in the myriads of stinging capsules (nematocysts) that form in the outer tissue layer (ectoderm) (Fig. 25). While large forms, as we have seen, are able to trap, paralyse and feed on sizeable fishes, the overall success of the cnidarians would surely have been much less had they not evolved the right kind of nematocysts for securing the abundant supplies of proteins etc. below the armour of the smaller crustaceans. In *Life in the Sea* by Gunnar Thorson (1971) there is a striking illustration of a small hydroid (*Hydractinia*) swallowing and entirely digesting a copepod, which is about as voluminous as the relaxed hydroid, in the space of 26 hours. Copepods are strong and active for their size and must be poisoned and paralysed before they can escape. Paralysis requires the injection of toxins below the cuticle of the prey. In a fish, on the other hand, the sensitive epidermis lies exposed over the scales and bones.

Nematocysts, the capsules of which range from ellipsoidal to globular in form and in size from 5 to 1,000 microns, are set on the tentacles and certain other parts of cnidarians. Each capsule contains a coiled-up thread with very thin walls (Fig. 25). When stimulated, and some nematocysts are triggered by a sensitive hair, the thread is forcibly and very rapidly everted. There are at least three main types of nematocysts: those that coil tightly and cling to the bristles on the surface of small crustaceans; those that fire a lassooing thread, sticky or covered with barbed spines; and those that pierce as they explode, so injecting the sting of their toxic contents. The threads of the last type are barbed to the tip and they rotate as they explode. 'Rotation of the tip must confer lateral and longitudinal stability on the advancing thread; and it is striking to see how straight are the tracks of discharges in distilled water, though the discharged tube is a most delicate structure. Rotation also imparts a cutter-like action to the barbs at the tip, as they pass to the barb series of the everted tube: this is probably unimportant in the lassooing type of capsule, but must determine the cutting and cuticle piercing powers of the injection type of stinging capsule' (Picken, 1957). There is much we do not know of the part played by nematocysts in the life of cnidarians, but we can at least begin to appreciate why they have held their place in the ocean from Cambrian times.

We have seen elsewhere that the outer tissue layer (ectoderm) of a cnidarian is separated from the inner layer (endoderm) by the gelatinous layer (mesogloea). The endoderm lines the body cavity and contains the digestive cells. During development the mysteries of morphogenesis are such that these three body-wall layers are fashioned around the sea-filled body cavity into medusoid or polypoid forms (Fig. 26). The adaptations of the gelatinous medusoid form for a planktonic existence should now be evident: polypoid forms, such as hydroids, sea-anemones and corals which have much less mesogloea, are primarily organized for a sedentary life and will be considered later with the other dwellers on the deep-sea floor. But both forms are not only essentially similar in body plan but are also radially symmetrical, particularly in the placing of their tentacles. As Buchsbaum and Milne (1960) say, 'Their elegant symmetry is an effec-

tive design for snaring prey from any direction and passing it on to a
centrally placed mouth.' The main exceptions to the rule are the
numerous siphonophores with tentacles, digestive and reproductive
members dispersed along the trailing part of the organism.

There are three classes of cnidarians. In the Anthozoa, which include
octocorals, sea-anemones, stony corals and black corals, the life history
centres on a polypoid phase. The deep-sea forms are considered elsewhere (pp. 178–86). In the Scyphozoa (jellyfishes) the medusoid form is

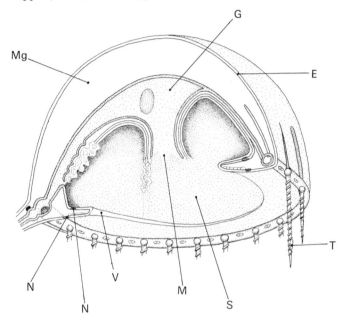

Fig. 26   Structure of a trachyline medusa, based largely on *Gonionemus*. E, ectodermal
covering of exumbrella; S, subumbrella; M, mouth; G, gastric cavity;
Mg, jelly-like mesogloea separating ectoderm from endoderm (gastrodermis)
lining the gastric cavity; V, velum; T, tentacle, bearing rings of nematocysts;
N, nerve rings. (Redrawn from Meglitsch, 1967)

predominant: the sessile polypoid stage produces young medusae, or develops directly to the adult condition, or is absent. The groups of Hydrazoa that mainly concern us here are the trachyline medusae, the siphonophores and chondrophores. Polypoid members of the important hydroid
group live on the ocean floor at all levels. Shallow-sea species may
have a medusoid phase (hydromedusae) in the life history. Hydroid
medusae also occur in the deep sea. For instance, during the 39th cruise
of the Russian research vessel *Vityaz* in the Kurile–Kamchatka Trench

area, such species as *Pandea rubra, Meator rubatra* (Bougainvillidae) and *Aequorea aequorea* (Campanulinidae) were taken. Their relatives in the orders Trachylina and Siphonophora play a more prominent part in deep-sea life. Hydrazoan medusae, which range from disc-like to bell-like forms, differ from the jellyfishes, inter alia, in developing a velum. The velum is a shelf-like inwards extension of the body wall around the rim of the medusa so as to restrict its opening to the sea. In the ectoderm of both the velum and the under surface (subumbrella) of the medusa is a circular band of striated muscle fibres, which act together to shape the swimming movements. Quick contractions of these muscles purse the disc or bell to force water through the yielding and narrowing velar ring in a propellant jet. When the muscles relax, the elastic, viscous jelly and the inflow of water to the subumbrella space restore the medusa to its resting form in readiness for the next pulse. Swimming movements may be desultory or repeated rhythmically. The rate of pulsation changes with ambient influences such as the presence of food, the intensity of light and the temperature.

Size for size, bell-shaped medusae must be faster swimmers than disc-like types, for during pulsations a bell will approach a streamlined form and produce a narrow, relatively powerful, velar jet. The form and structure of the sea-wasps (Cubomedusae), which are easily the fastest of the jellyfishes, support this contention. The bell-like body, which is disc-shaped in most jellyfishes, has four flattened sides and the margin of the bell is tucked in to form a false velum (velarium). Pulsations may be as much as 150 per minute and speeds of 1 knot or more have been estimated, making the fastest sea-wasps respectable members of the nekton. Thus, a bell-like form producing a concentrated jet and, of course, a well-developed neuromuscular system would seem to make for speed in medusae and jellyfishes. Close comparison of the fishing techniques of bell-like and disc-like species should be interesting. Some disc-like forms (e.g. *Cyanea*) fish as their tentacles stream outwards during a slow parachute descent. Hydroid leptomedusae tend to be disc-shaped and have balancing organs (statocysts) round the margin of the disc. The other type (Anthomedusae) are bell-shaped with light-sensitive organs (ocelli). Others troll long tentacles as they pulse in a drifting sea. Do the bell types, which presumably rely more on their own movements to carry out their fishing sweeps, catch much of their prey between pulsations, when their tentacles have chance to spread?

Most of the trachyline medusae are high-seas forms with headquarters in the warmer oceanic regions at depths between the surface and bathypelagic levels. Bell widths vary from a few millimetres to 10 centimetres. The species of one group (Trachymedusae) have a plain bell margin and include *Aglantha digitale, Halicreas minimum* and *Crossota brunnea*, all virtually cosmopolitan forms (Fig. 27). The first, with a transparent bell, sometimes with rosy tints, is most abundant in the epipelagic zone: the populations of the other two, which have bowl-shaped bells, are centred at mesopelagic levels. In *Halicreas* the radial canals and tentacles are tinged with bright red pigments while the bell of *Crossota* is darkly coloured. Species of the other group (Narcomedusae) have firm glassy bells with margins divided into lobes by the bases of the tentacles. A large round mouth leads to a capacious stomach. Forms that occur at epipelagic to deep-sea levels include representatives of the genera *Aegina*, *Aeginura* and *Solmissus*. In the Adriatic Sea the daily vertical migrations of *Solmissus albaescens* have been assessed by over 100 net samples taken between 1,000 metres and the surface. By day the populations are centred between depths of 400 and 600 metres, and during the afternoon individuals start to rise to higher levels. By night if the moon is full the medusae stay below a depth of 100 metres but on dark nights swim nearer the

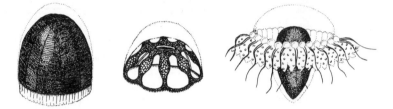

Fig. 27   Three deep-sea medusae. From left to right, *Crossota brunnea*, *Halicreas rotundatum* and *Atolla*.

surface. This species, which has a well-developed velum with strong circular muscles, was estimated to ascend at a rate of 50 metres per hour. Copepods form the main food, and the investigator thought that the medusae may have been following their prey (Benovic, 1973).

Trachyline medusae which range from a few millimetres to 100 mm in width, are smaller than jellyfishes, but they evidently compensate for their limited prey-snaring radii by relatively high population densities, at least in highly productive waters. In the North Pacific (Kurile–Kamchatka Trench), for instance, maximum numbers of *Aglantha digitale* were over 1,000: lower down in the less food-rich mesopelagic zone the comparable figure for *Crossota brunnea* was 50 (Vinogradov, 1968).

The Siphonophora are pelagic hydrazoans formed of an interconnected complex of medusoid and polypoid members that serve separate functions and are integrated as a colonial organism. The polypoid members are modified as: digestive zooids (gastrozooids), with a mouth and usually a single tentacle; protective and prey-catching zooids (dactylozooids), slender, mouthless and armed with a tentacle full of nematocysts; and reproductive zooids (gonozooids), producers of the medusoid gonophores that form either eggs or sperm but never develop into complete medusae and are seldom set free. Beside their modification as gonophores, medusoid members form swimming bells (nectocalyces), with velum, four radial canals and ring canal, and bracts—gelatinous, leaf-like appendages that have buoyant properties. It is as though the separate medusoid and polypoid phases of the hydroids have been fused into one polymorphic organism.

The 150 or so species of siphonophores are predominantly inhabitants of the oceanic province from epipelagic to deep-sea levels. In two groups (Cystonectae and Physonectae) the colony is topped by a gas-filled pneumatophore or float and they are mainly distinguished by the presence or absence of swimming bells: physonects have one or more swimming bells below the pneumatophore, while cystonects have none. Members of the third group (Calycophorae) have one or more swimming bells at the head of the colony but are without a pneumatophore (Fig. 28).

Cystonects include the Portuguese man-of-war, *Physalia physalis*, a prominent member of the neuston (p. 42), and species of *Rhizophysa*. *Physalia*, which lives on the warm ocean but may drift into temperate waters, has a sail-topped float below which are tidily attached the tentacular, digestive and reproductive members. In *Rhizophysa* these members are attached at intervals to a long stem that trails below a sizeable pneumatophore.

Nearly all siphonophores are physonects or calycophorans and most species resemble *Rhizophysa* in that the tentacular, digestive and reproductive members are hung on a long stem below the head of the colony. More precisely, the gelatinous bracts cover these members to form separate clusters (cormidia) that are repeated regularly along the stem which is very much alive, especially in its remarkable powers of extension and contraction. Even in species with a float, much of the buoyancy of the colony resides in the bracts (p. 319).

Physonects, such as *Nanomia* and *Halistemma*, draw their line of trailing members through the sea by pulsing a relatively small number (15) of swimming bells. In more complex forms (e.g. *Agalma*) the bank of bells consists of more numerous units. Certain forms (e.g. *Physophora* and *Discolabe*) have the nutritive and reproductive zooids clustered radially about a short stem below the pneumatophore and swimming bells. *Nanomia* (Fig. 28) has been well studied. Underwater films from submersibles show it capable of lively aquabatics. Evidently the activities of the various zooids are coordinated by the nervous system. Swimming bells pulse in unison to move the colony along a given line but are equally capable of asymmetrical action to change course suddenly. Contractions

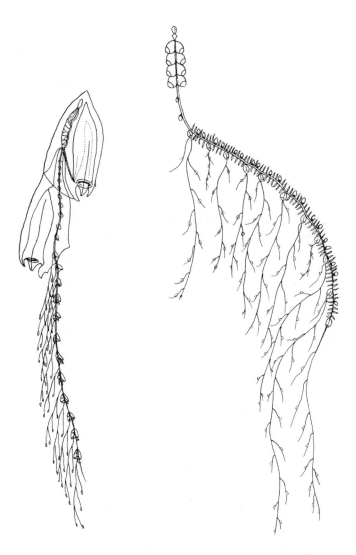

Fig. 28   Representatives of the two main groups of siphonophores. Right, *Nanomia bijuga* with a small gas-filled float (pneumatophore) topping the colony just above the swimming-bells (physonectae), below which buoyant gelatinous bracts provide lift for the prey-catching zooids, digestive zooids and reproductive zooids. Left, *Diphyes* with the two swimming-bells that provide lift and locomotion for the trailing part of the colony.

of the tentacular members lead to lively activity of the neighbouring digestive members, and in turn digestive products are pumped to other parts of the colonial organism, surely one of the most fascinating forms of animal life.

Most species of siphonophores belong to the Calycophorae, some of which are the most widespread and abundant in the order. When mature the cormidia of most calycophorans break loose to become free-living sexual forms (eudoxids), and they are less prevalent than the asexual, colonial stages. This is to be expected if the production by gonophores of sexual cells, and their eventual union to form planula larvae, is seasonal. Also the eudoxids are sometimes concentrated at deeper levels than their parent colonies. Species range in size and activity from small darting diphyids to large and sluggish prayids. The latter forms have two or more large swimming bells that are highly gelatinous and head a long line of trailing members extending for several metres. As meant by their name, diphyids have two swimming bells, which are usually similar and one is just before the other (Fig. 28). One species, *Chelophyes appendiculata*, may well be the commonest and most widely distributed species of siphonophore (Pugh, 1974).

Most siphonophores live in the warmer waters of the open ocean, where numerous species are widely distributed. Species of *Muggiaea* and *Nanomia* seem to be more neritic than oceanic. The well-known species *Dimophyes arctica* is a cold-water species that lives at epipelagic levels in arctic and antarctic areas but is centred in cold, deep waters in warmer intermediate regions (Ekman, 1953). It is one instance of equatorial submergence (see p. 435). In depth the populations of many species are centred in the epipelagic zone and some extend to mesopelagic levels. Certain species are predominantly mesopelagic and a few are even concentrated at upper bathypelagic depths. Catches off the Canary Islands have given intriguing indications of such vertical stratification, especially of how the three most abundant species of *Lensia* seem to have partitioned the water column. In each species about two-thirds of the catch are centred at definite and descending levels: *Lensia subtilis* 150 metres; *L. multicristata* 250–450 m; *L. achilles* 570–625 m. There is also evidence that mesopelagic as well as epipelagic species undertake daily vertical migrations (Pugh, 1974).

Siphonophores, especially the smaller species of diphyids, are among the most abundant of the larger forms of zooplankton (Fig. 28). They are so transparent and elusive that it takes some little time before one realizes that a catch is full of the smaller kinds. In their natural surroundings they must be invisible. When they luminesce, a bluish glow lights up the glassy bells of the colony.

The purple sail or by-the-wind sailor (*Velella*) and its relatives (*Porpita*, *Porpema*), which, like the Portuguese man-of-war (*Physalia*), are members of the pleuston (p. 42), were once classified as siphonophores but are now given their own order (Chondrophorae). The units of the colony hang from a circular or oval, gas-filled float that is strengthened by chitin and in *Velella* bears an upright sail. There is a large central digestive member (gastrozooid) surrounded by many reproductive members (gonozooids) that produce free medusae and in turn are fringed by a circle of prey-catching members (dactylozooids). Like *Physalia*, the chondrophores are purple to deep blue in colour and have their sailing headquarters over warmer oceanic regions.

Nearly all the deep-sea species of the jellyfish class (Scyphozoa) seem to lack a fixed polypoid stage in their life history (see also pp. 178–9). Nearly all belong to the order Coronatae and range from flattened to dome-like forms that vary in diameter from a few to 25 centimetres. The ordinal name refers to the groove that crowns the umbrella. Just below this coronal groove the jelly of the bell is fashioned into a circlet of thick pedalia, some or all of which bear a single solid tentacle. The margin of the bell is scalloped into lappets that alternate with the pedalia. As Buchsbaum and Milne (1960) say, 'The beautiful sculpturing of these masses of jelly remind one of some of the gelatin desserts that have been shaped in grooved, dome-like metal moulds.' The predatory powers of *Periphylla hyacintha*, which has a dome-like, purple-washed bell and 12 tentacles, has already been stressed (p. 81). Hardy (1956) remarks that this species is usually illustrated '... with its tentacles hanging limply down; in life they are extended in all directions feeling outwards for possible prey'. The saucer-like umbrella of *Atolla* has brown to dark red markings and bears two alternating sets of pedalia with 16–32 tentacles (Fig. 27). The disc-like *Nausithoe rubra* is a 'beautiful dark madder-brown' (Hardy, 1956). Representatives of these and other coronate genera are commonest at mesopelagic levels in the subtropical and tropical belts but range into colder regions, where they may be taken at epipelagic levels.

In the other groups of jellyfishes, *Pelagia noctiluca* (order Semaeostomae), with purple-rose or lilac colouring, is one of the few species that are truly oceanic. It is mainly an epipelagic form, though like the deep-sea species, luminescent. Box jellies (Cubomedusae) and many-mouthed jellyfishes (Rhizostomeae) are predominantly inhabitants of warm in-shore waters but certain species live in the upper waters of the open ocean. Kramp (1956b) makes special mention of the 'beautiful brown and violet *Crambionella orsini*, a rhizostome jellyfish that was often taken by the Danish Research Vessel *Galathea* during her passage from Africa to India. Females of the pelagic octopus, *Argonauta*, the paper nautilus, sometimes attach themselves to this jellyfish. In Kramp's words, 'Between Mombasa and the Seychelles we saw the remarkable sight of a number of *Argonauta* ensconced on top of jellyfish, *Crambionella orsini*, which drifted in large numbers round the ship. To the best of our knowledge this had never been seen before.'

CTENOPHORA (COMB-JELLYFISHES)

Comb-jellyfishes are organized in much the same way as cnidarian jelly-fishes. The outer (ectodermal) and inner (endodermal) tissues are separated and supported by a massive gelatinous matrix (mesogloea) that is strengthened by connective tissue fibres and shapes the organism. As in jellyfishes, the mesogloea contains amoeboid cells, but there are also muscle fibres, and presumably nerve cells. The two groups are also alike in that the only body space is the endoderm-lined digestive cavity, which has a mouth but no anus. This similarity in degree or organization and the disposition of the body plan in radial symmetry is recognized by placing the two groups in the grade Radiata.

The other grade of animals with definite tissues and organ systems (eumetazoans), which comprises most of such animals, are bilaterally symmetrical and called Bilateria.

More precisely, the ctenophores are biradially symmetrical Radiata (Fig. 29). Most species (Tentaculata) have mobile tentacles that are densely studded with adhesive cells (colloblasts); cells unique to cteno-phores and quite unlike the nematocysts of cnidarians (Fig. 25). One ctenophore, *Euchlora rubra*, has nematocysts (like those of Narcomedu-sae) but no colloblasts. 'The protruding heads of the adhesive cells are very sticky and cling to prey. At their inner ends they are attached to spirally coiled contractile filaments that yield to the pull of struggling prey but cannot be wrenched loose' (Buchsbaum and Milne, 1960). Ctenophores without tentacles (Nuda) just swallow their prey. There are about 80 species, all marine.

No ctenophore is built for propulsion by jet reaction—most species make their slow way by the sequential beating of ciliary plates that are disposed in radial rows over the animal. Each ciliary plate is made of numerous long cilia that are fused at the attached end and so look like the teeth of a comb, hence Ctenophora, meaning 'comb-bearers'. The play of light on a moving ctenophore may be the only indication of its presence. As Buchsbaum and Milne (1960) say, 'The delicate trans-parency of comb jellies makes them all but invisible in the water, so that often they reveal themselves only in the rippling iridescence of the rows of beating combs as they diffract the light.' At night, though, ctenophores are brilliantly luminescent, and they flash along the comb rows when disturbed by the sea.

Lighthill (1969) implies that movement by ciliary activity is appropriate only in small organisms. At first sight, ctenophores seem to be con-spicuous exceptions, for species that move by beating ciliary combs range

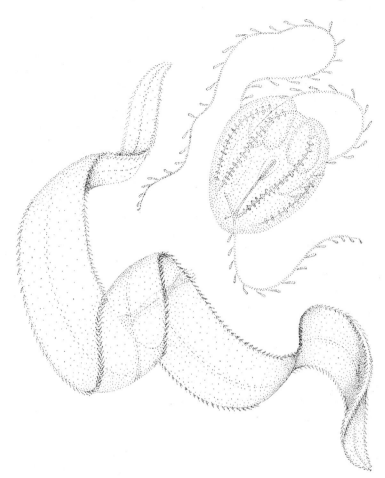

Fig. 29 Two ctenophores, Venus' girdle (*Folia*) and the sea-gooseberry (*Pleurobrachia pileus*). (Redrawn from Marshall, 1971)

in size from a few millimetres to about 20 centimetres. First, though, the beating is by sizeable ciliary paddles not by tracts of discrete cilia. Moreover, the paddles beat in effective metachronal rhythm (each paddle in a row is later in stroke than the one immediately in front), more precisely in antiplectic metachronism, which differs from symplectic metachronism in that the lag in beat between one ciliary unit and the next is greater, and indeed sufficient to give each paddle greater freedom of movement during its effective stroke. Studies of the ciliated protozoan *Paramecium* and the ctenophore called the sea gooseberry (*Pleurobrachia*) showed that this kind of metachronal rhythm is more effective than the other in propulsive power. Even so, most ctenophores are slow swimmers. But members of the tentaculate group Lobata, such as *Ocyropsis maculata*, move quickly by strong down beats of their large and wing-like oral lobes. Venus' girdle (*Cestum veneris*), a ctenophore that is compressed in the tentacular plane to a beautiful band-like form and reaches a length of $1\frac{1}{2}$ metres, swims by undulations of its body (Fig. 29).

Apart from a few bottom-dwelling forms, ctenophores are pelagic animals par excellence. Species such as *Beroe cucumis* are cosmopolitan in the widest sense and in cold waters may be tinged with pink or lavender colours. The oceanic species are predominantly epipelagic, but there are two deep-water genera (*Bathyctena* and *Aulacoctena*).

TUNICATA

Much as pelagic cnidarians have evolved, as it were, around their hydrostatic skeleton to become predominant among gelatinous forms of carnivorous zooplankton, so have pelagic tunicates attained comparable success among gelatinous forms of herbivorous zooplankton (see also pp. 80–1). There are over 2,000 species of tunicates, most of which are ascidians and benthic in habit from intertidal to deep-sea levels. Deep-sea ascidians are considered in the next chapter.

Since the organization of pelagic cnidarians is easily run (pp. 82–3), it is understandable that some species live at deep-sea levels well below the productive surface zone. Though they may be taken below this zone, and there are a few deep-sea species, pelagic tunicates are primarily forms of neritic waters and the epipelagic zone of the open ocean. Again this might be expected if one considers their dependence on species of phytoplankton. Like their benthic relatives, pelagic tunicates trap such food (and organic particles) by means of mucous secretions that come, as already stated, from the endostyle, which is a glandular part of the pharynx.

The Larvacea are so called because in certain respects they are like the larval stages of ascidians. Most of the eighty-odd species are tiny and transparent with a trunk barely exceeding 1 millimetre, and a tail usually not more than three times the length of the trunk. Special cells (oikoblasts) on the forward part of the trunk rapidly secrete a transparent and gelatinous filter-chamber (house) that varies in size and complexity from species to species. The house is much larger than its maker, which draws water through the filters by lashing its tail. In most species of one family (Oikopleuridae) the water enters through a fine sieve and then circulates inside through a very fine filtering system that funnels to the animal's mouth (Fig. 30). Inside the mouth the food particles are trapped in a conical sieve of mucus that is maintained and formed by endostylar and ciliary activity. A twisted cord of mucus and food enters the stomach. During feeding water pressure builds up in the house and eventually a hinged door by the exit to the house suddenly opens to release a jet of water, and as one jet follows another these extraordinary animals move

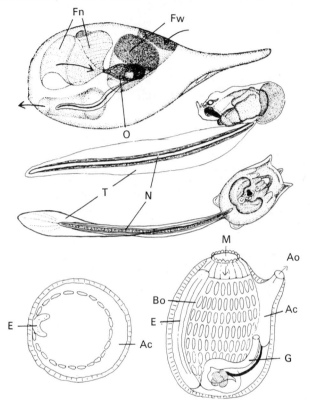

Fig. 30   Top, a larvacean *Oikopleura* (O) in its house, and below, the animal enlarged. Bottom, a tadpole larva of a sea-squirt (ascidian). Bottom right, the structural plan of a sea-squirt; bottom left, a cross-section through the pharyngeal region of a sea-squirt. Ac, atrial chamber; Ao, atrial opening; Bo, row of branchial openings; E, endostyle; Fw, filtering windows; Fn, filter-net; G, gut; M, mouth; N, notochord; T, tail. (From Marshall, 1971, *Ocean Life*)

among the finest pastures of the sea. Species of another family (Fritillarii-dae) secrete a food trap that is pushed by the animal when it is feeding and suspended outside its filter. By such means larvaceans trap the smallest kinds of phytoplankton down to one micron in size (see also p. 224).

Larvaceans are very common in all parts of the ocean. Species diversity is greatest in subtropical and tropical waters. More than is realized at present, these minute tunicates must be an important source of food for young stages of deep-sea animals and for vertically migrating mesopelagic animals, such as lantern-fishes. The largest larvacean, *Bathychordaeus*, which reaches a length of more than 77 mm and builds a very voluminous house without filtering windows, is a deep-sea (mesopelagic) form. *Bathychordaeus*, which is believed to feed on detritus and small planktonic organisms, has an enlarged sac-like oesophagus and stomach.

Pyrosomids, salps and doliolids form the other group (Thaliacea) of pelagic tunicates (Fig. 31). Inside their gelatinous tunic, they resemble ascidians in their general organization, but are modified for pelagic feeding and locomotion in that the atrial cavity is behind the pharynx and discharges water through a posterior opening.

Pyrosomids develop into colonies '... consisting of numerous individuals embedded in the wall of a gelatinous tube, which is usually long, more or less cylindrical, closed at one end and open at the other. The zooids have their branchial siphons at the outer surface of the tube, the atrial siphons opening into the common cloacal cavity. Each accordingly passes a water current from the exterior to the interior, where the individual currents fuse into a powerful stream emerging from the open or posterior end of the tube. Locomotion is accordingly a continuous jet propulsion. A diaphragm at the open end narrows the aperture and increases the velocity. Colonies are often colourless and transparent, but may have a slightly yellowish, blue-green or grey colour. They may be soft and gelatinous or rigid and cartilaginous in texture. The outer surface is never quite smooth' (Berrill, 1950).

Though predominantly warm-water forms, pyrosomids (Fig. 31) are found in all oceanic areas apart from the Arctic Sea. Colonies range in length from a few centimetres to over 4 metres. As their name suggests, pyrosomids are luminescent; indeed, they may well produce the brightest light of all luminous marine organisms. The light comes from photogenic bodies on each side of the pharynx and has been said to range in colour through red, orange, greenish, sky blue, greenish-blue and azure. But one set of careful measurements showed that most of the light energy is blue-green (482 mμ) with a secondary greenish peak (525 mμ) (Boden and Kampa, 1964). Luminescence has also been observed in the larvacean genus *Oikopleura* and in salps and doliolids.

Though they are rather simply organized, doliolids (Fig. 31) have a complex life cycle. Berrill (1950) remarks that their '...method of attainment of a sexually mature form is probably the most intricate in the whole of the animal kingdom'. This sequence of events is less complicated in salps. Essentially both groups have an asexual and a sexual phase in the life history. In salps solitary individuals with no gonads (oozooids) have a budding stolon producing a chain of individuals that form the

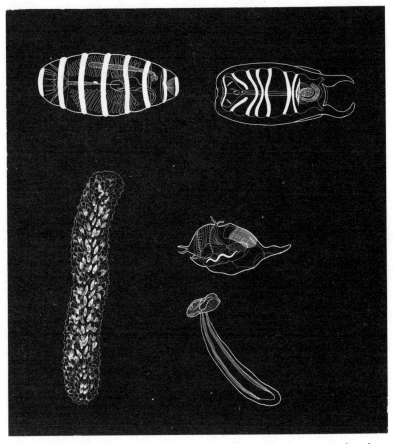

Fig. 31   Pelagic tunicates. Top left and right; *Doliolum nationalis* (gonozooid) and a
salp, *Salpa* (*Thalia*) *democratica*; below left, a small pyrosomid, *Pyrosoma
spinosum*; below right, larvaceans, above *Oikopleura albicans* (indicated by
serpentine tail) in the filtering 'house'; below *Oikopleura fusiformia*. (From
Marshall, 1954)

aggregate phase and produce ova and sperm. The fertilized eggs produce
oozooids. Doliolids have solitary sexual individuals (gonozooids) and the
fertilized eggs produce tailed larvae that grow into asexual oozooids.
Each oozooid has a ventral budding stolon and a dorsal spur at the posterior end. Buds formed by the stolon migrate up and become attached
to the dorsal spur where they grow into multiple lateral rows of feeding
individuals (trophozooids) and a median row of phorozooids that eventually become detached and grow into gonozooids.

In both groups asexual and sexual forms have an individual structure, but in essence the means of propulsion and food procurement are similar. Propulsion is centred in the muscle bands that entirely or partly encircle the body wall just below its substantial gelatinous tunic. As these muscles relax and contract water is taken into the branchial chamber and expelled in jets from the atrial chamber. Bursts of locomotor activity are interspersed with longer and irregular pauses. Much as in larvaceans, phytoplankton organisms and particles that enter the pharynx are trapped in a mucous net that is secreted by the endostyle and drawn across the pharynx by ciliary activity. In the aggregate form of the salp *Pegea confoederata*, for instance, the mucous net is conical and continuously renewed and is evidently able to retain particles as small as 1 micron. The remains of flagellates, coccolithophorids, diatoms, tintinnids, foraminiferans and radiolarians were found in faecal material.

Salps (Fig. 31) are larger than doliolids. Solitary salps vary in length from a few to 200 millimetres. Chains of aggregate forms may reach lengths of 2 metres or more. Single doliolids span from a few to 30 mm and the chains of zooids on the nurse oozooid reach 40 centimetres. Like other pelagic tunicates, salps and doliolids are mainly cosmopolitan in the subtropical and tropical belts. A number of species extend into temperate and polar waters. In equatorial regions much of the planktonic biomass in the summer months consists of salps, doliolids and pyrosomids. Sometimes the sea is thick with tunicates and at night the light displays of pyrosomids is a sight not to be forgotten. Apart from turtles, fishes and numerous invertebrates, notably heteropods, their main predators are various kinds of seabirds.

Pelagic snails (pteropods and heteropods)

In two groups of pelagic gastropods the muscular foot, which is one of the key features that contributes to the outstanding success of gastropods in benthic habitats, has been modified into wing-like appendages. Hence they are known as pteropods (winged-feet). More precisely, the wings are outgrowths of the sides of the foot and form the propulsive organs of the shelled pteropods (Thecosomata) and the naked pteropods (Gymnosomata), which both belong to the opisthobranch group of gastropod molluscs but are not closely related.

In one family of shelled pteropods (Limacinidae) the wings and head emerge from a transparent shell measuring a few millimetres in length or width. Morton (1958) writes, 'The foot has a broad sole and bears an operculum, and from its sides are produced long parapodia, or "wings", muscular at the bases and thin towards the tips. *Limacina* rows itself upwards in a broadly spiral course, using these as oars, and drops as a dead weight by holding them together motionless above the body.' Species of the family Cavoliniidae have a bilaterally symmetrical

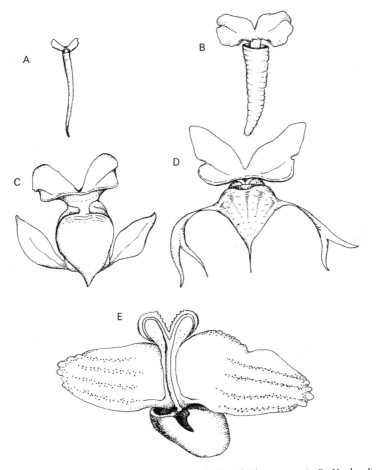

Fig. 32   Oceanic pteropods. A, *Creseis acicula* (length about 10 mm); B, *Hyalocylis striata*; C, *Cavolinia tridentata*; D, *Diacria trispinosa*; E, *Gleba cordata* (span of 'wings' about 75 mm). (Redrawn from Gilmer, 1974)

shell, which may be long and conical (*Creseis*), dorso-ventrally flattened (*Clio*) or inflated and flask-like (*Cavolinia*) (Fig. 32). *Cavolinia tridentata*, which is about 22 mm long, bears a pair of greenish, leaf-like appendages, each strengthened by a central vein that emerges from a rear angle of the shell (Fig. 32). Presumably these appendages enable the animal to plane through the water, and not lose too much height, between bursts of swimming activity. Members of another family (Cymbuliidae) are large (up to 50 mm across the wings), gelatinous and close to perfect buoyancy. They have no true shell but lie in a boat-like pseudoconch made of chitin. 'In *Cymbulia* a pair of broad based wings are attached along the sides. They are flapped like a butterfly's wings with an up and down rowing movement; and they show also an undulation from the front backwards, which provides a forward thrust. In *Gleba* (Fig. 32) and *Corolla* the pseudoconcha is a wide saucer and the wings have grown together in a heart-shaped fringe which is moved gracefully in swimming like a skirt or the bell of a medusa' (Morton, 1958).

The shelled pteropods feed on micro-organisms, particularly on the minute plants of the nanoplankton. Species of Limacinidae and Cavoliniidae set up ciliary currents on the wings and mantle, so drawing in organisms and food particles that are trapped by a film of mucus. Cymbuliids secrete a large mucus web. *Gleba cordata* (Fig. 32), which has a wing span of about 50 mm, spins a 2 metre-span web by means of mucus glands at the tips of the wings. 'Gently manipulated by the cilia in the animals' elongated mouth, the free-floating membrane slowly sinks, collecting a variety of tiny plankters on its sticky surface. Eventually, *Gleba* draws most of the web down its proboscis and consumes it' (Hamner, 1974).

Like other members of the zooplankton that are predominantly herbivorous, the shelled pteropods are both common and abundant. In the Beaufort Sea region of the arctic, the cosmopolitan species *Spiratella helicina* appears to be the cause of a sonic scattering layer at a depth of 50 metres (Hansen and Dunbar, 1970). Nearly all pteropods are epipelagic and most species live in the warmer oceanic belts, where a number are circumtropical. Shells of pteropods may be found in the sediments, but pteropod ooze occurs in significant amounts only in the Atlantic Ocean, mainly in the central South Atlantic. Cymbuliids, with their large digestible cast-nets of mucus, are evidently very efficient herbivores and probably much more abundant than is appreciated at present.

The naked pteropods (Gymnosomata) are much less abundant than the shelled species, which is not surprising in view of their carnivorous habits. The body is more or less spindle-shaped and may attain a length of more than 40 mm. Gymnosomes move quickly on powerful wings. 'The animals move upwards or forwards by sculling with them, twisting them at the "wrist" to allow a power stroke at both the upward and downward beat' (Morton, 1958). The prey, much of which consists of shelled pteropods, is taken by a formidable buccal armature consisting of various combinations of prehensile hooks, sucker-studded tentacles and adhesive oral papillae. One species, *Clione limacina*, which, like its prey *Spiratella helicina*, is sometimes an important item in the diet of baleen whales in arctic waters, is also similar in its 'bipolar' distribution. But gymnosomes are most diverse in warmer oceanic regions.

Heteropoda, known by some 25 species, belong to the large prosobranch group of snails. Like many other planktonic animals they are virtually transparent. In one family (Atlantidae) comprising the smallest species (up to 25 mm long), the coiled transparent shell is so compressed and keeled that the snail stays upright as it swims, which, as in all heteropods, is effected by undulations down a fin-like part of the foot. The shell is much reduced in the Carinariidae (length 2·5–5·0 cm) and entirely absent in the Pterotracheidae (2·5–10·0 cm). Members of both families have an elongated gelatinous body that is close to neutral buoyancy. When these heteropods swim quickly, which is aided by serpentine movements of the body in pterotracheids, the fin side is uppermost.

All heteropods have large tubular eyes moved by special muscles, which may well help them find their prey. Morton (1958) writes, 'The fast swimming Heteropoda, with movable telescopic eyes, are active predators, catching medusae, small fish, copepods, and pteropods and themselves falling prey to larger heteropods. The large *Pterotrachea* and *Carinaria* carry the buccal bulb at the end of their flexible "trunk" periodically thrusting out the sharp-toothed radula to seize prey, or even to attack the human finger-tip.' Heteropods swallow their prey entire, which also includes hyperiid amphipods, salps and arrow-worms.

The headquarters of these lively pelagic snails is in the epipelagic zone of the subtropical and tropical belts, where numerous species are more or less cosmopolitan. But their place in the ocean is limited by comparison with that of the arrow-worms, which are also predators and have much the same range of lengths.

Chaetognatha (arrow-worms)

Three groups of worm-like invertebrates are predominant in marine ecosystems. In the zooplankton of both the neritic and oceanic provinces, the Chaetognatha or arrow-worms are much more ubiquitous and abundant than the pelagic species of polychaete worms, which, together with the nematode worms, are outstanding among the worm-like invertebrates in benthic environments.

Arrow-worms maintain their place in the ocean, which goes back to Cambrian times, with relatively few (about 50) species. Most are below 40 mm in length and the entire range is from a few to over 100 millimetres. A rounded head, armed on each side with a set of grasping spines, passes evenly into a transparent, torpedo-shaped body that extends in the horizontal plane into one or two pairs of lateral fins and a terminal tail fin. The fins are immobile and supported by a double set of delicate fin rays (Fig. 33). Apart from a few benthic species of the genus *Spadella*, all arrow-worms are planktonic and most species are placed in the genus *Sagitta*.

In most arrow-worms the body is more or less stiff and turgid and is held straight between periods of activity. Movement comes from the quick contractions of substantial bands of muscles that run down the body. Arrow-worms seem to spend much of their time slowly gliding on their fins, '...eventually however, they begin to sink, whereupon they resort to swimming to maintain their level. Swimming consists of short swift forward darts, each covering a distance of around 5 cm; this is followed by gliding on the momentum of the darts and a return to floating. The forward dart is so rapid as to be difficult of analysis but is believed to result from the alternate contractions of the dorsal and ventral longitudinal muscle bands of trunk and tail. This would seem to produce up-and-down body waves and tail flirts, but the movement is so rapid that

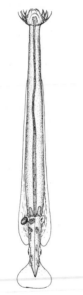

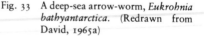

Fig. 33   A deep-sea arrow-worm, *Eukrohnia bathyantarctica*. (Redrawn from David, 1965a)

it appears to the naked eye like a trembling' (Hyman, 1959). Arrow-worms may dart to elude their predators or take their prey, both of which are detected and located by groups of bristle-bearing sensory cells that remind one of the neuromast organs in the lateral line system of fishes (Marshall, 1971). The prey is seized in the spine-like jaws and swallowed whole. Copepods are the most frequent prey, but other chaetognaths, euphausiids, medusae, pelagic tunicates and young fish are also taken. Perhaps chaetognaths feed on most kinds of zooplankton that they are able to swallow.

Unlike most other groups of planktonic animals, chaetognaths have no larval phase. They are hermaphrodites and the ovaries mature after the coelomic cavities of the tail are full of ripe sperm. In *Sagitta* self-fertilization is said to be common. The eggs hatch into young looking rather like the adults. Wherever they live, and in oceanic waters arrow-worms are world-wide in occurrence at epipelagic and deep-sea levels, fecundity and the cycles of reproduction are such as to maintain relatively abundant populations. Between New York and Bermuda the displacement volumes of epipelagic arrow-worms were usually second to those of the copepods and averaged 13 per cent to the total volumes of zoo-plankton (Grice and Hart, 1962). At levels from the surface to 1,000 metres in the western North Pacific the biomass of arrow-worms was about 10 per cent of the total biomass of zooplankton and micronekton irrespective of the depth (Aizawa, 1974). In absolute terms the number of individuals—and the number of species—per unit volume of water are greater in the upper 200 metres than below this layer. Species diversity is greatest in warmer oceanic waters. Thus, in the subtropical and tropi-cal waters of the Indo-West Pacific faunal province, which extends from the Red Sea and east coast of Africa across the Indian Ocean to the Hawaiian Islands and Polynesia in the western Pacific, about 85 per cent of the known species have been recorded, some of which are circumtropi-cal. Species such as *Sagitta bipunctata*, *S. enflata*, *S. hexaptera* and *Ptero-sagitta draco* are cosmopolitan at epipelagic levels in both temperate and warm oceanic waters. Species with populations centred at mesopelagic levels include *Eukrohnia* spp. (Fig. 33), *Sagitta lyra*, *S. decipiens* and *S. macrocephala*. *Bathyspadella edentata*, which may be bathypelagic, is without jaw spines. Certain mesopelagic species may be tinged in varying degrees with orange or red pigments. Except for the bluish pigmentation of neustonic forms, other chaetognaths are virtually transparent.

Diurnal changes of level associated with vertical migrations are well marked in certain species (e.g. *Sagitta elegans*, and *S. bipunctata*) in neritic waters, where populations tend to be concentrated near the sur-face during the night. Oceanic species may be partial migrators (Angel and Fasham, 1974) or non-migrators. In the Southern Ocean, where *Sagitta gazellae* is one of three endemic species, there was no clear evi-dence that *gazellae* is a diurnal migrator. But there are seasonal changes of level associated with the life cycle. David's (1965a) analysis suggests that maturing animals gradually sink and eventually release their eggs at about a depth of 1,000 metres. The eggs rise and probably hatch at about 250 metres, the young growing into juvenile animals that mass in the upper 100-metre layer. These growing populations migrate to the 100–250-metre layer in the winter months, but rise and resume life in the surface layer in the spring.

Polychaete and nemertean worms

Like the arrow-worms, planktonic species of polychaete worms are predatory and known from relatively few (about 60) species. Most belong to four families (Alciopidae, Tomopteridae, Typhloscolecidae and Phyllodocidae) of which the first two are entirely comprised of pelagic forms and best known to marine biologists. Fauvel (1959) has caught their attraction, '*Les Tomopteris et les Alciopiens pélagiques sont transparents comme du cristal et seraient à peu près invisibles n'étaient les taches vivement colorées des parapodes des premiers et les enormes yeux rouges*

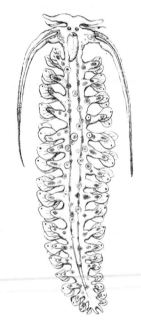

Fig. 34 A pelagic polychaete worm, *Tomopteris kefersteinii*. (After Greeff)

des seconds.' ('The pelagic tomopterids and alciopids are as transparent as crystal and would be almost invisible if it were not for, firstly, the vividly coloured spots on their parapodia and, secondly, their huge red eyes.') Some of the brightly coloured spots on the parapodia of tomopterids are traceable to rosettes of yellow cells that luminesce through large surrounding hyaline cells (Fig. 34). The large projecting parapodia end in paddle-like expansions, and it is by sequential strokes down the parapodial series that tomopterids row swiftly through the sea. Alciopids have shorter (uniramous) parapodia but have lively motions as they seek their zooplankton prey, aided no doubt by their large camera-type eyes.

Most species of pelagic polychaetes live in warmer oceanic regions and as Friedrich (1969) remarks, '...they often occur sporadically and solitarily, so that the role of these predatory forms in the biology of the pelagial is of secondary importance in comparison with that of the benthonic polychaetes which play a very important part in the biology of the benthos'. One species, *Tomopteris septentrionalis*, lives at epipelagic levels in artic and antarctic regions but at deeper levels in the warmer oceanic belts. Certain deep-sea species of *Tomopteris*, as Sir Alister Hardy (1956) illustrates so well, are almost entirely red.

Some of the pelagic nemertean worms, all of which seem to live at deep-sea levels, also develop red pigmentation. The body, which ranges in length from 6 to 20 cm and is usually widened, sometimes with undulating edges, has a gelatinous mesenchyme. Species of *Pelagonemertes* have a transparent body and a branched, orange or red-coloured gut: others, such as *Bathynemertes hardyi*, are scarlet or plum-coloured. *Nectonemertes* has been observed to swim rapidly with strong undulations. Pelagic nemerteans presumably use their long protrusible proboscis to take their prey in the zooplankton.

Nekton (and micronekton)

Nektonic animals have greater swimming powers than those that comprise the zooplankton, and much as one may usefully distinguish a micro-zooplankton from the larger forms of zooplankton (p. 68) so may the smaller kinds of nekton be considered as micronekton (p. 42). The members of each group have a size and activity range that may be sensed by the dimensions, mesh size, and towing speed of the nets that have been designed to catch them most effectively. A small conical net with a mesh size of 200 microns and towed at a speed of about half a metre per second is used to sample the micro-zooplankton (p. 68). A larger conical net with a mouth area of 1 square metre (70 cm diameter) and mesh size from 300 to 500 microns, which is usually towed at something like 1–3 metres per second, samples the larger members of the zooplankton. The net designed by the Institute of Marine Resources, La Jolla, California, to sample the micronekton of eastern Pacific waters has a mouth area of 2·3 square metres and oblong meshes of 5·5 × 2·5 mm. Towed at about 5 metres per second, this net takes the larger crustaceans, small cephalopods and fishes of a size range from about 1 to 10 centimetres (Strickland, 1972). Nets such as the Isaacs–Kidd midwater trawl and the modified Tucker net used by the Institute of Oceanographic Sciences with a mouth area of several square metres, and smallest (cod-end) meshes of 77 meshes/inch, are towed at speeds of 1½ knots (0·77 metres per second) or more and catch micronekton and smaller fish and squid members of the nekton (> 10 cm in length). Still larger midwater nets increase the size range and catches of nekton, but there are still limitations. As Clarke (1971) observes, 'Larger commercial midwater trawls are regularly used for biological sampling, and here the mouth is held open by otter boards, floats on the headline and large weights on the footrope. Examples are the British Columbia midwater trawl, the Engels midwater trawl and the four-door Cobb midwater trawl. These trawls are difficult to operate from multi-purpose oceanographic vessels. However, they are valuable in sampling animals up to about half a metre in length. The catch is not easy to use in comparative studies because the mouth area and shape change with speed, and graduation of the mesh size along the net makes interpretation difficult. Even with nets of this enormity with mouth areas of 2,800 square metres and 750 square metres (less when fishing because they close up due to drag) the larger animals, known to be present by observation, are not caught.' Such observation may be made by examining the food eaten by sperm whales. Elsewhere, Clarke has calculated that a sperm whale of eleven tons eats no less than 130 tons of squid in a year. Many of these squid will be over a metre in length and some, such as *Architeuthis*, span well over 10 metres.

Apart from their size and activity, micronektonic animals differ from those of the zooplankton in form and organization. The shrimp-like mysids and euphausiids and the deep-sea prawns, like the squid and fishes, have more or less streamlined spindle-shaped bodies (Marshall, 1971). In keeping with their size and complex organ system, all these animals have elaborate gills. In the twilight world of mesopelagic levels eyes are very well developed and many species have complicated constellations of light organs. Size, form and activity are correlated with massive muscular systems. The jointed exoskeleton of crustaceans is associated with a complex system of muscles, which are very highly developed in micronektonic forms. In a mysid, euphausiid or deep-sea prawn the weight of muscle must be well over 50 per cent of the total weight. During cruising movements the most active part of this muscle is that moving the swimming legs (pleopods), but during escape movement the whole of the powerful abdominal musculature propels the animal backwards by a series of rapid tail-flips. Muscle and skeleton make up much of the dry weight of such forms. For instance, the dry weight as a percentage of the wet weight reaches 15 per cent or more in amphipods, mysids and euphausiids. In siphonophores, medusae, ctenophores, arrow-worms and tunicates the corresponding figure ranges from 0.3 to 10.0 (Parsons and Takahasi, 1973). The mantle muscle of squids, which expresses propellant jets from the mantle cavity, is about 30 per cent of the body weight in active species (Trueman, 1975). In active fishes the segmented lateral muscles used in swimming are 40 per cent or more of their total weight. Lastly, a high percentage of micronektonic forms are carnivores, whereas many zooplankton species, forming much of the total biomass, are herbivores. In other words, compared to the zooplankton most organisms of the micronekton may be regarded as predatory-behaviour-machines that are quite expensive to maintain.

The greatest contrast between the micronekton and nekton is in the epipelagic zone. Here is the swim of baleen whales, sperm whales, dolphins, turtles, whale-shark, basking-shark, isurid sharks, grey sharks, thresher-sharks, sailfish, marlin, spearfish, swordfish, tuna, flying-fishes and various squid (ommastrephid and onychoteuthid). This is Isaac's (1969) vivid appreciation, 'The pelagic region contains some of the largest and most superbly designed creatures ever to inhabit this earth: the exquisitely constructed pelagic tunas; the multicoloured dolphinfishes, capturers of flying fishes; the conversational porpoises; the shallow- and deep-feeding swordfishes and toothed whales, and the greatest carnivores of all, the baleen whales and some plankton-eating sharks, whose prey are entire schools of krill or small fishes. Seals and sea lions feed far into the pelagic realm. In concert with these great predators, large carnivorous sharks await injured prey. Marine birds, some adapted to almost continuous pelagic life, consume surprising quantities of ocean food, diving, plunging, skimming and gulping in pursuit. Creatures of this region have developed such faculties as advanced sonar, unexplained senses of orientation and homing, and extreme olfactory sensitivity.' Compared to all these large and active members of the nekton, the crustaceans and small fishes of the micronekton are indeed shrimps and so much whitebait.

One need only contrast this state of affairs with conditions immediately below in the mesopelagic zone (150–1,000 metres), which is the most productive part of midwater space, to see how much life is scaled down in the deep sea. If the size range of micronektonic animals is taken to be about 1–10 cm, then virtually all of the larger crustaceans and many of the cephalopods and fishes are included. Some of these animals, such as certain species of euphausiids, prawns, squid and lantern-fish, swim up to the epipelagic zone each day, where they form part of the micronekton during the night. The nekton of the mesopelagic zone consists of a few large species of fishes (e.g. the larger gempylids, trichiurids and alepisauroids) and squid (e.g. certain histioteuthids). Though more is known now of the larger forms of pelagic deep-sea animals, Hjort's (1912) description of this fauna as Lilliputian still seems apt.

Euphausiacea

If their near ubiquity, trophic range and numerical abundance are considered together, the euphausiid shrimps emerge as the 'copepods' of the micronekton. Considering just the oceanic species, euphausiids live from epipelagic to bathypelagic levels throughout the four oceans. Species fall into three main trophic types, as in copepods (pp. 232–3): herbivores (when conditions are favourable), omnivores and carnivores. In one measure of abundance, the mean displacement volume of euphausiids was about 5 per cent of the total catch figure in a series of net hauls in the epipelagic zone between New York and Bermuda (Grice and Hart, 1962). As already stated (p. 78), the corresponding figure for copepods was about 50 per cent, and one must always remember that copepods are much easier to catch than euphausiids. At levels between the surface and 1,000 metres from stations in the western North Pacific, the biomass of euphausiids was 8 to 15 per cent of the total catches of zooplankton and micronekton (Aizawa, 1974). During the antarctic summer, vast swarms of euphausiids, nourish baleen whales, crab-eater seals, penguins and fishes.

The 85 species of euphausiids (order Euphausiacea) are closely related to the large order of decapod crustaceans which includes shrimps, prawns, crayfish, lobsters and crabs. Indeed, certain authorities class them with the Decapoda. All species are shrimp-like in appearance (Fig. 35) and have well-developed antennules and antennae that bear olfactory and tactile receptors. Just behind the mouth the paired mandibles, maxillules and maxillae collect, macerate and manipulate the food just before it is swallowed. The outer branches (exopodites) of the 6–8 pairs of thoracic limbs, to which the gills are attached, produce currents of water from which food particles are screened by the setae on the inner branches (endopodites) of these limbs.

Euphausiids cruise by the sequential paddling motions of the two pairs of swimming limbs (pleopods) on each of the first five segments of the abdomen. The sixth and last segment has no such appendages but articulated at the end of this segment is a slender, tapering telson, on either side of which is a flattened biramous limb (uropod). Telson and uropods form a tail fan, which is brought powerfully into play by a series of quick flexures of the muscular abdomen, when the animal shoots backwards, perhaps to evade a predator.

Species range in length (tip of rostrum to end of telson) from about 10 to 150 mm. The largest of the epipelagic species is *Euphausia superba* (50–60 mm), the whalers' krill of oceanic areas south of the Antarctic Convergence.

Smaller species of *Euphausia* and various smallish species of other genera (e.g. *Stylocheiron* and *Nematoscelis*) are typical of the warmer epipelagic waters (Fig. 35). Other species of these three genera and representatives of such genera as *Thysanopoda* (Fig. 35) and *Nematobrachion* live at mesopelagic levels. Large species of *Thysanopoda* (adults of *T. cornuta* and *T. spinicaudata* may exceed 100 mm in length) and *Bentheuphausia amblyops*, which are widely distributed, are typical of bathypelagic levels. These species of *Thysanopoda* may be largely red in colour:

other species may be touched and spotted with reddish pigments but are otherwise virtually transparent.

Unlike other euphausiids, *Bentheuphausia amblyops* has regressed eyes and no light organs. Most species have ten light organs: one in each eye stalk, one pair at each of the bases of the second and seventh thoracic limbs and a single organ lying between each of the first four pairs of

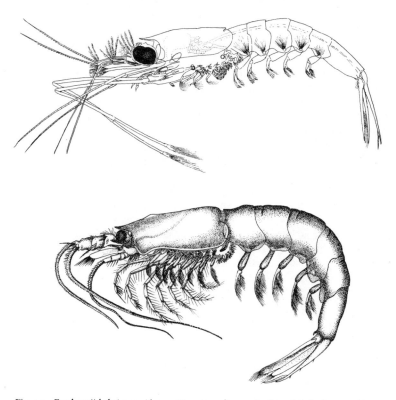

Fig. 35   Euphausiid shrimps. Above, *Nematoscelis mantis*. Note bilobed eye and long second thoracic legs. Below, *Thysanopoda acutifrons*.

abdominal swimming legs (Mauchline and Fisher, 1969). Each photophore has a lens and a group of light-producing cells (see also pp. 357).

The Antarctic krill *Euphausia superba*, which feeds on the abundant crops of phytoplankton during the summer, is a seasonal herbivore. At other times it is an omnivore like most of the euphausiids. Carnivorous species, which prey on copepods and other suitable forms of zooplankton, belong to the genera with bilobed eyes: *Stylocheiron, Nematoscelis* and *Nematobrachion*. These forms also have a very long pair of anterior thoracic limbs ending either in a few large spines or small chelae, presumably for grasping captured prey (Fig. 35). The functional design of the feeding and alimentary devices of euphausiids and the part these animals play in oceanic food chains and economy are considered later (pp. 236–9.) Each day certain mesopelagic species migrate between their daytime levels and the surface water where they find much of their food while other species, both epipelagic and mesopelagic, stay at one level. Recent survey shows that migrant species of the same genus overlap in their daytime vertical ranges, whereas there is vertical segregation between non-migrant species (Baker, 1970, and p. 236).

Euphausiids produce small eggs and after fertilization each eventually hatches into an ovoid nauplius larva a millimetre or less in length. The nauplius has three pairs of swimming legs which later form the two pairs of antennae and the mandibles. The nauplii develop after moulting their exoskeleton into second stage nauplii. A series of growth stages and moults takes the nauplius through a metanauplius stage (in which the future abdomen is seen), calyptopis stages (with a hooded carapace and more definite abdomen) and furcilia stages, which become more and more like the adult. The last furcilia stage moults, forms and changes into an adolescent. Deep-sea species apparently spawn at depth and the sequence of events in some mesopelagic species, at least, is that the early larval life is passed in the productive surface water. If all the larval stages of bathypelagic species remain within the depth range of the adults (Brinton, 1962) one wonders how the smallest larvae manage to feed themselves (see also p. 433).

Deep-sea prawns

Deep-sea prawns (or 'shrimps' if you are an American or Japanese marine biologist) are often prominent in the catches of large nets fished from epipelagic to bathypelagic levels in temperate to tropical reaches of the ocean.

In biomass the catches of such prawns are close to those of the euphausiids and generally comprise about 10 per cent of the total biomass in samples taken between the surface and a depth of 1,000 metres. Deep-sea prawns, euphausiids and fishes make up most of the micronekton and their contributions respectively range from 15–25, 35–50 and 20–45 per cent of the total biomass (Omori, 1974). As a rule euphausiids are smaller than prawns, so the former easily outnumber the latter. Moreover, the part played by prawns in the pelagic life of subpolar and polar waters is insignificant compared to that assumed by the euphausiids.

Shrimps and prawns form the Natantia, or swimming decapod crustaceans. Lobsters, crayfish and crabs, which are predominantly benthic and considered in the next chapter, comprise the Reptantia, or crawling decapods. The characteristic features of natant decapods are: the body is laterally compressed; the base of each antenna is articulated to a movable, plate-like scale, the thoracic legs are slender, and there is a full complement of well-developed swimming legs (pleopods) on the abdomen. In nearly all species the carapace juts forwards as a prominent rostrum, which is often saw-toothed and compressed. During quick movements, when prey may be seized, the compression of the body and rostrum will tend to keep a prawn from veering off a line that is guided by the antennal scales, which act as rudders. Compensatory movements

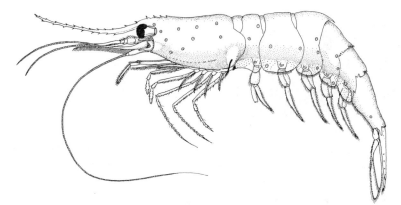

Fig. 36   A deep-sea prawn, *Systellapsis debilis* (5–6 cm). Note the luminescent organs. (Redrawn from Herring and Clarke, 1971)

of the scales may also help to control rolling and pitching. Deep-sea prawns thus seem to be well fitted for an active, pelagic life. They also have a much lighter, less well calcified exoskeleton than the bottom-dwelling, crawling decapods. The latter never have a laterally compressed body, antennal scales are vestigial or absent, and the pleopods which are not used in swimming, are often reduced.

Swimming decapods fall into two main groups, Penaeidea and Caridea, which are readily distinguished. Carideans have expanded side-plates (pleura) on the second abdominal segment that overlap those of the adjacent segments, whereas there is no such overlap in penaeids. Furthermore, most penaeids have a small claw (chela) at the end of the third thoracic leg, which is not developed in carideans. Deep-sea penaeids such as *Gennadas*, *Funchalia* and sergestids (*Sergestes* and *Sergia*) have extremely long whip-like antennae and the lash is thickly and regularly set with setae, many of which are presumably tactile or olfactory in function. These prawns are rather slender in form but they have powerful

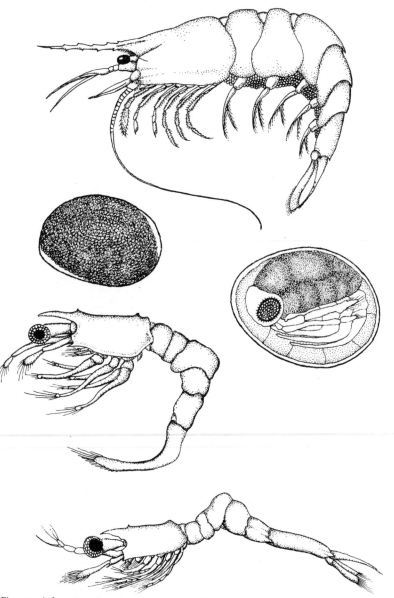

Fig. 37   A deep-sea prawn *Acanthephyra quadrispinosa*. Above, adult female carrying eggs and immediately below, left, egg (1 mm long), right, eyed-embryo in egg. Below egg, second-stage larva; bottom, fifth-stage larva. (Redrawn from Aizawa, 1974)

pleopods and in some the rearmost pairs of thoracic legs have become strong jointed oars. One has the impression of sudden powers to dart on prey detected by the searching sweeps of the long sensory antennae (Fig. 38). Deep-sea carideans also have strong swimming legs. Species of the family Oplophoridae (e.g. *Acanthephyra*, *Notostomus* and *Hymenodora*) are stoutly made: those of *Notostomus* have an 'inflated' carapace.

There are over 200 species of pelagic prawns, most of which range from 10 to 100 mm in length (Figs. 37–39). The largest (e.g. certain species

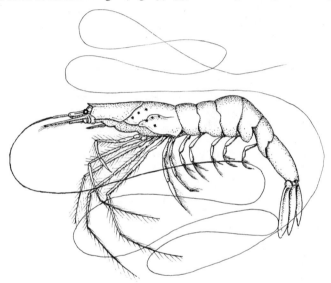

Fig. 38   The sergestid prawn *Sergestes corniculum* (6 cm). When this prawn swims the long lash of each antenna is probably streamed parallel to the long axis of the body. (After Sund)

of the genera *Acanthephyra* and *Notostomus*) reach lengths from 130 to 180 mm and often seem to live near the deep-sea floor (Omori, 1974). Nearly all species live in the subtropical and tropical belts and some extend into temperate waters. As already indicated, deep-sea prawns are poorly represented in the coldest oceanic regions, where, for instance, in antarctic waters five species have been recorded (Yaldwyn, 1956). There are parallel differences in the latitudinal diversity of euphausiids, but the six antarctic species, especially *Euphausia superba* (p. 437), are key forms in the economy of life.

Euphausiids are also more prominent than the prawns, both in individuals and species, in the temperate to tropical reaches of the epipelagic zone. Of the mesopelagic prawns, those that live around 500–700 metres or above by day (upper mesopelagic) are transparent, or translucent with scattered red chromatophores, or half red (*Sergestes* spp., *Funchalia* and *Parapandalus*). All species that occur below 500–700 metres by day (lower mesopelagic) have numerous small chromatophores and a heavily pigmented exoskeleton that ranges from deep red to scarlet in colour (e.g. *Sergia*, *Acanthephyra*, *Notostomus*). Many of the mesopelagic species undertake daily vertical migration, but in bathypelagic forms (e.g. *Acanthephyra stylorostralis*, *Hymenodora glacialis* and *Physetocaris microphthalma*) the main part of the population lives below 1,000 metres throughout the entire day (see Foxton, 1970, and Omori, 1974). Certain bathypelagic species seemingly extend down to depths of 5,000 metres or more (Vinogradov, 1968).

Some deep-sea prawns have light organs (Fig. 36). The most remarkable are modified liver tubules (Organs of Pesta), which may have reflectors and lenses to enhance and concentrate the down-welling light. These organs are developed in both penaeids (*Gennadas*, *Sergestes* and *Sergia*) and carideans (*Hymenodora* and *Parapandalus*), but in the former group, for instance, they occur only in the semi-transparent upper mesopelagic species (e.g. *Sergestes*). The lower mesopelagic species of *Sergia* are a uniform red colour and there are no Organs of Pesta. Dermal photophores may also be present, and those that have lenses (*Sergestes challengeri* etc.) also live at upper mesopelagic levels. The all-red species develop dermal photophores without lenses and these live, as already stated, at lower mesopelagic levels. Bathypelagic species, such as *Sergia japonicus*, are without photophores. There are thus correlations between type of pigmentation, photophore development and depth ranges. Indeed, as Foxton (1970b) has suggested, and is considered elsewhere (p. 360), these correlations remind one of parallel developments in certain cephalopods and fishes.

Deep-sea penaeids produce numerous small eggs that are shed into the water. After fertilization they hatch into minute nauplius larvae with the usual three pairs of swimming legs that later become the antennules, antennae and mandibles. The nauplii change after moults into protozoea stages, which have functional mouthparts. Then comes a zoea stage (called acanthosoma in some sergestids) in which some or all of the thoracic limbs function in swimming, but the abdominal limbs are absent or rudimentary. The succeeding post-larval stages (mastigopus of sergestids) look more and more like the adults and the abdominal segments acquire their full complement of setose swimming legs. In some species of *Sergestes* and *Sergia* the carapace of the protozoea and zoea larvae bristle with long spines that may have a protective function as well as helping to slow down the sinking rate of these larvae.

Whereas the eggs of sergestids are a fraction of a millimetre in diameter, deep-sea carideans produce relatively few, large eggs which range from about 1 to 4 mm (along the longest axis). The eggs are carried by the female on long setae of the abdominal limbs where they stay until they hatch into relatively advanced stages (protozoea or zoea). After a few post-larval stages the adult form is attained. The early larval stages are passed within the egg, and as Omori (1974) says, the advanced young that are produced have well-developed swimming and feeding powers that give them considerable safety from predators and independence of supplies of minute food. The penaeids rely on safety in numbers. A comparative survey of such reproductive and larval strategies in the deep-sea will be found elsewhere (Chapter 12).

Peracarid crustaceans

In euphausiids, shrimps, prawns, lobsters, crayfish and crabs, which are classed as eucarids (Greek *eu*, true, and *karis*, shrimp), the carapace is

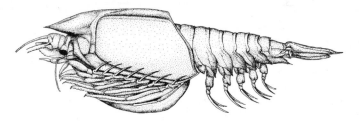

Fig. 39   Lateral view of a large female of the mysid shrimp *Gnathophausia ingens* (35 cm). (Redrawn from Clarke, 1961)

large and covers the thorax. The eggs, if carried, are attached to the pleopods. Peracarids (Greek *pera*, pouch), as implied by their name, have a brood pouch in which the female bears the rather large eggs until they hatch. The carapace, if present, is not fully developed and leaves at least four of the thoracic segments free. Pelagic peracarids belong mainly to the opossum shrimp group (Mysidacea) and the Amphipoda. Cumaceans, tanaids and isopods, which are entirely or largely benthic, are considered in the next chapter.

Deep-sea mysids have been taken in nets fishing below 6,000 metres (Vinogradov, 1968). But they are commonest at mesopelagic levels, where live species of such genera as *Gnathophausia*, *Eucopia* and *Lophogaster*. Species of *Gnathophausia*, which are red and look like prawns, are also prawn-like in size (Fig. 39). Most mysids are from 20 to 30 mm in length, but a female *Gnathophausia ingens* measured 351 mm (from tip of rostrum to tip of telson) and other large specimens of this species range from 145 to 210 mm (Clarke, 1961). Species of *Eucopia*, which have long, frail-looking thoracic legs, are also red. There is a fine painting of a scarlet

E. unguiculata by Sir Alister Hardy (1956). Many mysids are filter feeders, but the deep-sea forms are likely to be omnivores.

Amphipods have sessile eyes, no carapace, and in nearly all species the body is flattened from side to side and presents an arched profile. Representatives of the subgroup *Hyperiidae* live at epipelagic and mesopelagic levels over most of the ocean. *Hyperia*, *Themisto* and *Parathemisto* species are most common at epipelagic levels, and individuals of one (*H. galba*) live commensally it seems, with large jellyfishes (e.g. *Cyanea*).*
The transparent frail-looking Diogenes shrimps (*Phronima*) with enormous eyes on a hammer head, have more drastic designs on gelatinous organisms. Adult *Phronima*, which may be common at mesopelagic levels and range from about 20 to 30 mm in length, will take over the swimming bell of a siphonophore or the test of a salp or pyrosomid and then proceed to eat away the inside of such structures until a gelatinous barrel is left. The *Phronima* lives in the barrel and moves it along by limb movements that draw water through its shelter, which also serves as a brood chamber and a nursery for the young. *Cystisoma*, another fragile but larger (15 cm) amphipod with a near crystalline transparency, is seen from time to time in nets fished at upper mesopelagic levels. Of one fine photograph, the Angels (1974) remark, 'The only colour in this specimen is the interference colour of the eyes and some red pigment from the animals' last meal.' But certain mesopelagic hyperiids are all red, and again we are indebted to Sir Alister Hardy (1956) for fine paintings (of *Lanceola* and *Scypholanceola*). Representatives of the largest group of amphipods (Gammaridea) also form part of the mesopelagic fauna. Even so, the peracarid crustaceans are less prominent in the economy of pelagic life than their eucarid relatives, the euphausiids and prawns.

Cephalopoda

When a research vessel is on night station, squids and fishes may be attracted to the illumination cast into the sea by the deck lights. The squid may also be drawn to the fishes, though it is difficult to tell predator from prey. One sees dark spindle-shaped bodies and glinting eyes moving just below the surface. Even in brighter surroundings a darting squid may be momentarily mistaken for a fish. Through such illusions, an observer is made aware of the extraordinary similarities between cephalopods and fishes. Beside the streamlined shape of active species and their prominent camera-like eyes, the two groups resemble each other in numerous other respects of their gross and fine structure (Marshall, 1971; Packard, 1972).

---

* Recent investigations indicate that most, if not all, hyperiid amphipods are associated with gelatinous zooplankters during some part of their life (Madin and Harbison, 1977; Harbison, Briggs and Madin, 1977).

These resemblances form just one striking illustration of the brute fact that during their evolutionary history distantly related kinds of plants and animals have converged in functional design in response to the pressures of natural selection in a given medium, or to subdivisions of a medium ranging from broad adaptive zones to narrow ecological niches. Though this is not the place to consider convergence, it must be said that this aspect of evolution is prominently displayed in the deep sea.

Many of the 700–800 species of modern cephalopods live in the deep sea. Pelagic species are mostly squids (Teuthoidea) together with a number of octopods, a few sepioids (cuttle-fish group) and the vampire squid (*Vampyroteuthis*). Most of the octopods and sepioids are bottom-dwellers. Epipelagic species, notably the ommastrephid squid, have firm torpedo-shaped bodies and strong swimming powers, which reside in their thick, muscular mantles. One species (*Ommastrephes bartrami*) and a hooked squid (*Onychoteuthis banski*), which have large triangular fins, have power enough to dart out of the water and fly through the air. After take-off, flying-squid are sustained largely, it seems, by continued expulsion of jets from the funnel, and they may cover distances of 50 metres or more (Packard, 1972). When a small shark approached two *Ommastrephes*, both were observed to turn a dark red and shoot out of the sea to a height of 2 metres. Though primarily epipelagic, ommastrephid squid have been photographed by baited cameras at mesopelagic levels. As in fishes (p. 43), epipelagic species are considerably larger than most of their relatives of lower levels. The flying-squid (*O. bartrami*) reaches a length of about 1 metre, while *Dosidicus gigas*, an ommastrephid of Chilean waters, grows to well over 2 metres.

There are smaller and more delicate squid in the epipelagic zone, especially members of the family Cranchiidae. Species of *Cranchia* and *Liocranchia* (Fig. 40), which may be quite abundant, have soft, globular bodies, large eyes, short arms and rather long tentacles. Apart from certain internal organs, these squid are virtually transparent. The liver is a particularly opaque organ, but in *Liocranchia* it is silvered over, presumably to hide it from predators (see also pp. 360–1). *Cranchia scabra* attains a body (or mantle) length of about 50 mm, which is less than half the length of an adult *Liocranchia reinhardti*.

Concerning the cephalopods of deeper levels, closing midwater nets fished accurately from RRS *Discovery*, the research vessel of the Institute of Oceanographic Sciences, UK, have enabled Malcolm Clarke and his colleague C. C. Lu to determine the vertical ranges of the commonest North Atlantic species throughout their growth stages and where possible to trace the daily changes in level of the species that undertake vertical migrations. They also found a latitudinal gradient in species diversity. From 60°N to about 20°N the number of species taken roughly doubled for every ten degrees of latitude (Lu and Clarke, 1975). Thus, like other

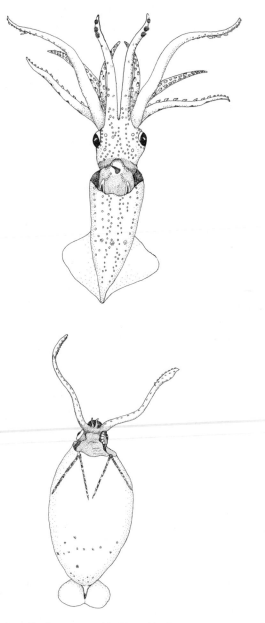

Fig. 40 Two juvenile deep-sea squid. Top, *Abralioptsis* sp. (7 mm); bottom, *Liocranchia reinhardti* (12 mm). (Redrawn from Roeleveld)

groups of marine organisms, cephalopods are most diverse in the warmer parts of the ocean (see also p. 117, and Clark and Lu, 1974 and 1975; Lu and Clarke, 1975). Much of what follows is taken from their work.

Mesopelagic squids of the upper 700 metres '... are in general small squids rarely exceeding 30 cm in overall length excluding the tentacles and averaging only about 15 cm with large numbers adult at 4–5 cm' (Voss, 1967). In the North Atlantic the commonest small squid of mesopelagic levels include *Abraliopsis affinis*, *Histioteuthis* spp.; *Pyroteuthis margaritifera* and *Mastigoteuthis schmidti*. In *Abraliopsis* and *Histioteuthis* most of the numerous light organs are on the lower side of the animal. The few jewel-like light organs of *Pyroteuthis* are set around the eyes, at the tips of its tentacles and in the body. Indeed, an outstanding feature of mesopelagic squid '... is the possession of numerous small to large light organs or photophores distributed chiefly upon the vertical surface of the mantle, funnel, head, arms and the ventral periphery of the eyeball' (Voss, 1967). Many mesopelagic fishes also have numerous light organs along their underparts, and, as in squid, some of these organs are of elaborate design (Fig. 41). Again, mesopelagic squid and fishes have large and very complex eyes, but no fish has eyes such as develop in histioteuthid squid, which have the left eye much larger than the other (pp. 394–5).

Two small sepioid squid, *Spirula* and *Heteroteuthis*, are also common in parts of the Atlantic at mesopelagic depths. *Spirula* floats head downwards, buoyed by an internal, gas-filled, spirally chambered shell, and shining upward at the end of the body just above the shell is a light organ. *Spirula* (body length 7 cm) is the only deep-sea cephalopod with a gas-filled buoyancy system. In diverse squid (e.g. cranchiids and histioteuthids) neutral buoyancy is achieved by buoyancy chambers filled with relatively light ammoniacal liquids (pp. 324–5). Such a system is found also in certain deep-sea prawns (e.g. *Notostomus*). Fishes manage well, if need be, with a gas-filled swimbladder (pp. 326–35). The other sepioid, *Heteroteuthis* (body length 4 cm), looks rather like its bottom-dwelling relatives. The ink it discharges on being disturbed is mixed with a luminous secretion, which reminds one of the clouds of luminous material emitted by certain deep-sea prawns and fishes (pp. 349–51). As already implied, convergent evolution is rife in the deep sea, especially in visual, luminescent and buoyancy systems.

In the catches from their North Atlantic stations, the two commonest midwater octopods studied by Clarke and Lu were *Vitreledonella richardi* and *Japetella diaphana*, which both grow to a mantle length of about 60 mm. Both species are virtually transparent (*Japetella* is sometimes reddish) and have the gelatinous consistency and something of the appearance of a jellyfish. The first is mesopelagic at all stages of growth, whereas half-grown and adult *Japetella* were taken at upper bathypelagic

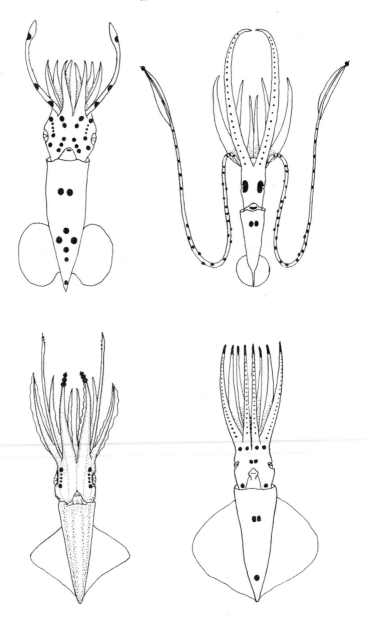

Fig. 41   Deep-sea squid. Top left, *Pterygioteuthis*; top right, *Chiroteuthis*; bottom left, *Abraliopsis*; bottom right, *Octopoteuthis*. The black markings represent light organs. (Redrawn from Young, 1972)

levels (1,000–1,500 m). A third gelatinous midwater octopod, *Amphitretus pelagicus*, has upturned tubular eyes that are reminiscent of similar organs in certain mesopelagic fishes (Fig. 12).

The finned octopods, such as *Cirrothauma* and *Cirroteuthis*, once thought to be bathypelagic, have been photographed near the deep-sea floor (pp. 141–2). They are also gelatinous and the large webs between their arms make them look even more like jellyfishes. The paddle-like fins project from a relatively small body. *Cirrothauma* is blind and certain species of *Cirroteuthis* have regressed eyes. No deep-sea octopod, or indeed any kind of octopod, is know to be luminescent.

The vampire squid, *Vampyroteuthis infernalis*, which is more or less cosmopolitan between lower mesopelagic and bathypelagic levels, is neither a squid nor an octopus (Fig. 12). With its eight broad-webbed arms, paddle-like fins and soft gelatinous body it looks rather like a finned octopod, but there are certain outstanding differences. Actually the vampire has ten arms, two being modified into long, mobile sensory filaments that are withdrawable into special pockets. Thus, it is also quite distinct from the squids in which two of the ten arms develop into special, prey-catching tentacles. *Vampyroteuthis* is placed in a separate order called the Vampyromorpha. It is also unlike any octopod in its well-developed set of light organs. Besides the many small organs that stud the skin, there are two clusters of luminous nodules on the back of the 'neck' and two large lights, each with an 'eye-lid', just behind the fins. Males reach a length of about 125 mm, females about 200 mm.

Some kinds of mesopelagic squid migrate daily between their daytime levels and overlying waters, where they swim by night, and presumably find much of their food. Among the species that are most common in the North Atlantic, Clarke and Lu found reasonable evidence that *Abraliopsis affinis*, *Bathyteuthis abyssicola* and *Pyroteuthis margaritifera* appear at epipelagic levels during the night. Other common squid, such as *Heliococranchia pfefferi* and *Mastigoteuthis schmidti*, are non-migrators, as are the octopods *Japetella* and *Vitreledonella*.

There are also ontogenetic migrations during the life history. The catches from *Discovery* stations suggest that most deep-sea cephalopods start their young life in the epipelagic zone and assume deeper levels as they grow to maturity. There is uncertainty as to where the eggs are laid (pp. 439–40).

In 1895, when the *Princesse Alice 1*, the research yacht of Prince Albert 1 of Monaco, was working near the Azores the local whalers harpooned a sperm whale, which happened to die under the bottom of the yacht. In its death flurry it spewed up parts of great squids new to science. On returning to Monaco the Prince had his yacht fitted as a whale-catcher and began a series of investigations on the sperm whale and its food, showing that it feeds mainly on squids, some with tentacles as thick as a man's arm and studded with great suckers bearing cat-like claws. But for these and subsequent investigations, and for the occasional stranded specimen, our knowledge of squids, particularly of their size range, would be very restricted. Sperm whales feed on medium-sized to the largest squid, which must be more abundant and larger than appears from the catches of midwater nets (Clarke, 1977b).

Fishes

Small fishes, euphausiid shrimps and deep-sea prawns make up most of the micronekton (pp. 105–7). Why are small squids not a major constituent, or are sampling devices at fault? Perhaps squids are the hardest to catch, though the elusive powers of a fish would seem to be much the same as those of a squid of the same size and shape. But the smallest pelagic squid are larger than the smallest fishes. Most of the prolific species of midwater fishes, which nearly all belong to the stomiatoid and lantern-fish groups, range from 25 to 70 mm in length when mature. If the arms are included in the length, hardly a small-sized species of squid falls into the above size span, which suggests that the smallest squids are more elusive than the smallest fishes. But are the latter more productive than the former, bearing in mind that animals feeding nearest the primary producers in a food chain (or pyramid) are the most productive (see also pp. 227–30). Are small fishes better able than their squid counterparts to take advantage of the wealth of small (mainly copepod) food in the ocean? Small stomiatoids and lantern-fishes certainly feed on eggs, larval stages, copepods, ostracods, arrow-worms, euphausids and so forth, but less is known of squid food. Prey caught by a squid or octopus is conveyed by the arms to the beak-like jaw and there chewed, if necessary, into pieces small enough to enter the oesophagus, which is restricted by its passage through the brain and cephalic cartilages. Diverse predatory fishes are able to swallow prey up to half of their own size. Small species of midwater squid, including migratory species, also feed on copepods and euphausiids (Clarke, 1966) and certain mesopelagic and bathypelagic cephalopods have suckers small enough to grip small planktonic animals. Further observation and investigation is needed. In the meantime one suspects that the fishes are able to take a wider range of small prey organisms, and thus feed closer than the cephalopods to the photosynthetic base of the food pyramid. For instance, it is improbable that any genus of deep-sea cephalopod has species to match the abundance and ubiquity of those in the stomiatoid genus *Cyclothone*.

The most diverse mesopelagic fishes, both in numbers of species and individuals, are stomiatoids and lantern-fishes (Myctophidae). There are over 300 species in the second group (Fig. 42) and about 250 in the first (Figs. 43 and 44). Both groups have a few representatives at bathypelagic levels near 1,000 metres and below, where, indeed most fishes belong to the dark-coloured species of *Cyclothone*. But in diversity alone the bulk of the bathypelagic fauna consists of ceratioid angler-fishes, which are known from about 100 species.

Stomiatoid fishes, which seem most closely related to the smelts, typically develop four rows of prominent and complex light organs along the

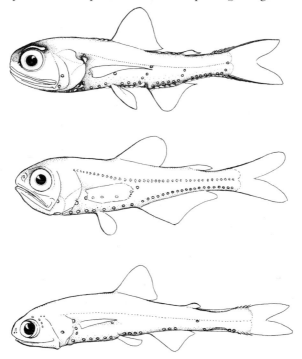

Fig. 42   Lantern-fishes (Myctophidae). From top to bottom; male *Ceratoscopelus townsendi* (43 mm), male *Electrona antarctica* (66·5 mm); *Taaningichthys bathyphilus* (ca. 60 mm). (Redrawn from Wisner, 1974)

underparts of the head and trunk and two rows at the same level along the tail. Small silvery kinds, such as *Vinciguerria*, *Valenciennellus* and *Bonapartia* (Gonostomatidae) and some of the hatchet-fishes (Sternoptychidae) are common at upper mesopelagic levels (200–600 m), but they are usually outnumbered by small transparent fishes, again of the genus *Cyclothone* (e.g. *C. signata* and *C. braueri*) (Plates I and II).

The smaller species of stomiatoids have relatively long jaws bearing numerous pointed teeth, part of a prey-taking system to deal with the range of food organisms already listed (p. 122). Most of the larger species

Fig. 43 Stomiatoid fishes. From top to bottom; *Valenciennellus tripunctulatus* (ca. 30 mm), the hatchet-fish *Argyropelecus affinis* (ca. 60 mm), the hatchet-fish *Polyipnus laternatus* (ca. 40 mm). (Redrawn from Weitzman, 1974; *Bull. Amer. Mus. Nat. Hist.* Vol. 153, Art. 3, New York)

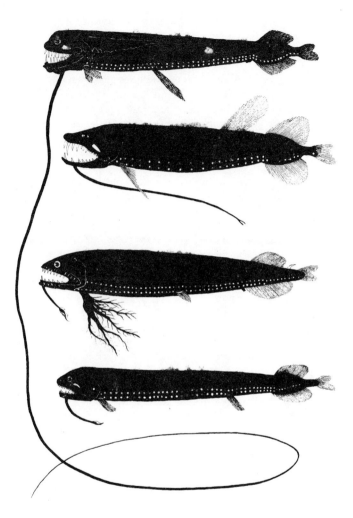

Fig. 44 Stomiatoid fishes (family Melanostomiatidae), showing variation in length and form of luminous chin barbel. From top to bottom; *Grammatostomias flagellibarba* (length of fish ca. 200 mm), *Bathophilus longipinnis* (ca. 50 mm), *Chirostomias pliopterus* (ca. 200 mm) (note the luminous tracery on the pectoral fin), *Melanostomias spilorhynchus*. (Redrawn from Marshall, 1954, after Beebe and Crane)

(about 100 mm or more in length) belong to the families Astronesthidae, Chauliodontidae, Stomiatidae, Melanostomiatidae, Idiacanthidae and Malacosteidae, and they all live at various levels in the mesopelagic zone. These fishes also have large jaws and some of the foremost teeth are fang-like. Tine-like teeth on mobile pharyngeal bones at the back of the throat grip and convey sizeable prey to the gullet. Such prey may consist of euphausiids, prawns, squid and fishes, and the gullet, stomach and wall of the abdomen are distensible enough to accommodate large meals. If the predator is stealthy and quick enough food may be a single fish or squid that is at least half the length of its captor.

Besides the ventral rows of light organs, the larger and rapacious kinds of stomiatoids have large photophores below and behind the eyes. The latter cheek lights may be very large and in some species better developed in males (pp. 367–8). Many species have thousands of small lights all over the skin and in the clear gelatinous tissue that may invest the body (as in *Chauliodus* and *Stomias*). Apart from *Chauliodus* spp., nearly all species have a chin barbel bearing beads and inlays of luminous tissue that may be very elaborate. In *Chauliodus* the second ray of the dorsal fin, which is very long, has a luminous tip. Perhaps these are angling devices with a luminous bait (p. 367).

The species of *Chauliodus* ('viper-fishes') and *Stomias*, which are ser-pentine fishes with a protrusible upper jaw, are silvery-sided with a honeycomb scale pattern, framing, as it were, the body lights. The largest species reach lengths of 300 mm or more. The black or bronze-coloured astronesthids ('star-eaters'), which range from about 40 to over 300 mm in length, have a conventional fish form and a fin pattern much like that of a smelt. The melanostomiatids ('black dragon-fishes') are much more diverse in form, ranging from long slim types (e.g. *Eustomias*) through fusiform types (e.g. *Echiostoma*) to short squat types (e.g. certain species of *Bathophilus*). Female idiacanthids copy, as it were, the black skin and serpentine form that is so common in melanostomiatids, but the males are not only pale-coloured but much the smaller sex (see also pp. 367–8). Though also black, the malacosteids have evolved in other directions. The suspension of the jaws is so angled backwards that the gape is very wide. There is no floor to the mouth cavity and the bare lower jaws with their strong pointed teeth remind one of the lethal mechanism of a rat-trap. The rat-trap fishes are also unusual in that the cheek photophores pro-duce red light, which is true also of *Pachystomias*, a melanostomiatid. All other luminescent organisms emit blue-green light (see Chapter 10).

Like the smaller kinds of stomiatoids, lantern-fishes are so successful that a sizeable net towed at mesopelagic levels almost everywhere in the ocean is almost bound to catch them. If, as seems very likely, they have not colonized the Arctic Sea, then it is improbable that other groups of midwater fishes will have gained a footing.

All lantern-fishes are more or less spindle-shaped and bear much the same fin pattern (Fig. 42). The jaws bear rows of small teeth, and though some kinds (e.g. certain species of *Lampanyctus* and *Gymnoscopelus*) are larger and wider mouthed than others (e.g. *Goniichthys* and certain species of *Diaphus*), there is no marked contrast between small plankton-eaters and larger predators on the micronekton, such as obtains in stomiatoids. While large lantern-fish are naturally able to accommodate somewhat larger food organisms than small species, the diet of all kinds is much as that already stated. The entire size range in the family is from about 25 to 250 mm.

While the number and disposition of the lights on the underside of the head and abdomen is more or less standard throughout the group, the constellation of lights that may be seen on a fish in profile seems to be unique in each of the 250-odd species. When lantern-fishes display these lights, are the individuals of a particular species exchanging recognition signs (p. 369)? There are also differences between the sexes in the development and placing of certain luminous organs (pp. 368–9). Most of the organs on the head and body look like minute pearl-buttons. The structure and function of lantern-fish lights are considered elsewhere (p. 369).

If the stomiatoids have evolved from smelt-like ancestors, their nearest deep-sea relatives are the argentinoid fishes, which are related to the salmonoid groups. The argentinoids are small-mouthed fishes of diverse body form ranging from short squat types (*Opisthoproctus*) through conventional types (*Winteria* and *Argentina*) to elongated types (*Dolichopteryx* and *Bathylychnops*) (Plate I and Fig. 133). Apart from *Argentina* and its near relatives, which live near the bottom over slope areas, these are fishes of the mesopelagic zone. Their adaptation to this twilight world is evident in their large and sensitive eyes, which are considered elsewhere (pp. 385–7). Here we need only note that tubular eyes are prevalent in this group (as in the above genera except for the third and last). Such eyes, with parallel optical axes that may be directed upward or forward, have also evolved (convergently) in certain stomiatoids (e.g. the hatchet-fish, *Argyropelecus*), two families of alepisauroid fishes (Scopelarchidae and Evermannellidae), giganturoids and the free-living males of the ceratioid angler-fish genus, *Linophryne*. In *Opisthoproctus* and certain other deep-sea argentinoids there is a well-developed anal light organ containing luminous bacteria (see also p. 357).

A unique kind of light organ has evolved in fishes of the family Searsidae, which are also mesopelagic and have large and very sensitive eyes (Fig. 45). Each organ of a pair, which is set on the shoulder region above the pectoral fin, forms photogenic cells in a darkly pigmented sac that opens backwards through a pore. In *Searsia*, at least, these shoulder organs discharge a bright cloud of luminous particles (Nicol, 1958). Some species have other (round, crescentic or strip-like) luminous organs, most of which are set on the underside of the fish (Fig. 45). Lengths range from about 50 to 210 mm.

Apart from *Platytroctes* and *Platytroctagen*, which have a compressed and keeled body, searsids are fishes of fusiform appearance. The closely related aleopcephalids ('slick-heads') are much more diverse in form and fin pattern. For instance, *Aulostomatomorpha* has a long tubular snout, *Leptoderma* an elongated body with a long anal fin, *Xenodermichthys* a fairly elongated compressed body and *Alepocephalus* is more or less fusiform with opposed and set-back dorsal and anal fins. Species of *Alepocephalus*, which may reach more than 0.5 m in length, live near the

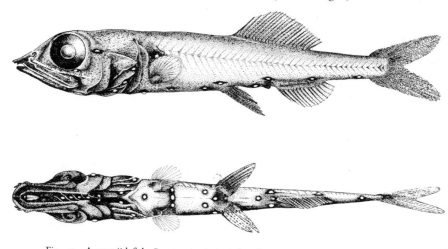

Fig. 45   A searsiid fish, *Persparsia tåningi* (length ca. 60 mm). (Redrawn from Parr, 1960)

bottom at depths down to 2,000 metres or more. Some alepocephalids are mesopelagic to bathypelagic. Genera such as *Xenodermichthys* have light organs while others have none.

There are over 1,000 species of deep-sea midwater fishes and at least one half are stomiatoids and lantern-fishes. The next most diverse groups are the deep-sea anglers (ceratioids) and alepisauroids. Lantern-fishes and alepisauroids are placed in the same order (Myctophiformes), but they are very different in functional design and habit. Most of the 100 odd species of alepisauroids belong to the family Paralepididae, which have been called barracudinas because some species remind one of those predatory fishes par excellence, the barracudas. Indeed, alepisauroids, which live from mesopelagic to upper bathypelagic levels, seem to prey largely on other fishes, and they are well equipped to make a living. Whether spindle-shaped or long-bodied, and there is a marked trend in the group for the evolution of elongated types, the jaws and inner parts of the mouth bear large stabbing teeth (scopelarchids, evermannellids, *Alepisaurus, Omosudis* and *Anotopterus*) or a set of prominent pointed teeth (paralepidids) (Figs. 46–48). The dentition consists also of many small, pointed teeth that help to grip the prey. Like the predacious stomiatoid fishes, alepisauroids have an expandable sac-like stomach, and even in species with body walls of limited distensibility (e.g. scopelarchids) prey up to one-third of the length of the swallower can be accommodated. Nearly all species are non-luminous.

Fig. 46   A lancet-fish, *Alepisaurus ferox*. (Redrawn from Maul, 1946)

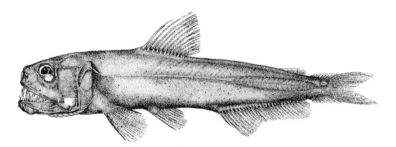

Fig. 47   An evermannellid fish, *Coccorella atrata* (ca. 45 mm). (Redrawn from Rofen, 1966)

Certain of the alepisauroids are among the largest species of midwater fishes. *Alepisaurus ferox* (Fig. 46), the lancet-fish, grows to about 2 metres, *Anotopterus* to about 1 metre and the largest barracudinas (e.g. *Notolepis coatsi* and *Paralepis atlantica*) to about 50 centimetres. The main range is from a few to 30 centimetres. Though larger than most lantern-fishes, few alepisauroids undertake daily vertical migrations that so occupy so many of their relatives. Indeed, the alepisauroids and other sizeable midwater fishes that are adapted to take large meals seem to be non-migrators. Most of the migrators are fishes of modest size with moderate capacities for relatively small food organisms. If, as seems very likely, vertical migrations are feeding sorties, are the migrators

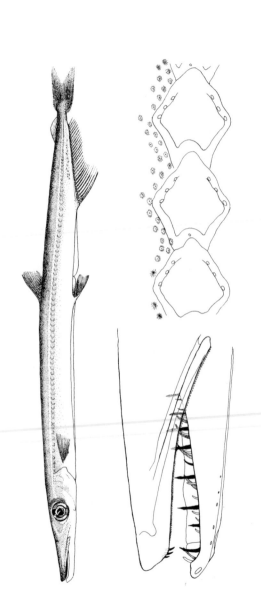

Fig. 48   A barracudina, *Lestidium atlanticum* (length ca. 175 mm). Below, side view of snout and main jaw teeth, and three lateral line scales. (Redrawn from Rofen, 1966)

'compelled' upwards to find the best feeding grounds or are there other factors? Are the non-migrators able to stay at one level because of their trophic opportunism? Such questions are considered elsewhere (pp. 247–8).

Giganturoid fishes and the gulper-eel *Saccopharynx* (Fig. 49) are also adapted to take a wide size range of food organisms. The giganturoids are elongated to long slim fishes with a wide mouth, a fearsome dentition and large forwardly-directed tubular eyes. Their scientific name refers to the very prolonged lower lobe of the tail fin. Like many other mesopelagic fishes, they are silvery.

The species of *Saccopharynx* and the related genus *Eurypharynx* (Fig. 145) are very elongated, dark-coloured fishes of lower mesopelagic to bathypelagic levels. Much of their length is taken up by a long tapering tail that is tipped with luminous tissue. The well-named *Eurypharynx pelecanoides*, with extremely long jaws that remind one of a pelican's bill, grows to a length of about three-quarters of a metre: the species of *Saccopharynx* reach lengths of between 1 and 2 metres. Though *Eurypharynx* has a much larger mouth than *Saccopharynx*, the latter has a more formidable dentition, a much larger stomach and a longer abdominal cavity. Indeed, *Saccopharynx* can swallow very large fishes but *Eurypharynx* feeds on relatively small prey, such as the young stages of prawns and small fishes.

Though the gulper-eels are highly modified fishes the fact that *Eurypharynx*, at least, produces transparent, leaf-like leptocephalus larvae may well mean that the true eels (Anguilliformes) are their nearest relatives. Midwater eels include the snipe-eels (Nemichthyidae), the bob-tailed snipe-eel (*Cyema*) and *Serrivomer*. The snipe-eels have long, pliable beak-like jaws, set with many minute teeth, and a very elongated body that may reach a length of nearly a metre. They have relatively large, wide-open eyes, whereas the bob-tailed snipe-eel, like other bathypelagic fishes, has small, reduced eyes (see also p. 390). *Cyema* (Fig. 145), which has a short tail 'arrow-feathered' with opposed dorsal and anal fins, may reach a length of about 125 mm.

The whale-fishes (Cetomimiformes), which also have small or reduced eyes, are also bathypelagic. Whale-fishes have large, small-toothed jaws and a considerable capacity, expandable if need be, for copepods and other small zooplankton. One species, *Barbourisia rufa*, which grows to a span of at least 20 centimetres, is one of the few reddish-coloured midwater fishes. The other dark-coloured whale fishes are sometimes tinged with red (Fig. 137).

Like other midwater species, whale-fishes are widely distributed in the ocean, but they are not commonly caught. Out of some 10,000 fishes representing well over 100 species that were taken off the Canary Islands, Badcock (1970) found two specimens and species of whale-fishes. When

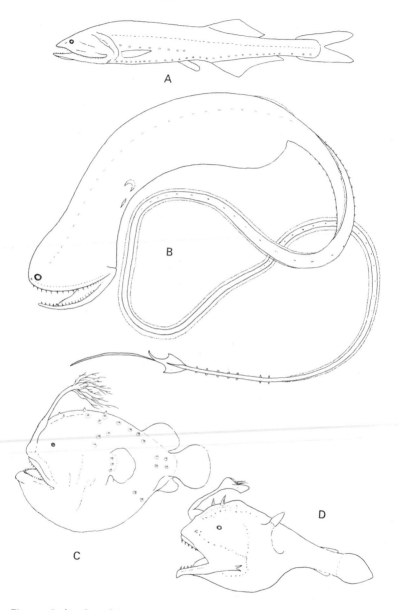

Fig. 49   Bathypelagic fishes. A, *Cyclothone*; B, *Saccopharynx harrisoni*, a gulper-eel; and two female ceratioid angler-fishes, C, *Himantolophus groenlandicus* and D. *Danaphryne nigrifilis*. (C and D are redrawn from Bertelsen, 1951) (redrawn from Marshall, 1951)

the whale-fishes have been properly revised they may well prove to be as diverse as the Melamphaidae, a family (known from about 30 species) with which they have some affinity. Melamphaids, with their round snout, short body, large mid-set dorsal fin and bands of minute teeth in the jaws remind one rather of lantern-fishes (Fig. 157). They also have about the same size range (20–200 mm). But their thoracic pelvic fins, each with a spine and 6 to 8 jointed rays, are part of a character complex that relates them to the 'lower' spiny-finned (berycoid) fishes. Melamphaids, which are mesopelagic to bathypelagic fishes with brownish to dark colouring, live in all major regions of the ocean, except for the Arctic and Mediterranean Seas. They are without light organs. In one genus (*Melamphaes*) there are 'dwarf' and 'giant' species. The former are evidently adapted for life in the impoverished central water masses, the latter for more productive peripheral waters (see pp. 480–3).

Fish faunas of neritic and coral-producing seas are dominated by 'higher' spiny-finned (perciform) fishes, but representatives of this great teleost group are in a minority in the deep sea. Midwater forms include the giant swallowers (Chiasmodontidae) and the trichiuroid fishes, predatory types with marked piscivorous ways. *Chiasmodon niger* can swallow fishes larger than itself. The trichiuroids are diverse in form, ranging from elongated types, such as *Paradiplospinus*, *Gempylus*, the snake mackerel and *Aphanopus*, the black scabbard fish, to those with a fusiform body, e.g. *Ruvettus*, the oil fish and *Nealotus*. Trichiuroids, which are somewhat reminiscent of alepisauroid fishes in form, fearsome dentition and sleek appearance, are represented also by a number of large species. The snake mackerel grows to about 1 metre, the black scabbard fish to 1·25 metres and the oil fish to over 1·5 metres. There is a deep line fishery for the second species off Madeira.

The depth of transition from the twilight of the mesopelagic zone to the sunless waters of the vast bathypelagic zone is near 1,000 metres. Dark-coloured species of *Cyclothone*, whale-fishes, gulper-eels and the deep-sea angler-fishes (Ceratioidea), which are related to the angler-fishes or monkfishes (*Lophius* spp.) and frog fishes (Antennariidae), form most of the bathypelagic fish fauna (Fig. 49). Angler-fishes (Lophiiformes) bear a rod and bait (evolved from the foremost dorsal ray) on top of the head, but in the ceratiods it is developed only in the females. In all but one form (*Caulophryne*) the bait is luminous and differs in structure from species to species (Bertelsen, 1951; Pietsch, 1974).

In each species the individuality of the lure may be seen in its shape, photogenic complex and dressing with dermal appendages, but there is no evidence yet that particular lures attract particular kinds of food organisms (Pietsch, 1974). It seems more likely that female angler-fishes, which have a wide, well-armed mouth and distensible accommodation, are trophic opportunists, able to take a wide size range of prey (see pp.

255–6). Perhaps the males are drawn to a proper partner after receiving the 'right' sign stimuli from her lure. But the males of a few species (e.g. *Centrophryne*) have tiny eyes, though their olfactory organs, like those of most male ceratioids, are highly developed. Perhaps the females with keen-nosed mates lay scent trails during the breeding season (pp. 404–8).

The males are much the smaller sex and their dwarfing is but one aspect of a complex and trenchant sexual dimorphism (pp. 261–2). After metamorphosis the males are about 10–40 mm in length, and there is little subsequent growth. They are active more or less fusiform fishes but in nearly all species the females are globular or pear-shaped. In certain ceratioids (e.g. *Edriolychnus* and *Cryptopsaras*) the male becomes parasitic on the female and in such forms she is from three to more than ten times the length of her mate. The free living males of *Himantolophus*, which is one of the ceratioids that seem to have other mating arrangements (p. 453), grow to about one-thirteenth the length (600 mm) of the largest known female (Bertelsen, 1951).

Though their means of reproduction are unusual, the life-history patterns of deep-sea angler-fishes are much like those of most other midwater fishes, whether mesopelagic or bathypelagic. Their eggs, which must be very numerous but have yet to be found, are probably shed in the depths and develop as they float upwards. During metamorphosis to adult form there is a descent to the levels of the adult living space. Here a few of the young fishes will eventually reach maturity.

Benthopelagic animals

If the transition from mesopelagic to bathypelagic zones is around a depth of 1,000 metres, it is evident that living space adequate for bathypelagic faunas exists only over regions of the deep-sea floor well below this transition level. Even so, the mean depth of the ocean is about 4,000 metres. Though the living space of benthopelagic animals is *near* the bottom, such faunas may well merge with mesopelagic faunas over the upper continental slope at depths from 200 to 1,000 metres and with the most abundant species of the bathypelagic faunas at depths lower down the slope between 1,000 and 2,000 metres (pp. 272–3). It is possible, of course, that competition between midwater and near-bottom faunas tends to exclude overlap between the two. But there is evidence that certain mesopelagic animals live near the bottom for part of their adult life. Large and heavy bodied species of deep-sea prawns (e.g. of the genera *Acanthephyra* and *Parapandalus*) may live on or near the deep-sea floor as adults (Omori, 1974). If their habit, as photographs suggest, is to rest on the bottom they are better regarded as benthic animals. But certain lantern-fishes, which have been seen swimming near the bottom and are consistently caught with nets fishing on or near the bottom, would seem to have much the same habit as benthopelagic fishes. Indeed, it looks as though the older individuals of such lantern-fishes take to a benthopelagic existence (pp. 272–3). Large individuals of the stomiatoid fish *Yarrella blackfordi* are also frequently taken in bottom nets (Grey, 1964).

Evidence that they have fed on or near the bottom may sometimes be found in midwater fishes. Black scabbard fishes (*Aphanopus*) caught by Madeiran fishermen occasionally contain halosaurs, which, as we shall see, are benthopelagic in habit. Large deep-sea cod (*Halargyreus*), also benthopelagic, have been found in gulper-eels of the genus *Saccopharynx* and there is also the remarkable record of sea-urchin remains in the related *Eurypharynx*. Thus, bathypelagic as well as mesopelagic fishes sometimes seek the interface between sea and land, and in all instances the attraction is presumably to good feeding grounds near or on the bottom. The extent of this attraction is unknown but if it is strong, as in regions of high productivity (p. 272), one might occasionally expect to see individuals of large midwater fishes in photographs taken near the bottom. There seem to be no such records. In the horizontal plane, however, diverse midwater fishes grow large but not to maturity in feeding grounds well removed from the fertile areas of their distributions (see p. 455).

Deep-sea cameras have taken many fishes in exposures near the bottom in all parts of the ocean (Heezen and Hollister, 1971). Benthic species, such as tripod-fishes and their relatives, rest on the sediments and will be considered in the next chapter. Of the fishes that habitually swim near the bottom the most often photographed are the rat-tails or grenadiers (Macrouridae). Pictures of other benthopelagic fishes include squaloid sharks, chimaeroids, alepocephalids, eels, halosaurs, notacanths, deep-sea cods and brotulids. Species of all these groups have been consistently taken with bottom nets, but photographic evidence was needed to confirm or refute conjectures about habit based on diet and functional design, as well as on mode of capture.

Evolution in benthopelagic fishes has tended strongly to produce elongated forms, and in nearly all the teleosts the dorsal and anal fins are long based and sustained by many fin rays. Such fishes, particularly the rat-tailed types, are evidently well adapted to life near the bottom over a wide range of depths (pp. 276–7). Benthopelagic fishes are also neutrally buoyant. The sharks and chimaeras have large buoyant livers charged with squalene, a hydrocarbon of relatively low specific gravity. Species of nearly all the other groups develop a capacious gas-filled swimbladder and there is evidence that the buoyant properties of this organ are maintained by its gas-secreting complex against the pressures exerted by the ocean (pp. 326–35). The alepocephalids, which in productive areas

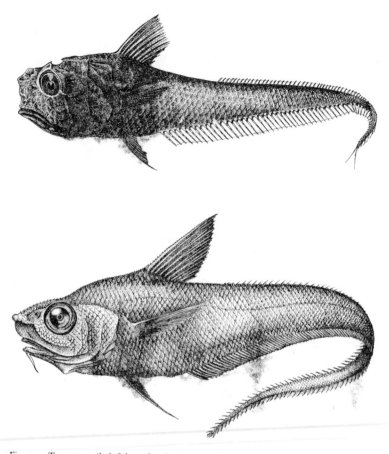

Fig. 50 Two rat-tailed fishes (family Macrouridae). Above, *Hymenocephalus aterrimus*; below, *Coryphaenoides colon*. (Redrawn from Marshall, 1973)

may be both diverse and common (Golovan, 1974), have no swimbladder. But they are likely to be close to neutral buoyancy, particularly because of the weak development of their skeletal and muscular systems (p. 337).

Benthopelagic fishes thus have no need to expend energy to keep their level. They are free, as cine-films show, to pursue their slow undulatory ways over the sediments as they search for food.

Like other benthopelagic fishes, rat-tails (Macrouridae) are most diverse over the upper continental slope from warm temperate to tropical belts of the ocean (Fig. 50). Their relatively frequent appearance in deep-sea photographs is not because other benthopelagic fishes are camera-shy. A series of trawl hauls down the continental slope and rise to depths near 5,000 metres nearly always nets more individuals and species of rat-tails than of any other comparable group. Indeed, if overall oceanic diversity could be expressed in numbers of individuals and species, rat-tails would surely emerge as the most diverse family of benthopelagic fishes. Brotulids, known from about 200 species, come next to the 250 or more species of rat-tails, but one has only to consult the reports of major oceanographic expeditions to find that many more fishes of the second family were caught. Ninety-five per cent of rat-tail species live at depths from 250 to 2,000 metres in all major oceanic regions. Most of the species with centres of abundance between 2,000 and 6,000 metres belong to the genera *Chalinura* and *Lionurus* (Marshall, 1965, 1973).

Rat-tails, which are related, inter alia, to the deep-sea cods (Moridae), have long, many-rayed dorsal and anal fins that meet at the end of a long, tapering tail. The species of one subfamily (Bathygadinae), with

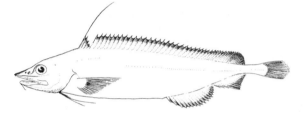

Fig. 51   Deep-sea cod, *Antimora rostrata*. (After Goode and Bean)

a wide terminal mouth, fine teeth in the jaws and long numerous gill rakers, are apparently adapted to feed over the sediments, particularly on copepods and other small crustaceans. Most species (subfamily Macrourinae) have relatively few, stumpy gill rakers, a restricted first gill slit, and the front of protrusible jaws (with a few exceptions) just behind to well behind the tip of the snout. These rat-tails take food organisms that live over and on the sediments. Species with a prominent, well-armed snout also root in the oozes, sieving swallowed deposits to retain small invertebrates (e.g. polychaete worms) on their gill rakers (Marshall, 1973). The largest known species, *Coryphaenoides pectoralis*, lives in the North Pacific and reaches a length of 1·5 metres: the smallest (of the genus *Hymenocephalus*) span about 20 centimetres.

The fifty-or-so species of deep-sea cod (Moridae) are lower shelf to slope dwellers to depths of about 3,000 metres. While like the cod-fishes (Gadidae) in many respects (Fig. 51), their unique feature is the development on the anterior chamber of the swimbladder of 'ear-pads' that fit closely against the auditory capsules. Such a hydrophonic arrangement in other fishes confers extra hearing powers, but do deep-sea cod hear each other? Unlike most of the macrourids they do not seem to have special means of sound making (p. 425). But certain morids (e.g. of the genera *Physiculus* and *Brosmiculus*) have converged with diverse macrourids in evolving mid-ventral light organs that contain luminous bacteria and open to the outside through a duct near the anus (p. 373). The largest deep-sea cod, *Antimora rostrata* (length to 65 cm or more), is cosmopolitan between depths of 400 and 3,000 metres.

Considering their success as adults, it is curious that so few young stages of rat-tails have been taken (see Marshall, 1965). Eels, halosaurs and notacanths ('spiny-eels') all produce leptocephalus larvae (p. 449), and they are also alike in other features, such as the structure of their swimbladder (Marshall, 1962). The best known and probably the most successful of the benthopelagic eels are the synaphobranchids, common over the upper continental slope but extending to depths of 3,000 metres or more. Halosaurs, which have much the same depth range, may be quite abundant over parts of the slope. One species (*Aldrovandia rostrata*) was caught by HMS *Challenger* at a depth of just over 5,000 metres (Fig. 52). The largest species grow to a length of more than half a metre. Notacanths are elongated, rather eel-like fishes with a row of spiny rays along the back (Fig. 52). As in halosaurs, the snout projects well forward of the inferior mouth and the long tapering tail bears a many-rayed anal fin down the whole of its span. Notacanths are small-scaled, generally greyish to dark-coloured fishes. The halosaurs develop relatively large scales and background colours varying from chocolate brown to greyish white or pink. Like rat-tails, fishes of these two groups tend to move nose-down over the sediments. The food of halosaurs seems to consist largely of crustaceans taken close to the bottom, whereas species of *Notacanthus* snip off pieces of sea-anemones, sea-pens and bryozoans (see also pp. 275–6). The deepest living notacanths (of the genus *Polyacanthonotus*) range to levels of about 3,000 metres. *Notacanthus chemnitzi*, the largest species, grows to a length of a metre or more.

Though deep-sea fishes live in environments that are much vaster and less productive than those over the continental shelves, their populations evidently tend to produce enough mature individuals for the two sexes to meet and reproduce. If, after one meeting, the female is provided by the male with enough sperm to cover her reproductive life, and if she later gives birth to living young that are advanced enough to have a good start in life, then one might think that here are reproductive adaptations that would be particularly apt for deep-sea fishes. But in all parts of the ocean egg-laying (oviparous) species of fishes far outnumber the viviparous kinds. In the deep sea some of the latter are squaloid sharks, though the only kinds to produce stores of sperm (spermatophores) that can be transferred through an intromittent organ to viviparous females are some of the brotulid fishes, the aphyonids and certain mesopelagic eel-pouts (see also pp. 457–8).

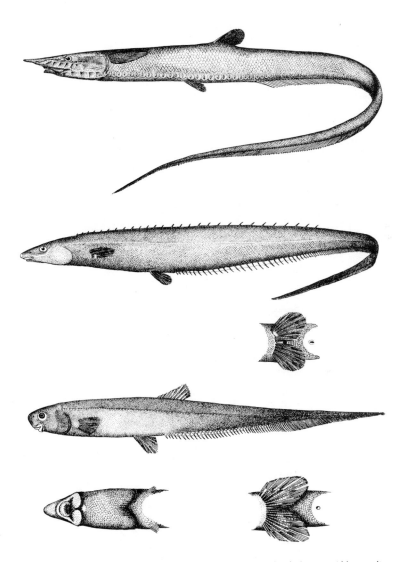

Fig. 52 Notacanthiform fishes. From top to bottom; the halosaur *Aldrovandia rostrata*; the notacanth *Polyacanthonotus rissoanus* and pelvic fins; *Lipogenys gillii* and below, ventral surface of head, showing sucker-like mouth, and pelvic fins. (Redrawn from McDowell, 1973)

There are over 70 species of viviparous brotulids and about half live along the shore, in coral reefs and in fresh waters. The remaining deep-sea species range between depths of 150 to 2,000 metres (Mead, Bertelsen and Cohen, 1964). The one hundred or more species of oviparous brotulids, which have no intromittent organ, are all deep-sea forms and several are known from depths exceeding 5,000 metres.

Brotulids are more or less elongated fishes with long based dorsal and anal fins that end near, or are confluent with, the small caudal fin (aphyonids have a shorter tail). Most species are between 10 and 30 cm when mature (Fig. 53).

Males of the oviparous species, like those of diverse rat-tailed fishes,

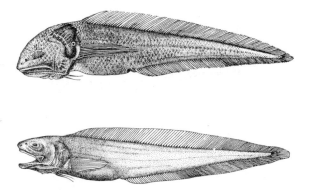

Fig. 53   Two abyssal brotulids. Above, *Abyssobrotula galatheae* (length 130 mm), from the Kermadec Trench, 5,230–5,340 m; below, *Bassogigas* (170 mm). (Above, redrawn from Nielsen, 1977; below, redrawn from Bruun, 1956)

have a drumming mechanism associated with the swimbladder. The sound-producing devices of viviparous brotulids are less well known, but in certain species both sexes have drumming muscles attached to the swimbladder and its skeletal suspension. The possible role of sound making in the life of these two dominant groups of benthopelagic fishes is considered elsewhere (pp. 425–7).

Shark members of benthopelagic fish faunas, as already implied, belong largely to the suborder Squaloidea, which have two dorsal fins with or without spines but no anal fin. They are dark skinned and some, such as *Etmopterus*, are luminous. *Etmopterus hillianus*, reaching a maximum length of 31.5 cm, is one of the smallest sharks. The deep chocolate-brown *Centroscyllium fabricii*, also luminescent, lives on both sides of the North Atlantic down to 1,500 metres and grows to about a metre in length. *Centroscymnus coelolepis* has a rather similar distribution to depths of 2,700 metres) and size but is without luminous organs. The largest species, which have appeared in bottom photographs, are sleeper sharks (*Somniosus*), *S. microcephalus*, the Greenland shark, growing to 7 metres or more. In general deep-sea sharks are most common over the upper continental slope between depths of 200 and 1,500 metres. Their food seems to consist largely of fishes, cephalopods and the larger crustaceans.

Throughout the ocean, down to depths of 5,000 metres or more, populations representing close to 1,000 species of deep-sea fishes move over the sediments. They are most diverse over the upper continental slope. Much of their food consists of benthic invertebrates but zooplankters, particularly copepods, may be found in their alimentary systems.

There would thus seem to be a benthopelagic zooplankton, and recent research certainly indicates that special kinds of calanoid copepods live near the deep-sea floor. During dives in the Woods Hole Oceanographic Institution's submersible *Alvin*, Grice (1972) used a specially designed pair of nets, each of which could be closed after towing. In the net he fished just above the bottom (20–50 cm above) there was a fauna of copepods that was not present in the net fished well above the bottom (120 and 260 metres over depths of about 1,000 and 1,500 metres respectively). The near-bottom copepods (7 species belonged to the genus *Xanthocalanus*) ranged in length from about 1 to 3·25 mm. Besides their small size, they had other characters previously considered to be typical of bottom-dwelling copepods, such as a plump body and the presence of long spines on the outer segments of the first swimming legs. Such copepods may not only be found in the gut of diverse benthopelagic fishes, but also in tripod-fishes, which stand clear of the bottom on their fin-ray under-carriage (Marshall and Merrett, 1977; see also p. 305). Grice (1972) calls them planktobenthic copepods, and he wonders whether they may live in the bottom sediment or flocculent zone just over the sediment. Perhaps they alternate between a sedimentary and free-swimming life, as may some of the amphipod and isopod crustaceans, many of which are ooze-eaters.

If their habits are like those of their cuttlefish relatives, small sepioid squid (e.g. *Rossia* and *Sepiola*) of the continental slope will spend part of their time buried just below the ooze, and part hunting for food above. But this hardly applies to cirrate octopods, which are evidently benthopelagic rather than bathypelagic animals. To quote Roper and Brundage (1972), 'Twenty-seven photographs from seven deep-sea localities in the North Atlantic reveal cirrate octopods in their natural habitat. The photographs demonstrate that these octopods are benthopelagic, living just above the bottom at depths of 2,500 to greater than 5,000 m. Typical cephalopodan locomotion is exhibited as well as a drifting or hunting phase, and possibly a pulsating phase. Animals range in size from approximately 10 to 128 cm in length, and up to 170 cm across the out-stretched arms and webs.'

Lastly, a physonect siphonophore, *Stephalia coronata*, may well live close to the sediments, for the few specimens known have come from bottom nets. The Angels (1974) suggest that '... the animal's tentacles trail along the bottom, presumably catching small animals from off the mud surface'.

It is reasonable, then, to conclude this survey of pelagic animals in the deep sea by realizing that organisms ranging in size from small cope-pods to metre-long fishes find a living near the sediments. Exploration of this benthopelagic fauna has barely begun, but should lead, inter alia, to closer appreciation of life on and in the oozes. To this benthic life we now turn.

Life on the deep-sea floor 5

Pelagic life is partitioned and maintained in dynamic but stratified stability between the interfaces of the ocean with the air and earth. Life on the deep-sea floor is at rest, as it were, in two-dimensional space, little or no energy being needed to counter the pull of gravity. The same is true, of course, of benthic animals in coastal seas, but during their early life most of these forms become part of planktonic space. In the deep sea, benthic invertebrates with typical planktonic larvae are in the minority. The young life of most species seems to have contracted towards the interface existence of their adult phase (pp. 460–3).

Deep-sea benthos, like that of the neritic province, is divisible into an epifauna and an infauna. Species of the infauna live in the sediment and range from large semi-burrowing forms, such as acorn-worms, to millimetre-sized copepods and even smaller young stages that move between the grains of ooze. The epifauna, which, according to Thorson (1971), comprises about four-fifths of all large benthic animals, are animals that move over the sediment or are attached to some suitable surface.

The infauna

The smallest animals of the infauna, from fractions of a millimetre to about two millimetres in length, are known collectively as meiobenthos (Gr. *meion*, smaller) (Fig. 54). Investigation of the deep-sea forms is now under way. In the Walvis Ridge area off southwest Africa samples from ten stations between depths of 1,200 and 5,170 metres were dominated by nematodes (94%), followed by harpacticoid copepods (2·3%) and polychaete worms (1·5%). There were also numerous nauplius stages (presumably of copepods), and a few tardigrades,* kinorhynchs† and ostracods (Dinet, 1973). Subsamples taken at different levels down the cores of sediment showed that conditions for meiobenthic life were favourable only in the uppermost 5 cm of sediment. Indeed, nearly 60 per cent of the animals come from the top centimetre. In a table comparing his results with others from the Indian Ocean, the western Mediterranean, the Iceland–Faroe region and the Iberian Sea, Dinet found that the density of the meiobenthos (individuals per 10 cm² of sediment) ranged from 24 to 1,554 at depths from 1,400 to 5,300 metres. Densities were lowest (33–78/10 cm²) in the Mediterranean, which is one of the most impoverished parts of the ocean.

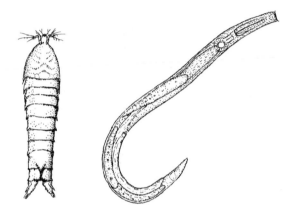

Fig. 54   Representatives of the meiofauna. Left, a harpacticoid copepod *Paramphia-cella* and, right, a nematode worm *Ethmolaimus*. (Redrawn from Parsons, Takahashi and Hargrave, 1977)

In a series of hauls between Cape Lookout, North Carolina, USA, and Bermuda at depths between 14 and 5,165 metres, Coull (1972) also found that nematodes and harpacticoid copepods accounted for most (90%) of the catch of meiobenthos. He even remarks that harpacticoids are the most abundant arthropod encountered in the deep sea. Sediment-dwelling species tend to be elongated and modified in various ways for dealing with their environment. For instance, Coull remarks that some of the deep-sea forms are made for digging into the ooze.

While the number of copepods taken decreased with depth, the number

---

*Tardigrada ('Water Bears') are small (50–1,000 μ) animals with a bilaterally symmetrical body formed of four segments, each with a pair of stumpy appendages bearing minute claws.

†Kinorhyncha, also minute, have a bilaterally symmetrical body of 13 or 14 segments: the first is the head, which bears circlets of curved spines.

of species (species diversity) increased markedly from shelf to abyssal levels. Samples from the shelf were always dominated by a few species, whereas below 1,000 metres many species of the more diverse deep-sea fauna proved to be new. Like the earlier findings of H. L. Sanders and his colleagues on the smallest members of the larger (macrofaunal) benthos, Coull's investigations seemed to support the idea that, though impoverished, the stable ('predictable') conditions on the deep-sea floor have favoured the evolution of diverse and largely infaunal, benthic fauna. His most intriguing discovery was of the marked individuality of the copepod fauna from one deep-sea station to the next. After examining nearly 700 copepods from 18 stations at depths greater than 400 metres, he found that very few species were duplicated from sample to sample. Indeed, over 50 per cent of the fauna is undescribed.

There are also bound to be many new species of nematode worms and when this part of the fauna is properly studied, it will be interesting to see if it follows copepod patterns of distribution. Tietjens' (1971) work on the ecology and distribution of the deep-sea meiobenthos off North Carolina shows that nematodes were dominant from 50 to 500 metres and, together with foraminiferans, were the only members of the fauna in large numbers between 800 and 2,500 metres. During a more recent survey in the same region Coull and his colleagues (1977) found that individuals representing species of the same two groups comprised 75 per cent or more of the total numbers of meiobenthic organisms, which were highest (mean 892/10 cm^2) in very fine silt at 4,000 metres. At all depths (400, 800 and 4,000 metres) most of this fauna came from the upper 3 cm of sediment, but at deeper levels (> 4 cm) the typical residents were nematodes and foraminiferans, asserting, as it were, their dominance.

No doubt there are both epifaunal and infaunal species of foraminiferans in the deep sea, just as in neritic waters. Gunnar Thorson (1971) writes, 'In the deeper waters of the Arctic seas, and in certain areas of the North Sea and Kattegat, one finds around fifty foraminifers of the genus *Astrorhiza* per square metre. For unicellular animals these are large—about 5 mm in diameter—but their shells, which are composed of sand grains cemented together, take up very little room in a bottom sample. In life, however, long shiny threads of protoplasm emerge from a multitude of holes in between the sand grains, threads which wreathe around each other to form an irregular network over the bottom.... *Astrorhiza*'s net can cover an area with a radius of 6–7 cm. In other words, even if only fifty of these inconspicuous creatures are found per square metre, over half the bottom surface will be a danger zone for the passage of many small crustaceans and young bottom animals.' Thorson considers *Astrorhiza* to be a member of the infauna. Epifaunal forams of coastal seas crawl over the bottom and weeds by means of their pseudopodial complex, feeding on diatoms, flagellates, zoospores of algae, etc. The deep-sea counterparts of these species presumably feed, at least in part, on digestible organic particles in the oozes. If the pseudopodia of a foram are probing the oozes while its shell rests on the oozes, it seems pointless to ask whether it is an epifaunal or infaunal species. And we have already seen that forams are among the dominant groups of the meiobenthos.

The general rule that there are more benthic than pelagic forms of animals in the ocean is well exemplified by the Foraminifera, which are represented today by over 1,000 bottom-dwelling and about 30 planktonic species (pp. 69–70). Russian investigations in the Pacific Ocean show that benthic species with a calcareous shell (or test) live at depths down to about 4,500 metres. Agglutinating species, which use sand grains and other materials (fragments of mica, magnetite, granite, topaz, sponge spicules, coccoliths, etc.) to form their tests, live at all depths down to the very deepest reaches of the ocean (Saidova, 1966) (Fig. 55). Most of the deep-sea agglutinating forms are ammodiscids, some of which have a long tubular test coiled like a ship's rope or in less regular fashion (measuring from about 0·5 to 1·5 mm in diameter). Of the calcareous species, buliminids, with a many-chambered test, are common over the upper continental slope. Presumably the lower limit of occurrence of calcareous species is related to the increasing solubility of calcium carbonate with depth (and hence a decreasing availability for the secretion of shells). More precisely there is a critical depth called the calcium carbonate compensation depth, where the rate of solution of $CaCO_3$ exceeds the rate of its deposition. (The solubility of $CaCO_3$ rises as the temperature drops, while increasing pressure increases the dissociation of $CaCO_3$ and CO_2 in sea water.) In the Pacific Ocean this critical depth lies between 4,000 and 5,000 metres and beyond it the $CaCO_3$ concentration in deep-sea sediments decreases rapidly with depth.

Deep-sea benthic forams, like the harpacticoid copepods, increase in diversity with depth. In the western North Atlantic (off the eastern United States) the number of species increases from about 20 between 200 and 400 metres to between 60 and 80 at levels just over 4,000 metres. Compared to species at depths below 100 metres the deep-sea species have smaller population densities (Buzas and Gibson, 1969). Both the number of species and the relative abundance of each are needed to obtain an ecologically meaningful measure of diversity but such matters are best considered elsewhere (Chapter 13).

It seems natural that coral reefs, which are so productive and diverse in ecological niches, should provide a living and lodging for many different kinds of animals. Until the early 1960s it also seemed natural that the monotonous, food-poor sediments of the deep-sea floor should support fewer and fewer species and individuals of animals as the depth increased. This is certainly true of the larger species of epifaunal inverte-

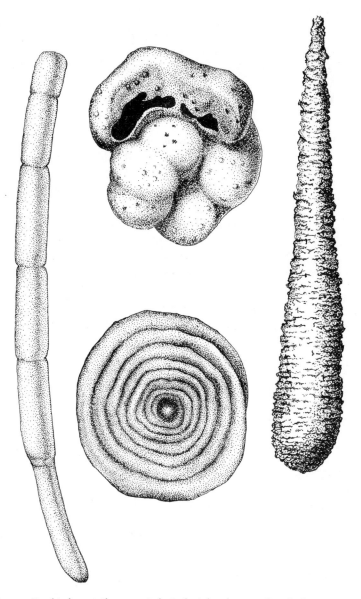

Fig. 55   Benthic foraminifera tests. Left, *Bathysiphon lanosum* (length about 25 mm);
upper centre, *Cribrostomellus apertus* (diameter about 10 mm); lower centre,
Ammodiscus profundissimus (diameter ca. 5 mm); right, *Rhizammina alta*
(length about 12 mm). (Redrawn from Saidova, 1970)

brates, as the results of deep-sea expeditions showed so well. But by using specially designed nets and sieves fine enough to retain the smaller kinds of invertebrates H. L. Sanders and his colleagues (Hessler and Sanders, 1967; Sanders and Hessler, 1969), who fished from shelf to abyssal levels (7,400 metres) between New England and Bermuda, discovered an astonishing diversity of such animals in the sediments.

Their 'epibenthic sled' is designed to skim the surface of the oozes (it also has a blade that can be set to bite into the top few centimetres), and it takes both epifaunal and infaunal invertebrates. The anchor dredge has a leading edge that bites to a specific (10 cm) depth and after filling to capacity, takes no more deposit. It is useful only on relatively smooth and soft bottoms and its catch consists largely of infaunal species. The samples are carefully washed through a sieve with a mesh size of 0·42 mm. Russian investigators evidently use a mesh size of 0·5 cm and they class their catch as meiobenthos. To quote Sokolova (1970), 'By meiobenthos one understands small invertebrates not larger than 0·5 cm (5,000 μ) and not smaller than 500 μ. The minimum size is determined by the mesh size of gauze No. 140 through which the bottom-dredged sample is washed. The maximum size is taken arbitrarily.' The smallest members of the 'Russian meiobenthos' would thus be the largest members of what is regarded as meiobenthos by other investigators.

Sanders and his colleagues found that the miniature infauna on the deep-sea floor between New England and Bermuda is dominated by poly-chaete worms (of a few millimetres) (40–80%), crustaceans (3–50%) and bivalve molluscs. Numerous species proved to be new to taxonomists.

The catches of bivalves showed that eulamellibranch species were domi-nant over the outer continental shelf and upper continental slope, but as the depth of sampling increased the protobranchs became more and more prominent. Indeed, these bivalves and other kinds, such as septi-branchs, form about 10 per cent by number of the abyssal benthic fauna in the Atlantic (Allen and Turner, 1974). Most bivalves are eulamelli-branchs and specialized for (suspension) feeding on micro-organisms and organic particles, which are collected by tracts of cilia on their large gills. Food channels along the lower edges of the gills convey the food to small labial palps, which bear a ciliary system that sorts the food particles before passing them to the mouth. The replacement in the deep sea of such bivalves by protobranchs would seem to indicate the lack of particu-late food of the right kind (and quantity).

Protobranch bivalves have simple gills formed of small plate-like fila-ments and in most species, including those in the deep sea, they aid in respiration and have at most a subsidiary food-collecting function, which is centred mainly in their large labial palps. Such forms of protobranchs are deposit feeders and when buried or half-buried in the sediment, the palps, which in some species have proboscis-like extensions, work over

148

the oozes gathering food particles that are eventually conveyed by cilia to the mouth. Deep-sea species rarely exceed 5 mm in shell length.

While protobranchs are the predominant bivalves in the soft oozes of the deep sea, septibranch species form a small but constant part of the fauna (Allen and Turner, 1974). These authors have studied the functional morphology of abyssal (2,000–6,000 metres) Atlantic species of the deep-sea family Verticordiidae, which have shells measuring a few millimetres across the major axis. The gills of verticordiids are more or less like those of eulamellibranchs, but in many other characters they look like typical septibranchs (Fig. 56).

Water enters and leaves a verticordiid bivalve through short (inhalent and exhalent) siphons, which bear special tentacles, whose function is reconstructed by Allen and Turner thus, 'The region of inhalent and exhalent tentacles is richly supplied with papillae that secrete an adhesive fluid. The secretion has two functions; to assist the animal to maintain a position at the surface of the very soft abyssal sediment and to capture food organisms such as copepods. The sticky tentacles are spread out over the surface of the sediment. On capture the food is "licked" off the tentacles by an inhalent valve and falls, or is carried, to a posterior facing buccal funnel formed by the expanded lips. The palps are extremely reduced in size. Oesophagus and stomach are muscular and the stomach is lined with a scleroprotein and is a crushing organ.' The species studied came from depths between about 450 and 5,000 metres. Though the pelagic life of the young is likely to be short (pp. 464–5), Allen and Turner found that individual species were widespread and present on both sides of the Atlantic.

Tusk shells (Scaphopoda), which live in a long tapering tube and are expert burrowers in sands and muds of coastal seas, are also likely to be part of the infauna of the deep-sea floor, where they exist down to a depth of nearly 7,000 metres. These molluscs may be rather numerous in deep-sea nets, and concerning one species, *Dentalium megathyrus*, known from depths between 1,450 and 4,100 metres off the entire Pacific coast of America, Wolff (1961b) writes, 'We may presume this large species (up to 7 cm long) is also buried in an oblique position similar to the shallow water forms (only the pointed lip of the shell with a hole for the respiration water being visible above the bottom), and that it also feeds on foraminifers, which are caught by a large number of their sticky tentacles, situated on the head and swollen at the tips.'

Since most of the life and organic material in deep-sea oozes is on or just below the surface, tusk shells need only make shallow burrows. Presumably the same is true of the worm-like species of aplacophoran molluscs, which feed on organic muds and ooze and live from coastal waters down to depths of more than 8,000 metres (Filatova, 1971, p. 48). In the deposit feeders, *Chaetoderma* and *Crystallophrissa* (Fig. 57), the foot is

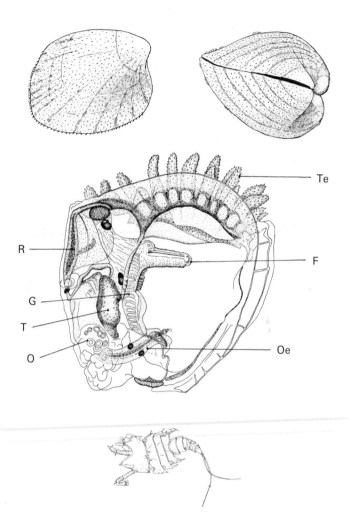

Fig. 56   An abyssal verticordiid bivalve mollusc, *Lyonsiella abyssicola*. Above, lateral
and anterior views of shell (length ca. 4·5 mm). Centre, general anatomy of
animal; F, foot; G, gill; O, ovary; Oe, oesophagus; R, rectum; T, testis;
Te, tentacle. Below, a cervinid copepod (length ca. 1 mm) removed from the
stomach of a *Lyonsiella abyssicola*. (Redrawn from Allen and Turner, 1974)

absent and the mantle forms a cylindrical tube around the worm-like body, which bears a pair of gills (ctenidia) at the posterior end. Burrows are made so that the gill-bearing end of the animal is left near the surface of the sediment. Species of worm-like molluscs, particularly of *Chaetoderma*, are quite common on the deep-sea floor (see, for instance, the faunal lists from stations worked by the Russians in the Scotia Sea and Atlantic sector of the Southern Ocean as described in Vinogradova et al., 1974 and Wolff, 1961). It will be interesting to know more of their biology.

On the deep-sea floor, detritus-eating bivalves (protobranchs) are more successful than suspension feeders (eulamellibranchs) and largely carnivorous kinds (septibranchs). Of the dominant groups of small crustaceans, most of the amphipods, isopods and tanaids are detritus feeders. The same is true of the small polychaete worms (Sanders and Hessler, 1969). As Birstein and Vinogradov (1966) stress, the deep-sea members

Fig. 57   An aplacophoran mollusc *Crystallophrissa indicum*.

of these and other deposit-feeding kinds of invertebrates use the same feeding methods as their shallow-water relatives. 'The conquest of the abyssal did not compel them to develop new feeding habits.' It is reasonable to argue that the three groups of crustaceans and their relatives the cumaceans were also preadapted reproductively for life on the deep-sea floor. The isopods, amphipods, tanaids and cumaceans are classed in the Peracarida, a group in which the eggs develop in a brood pouch formed by the females. The eggs hatch at an advanced stage, and getting a good start in life, as we shall see (pp. 463–5), is an outstanding reproductive adaptation of many deep-sea invertebrates.

Isopods live from the shallowest to the greatest depths of the ocean. Many of the 4,000-odd species have a dorso-ventrally flattened body, but there is a wide variation from the extremely flattened serolids to forms with more or less cylindrical bodies. In the deep sea, for instance, species of *Eurycope* have a short depressed body, whereas some species of *Macrostylis* have a rather slender body that is rounded in cross-section (Fig. 58). Most of the deep-sea isopods range in length from about 2 to

10 mm and beyond a certain level that varies from place to place (about 400 metres in the Arctic, 1,000 metres in the northwest Atlantic and Antarctic and 3,000 metres in Peru–Chile Trench) the isopod fauna (including *Eurycope* and *Macrostylis*) consists largely of species belonging to the subgroup Asellota. Isopods of the upper continental slope are well pigmented, particularly with red and black markings, but abyssal forms, besides being blind, lose their pigment and are usually white or dull grey in colour (Menzies et al., 1973).

Nearly all of the deep-sea amphipods are gammarids, forms in which there is a strong tendency for the (arched) body to be compressed from

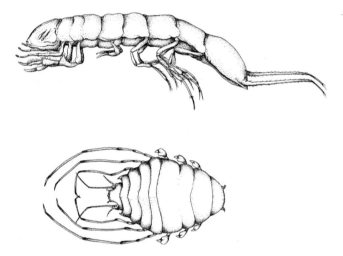

Fig. 58   Peracarid crustaceans (isopods). Above, *Macrostylis galatheae* from the Philippine Trench, 9,790 m; below, *Eurycope* sp. (Above redrawn from Wolff, 1956)

side to side (Fig. 59). Like the isopods, they range from intertidal levels to the very bottom of the ocean. Most of the deep-sea species measure from 2 to 25 mm in length. The tanaids (Tanaidacea) are known from all levels down to about 8,000 metres (in the Kermadec Trench). They have rather slender, more or less cylindrical bodies and range from about 1·5 to 25·0 mm in length (Fig. 59). Nearly all of the larger species (> 10 mm) have been taken in relatively shallow waters in antarctic and temperate regions or at slope or abyssal levels (Wolff, 1956).

Where do all of these peracaridan crustaceans live? Deep-sea cameras, which have given insight to the lives of many bottom-dwelling animals, are of little help, for all tanaids and nearly all amphipods and isopods are smaller than the present limit of resolution of submarine photography. The giants among isopods belong to the genus *Bathynomus*, which attain lengths of 40 cm and are known from shelf to slope levels in the Gulf of Mexico, the Sea of Japan and the Bay of Bengal. Serolids such as *Serolis bromleyana* reach a length and width of over 5 cm. Indeed it is evident that they grow to even larger size, for there are two photographs of giant serolids resting on the sediments of the Falkland

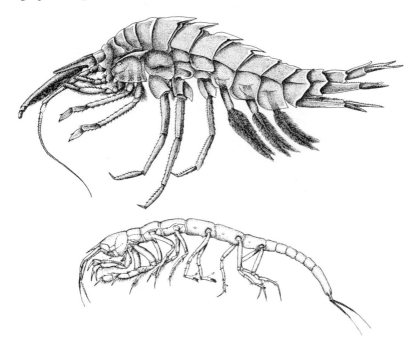

Fig. 59   Peracarid crustaceans (gammarid and tanaid). Above, a gammarid *Parargissa galatheae*, male, length 50 mm; below, a tanaid, *Apseudes galatheae*, female, length ca. 18 mm from the Kermadec Trench 6,770 m. (Above redrawn from Barnard, 1961, below redrawn from Wolff, 1956)

Plateau—3,181 metres—and the Scotia Sea—2,976 metres (Heezen and Hollister, 1971). The largest known amphipods have yet to be caught. In a sequence of photographs taken by the Monster Camera at a depth of 5,300 metres in the eastern Pacific a number of giant gammarids (of about 28 cm) were shown around the bait (Hessler, 1972a). It is also clear that there are populations of scavenging and sizeable gammarids at great depths in other parts of the ocean. For instance, the same camera revealed a mass of gammarids swarming round the bait can at a depth of 7,000 metres in the Peru–Chile Trench (Isaacs and Schwartzlose, 1975). Even so, most of such deep-sea peracaridans are small and it may well be that they spend much of their life seeking detrital and other food either just above or just below the surface of the sediments. Indeed, during a dive to 4,160 metres off Madeira in the French bathyscaphe *Archimede*, Wolff (1971) observed, '... an incredible number of isopods and amphipods swimming close to the bottom (from 0 to 1–2 m above it)'. Concerning the tanaids, the members of one family (Neotanaidae), unlike others of their group, do not build tubes. Gardiner (1975) suggests that they are epifaunal and probably feed on the organic detritus lying on the sediment. He adds that it is unknown whether they are able to burrow, but that they probably can when necessary.

As already stated, small polychaete worms are also prominent members of the infauna. Russian investigations in deep subtropical and equatorial regions of the Pacific Ocean revealed that diverse polychaetes less than 5 mm in length were usually well represented in their samples, which were dredged between depths of 4,260 and 6,000 metres. In all parts, from a relatively rich (eutrophic) equatorial area to food-poor (oligotrophic) subtropical areas both north and south of the Equator, the biomass of small polychaetes, and they are described as detritus-feeders, ranged from 12 to 55 per cent of the total amounts (Sokolova, 1970). Larger species of deep-sea polychaetes are evidently members of the epifauna and will be considered later (pp. 173–4).

Diverse soft-bodied invertebrates, mainly molluscs and worm-like forms, burrow in the sediments and they are well adapted, particularly in the mobile interplay of their neuromuscular system and fluid skeleton, for such a mode of life (Clark, 1964, Trueman, 1975). In inshore waters the sipunculid and priapulid worms depend on successive eversions and introversions of a large proboscis to force them through sand and mud. Representatives of both groups range into the deep sea and in the Kurile–Kamchatka Trench have been taken at depths beyond 6,000 metres (Murina, 1974) (Fig. 60).

Shallow-sea sipunculids swallow considerable quantities of sediment and presumably digest any contained organisms and detrital material. Priapulid worms are said to feed on living prey, micro-organisms and detrital particles. In the deep sea, meiofaunal organisms and detritus are concentrated just below the surface of the sediments (p. 144) and it may well be that these worms move over the bottom rather than burrow (see also p. 303). This is certainly true of deep-sea species of acorn-worms (Enteropneusta), which in inshore waters also burrow quickly by means of a mobile proboscis. Photographs taken by deep-sea cameras show that acorn-worms '... either burrow shallowly by means of a soft proboscis or crawl along the bottom'. These worms which swallow the sediment as they move in spiral paths, and may be traced by the coils of remoulded sediment '... are quite frequently seen in the equatorial and

South Pacific. On a research cruise between Wellington, New Zealand and Tahiti, the camera recorded them at nearly every station made in depths greater than 4,000 metres ... Later they were photographed in over 10 per cent of the more than 500 stations made in the southeast Pacific and Scotia Sea. Still more spirals have been found in pictures from

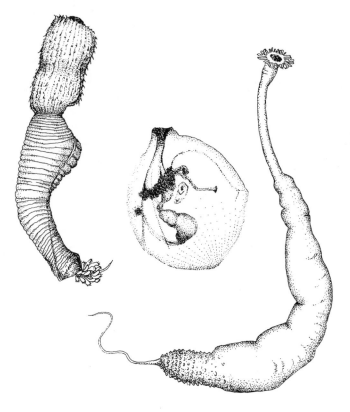

Fig. 60   Priapulid and sipunculid worms. Left, *Priapulus abyssorum*, length ca. 200 mm; right, *Golfingia flagrifera*, a sipunculid; centre, a sipunculid *Pelagosphaera* larva. (Left, after Menzies, 1959)

the North Atlantic, Indian Ocean and North Pacific. But they appear most often in pictures from the high southern latitudes and they are especially common in the South Pacific' (see Fig. 1). At one South Pacific station, the acorn-worm at the head of a large coil of sediment was estimated to be about 1 metre long and 5 cm thick (Heezen and Hollister, 1971).

Inshore species of echiurid worms live in burrows or some other kind of shelter, including the shells and tests of other animals. An extensible proboscis, which emerges from the trunk region, secretes a funnel or covering of mucus for gathering food. In non-tube-dwelling species small organisms and particles stick to the mucus on the forepart of a mobile, searching proboscis and are then passed to the mouth along a ciliated groove.

Russian expeditions found that echiurid worms, all with a long proboscis, range into the deepest parts of the Pacific Ocean. For instance, *Vityazema* is known from depths between 5,500 and 10,000 metres and *Alomosoma* between 500 and 8,500 metres. Professor Zenkevitch pictures one deep-sea form (*Tatjanella grandis*) with its trunk below the

Fig. 61 An echiurid worm, *Tatjanella grandis*, habit as reconstructed by L. A. Zenkevitch.

ooze and its proboscis extending over the surface (Fig. 61). After gathering food in one direction the proboscis can be extended along a neighbouring radius, eventually leaving a series of spoon-shaped marks in the ooze that radiate from the burrow (Fig. 61).

Finer and more numerous impressions around a central burrow may well be the work of tube-dwelling polychaete worms, such as terebellids. Some terebellids stretch their long tentacles over the deposit and feed in much the same way as echiurid worms. Towards the end of each tentacle a flattened zone becomes attached to the sediment, so leaving an outermost section to explore for detrital material, which is channelled to the mouth by the cilia and muscles associated with a groove in the tentacle. Investigations on Russian research vessels in the Pacific down to depths of 6,000 metres or more have shown that terebellids and other detritus-feeding polychaetes are quite abundant in sediments with a relatively high content of organic matter.

Deep-sea cameras may sometimes catch the burrowers in their excavations. Sea-anemones which have been photographed and dredged from all depths in the ocean, include species that burrow in the ooze and may be responsible for some of the larger craters, mounds and burrows seen in photographs (Heezen and Hollister, 1971). Their relatives, the cerianthids, are active burrowers and can be seen with their slender columns thrust well into the sediment. Of the echinoderms, certain brittle-stars bury themselves so that their rays are the only parts visible above the surface of the ooze. Starfishes, some of which look like porcellanasterids, may also be seen half buried in the ooze. As we shall see later, certain species of both groups feed on detritus.

In coastal waters many kinds of fishes live in burrows, but in the deep

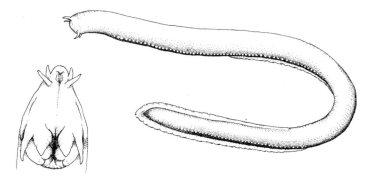

Fig. 62   A hagfish (*Myxine glutinosa*) and, left, the lower view of the head. (Redrawn from Bigelow and Schroeder)

sea the only kinds that may have such a habit are the hagfishes (Myxinidae) (Fig. 62). Most species have been taken from shelf to slope levels in temperate latitudes, both north and south. Though without true jaws, they use laterally opposed and mobile tooth plates to feed on dead fishes and capture live food, such as worms and crustaceans. Hagfishes frequently swarm around the bait of automatic cameras. Isaacs and Schwartzlose (1975) made the following observations, 'Hagfish thrive at depths from 200 to nearly 2,000 metres. At first we were puzzled by the reluctance of other fish to penetrate the Gorgon's-head tangle of hagfish and feed on the bait. Close-up motion pictures gave the answer: the hagfish enclose the bait in a thick cocoon of slime that other fishes find distressing. On a number of occasions fish were observed to emerge from the feeding mass making frantic attempts to clean their gills of slime. The spectacular ability of hagfish to exude slime has long been known: the exudation of a single hagfish can convert a large container of water into a slimy gel. Their employment of this defence mechanism to sequester food, however, had not been suggested.' Like all good 'bonanza strategists' (Wilson, 1971) hagfishes have the means of keeping much of the spoils for themselves.

The epifauna

Attached animals

Conditions of life in the deep sea are such that both attached and freely moving kinds of animals have penetrated to the deepest reaches of the ocean. Attached forms of invertebrates fasten themselves to the sediments, rocks and other suitable surfaces, which, beside each other, include some of the following, '... garbage, paper, radioactively contaminated tool and waste, junk, wrecks from peace and war; artillery projectiles; clinkers from coal-burning steamships; ballast; telephone and telegraph cables; beer bottles, tin cans and much more, imaginable, virtually unimaginable and sometimes, unmentionable' (Heezen and Hollister, 1971).

While some of the most closely studied kinds of inshore life are the invertebrates attached along certain rocky shores, the fauna fastened to the rocky mountains of the deep sea is not only remote but very difficult to investigate. The rocky mountains of the mid-ocean ridges, which are from 1,000 to 4,000 kilometres in width, unfold under the ocean over a total distance of about 40,000 miles. The peaks of the mountains rise 2 to 4 kilometres above the ocean floor. The ocean-basin floor, which lies between the continental margins and the oceanic ridge system, and over which is spread most of the expanse of oceanic sediments, makes up one-third of the entire ocean-basin systems in each of the Atlantic and Indian and three-quarters of the Pacific Ocean basins. The oceanic ridges and their mountains are thus vast underwater features and the fauna attached to immense areas of bare rock is not, as already implied, easy to collect. To learn something of this fauna, the marine biologist should go to sea with a geologist and look in his rock dredge. He should also study all relevant photographs which might suggest, inter alia, ways of sampling the fauna.

Though deep-sea cameras will not show the smallest benthic animals, they have revealed intriguing glimpses of some of the larger invertebrates attached to the rocky surfaces of the oceanic ridges (Heezen and Hollister, 1971). Stalked sponges, sea-anemones, sea-fans (gorgonians), sea-pens (pennatulids) and sea-lilies live on the ridge mountains and on other rocky surfaces, such as sea mounts and isolated boulders. But these volcanic and rocky surfaces may be extensively encrusted with manganese dioxide, which may inhibit the settlement of sessile organisms. On the other hand, several groups of organisms are active in the construction of manganese nodules. Greenslate (1974) thinks that most of the organisms that are potentially important in the origin and development of nodules are microscopic (less than 0·5 mm across). Moreover, tubes of *Saccorhiza*, a foraminiferan, had formed on all the nodules he examined from Pacific basins. Other organisms on the surface of nodules form structures that contain large quantities of iron and manganese.

PROTOZOA

In an analysis of hadal life as seen on bottom photographs taken in five trenches in the southwestern Pacific Ocean, Lemche and his colleagues (1976) remark, 'The hitherto almost neglected group of protista, the Xenophyophoria, is among the dominating organisms, some with pseudopodia up to 12 cm long.' These epibenthic protozoans, which are known from the three oceans and antarctic regions from the littoral zone to very great depths, are rhizopods with a multinucleate plasmodium enclosed by a branched tube system. The pseudopodia emerge from the free ends of the tubes. A test, formed partly of agglutinated materials, such as the skeletal remains of foraminiferans, radiolarians and sponges, surrounds the plasmodium. The Xenophyophoria are by far the largest living protozoans. The largest known is 25 cm in diameter but it is leaf-shaped and only about 1 mm in thickness. The body is either rigid—disc-like, spherical, oval or star-shaped—or flexible—tree-like or flake-like (Tendal, 1972) (Fig. 63). In the above photographic survey many spherical forms 2–5 cm across (Psamminidae) were seen on soft bottoms and about a half were surrounded by a somewhat irregular stellate figure, interpreted as outstretched pseudopodia. Other forms were lumpy, more or less oblong and rounded (Cerelasmidae), and there were also branching and flake-like kinds (Stannomidae). Lemche and his colleagues suggest that particulate food on the oozes is collected by the pseudopodia. In certain areas they estimated that 30 to 50 per cent of the bottom was constantly covered by the pseudopodia, which may also catch living prey, as do some of the larger sublittoral foraminifera (p.145). The depths covered by the photographs range down to nearly 9,000 metres (in the North New Hebrides Trench).

There are also 'neglected' organisms in benthic communities under the central gyre waters of the Pacific Ocean. Hessler (1972b) describes them as '... small organisms composed of dendritically joined or anastomosing organic tubules with clay particles agglutinated to the surface' and they comprised '... the bulk of organic remains in samples from the central North Pacific'. The Russians call them *vitvistii kamochki*, or 'little branching clusters'. They are reminiscent of Xenophyophoria but Hessler adds that they are orders of magnitude smaller than any known representatives of the group. They are widely distributed and abundant

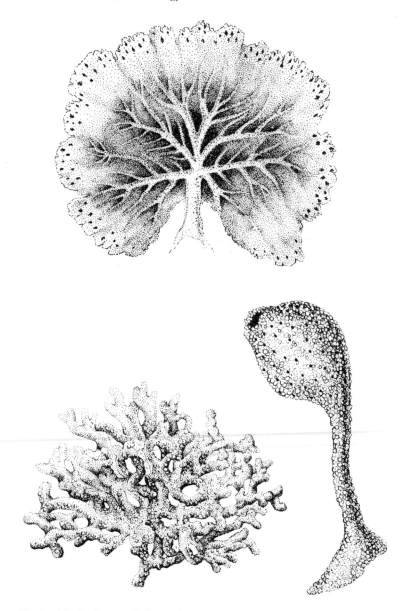

Fig. 63 Newly discovered rhizopod protozoans, the Xenophyophoria. Above, a branching form *Stannophyllum jenosum* (about 200 mm across); bottom left, *Stannoma coralloides* (about 35 mm across); bottom right, *Ammolynthus haliphysema* (height about 40 mm). (Redrawn from Tendal, 1972)

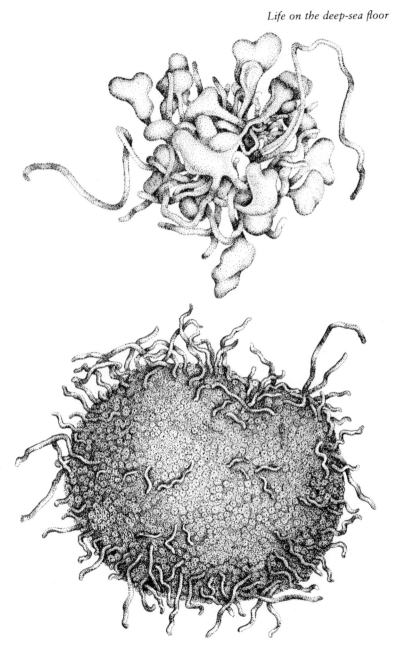

Fig. 64   A newly discovered group of foraminiferan protozoans, the Komokiacea. Above, *Normanina tylota* (0·5 mm); below, *Edgertonia argillispherula* (1 mm). (Redrawn from Tendal and Hessler, 1977)

over the deep-sea floor, and a recent study by Tendal and Hessler (1977) shows them to be agglutinating forams that must be placed in a new superfamily—Komokiacea (Fig. 64 and p. 291).

SPONGES (PORIFERA)

Sponges are the least complex, and among the most successful of the sessile many-celled animals in the marine benthos. In all but a few species the cells and their gelatinous matrix are supported by an intricate skeleton of spicules and/or fibres. Many species of the most diverse group (Demospongia) have a skeleton of siliceous spicules, others develop a composite skeleton of siliceous spicules and fibres of spongin or of fibres alone. The remaining sponges form a skeleton of six-rayed spicules of silica (Hexactinellida), or of calcareous spicules (Calcarea). Virtually all deep-sea sponges belong to the first two groups.

The metabolic life of sponges is centred on collar-cells (choanocytes), each of which bears a mobile flagellum that emerges through a filmy basal collar of protoplasm. The concentrated flagellation of these cells, which line the inner passages and chambers, circulates sea water through the sponge. Water is drawn in through small pores in the body wall, and after flowing through the canals and cavities, emerges through one or more exhalent openings (oscula) (Fig. 65). Oxygen is thus brought to the tissues and waste products eliminated. Since collar cells also ingest and digest the greater part of the food particles that may enter the sponge, it is fair to say that the direct and indirect effects of their activities are unsurpassed by any of the other kinds of cells in the animal kingdom. One might almost say that the simplicity of sponges is based on the versatility of their collar cells.

The vitreous appearance of the spicular framework of hexactinellids readily suggests their vernacular name, glass-sponges. Poincaré, the French mathematician and physicist (who was actually using the skeletal structure of a glass sponge in a metaphorical sense to illustrate the limitations of bare structural analysis in science), wrote of '... those delicate assemblages of siliceous needles which form the skeleton of certain sponges. When the organic matter has disappeared, there remains only a fragile and elegant lace-work. True, nothing is there except silica, but what is interesting is the form the silica has taken, and we could not understand it if we did not know the living sponge which has given it precisely this form' (quoted from Nash, 1963). Unfortunately, by the time a glass-sponge appears on the deck of a research vessel it is often far from being a living sponge. Venus' flower-basket sponges, *Euplectella* spp., are among the best known (Fig. 66). The walls and the sieve-cap of a *Euplectella* bear pores that lead to flagellated chambers, and it seems that water is drawn in and discharged all over the sponge. Apparently, each of the many flagellated chambers in the body wall maintains its own circulation, and perhaps this is true of all glass-sponges.

Hexactinellids are more or less radially symmetrical sponges with cylindrical, vase, urn, funnel or such-shaped bodies. A body wall of varying thickness encloses an inner cavity (spongocoel) that opens at the summit by a wide osculum. The osculum is bounded by upright spicules or covered over (as in *Euplectella*) by a sieve-plate of silica.

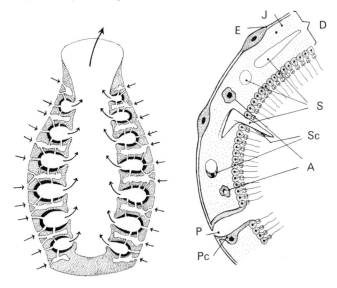

Fig. 65   Left, inner structure and circulation through a syconid sponge (after Wells, 1968). Parts of the chambers lined with collar cells are shown black. Right, cellular structure of a sponge, showing the three main types of cells; outer epithelial cells, amoebocytes and ciliated collar cells (choanocytes). E, epithelium; J, jelly; D, dermal layer; P. pore; Pc, porocyte; S, spicules; Sc, scleroblasts; A, amoebocytes. (From Marshall, 1971, *Ocean Life*)

Nearly all of the 440-odd species of glass-sponges live below the continental shelf, where they are most diverse and numerous at depths between 300 and 2,000 metres. In polar waters, particularly in the antarctic, glass-sponges, which in other parts of the world live at deep-sea levels, have colonized shallow waters. Certain members of the Rossellidae, for instance, are abundant in coastal waters of the antarctic at depths around 10 metres (Koltun, 1970). Abyssal species range to depths of 6,000 metres or more.

The glass-sponges are rooted in the sediments or fastened to rocks by basal extensions of the body that are supported by long spicules. In cup-shaped forms, such as *Pheronema*, these spicules form 'adventitious'

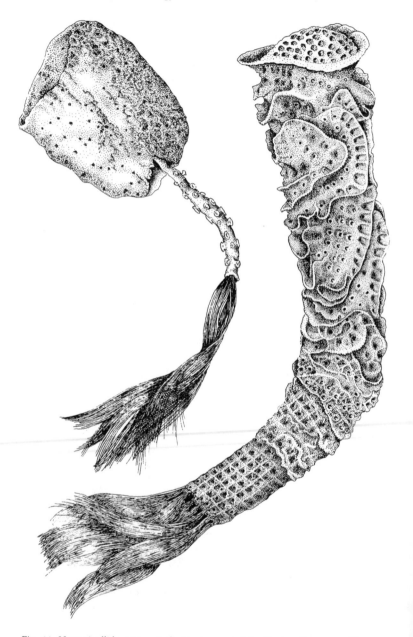

Fig. 66. Hexactinellid sponges. Left, *Hyalonema*; right, *Euplectellum aspergillum*.

roots. The tower-like species (*Euplectella, Holascus*) develop basal spicules that are twisted together to form a short stem and rooting system. The stem is long to short in the Leucopsacidae, which have a thick-walled cup or egg-shaped body. Long-stemmed forms include the tulip-like *Hyalonema* (Fig. 66), one of the best known glass-sponges, and the mushroom-headed *Caulophacus*. During a dive in the *Deep Star* Ron Church photographed a number of magnificent *Hyalonema* at a depth of about 1,000 metres in the San Diego Trough off southern California. The yellow heads, about 25 cm in width, were carried on metre-high stalks that were festooned with large plumose sea-anemonies (*Bolocera pannosa*). Church (1971) remarks, 'When a sponge dies, the glassy stalk remains. I have seen eerie forests of them rising from the sea floor like the stems of so many gigantic champagne glasses.' A species of Venus' flower basket (*Euplectella imperialis*) grows to a height of at least 60 cm, while *Monoraphis chuni* may attain 1 metre. Sizes range down to a few centimetres, but the size attained may well be related to the local productivity. Glass-sponges are rare on deep impoverished sediments (red clay or green clay) but may be numerous on diatom ooze (as in the Antarctic), radiolarian ooze, on gravels and on rocky bottoms.

There are about 4,000 species of sponges and some four-fifths belong to the Demospongia. Most of the sponges that live in littoral and shelf waters are representatives of this class. Some encrust rocks and other suitable surfaces; others grow into more or less irregular masses or branching forms. 'The same species may grow erect branches in quiet waters or cling matlike, moulded to the substrate, where the surf is strong.' There appear to be few encrusting sponges in the deep sea. 'The finger-like vaselike and fanlike sponges are characteristic of warm quiet seas or of the deep ocean bottom' (Buchsbaum and Milne, 1960). There are irregular, stumpy or clustered kinds of Demospongia over the upper continental slope, but with deepening water the species converge in form with the glass-sponges. The body becomes more or less radially symmetrical and in most species is borne on a stalk with a basal attachment to rocks or sediments. At depths of 3,000 metres or more the Demospongia are largely represented by small stalked species of the genera *Cladorhiza, Asbestopluma* and *Chondrocladia* (Lévi, 1964) (Fig. 67). The deepest record of sponges is of the species *Asbestopluma occidentalis* and *Chondrocladia concrescens* from depths between 8,175 and 8,840 metres in the Kurile–Kamachatka Trench (Koltun, 1972). Both species are known between wide depth ranges, the first between 820 and 8,840 metres (the shallowest record is off Vancouver), and the second between 200 and 8,660 metres (the shallowest records are in arctic and subarctic waters).

While deposit feeders, such as the elasipod sea-cucumbers, are poorly represented on the most deserted oceanic bottoms, sponges and other

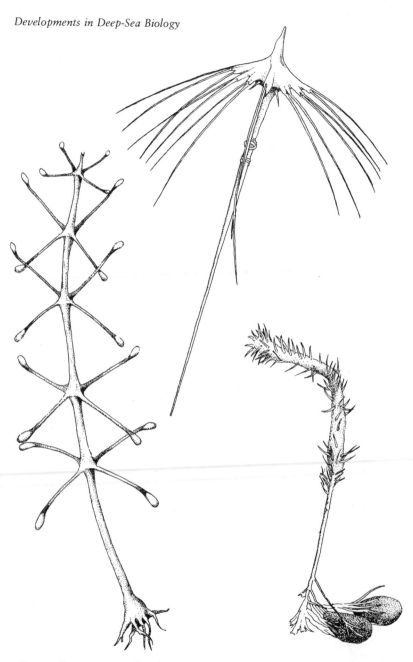

Fig. 67 Demospongia. Left, *Chondrocladia concrescens*; upper centre, *Cladorhiza longipinna* (height about 50 mm); right, *Cladorhiza septemdentalis* (height about 60 mm). (Upper centre and right redrawn from Koltun, 1972)

suspension feeders are better adapted for life under waters of low productivity (see Sokolova, 1972 and pp. 297–302). Species of Demospongia seem to be better at living in poor conditions than their hexactinellid relatives. For instance, in the northwest Pacific, a group of stations in waters of low productivity (around 150°E, 30°N) and at depths ranging from about 2,000 to 6,000 metres, yielded small sponges (3–6 cm) of the genera *Chondrocladia*, *Cladorhiza* and *Abyssocladia*. The glass-sponges, such as *Hyalonema apertum* and *Holascus* spp., were taken only at more northerly stations in waters of higher productivity (Koltun, 1972). One can only wonder whether this kind of success, and indeed whether the entire oceanic success of the Demospongia, is largely related to their being the most highly organized of sponges, especially in their elaborate (leuconoid) systems of flagellated chambers and passages.

Sponges, which are usually hermaphroditic, reproduce asexually by budding off new individuals from the parental stock and by freeing special buds known as gemmules. Fertilized eggs produce ciliated larvae. Buds and gemmules have been seen in deep-sea sponges but little is known of their sexual means of reproduction.

In systematic order the next sessile forms are hydroids, sea-pens, sea-fans, sea-anemones and so forth, members of the Cnidaria. Like the sea-lilies (Crinoidea), nearly all cnidarians are passive suspension feeders. Such animals wait for the movements of life or sea or gravity to deliver their food. Active suspension feeders, such as the sponges, bryozoans, brachiopods, ascidians and tubicolous polychaete worms, create their own local circulation, from which they filter suspended food. Barnacles comb the sea for food. These groups are now considered, but before turning to the passive suspension feeders, the beard-worms (Pogonophora), which seem to depend entirely on dissolved organic substances and are most closely related to the polychaetes, will be reviewed.

ACTIVE SUSPENSION FEEDERS
Bryozoa (*moss animals*)
Like some of the sponges and ascidians, certain bryozoans encrust rocks, weeds and other surfaces. There are also creeping, branching and leaf-like forms. Bryozoans are made of colonies of minute zooids, each housed in a chitinous or gelatinous case of its own secretion. Each zooid has a circular or crescentic lophophore with a series of slender, ciliated tentacles, which when thrust into the sea and unfurled, set up currents that draw food particles towards the gut.

Living bryozoans, which are known from about 4,000 species, are most diverse and abundant below the tide marks in coastal waters. The deepest record is of a species of *Bugula* from the Kermadec Trench at a depth of between 8,210 and 8,300 metres. Species of *Bugula* are tufted in form, which recalls that there is a marked tendency, as Schopf found off New England, for the proportion of species with erect colonies, as opposed to encrusting species, to increase with depth. At 10 metres less than a quarter of the species were erect, whereas below 100 metres more than half were. In an earlier paper Stark analysed the habit of species taken around 2,000 metres by the *Challenger* Expedition and found that

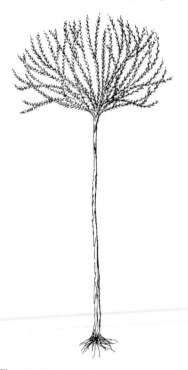

Fig. 68   An abyssal bryozoan, *Kinetoskias*.

only 9 per cent were encrusting or nodular, whereas nearly all the rest had some kind of erect form, especially arborescent. Evidently, encrusting species, like those of sponges and ascidians, are better suited to lively coastal waters than to the quieter reaches of the deep sea (Ryland, 1970).

The life of deep-sea bryozoans is well summarized by Ryland (1970), 'On the continental slope and beyond, rocky outcrops bear characteristic assemblages of bryozoans, together with brachiopods, corals and so on. Schopf (cited earlier) has noted that the maximum number of species per dredge station decreases from as many as 64 in 300 metres to 15 in

1,000 metres and 5 below 2,000 metres. Beyond the slope, absence of hard substrata probably restricts the occurrence of sessile epifauna. Here bryozoans are generally of the erect type and cellularioids, such as *Levin-senella* and *Himantozoum*, predominate. They form bushy colonies attached to the shells of foraminiferans by a stalk of kenozoids.' A colony of *Levinsenella magna*, taken at a depth between 4,540 and 4,600 metres in the North Atlantic by the Swedish Deep-Sea Expedition (1947–48) was about 18 cm in height (Silen, 1951). Colonies of *Kinetoskias*, which may be rooted in the oozes by rhizoids at the base of their long stem of fused kenozoids, are also known from abyssal depths (Fig. 68).

Ryland's concluding remarks on deep-sea bryozoans are as follows, 'Another characteristic feature of abyssal bryozoans appears to be that the tentacles are relatively longer than in comparable shelf species and can be protruded a very long way. Stomach contents from material preserved shortly after capture (Schopf) included detritus but no recognizable organisms, suggesting that the zooids filter detritus in suspension over the bottom.'

Brachiopoda (lamp-shells)
Like the bryozoans, the brachiopods have a 'true' body cavity (coelom) and a lophophore. The animal is enclosed in a bivalve shell (size range 5–80 mm) formed of upper and lower valves that are formed and lined by a mantle lobe of the body wall. The attachment to rocks, shells, corals and so forth is either direct or (in most species) by way of a stalk (pedicle). The lamp-shells are exclusively marine and the existing species (about 260) live in most parts of the ocean from the intertidal zone to depths

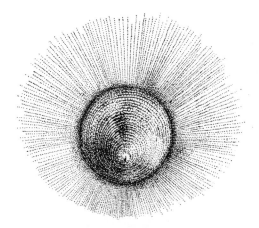

Fig. 69 A cosmopolitan brachiopod, *Pelagodiscus atlanticus*.

of 5,000 metres or more. Brachiopods, decapod crustaceans and flat-worms (Turbellaria) have yet to be dredged from the trenches (depths of 6,000 metres or more). Some species, such as *Terebratulina retusa* and *Macandrevia crania*, range from inshore waters to levels near 4,000 metres. There are also abyssal species that are unknown from moderate depths. *Pelagodiscus atlanticus* (Fig. 69), which is known from all ocean depths between 360 and 5,500 metres, is said to be the most cosmopolitan brachiopod. Its larvae (Fig. 164) have been netted near the surface off Brazil, Ceylon and Japan. Another deep-sea species, *Chlidonophora chuni*, known from off the Maldives at a depth of 2,253 metres, has a long pedicle with fine root-like outgrowths that attach to foraminiferan shells in the ooze. Species such as *Terebratalia tisimani* (63–235 metres) and *Diestothyris frontalis* (40–400 metres) seem to be confined to more modest depths.

Ascidiacea (sea-squirts)

Ascidians, which form a class of the Tunicata, are among the most flourishing of the attached forms of benthic life. They are much more diverse than their pelagic relatives, the Thaliacea and Larvacea (pp. 94–8). Well over three-quarters of 1,500-odd species are neritic forms but the entire class is represented at all benthic levels down to about 8,000 metres. Close to 50 species range to depths of 2,000 metres or more (Millar, 1959).

Ascidians are enclosed in a cellulose-based tunic that varies widely in texture and thickness. In nearly all species the most voluminous part of the body is the pharynx and its gill slits, which are usually subdivided to form an elaborate branchial basket. The pharynx is a ciliary-mucus food-gathering device much like that in the pelagic tunicates (p. 98). Cilia are so disposed over the basket that their concerted motions draw water (and food particles) through an oral siphon into the pharynx, which in turn leads to an exhalent atrial chamber. Ascidians, as we saw, are active suspension feeders. Indeed, some species are able to move several thousand times their body volume of water through the pharynx in a day. 'Mucus from the endostyle moves up across the pharyngeal bars and food particles stick to it. Strands of mucus are delivered to the dorsal lamina, which bends over to form a tube or groove along which the food moves to the oesophagus' (Meglitsch, 1967).

Ascidians are solitary or integrated in colonies that form massive or encrusting forms. Colonial species are largely confined to coastal seas, though certain polyclinids, for instance, extend into upper slope waters. Many of the deep-sea forms belong to the families Styelidae and Pyuridae, and in both there are species in which the individual is borne on a stalk with rooting processes that ramify in the sediment or form attachments to hard objects. *Dicarpa simplex* of the first family, and known from depths between 2,470 and 4,600 metres, has a short stalk, but in *Culeolus*

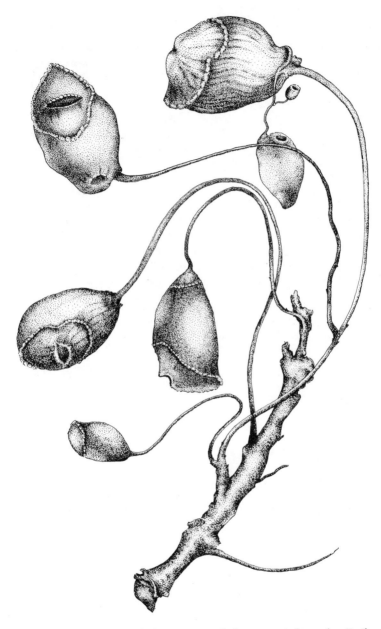

Fig. 70 Stalked ascidians of the species *Culeolus murrayi* from the Kurile–Kamchatka Trench. (Redrawn from Vinogradova, 1970)

spp. (204 to 7,295 metres and of the second family) the stalk is short to very long and slender (Fig. 70). In the latter genus, the test, which may be raised into small rounded swellings and decorated with a fringe of papillae around the atrial opening, is so attached to the stalk that the

Fig. 71 An abyssal molgulid ascidian, *Molgula immunda*, from the Kermadec Deep, 4,410 m. Intact animal on the right, animal removed from test on the left. (Redrawn from Millar, 1959)

animals stream into the current with the oral siphon foremost (see Vinogradova, 1970, and Fig. 70). Presumably the body is more or less neutrally buoyant. Other deep-sea members of these and other families (e.g. Molgulidae) are squat or globular in form and the test is usually covered with a coat of fine hairs that support the animals on the sediment (Millar, 1959) (Fig. 71). Nearly all are small with a body ranging in length from a few millimetres to a few centimetres.

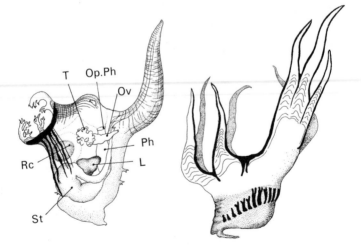

Fig. 72 Predatory deep-sea ascidians. Left, *Hexacrobylus indicus*, Op. Ph, opening to pharynx at base of oral siphon; Ov, ovary; Ph, pharynx; Rc, renal concretion; St, stomach; T, testis; right, *Octacnemus bythius*. (Redrawn from Millar, 1959)

Concerning their means of subsistence, Millar (1959) remarks that the ascidians '... evolved in comparatively shallow water and are adapted to life in waters rich in phytoplankton. The filter-feeding habit is related to these conditions and when certain ascidians started to penetrate great depths, the decreasing quantity of phytoplankton presented them with difficulties. Most deep water species, however, are still filter feeders and must find enough suspended food in the water.' The stomach contents of individuals of *Culeolus suhmi* taken near 5,000 metres contained the remains of phytoplankton cells and radiolarians. But the abyssal genera *Octacnemus* and *Hexacrobylus* (Fig. 72) which have lost the ciliated parts of the branchial sac and evolved a much-enlarged oral siphon, feed on copepods and the like. Millar suggests that other deep-sea ascidians with the same anatomical modifications are likely to have a similar diet (see also pp. 295–6).

Tube-dwelling polychaete worms

Polychaete worms are among the few dominant groups of the marine benthos. Except for a handful of pelagic forms (pp. 104–5) all of the 3,500 known species are well represented in either the epifauna or infauna of the ocean floor. While most species live in near-shore waters, particularly in subtropical and tropical regions, deep-sea forms have colonized all levels down to the greatest depths. Well over 150 species have penetrated to depths beyond 3,000 metres.

Polychaetes that form part of the miniature infauna, and burrowing terebellid polychaetes, were considered earlier (p. 154). The tube-dwelling species of the epifauna are our present concern, while the active (errant) polychaetes will be left until the moving members of this fauna are reviewed (pp. 201–3). Though not taxonomically tidy, such ecological division may remind us that in their ranging mastery of benthic living spaces, the polychaetes are surpassed only by the crustaceans.

Wolff (1961b) has given an interesting survey of the animal life from a trawling made on the *Galathea* at a depth of 3,570 metres off the Pacific coast of Central America. At this station, which was in a highly productive upwelling region, about 2,100 specimens of invertebrates and fishes were taken belonging to about 130 species. The polychaete worms, represented by about 310 specimens and 20 species, were third in order of relative abundance (after sea-cucumbers and brittle-stars). The dominant tube-dwelling species were maldanids that live in heavy tubes of black ooze but there were also tubicolous eunicids and serpulids (the latter secrete calcareous tubes). Members of these three families and of the sabellids and ampharetids extend into the deep sea well beyond a depth of 3,000 metres. Their lengths range from a few millimetres to 20 cm or more.

The classic types of suspension feeders among the tubicolous polychaetes are the sabellids and serpulids, known as feather-duster worms from the crown of pinnate tentacles ('gills') that emerge from the head. When spread and extended above the tube, these feathery appendages, which bear ciliary tracts and mucus cells, collect particulate food and convey it to the mouth. Other kinds of tubicolous polychaetes, such as the maldanids and ampharetids, browse over the oozes on detrital material etc.

The tube-dwelling worms that are most at home in the ocean are both small and feed on suspended organic particles. These are serpulids, and together with other suspension feeders, the sponges and scalpellid barnacles, have a much less restricted distribution throughout the deep ocean than the deposit feeders of the benthos (Sokolova, 1972). The most wide-spread deserts on the deep-sea floor lie below the great gyres that encircle the middle (subtropical) latitudes of the Pacific Ocean, both north and south. Here, on impoverished red clays, at depths down to about 6,000 metres, live small serpulid worms, together with such forms as foraminiferans (mainly agglutinating species), small sponges, nematode worms, small sipunculid worms, small bivalve molluscs, etc. (Filatova and Zenkevitch, 1966)). Serpulids and other tubicolous polychaetes also range into the trenches down to depths of 8,000 metres or more.

Cirripedia (*barnacles*)
Most of the 800 or so species of barnacles live in a calcareous shell of discrete plates and have six pairs of thoracic appendages that end in filamentous cirri (order Thoracica). Nearly all of the deep-sea species belong to this order, but the burrowing barnacles (Acrothoracica) and the Ascothoracica, which are partly or wholly parasitic on cnidarians and echinoderms, have a few deep-water representatives (Newman, 1974).

Nearly 50 species of barnacles have penetrated to depths beyond 3,000 metres, and most either belong to the stalked barnacle genus, *Scalpellum*, or to the unstalked, asymmetrical genus, *Verruca* (Nilsson-Cantell, 1955). Indeed, about 40 of the 175 species of *Scalpellum* have been taken below a depth of 2,000 metres. They attach themselves to the stems of glass-sponges, corals, scaphus-tubes of sea-anemones, the stalks of ascidians, the branches of bryozoans, the tubes of beard-worms, shells, pumice-stones, clinkers, telegraph cables and so forth. The length of the 'head' (capitulum) of the deep-sea species ranges from a few millimetres to about 6 cm (Fig. 73).

Deep-sea barnacles doubtless use their filamentous thoracic appendages to comb the sea for particles of organic matter and small organisms. While *Scalpellum* species, as has already been stated, have a relatively unrestricted distribution in the deep ocean, they are not found, unlike the serpulid worms, on the organically poorest stretches of red clay in the deep-sea basins (Sokolova, 1972). The deepest record is of a *Scalpellum* sp. from a depth of about 7,000 metres in the Kermadec Trench.

Fig. 73 A deep-sea barnacle, *Scalpellum regium*. (Redrawn from Menzies et al., 1973)

Scalpellum vitreum, a short-stalked species, is known from the western Indian Ocean, the Indonesian area and the northwest Pacific, including the Sea of Okhotsk, between the wide depth range of 74 to 6,096 metres. Numerous species appear to be confined to upper slope levels (Zevina, 1972).

ABSORBERS
Pogonophora (beard-worms)
The beard-worms are long, slender animals that live in a closely fitting tube of their own secretion. The tube is of chitin and protein and may be plain and smooth, or formed of successive rings or funnel-shaped sections. The lower end of the tube is thrust into the sediment so that the emergent upper part is more or less erect.* A single tentacle or brush of tentacles springs from the short, foremost part of the body (protosome). Nearly all of a beard-worm's length is taken up by a long slender trunk with two girdles of small, toothed bristles in the middle. These bristles are so made to grip the wall of the tube that each half of the trunk on either side of the girdles is free to extend or contract independently of the other. The body ends in a short segmented region (the opisthosoma) with several rings of rod-shaped setae. Beard-worm and polychaete worm setae are very alike in morphology, chemical composition and mode of formation, which, together with other common features, suggests that the two groups are rather closely related (George and Southward, 1973) (Fig. 74). But the Pogonophora are still kept in their own phylum.

---

*On the tubes of beard-worms one may find foraminiferans, sponges, hydroids, scyphozoan polyps, alcyonarians, small sea-anemones, serpulid worms, bryozoans, barnacles, ascidians and even crinoids.

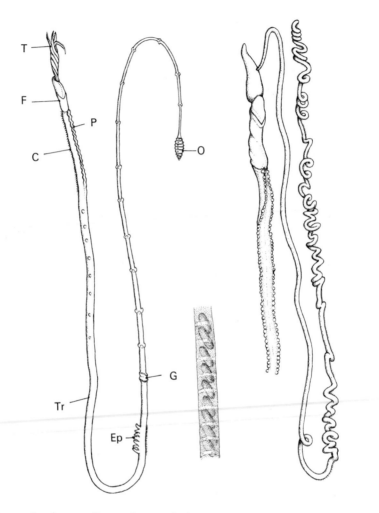

Fig. 74 Beard-worms (Pogonophora). Left, diagram of a typical pogonophore, very much shortened, showing the main body regions, T, tentacle; F, forepart; P, paired papillae; C, cilia; Tr, trunk; Ep, enlarged papillae; G, girdles; O, opisthosoma (with chaetae). Right, *Siboglinum caulleryi*, a species with a single long tentacle. Centre, eggs of a pogonophore in tube. (Left, redrawn and somewhat modified from George and Southward, 1973; centre, redrawn from Ivanov).

One of the largest beard-worms, *Lamellibranchia barhami* (with a tube from 600 to 700 mm in length and a maximum diameter of 9 mm) has about 2,000 long tentacles, which seem made for suspension feeding or at least for exploring the oozes to gather organic particles. But pogonophores are without a mouth, alimentary system or anus—they exist by absorbing organic substances through the epidermis, which has microvillae like the absorptive surfaces of other animals. Tests with radioactive organic compounds have shown that these worms can absorb amino-acids, glucose and fatty acids from dilute solutions and concentrate these substances many times in their blood tissues (Southward, 1971a).

Beard-worms are known from a hundred species (or so), taken between depths of about 25 and 10,000 metres. Their lengths range from a few millimetres to over 50 cm and their widths from 0·5 to 2·5 mm or more. One of the longest (and deepest-living) species, *Zenkevitchiana longissima*, about 35 cm long, lives in tubes that may reach a length of 1·5 metres. This species is evidently abundant in parts of the Kurile–Kamchatka Trench, even between depths of 8,000 and 9,000 metres, for Ivanov remarks that when the net came up from these levels, myriads of long white tubes were twisted round the frame of the dredge. Russian expeditions have taken numerous other species in this and neighbouring regions of the northwest Pacific. Since 1963 new species (27) have been taken in the northwestern Atlantic, Norwegian fjords, the eastern Pacific, Japanese waters, the Indian Ocean and the South Atlantic. Most kinds of beard-worms live in cool or cold waters (< 10°C), whether shallow or deep, but between Florida and Cape Hatteras there are species living in upper slope waters that vary in temperature throughout the year from 8 to 18°C (Southward, 1971b).

Even remembering that there is still much exploration to be done, it looks as if the beard-worms are largely confined to the more productive parts of the ocean. They are certainly not one of the typical groups of animals that live in the most deserted reaches (4,000–6,000 metres) of the ocean basins in middle latitudes (pp. 297–302). But in the Kurile–Kamchatka Trench, which is below very productive waters, they are very abundant at all levels. Kirkegaard (1956) quotes Ivanov as concluding that the food requirements of pogonophores must be great, since they produce considerable quantities of eggs and sperm and continue to grow throughout their life span. He himself concludes (from Pacific distributions) that they are restricted to areas where organic production is high. The Southwards (1967), who studied the quantitative distribution of pogonophores from a transect between Woods Hole, Massachusetts, and Bermuda and found that numbers in a square metre ranged from 2 to about 30, decided that '... pogonophores are likely to be found where there are comparatively large numbers of other animals and they are unlikely to be found where other animals are scarce'. It may well be that where the fauna is relatively numerous and diverse, local concentrations of dissolved organic substances, which may be released or excreted or arise from death and decay, are adequate to meet the requirements of resident beard-worms.

PASSIVE SUSPENSION FEEDERS

Wainwright (1976) defines passive suspension feeders as '... those animals that live attached to solid substrata and that rely on ambient water currents for food, gas exchange, sanitation and gamete dispersal. They feed by catching particles out of the water stream as it flows by.' The attached kinds of cnidarians and the crinoids, some of which look remarkably alike, are our present concern.

Cnidaria

All cnidarians develop stinging-cells (nematocysts) but not all depend exclusively on these cells to secure their food. In a number of the sea-anemones and diverse corals, ciliary tracts on the tentacles (or over the whole body in primitive anemones) form part of the feeding mechanism and are particularly effective in those species that depend on minute kinds of plankton. The soft coral *Alcyonium digitatum* (dead-men's fingers), will feed on suspensions of the diatom *Skeletonema*. Nothing is known of food procurement by benthic cnidarians in the deep sea, but it may well be that many species use cilia (and mucus) as well as nematocysts. This would ensure that if there were no living prey, organic particles could be utilized. Even so, deep-sea anemones have not turned themselves into sponges. Like carnivorous species of starfishes, sea-anemones tend to be distributed under the most productive parts of the ocean (Sokolova, 1972).

Though most diverse in species and colonies on the upper continental slope, deep-sea hydroids have been taken at all depths down to trench levels. The deepest record is of *Halisiphonia galatheae* from the Kermadec Trench between depths of 8,210 and 8,300 metres. *Aglaophenia tenuissima*, which was also taken in this trench (6,660–6,720 metres), is known also from the Great Australian Bight at depths between 293 and 585 metres (Kramp, 1956a). One of the best known of the hydroids is *Branchiocerianthus imperator*, a magnificent species that reaches a height of nearly 2 metres and has been taken at depths between 200 and 5,304 metres in the Arabian Sea, Gulf of Panama and the northwest Pacific.

Nearly all deep-sea jellyfishes seem to have lost the fixed polypoid phase in their life history (p. 85). The exceptions are evidently related to scyphoid polyps that are described under the generic name *Stephanoscyphus* and are widely distributed down to the deepest oceanic levels. The polypoid individuals live in long chitinoid tubes that attain a height of about 25 mm and are attached by a basal disc to pieces of wood, stones, rocks, pumice, sponges, gastropod and bivalve shells, tubes of polychaete worms and beard-worms and so forth. The best known species, *Stephanoscyphus simplex*, has been taken in all oceans between depths of 430 and 7,000 metres (Krämp, 1959) (Fig. 75).

One species, *Stephanoscyphus mirabilis*, produces medusae that belong to the genus *Nausithoë*, one of the coronate jellyfishes (p. 91). Kramp (1959), who found three recently liberated medusae in the tube of a specimen of *S. simplex* taken off the Philippine Islands, thought they looked like young *Nausithoë punctata* but was only prepared to assign them to the order Coronatae. Whatever their precise relationships, it is clear that scyphoid polyps have adaptable ways of making a living, for they are one of the characteristic groups of animals that live in the deepest and most deserted parts of the Pacific Ocean basins (Filatova and Zenkevitch, 1966).

If one thinks of their diversity (ca. 6,000 species), ubiquity and biomass (think only of coral reefs), the anthozoans are certainly the most successful class of cnidarians. Moreover, most of the biomass is centred on the fixed polypoid stage, whether solitary or colonial—there is no medusoid phase, the larvae, where known, starting life as small, free-swimming ciliated stages (planulae). There are two subclasses, Alcyonaria (octocorals), whose polyps have eight pinnate tentacles etc., and Zoantharia (hexacorals) with polyps bearing tentacles that are simple (rarely branched) etc. The first subclass includes three main orders: Alcyonacea (soft corals), Gorgonacea (horny corals) and Pennatulacea (sea-pens). The second includes four main orders: Actiniaria (sea-anemones), Madreporaria (true or stony corals), Antipatharia (black corals) and Ceriantharia (tube anemones). Representatives of all these seven orders have penetrated to deep-sea levels.

The soft corals (Alcyonacea), whose fleshy colonies are generally supported by a loose skeleton of calcareous spicules, are most typical of warm inshore waters, particularly in Indo-Pacific regions. Most species live between the tide marks and a level of 200 metres, but a few exist in deep-sea reaches to 3,000 metres or more.

The horny corals (Gorgonacea), also called gorgonians, sea-fans, and sea-feathers, develop into plant-like forms with slender branches that stem from a short main trunk with a fastening to the substrate by a basal plate or tuft of stolons (Fig. 76). The colony is supported by an axial skeleton of calcareous spicules, or of gorgonin or of both. The short polyps arise from the sides of the skeletal axis.

Most gorgonians live in warm inshore waters, especially in coral reefs and atolls. The deep-sea forms, which are often rooted to the sediments, are known down to depths of 5,000 metres or more. The deepest record (of *Primnoella krampi*) is from the Kermadec Trench at a depth of 5,850

Fig. 75 *Stephanoscyphus* (attached stage of Scyphozoa). Right, individual in tube with (above) a
number of strobilized medusae (height of tube
about 10 mm); below, medusa of *Stephanoscyphus simplex*. (Redrawn from Kramp,
1959)

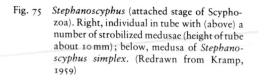

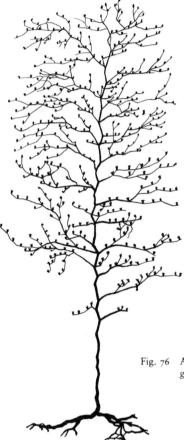

Fig. 76 A gorgonian deep-sea octacoral, *Chryso-gorgia*. (From Marshall, 1954)

metres (Madsen, 1956a). The deep-sea setting of gorgonians, sea-pens and corals has been well captured by Heezen and Hollister (1971), who write that these cnidarians '... which feed on tiny animals swept up by their deadly fingers, require a moderate to appreciable bottom current. They are seen on sea mounts, escarpments, the continental slope and in other active areas, but are not seen in a completely tranquil environment. The corals and gorgonians which usually are attached to rock, require the swiftest currents, while the pens attach themselves to the sediment and inhabit somewhat quieter areas.' Like some of their shallow-water relatives, certain deep-sea gorgonians, such as *Primnoella*, branch regularly and pinnately in one plane, which is kept at right angles to the direction of the current. Both the gorgonians and sea-pens seem to be patchily distributed over the deep-sea floor. Heezen and Hollister (1971) quote the pilot of the bathyscaphe FRNS–III as saying that he had seen these cnidarians '... only once at the bottom of the Setubal Canyon in the Atlantic off Portugal at about 1,800 metres. Dark red pennatulids and orange gorgonians covered the bottom at the rate of at least one colony per square metre. For once the bottom had lost its desert aspect, for these colonies of animals could easily be taken for finely serrated flowers.'

In the pennatulids the colony consists of one very long axial polyp, which bears on its sides many (dimorphic) lateral polyps. The lower end of the axial polyp forms a plain stem without a base or attachment processes—it is simply thrust into the sediment, usually by means of the

Fig. 77 Deep-sea pennatulid, *Umbellula*.

peristaltic contractions of a large end-bulb. The axis may be long and slender, even whip-like (*Chunella* and *Umbellula*), or at the other extreme thick and club-like (*Kophobelemnon*). The largest colonies reach a height of about one metre (Fig. 77).

Like the gorgonians, the pennatulids are most diverse and abundant in shallow inshore waters of subtropical and tropical regions, and they are also widespread in the deep sea (down to depths of about 6,000 metres, the deepest record—*Umbellula* sp.—coming from a depth between 6,000 metres and 6,730 metres in the Kermadec Trench). Even at considerable depths they may be quite abundant, the Swedish Deep-Sea Expedition taking 157 specimens of *Umbellula guntheri* at a depth between 5,033 and 5,044 metres in the equatorial Atlantic (Broch, 1957).

Umbellula is one of the classic animal types of the deep-sea floor, and so are the stalked sea-lilies (crinoids), both converging in their palm-like forms—forms that look well designed for passive suspension feeding. Bent over by currents, they look like palms in a breeze, though we shall see that the sea-lilies are not altogether pliant (p. 295). Some deep-sea pens develop a long axis with a fringe of many small polyps on either side (e.g. *Pavonaria* and *Funiculina*). There are striking photographs in Heezen and Hollister's (1971) book showing such kinds at a depth of 214 and 1,503 metres on the continental slope off New England and at 4,821 metres on a sea mount in the western Atlantic. One well-known species, *Funiculina quadrangularis*, is widely distributed over the upper continental slopes (from about 35 to 2,000 metres, or more). Sea-pens luminesce when stimulated, the light (blue-green, yellowish or violet) spreading over the colony from the site of the stimulus.

Three small sea-anemones attached to a rock trawled from a depth of 10,190 metres in the Philippine Trench provided the first evidence that many-celled animals exist in the deepest reaches of the ocean (Bruun, 1956). In general, as we have seen (p. 178), sea-anemones are widely distributed under the more productive waters of the deep ocean. In his report on the species taken by the Danish Deep-Sea Expedition (1950–52) at trench levels of 6,000 metres or more, Carlgren (1956) concluded that sea-anemones '... occur in the greatest depths of the oceans in rather great numbers, both as regards number of species and number of individuals. Reproduction must take place in these depths, as evidenced by the presence of well developed fertile organs, especially testes, as well as by the multitude of young individuals.'

Excellent pictures of sea-anemones have appeared in numerous photographs of the deep-sea floor (see Heezen and Hollister, 1971). Beside the burrowing forms, it seems that certain kinds are able to move over the oozes. Under very productive parts of the ocean, as at the 3,600-metre station worked by the *Galathea* off the Pacific coast of Central America (see also p. 173), sea-anemones can be quite abundant. At this station 95 specimens were represented by eight species. Most of them belonged to *Edwardsia* (26), a genus with burrowing species that are well known in inshore waters over much of the world, and *Chondrophyllia coronata* (13), a species recorded from the eastern Pacific and Atlantic between depths of 600 to 3,570 metres.

Of the anemone-like forms that are classed in the order Zoanthidea, most species habitually grow on other animals—they are epizoic. They are colonial or solitary and live from littoral to deep-sea reaches all over the ocean. Species of the genera *Epizoanthus*, *Parazoanthus* and *Zoanthus* attach themselves to sponges, hydroids, corals, gorgonians, bryozoans, worm-tubes and shells inhabited by hermit crabs. The stalk of the deep-sea glass-sponge *Hyalonema* is often studded by small polyps of *Epizoanthus*. This classic association is splendidly portrayed by Poul Winter in Plate IX of Wolff's (1961) report.

The true or stony corals (Madreporaria), many of which are associated in the formation of reefs, have as already stated, deep-sea representatives. There are colonial forms, such as the ivory corals (*Lophohelia*, *Amphihelia*) and solitary, cup-like types (Flabellidae, Microbaciidae, etc.) (Fig. 78). The ivory corals from stony thickets over the upper continental slope and in their branches live sponges, gorgonians, barnacles, bivalve molluscs and so forth. Extensive thickets of *Lophohelia* are found on both sides of the North Atlantic (at 200 to 1,000 metres) and off Argentina. Menzies et al. (1973) suggest that these deep-sea coral reefs '... are

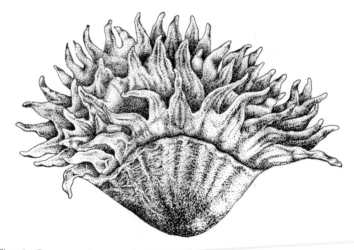

Fig. 78   Deep-sea solitary coral, *Flabellum goodei* (height about 40 mm). (Redrawn from Menzies et al., 1973)

probably world wide features which were once continuous but have been obliterated in some places because of slumping and erosion'.

One of the best known of the solitary deep-sea corals is *Flabellum*, the polyps of which live in more or less conical cups of calcite, with the apical end in the sediments (Fig. 78). One species (*F. apertum*) has been recorded between depths of about 400 and 2,800 metres from localities in all three oceans. Three genera (*Fungicyathus*, *Deltocyathus* and *Leptopenus*) are known from abyssal depths (> 2,000 metres) but only the first has penetrated beyond 5,000 metres. The microbaciid corals range from depths of 15 to nearly 4,000 metres. The deeper-living species have reduced skeletons and tend to be larger compared to those from shallower levels—the diameter of the coral-cups (coralla) range from

1·0 to 3·6 cm. Microbaciid corals have been found as fossils, and Squires (1967) suggests that they have been evolving for at least 60 million years '... during which they migrated into deeper waters of the ocean and in the course of this migration became adapted to life in an environment which is progressively more alien to the presence of calcium carbonate skeletons'.

The black corals (Antipatharia), which come largely from waters deeper than 100 metres in the subtropical and tropical belts, form slender, branching, plant-like colonies. The deepest recorded species, *Bathypathes patula*, has been taken in the Kurile–Kamchatka Trench between depths of 8,175 and 8,840 metres (Fig. 79). The thorny, axial skeleton

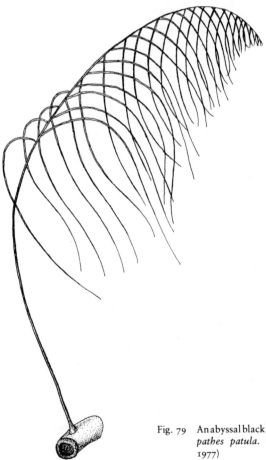

Fig. 79   An abyssal black (antipatharian) coral, *Bathypathes patula*. (Redrawn from Pasternak, 1977)

over which the polyps are disposed is made of a horny substance (anti-pathin) allied to gorgonin. Colonies range from a few centimetres to 2 or 3 metres in height.

Little is known of the life of black corals, but in favourable places, at least, they form luxuriant growths and attract many other animals. Attached to 'trees' of black coral (*Antipathes*) that Church (1971) saw at a depth of about 400 metres in the Coronada Canyon (off San Diego, California) were sea-anemones, sponges, starfishes, feather-stars, mol-luscs, barnacles and the egg case of a filetail shark, *Parmaturus xaniurus*.

Lastly, the Ceriantharia, which are long, solitary, anemone-like forms, live in mucus tubes thrust into the sediment. They have been con-sidered already under the infauna (p. 157), but there is a photograph of six long tube-dwelling anemones that are attached to a rock at a depth of 8,143 metres in the north wall of the Puerto Rico Trench. If, as seems very likely, they are cerianthids (see Heezen and Hollister, 1971), this is the deepest record of the group.

Echinodermata (Crinoidea)

The feather-stars (comatulids) and sea-lilies (stalked crinoids), which form one class, Crinoidea, of echinoderms, are represented today by about 550 and 80 species respectively—there are over 5,000 extinct species. Though their organization and mode of life diverges markedly from that of other echinoderms, their affinities are shown inter alia by: (1) a water vascular system with extensions into tube feet, (2) a radiate structure and (3) a skeleton made of separate plates of calcite, each com-posed of a single crystal. In crinoids, these plates fit closely together to cover the main part of the body (corona) in an underlying calyx that is roofed by a membranous or leathery tegmen around the mouth. The jointed arms, which are feathered with pinnules, arise at the junction of calyx and tegmen. To the base of the calyx is attached the stalk of the sea-lilies and the cirri of the feather-stars. The jointed, flexible stalk is formed by a series of hollow skeletal ossicles (columnals) and each of the mobile cirri consists of a flexible series of ossicles known as cirrals (Fig. 80). Actually, crinoids have yolky larval stages (vitellaria) that even-tually settle on the bottom as stalked (pentacrinus) stages. Later in life, feather-stars set themselves free from their attachment—free to swim with their arms or to use their cirri to creep, clamber or perch. When fully grown, feather-stars use their cirri, which vary greatly in number, length and form, to rest on the sediment or cling to rocks and other organisms, such as sponges, corals, tube-dwelling polychaete worms, beard-worms, bryozoans and so forth. If detached from their hold, most feather-stars will swim by quickly flexing their arms, but it seems that their habit, except for local adjustments of position, is to stay where they are.

Fig. 80   A sea-lily, *Bathycrinus carpenteri*. Left, lappets and tube-feet of food groove; centre, entire organism; right, a portion of an arm with pinnules, bearing tube-feet.

Feather-stars are most diverse and abundant in inshore waters, particularly in coral reefs of Indo-Pacific regions. Nearly all of the deep-sea species live at slope levels above 3,000 metres, the exceptions being certain members of the most diverse family (Antedontidae) that extend to depths of about 5,000 metres. In very productive waters at upper slope levels they may be very abundant, as, for instance, off Portugal, New-foundland and New England. Thousands of individuals of the species *Hathrometra tenella* came up in one haul of the dredge at 240 metres off Martha's Vineyard. The arm span of the largest species is about 600 mm, which is well above that of most deep-sea species.

Shallow-water feather-stars have two ways of intercepting food. Those exposed to currents deploy their arms in a fan-like array, the pinnules and tube feet of each arm being disposed in a plane at right angles to the direction of the flow. The tube feet, which are in groups of three, have inset mucus glands and bear papillae with sensitive hairs. When food particles impinge on the tube feet, threads of mucus are ejected, which are then wafted inwards to the food groove on each pinnule by the lashing movements of the longest tube feet. Ciliary currents in the food groove take over and eventually convey strands of food-laden mucus to the mouth. If the water is slack, such species may spread out their arms radially so as to intercept the settling of particles under gravity ('collecting-bowl' posture). Species that live in sheltered positions in a reef seem only to adopt the latter posture. Meyer (1973), from whose interesting paper the above is taken, refers to the observations by J. M. Pérès on deep-sea species. While one (*Heliometra glacialis*) was seen to hold its arms in a 'collecting-bowl' posture, other observations suggest that a vertical filtration fan may be used if there is current enough. The food of crinoids consists of phytoplankton, small zooplankters (such as radiolarians, foraminiferans, and the smallest crustaceans) and detrital particles.

Modern sea-lilies, which form stalks up to half a metre in length, are dwarfed by some of their fossil relatives, the largest of which had 20-metre long stems. During Carboniferous times the members of sea-lily communities grew to different levels above the bottom and it seems that '... an appropriate strategy for suspension feeding was to exploit the higher levels of the water column, to get the food first, so to speak, rather than wait for it to become incorporated into the sedimentary detritus' (Valentine, 1973). Most modern species are deep-sea forms and with increasing depth tend to increase their height and food-gathering powers by developing longer joints in the stem, arms and pinnules. If one scrutinizes deep-sea photographs (Heezen and Hollister, 1971) it certainly looks as though sea-lilies do spread out their arms in a 'collecting-bowl' posture if there is little or no current. The stem bends over in a current, but the arms, as already hinted do not follow the flow—they tend to be flexed and held against the current so as to form what must be filtration-fans.

Since the food-gathering postures of sea-lilies seem much like those of feather-stars, one wonders how far the possession of a stalk is related to their greater penetration into the deeper reaches of the ocean. At depths greater than 3,000 metres most crinoids are stalked and belong to the families Hyocrinidae and Bathycrinidae (Gislen, 1951). Species of these families have a plain stem whereas isocrinids, which are upper slope dwellers, have a relatively long stem encircled at frequent intervals by whorls of five cirri. Stalked crinoids live at depths down to about 6,000 metres over most of the ocean. Here they are cemented to hard substrata by a basal disc or attached to the sediment by way of a fine 'rooting' system (Fig. 80). Over the upper slope they may be thick on the ground in favourable areas. For instance, off Sand Key, Florida, a haul of the dredge at 550 to 700 metres contained so many heads and stems of *Rhizocrinus* that Alexander Agassiz pictured the net as passing through an underwater forest.

The errant epifauna
Echinodermata

Brittle-stars (Ophiuroidea), sea-stars or starfishes (Asteroidea), sea-urchins (Echinoidea) and sea-cucumbers (Holothuroidea), the four other classes of modern echinoderms, are outstanding among the benthic animals that move slowly over the deep-sea floor. Brittle-stars, which flex their mobile jointed arms to row themselves over the sediments, are the most active. Sea-stars slowly pick their way on their tube feet, as do sea-urchins, aided in some species by their spines. Most of the sea-cucumbers also have ambulatory tube feet, which leave characteristic footprints on the sediments. Limited by their radiate form and slow moving parts and weighed down by their calcareous skeleton, most echinoderms are bottom-dwelling animals par excellence. The exceptions, which spend all or part of their life moving freely along the bottom, are among the holothurians (pp. 197–201), which of all echinoderms are least encumbered by their skeleton and have evolved a considerable degree of bilateral symmetry.

OPHIUROIDEA (BRITTLE-STARS)

Libbie Hyman, after stating that there are about 1,600 existing species of brittle-stars, thought that this class '...may be regarded as the most successful echinoderm group living today, and this is probably to be attributed to the smaller size and greater agility of its members'. Their slender jointed arms, simple or branched, are sharply separated from a relatively small central disc, which has an underlying mouth but no anus.

Like crinoids, some of the brittle-stars use their arms as filtration nets. Indeed, the euryalous kinds, which are predominantly deep-water inhabitants, are said to feed almost exclusively on filtered planktonic organisms (Fell, 1966). Members of one family, the basket stars (Gorgonocephalidae), are thought to feed only by filtration, the branched arms (as in *Astrophyton muricalum*) being extended to form a very large filter fan (Magnus, 1967). Concerning *Gorgonocephalus*, Mortensen (1927) writes, 'The species of this genus live rather gregariously; on rockly ground swept by currents they may be found in great numbers, covering the ground, clinging to one another, their richly branching arms forming a dense network in which are caught the pelagic organisms (Copepods, Appendicularians, etc.) on which they feed. They are also often found clinging to Gorgonians. They reach a considerable size, up to ca 10 cm diameter of disk.' When uncoiled each branching arm may extend for 30 cm or more. Species of the main forms, *Asteronyx*, of another family, which have long and slender, unbranched arms, twine them around various kinds of colonial cnidarians. *Asteronyx lovenii*, virtually cosmopolitan between depths of 100 and 1,800 metres, clings to sea-pens and is said to feed on their polyps as well as on planktonic animals (Fig. 81).

Members of the main group, Ophiurae, which comprises the serpent stars, also use their arms (usually five and simple) as nets. For instance, *Ophiocoma scolopendrina*, a species common in the tidal zone of the Red Sea, will use its arms to collect food from the bottom during slack water, but in moving water it holds up its arms as filters (Magnus, 1967). Other inshore species have similar feeding habits, which are likely to obtain also in deep-sea species. Photographs show that the latter burrow into the sediment so that one or more arms are raised above the bottom and held into the current. Presumably they also extend their arms over the sediment to sweep up organic particles and small organisms. In general ophiuroids are carnivores and scavengers as well as particle catchers and the deep-sea species, like their carnivorous counterparts among the sea-stars, live under the most productive parts of the deep ocean (Sokolova, 1972). Out of six species of *Amphiophiura*, which is an abyssal genus, five were found to be carnivores, one a detritus feeder (Litvinova and Sokolova, 1971).

Brittle-stars live in all parts of the ocean from littoral to deep-sea levels of 6,000 metres or more. In one trawl haul between depths of 6,600 and 6,720 metres in the Kermadec Trench the *Galathea* Expedition obtained 177 individuals of *Ophiura lovenii*, which is otherwise known from all three oceans between the above depths and 2,500 metres (Madsen, 1956b). Deep-sea ophiuroids are most diverse and abundant over the upper continental slope, particularly in warmer oceanic regions. At modest

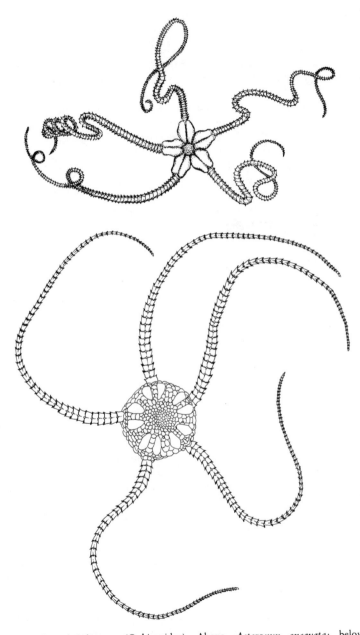

Fig. 81 Two brittle-stars (Ophiuroidea). Above, *Asteronyx excavata*; below, *Ophiomusium lymani*. (Redrawn from Wigley and Emery, 1967)

depths they may be seen covering the bottom like the repeated motif on a carpet. The most photographed and one of the best known brittle-stars is *Ophiomusium lymani*, which is widely distributed over the ocean between depths of 700 and 4,000 metres. It is one of the largest deep-sea species, with a disc diameter and arm length reaching 5 cm and 15 cm repectively (Fig. 81). On the American continental slope between New England and Cape Hatteras, 'Photographic evidence indicates that 80 per cent of the large *Ophiomusium* lie exposed on the sediment surface: 10 per cent are partially buried in the sediment; and 5 per cent are more deeply buried, with only the arm tips exposed. The remaining 5 per cent have the disc raised several cm above the bottom in a peculiar stance' (Wigley and Emery, 1967). This species is whitish, but *Ophiacantha* species, which are nearly all deep-sea, range from yellowish-white to dark brown hues. *Ophiacantha bidententa*, like numerous other deep-sea echinoderms, is luminescent (Herring, 1974b).

ASTEROIDEA (SEA-STARS)

Sea-stars (starfishes) have a more or less flattened, flexible body that ranges from stellate to pentagonal shapes. The brisingid sea-stars with long slender arms that are trenchantly distinct from a small central disc, look most like the brittle-stars: at the other extreme the arms are so short and broad that the entire animal is pentagonal in form.

There are about 2,000 species of living sea-stars, many of which are inhabitants of the deep-sea floor. The Brisingidae, already mentioned, are a deep-water family. The serpentine arms of *Brisinga endacacnemos*, a North Atlantic species known between depths of 200 and 2,000 metres, may attain a length of nearly 40 cm but the disc will be little more than 2·5 cm in diameter (Fig. 82). This is a handsome orange-red species and Mortensen (1927) remarks, 'When complete, probably the most magnificent of all sea stars.' Brisingids are widely distributed over the ocean. The deepest known species, *Freyella mortenseni*, was taken in the Kermadec Trench at depths between 5,850 and 6,160 metres (Madsen, 1956).

The cushion stars (Pterasteridae)—plump with five stout rays and a fin-like membrane along the edge of the body—are also well represented in the deep sea. A membrane over the upper side of the body forms a brood pouch, where the eggs develop. Where known, deep-sea-stars and other deep-sea echinoderms usually produce large yolky eggs, which probably hatch into advanced young—there seems to be no planktonic larval stages such as are found in most coastal species (see also pp. 464–5). Cushion stars, which are generally yellowish to red in colour, have been taken down to about 6,700 metres (in the Kermadec Trench, where *Hymenaster blegvadi* was taken by the *Galathea* at depths between 6,600 and 6,720 metres, Madsen, 1956b). The Benthopectinidae, with rather

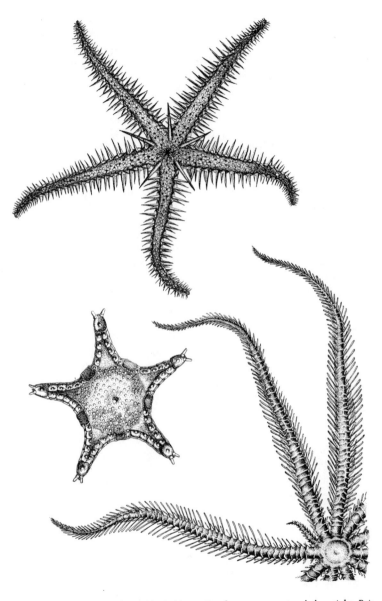

Fig. 82 Three sea-stars (Asteroidea). Above, *Benthopecten armatus*; below right, *Brisinga endecacnemos*; bottom left, *Porcellanaster coeruleus*. (Above and below right redrawn from Mortensen, 1927)

slender arms bordered by spines (Fig. 82), differing from all other sea-stars in developing a pair of dorsal muscles in each arm, are entirely deep-sea and well represented at abyssal levels beyond 2,000 metres. Two other families, the Astropectinidae (stellate and with conspicuous marginal plates) and the Goniasteridae (with large and distinct marginal plates), are also well known at deep-sea levels.

Like most of the shallow sea-stars, the deep-sea species of the above families are largely carnivorous. They take foraminiferans, worms, bivalves, other echinoderms, etc., but some seem to collect detritus by means of ciliary tracts on the arms. Members of the Porcellanasteridae, known from deep waters beyond 1,000 metres in all oceans, are ooze-eaters (Fig. 82). The ooze in the capacious stomach may contain the remains of foraminiferans, radiolarians, sponges, worm tubes, minute snails and plant debris (Madsen, 1961a). The family name comes from the porcelain-like marginal plates and they are also characterized by the presence of one or more cribriform organs in each of the five arm angles. In the living animal these organs appear as yellowish or orange vertical bands on pale greyish background colours. The porcellanasterids, as already stated (p. 157), are burrowers and in the opinion of Madsen (1961a) '... the function of the complicated cribriform organs ... is to create and maintain a circulation of water around the animal in its burrow, a channel presumably being kept open to the surface by means of the apical cone or appendage ... And no doubt the ciliary actions, besides supplying fresh water for respiration, also serve to draw down from above surface-matter to be caught in a secretion of mucus and led to the mouth.' These sea-stars are thus more infaunal than epifaunal, but they are considered here in contrast to their carnivorous relatives of the deep-sea floor. There are over twenty-five species and several are more or less cosmopolitan in distribution. Their size, expressed as the distance from the centre of the disc to the arm tips, ranges from about 35 to 100 mm. Though ooze eaters, they are apparently absent from the deeper parts of the trenches (> 7,000 metres), which is in sharp contrast to the greater depth ranges of some of their deposit-feeding relatives, the elasipod sea-cucumbers (see pp. 199–201).

ECHINOIDEA (SEA-URCHINS)

Sea-urchins are globular, egg-shaped or flattened echinoderms, most of which 'live in' a shell-like test made of closely fitting plates (the echinothuroids have a flexible test of imbricating or separated plates). The plates bear tubercles that form ball and socket joints with the bases of movable spines. The tube feet are in ten meridonial series. The mouth, as in brittle-stars and sea-stars, opens on the underside of the body. The intestine is long and coiled.

Regular sea-urchins develop a more or less globular body (Fig. 83). In coastal waters they feed largely on algae, but also take a variety of animal food. The test is usually elongate-oval in the irregular species, many of which are burrowers in the sediment of inshore waters. They '...eat all sorts of small bottom organisms, especially mussels, snails, Foraminifera, which they are able to pick out from the bottom material by means of their penicillate tube foot; or they fill their alimentary canal with all sorts of bottom material, absorbing the organic particles contained therein' (Mortensen, 1927). Representatives of both groups live in the deep sea.

Regular sea-urchins, which appear more frequently on deep-sea photographs than their irregular relatives, range to depths of more than 6,000 metres. Nearly all of the flexible sea-urchins (Echinothuriidae) are deep-sea forms (Fig. 83). Certain species with a test diameter of over 25 cm are the largest of all urchins. Mortensen (1927) remarks that some are gorgeously coloured (purples, reds or violets) and their poison-laden spines produce painful stings. At least five species are known from depths greater than 3,000 metres, one, *Kamptosoma abyssale*, reaching a depth of 6,235 metres (depth range 4,374–6,235 metres) (Mironov, 1971). Species living at upper slope levels often contain fragments of land plants and algae, sometimes mixed with animal remains (Lawrence, 1975).

Certain members of the family Echinidae, which are most diverse and abundant in coastal waters, penetrate to slope levels. When searching for the remains of the submarine *Thresher* at depths around 2,500 metres off Nova Scotia, Brundage et al. (1967) used an 'urchin scale' to estimate the size of objects in photographs of the deep-sea floor. The numerous echinids in their pictures measured about 4 to 5 cm in diameter (excluding the spines). Most of the slate-pencil sea-urchins (cidaroids) live at upper slope levels. These urchins may also contain plant fragments but invertebrate remains are much commoner (Lawrence, 1975). Comparison of their diet and distribution with those of the flexible sea-urchins would seem to be well worth while.

Of the irregular urchins, the pourtalesiids and urechinids are confined to the deep sea. Both have a thin fragile test, which is generally more or less bottle-shaped in the former and ovoid in the latter. As in the other spatangoid urchins, the mouth opens anteriorly and is without teeth and jaws.

Pourtalesiids bristle with long spines and evidently live on the surface of the oozes, judging by an excellent photograph of *Pourtalesia miranda* taken at a depth of 2,000 metres in the Cap Breton canyon off Santander (Southward et al., 1976). These urchins, which are particularly well represented in antarctic waters, range to depths over 6,000 metres, the deepest record (of a *Pourtalesia aurorae* at 7,250–7,290 metres) coming from the Banda Trench. Indeed, the pourtalesiids seem to have evolved more

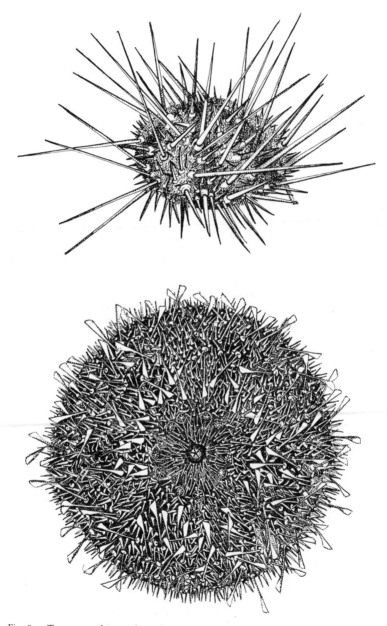

Fig. 83   Two sea-urchins (Echinoidea). Above, a regular sea-urchin, *Echinus affinis*; below, an echinothuroid urchin, *Hygrosoma petersii*.

abyssal species than any other comparable group of sea-urchins. Measured by the length of the test, sizes range from about 40 to 80 mm.

Like the regular urchins, the pourtalesiids will leave little more than the imprint of their spines as they browse over the oozes. But their relatives, the (amphisternous) heart-urchins, are prominent and perhaps the principal ploughers of the deep-sea floor. The activities of heart-urchins appear in photographs as broad meandering tracks at the head of which one may sometimes see the urchin shouldering its way through the surface sediment (Heezen and Hollister, 1971). Presumably the deep-sea heart-urchins feed rather like the shallow-burrowing species *Spatangus purpureus* which is known from tidal levels to about 1,000 metres in the northeastern Atlantic and the Mediterranean. Brush-like tube feet around the mouth pick up particles of sediment (and adhering organic particles) and convey them to the mouth, or such particles may be swept to the mouth by converging tracts of cilia. Perhaps the extent of the plough-tracks of deep-sea heart-urchins over a given period of time is inversely related to the organic content of the oozes. At all events, their tracks have been seen on photographs taken down to depths of 4,000 metres or more.

HOLOTHUROIDEA (SEA-CUCUMBERS)

In the eyes of deep-sea cameras the dominant animals of the benthos would seem to be sea-cucumbers. 'Holothurians feeding on bottom sediments mix and till the surface muds on an enormous scale, producing features more widespread and more visibly evident than those produced by any other animal on earth' (Heezen and Hollister, 1971). Even so, they thrive only on relatively rich sediments, being rare or absent over the most deserted parts of the deep-sea floor (pp. 297–301). Indeed in ubiquity, species diversity and numbers of individuals, if not in biomass, the dominant benthic animals may well prove to be free-living nematode worms.

In holothurians, the mouth, which is encircled by tentacles, is typically at or near the front end of an elongated, often flattened body with distinct ventral and dorsal surfaces. The anus is usually at the hind end. The skin, sometimes thick and leathery, contains a reduced skeleton of microscopic spicules or large imbricating plates or (in a few species) is entirely unprotected. The podia, which are absent in some forms (synaptids), are in the form of locomotory tube feet and typically disposed in five double series. But there are often multiple series, particularly on the ventral surface.

All five orders of holothurians have deep-sea representatives. Indeed, the Elasipoda are entirely a deep-water group. The Dendrochirota which, as their name implies, have tree-shaped tentacles, are predominantly shallow-water forms. They extend their sticky tentacles to catch plankton or detritus, which are passed to the mouth. Certain species of the typical sea-cucumber genus *Cucumaria*, extend to upper slope levels, and one or two are known only from deep waters (e.g. *C. abyssorum* from 2,000 to 4,000 metres in the eastern tropical Pacific and the Atlantic Ocean). Members of the genus *Echinocucumis*, thick-set forms with the mouth and anus at the end of upturned ends of the body, are largely confined to deep waters. Mortensen (1927) suggests that *Echinocucumis hispida*, which has been taken in the eastern North Atlantic between depths of about 50 and 1,500 metres, lies buried in the mud, with just the mouth and anus above the sediment.

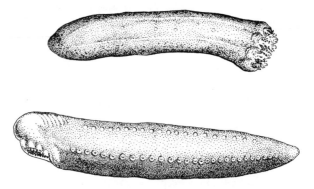

Fig. 84   Two deep-sea-cucumbers (Holothuroidea). Above, an apodan species, *Myriotrochus bruuni* (length ca. 8 mm) from the Philippine Trench 10,190 m; below, a molpadonian species, *Hadalothuria wolffi* (length 20–35 mm) from the New Britain Trench, 8,900 m. (Redrawn from Hansen, 1956)

Some of the Apoda, so-called because they are without tube feet, certainly burrow in the sediments of coastal waters, where they are most diverse and abundant. *Myriotrochus* is largely a deep-sea genus, and good photographs taken at a depth of 4,757 metres in the Bellingshausen Basin, South Pacific, show an apodous species, which looks very like a *Myriotrochus* lying on the sediments (Heezen and Hollister, 1971). The Danish Deep-Sea Expedition (1950–52) trawled over a hundred specimens of a new species, *Myriotrochus bruuni*, between depths of 10,120 and 10,210 metres in the Philippine Trench (Hansen, 1956) (Fig. 84). Members of this genus, which gets its name from the wheel-like ossicles, are known from littoral waters (in the arctic) to the greatest depths of the trenches (Belyaev, 1970). Lengths range from about 7 to 100 mm.

The Molpadonia, which are virtually apodous, have a thick sausage-shaped body that is generally prolonged into a tail-like appendage. Mortensen (1927) says that they all live more or less buried in the sediment and are deposit feeders. Indeed, except for the dendrochirotids, nearly all holothurians feed on organic material in the oozes, which are pushed into the mouth by the ring of tentacles. Most molpadonians live in deep waters. Two species, *Molpadia violacea* and *M. danialsenni*, are littoral in antarctic regions but extend northward into deeper waters at 1,000 to 3,000 metres. A new species, *Hadalothuria wolffi*, which is without a caudal extension, was taken in the New Britain Trench between 8,810 and 8,940 metres by the Danish Deep-Sea Expedition (Hansen, 1956) (Fig. 84).

The members of the two remaining orders shovel sediment into their mouths with shield-shaped (peltate) tentacles, but the Elasipodá are without respiratory trees, which are well developed in the Aspidochirota. Most of the deep-sea aspidochirotans belong to the family Synallactidae, some of which grow to a length of 20 cm or even more. Species of *Pseudostichopus* which have the habit of agglutinating foreign bodies (sponge spicules, pteropod shells, foraminiferan shells, etc.) to their skin, are more or less cosmopolitan mostly between depths of 2,600 and 5,300 metres (Mortensen, 1927). But over thirty specimens of one species (*P. villosus*) were taken at depths between 6,660 and 7,000 metres in the Kermadec Trench (Hansen, 1956). Even so, the main trench-dwelling holothurians are elasipods.

'The Elasipoda are a group of holothurians of bizarre appearance, being provided with large conical papillae, marginal rims, tail-like appendages and so on. They are often more or less flattened, with pronounced bilateral symmetry. The mouth, surrounded by ten to twenty tentacles of the peltate type, is generally displaced ventrally. The scanty podia are usually reduced to one or two rows on the flat ventral surface and lack terminal discs . . .' (Hyman, 1955). Sizes range from about 1 to over 30 cm and colours are not always drab grey. Some species are suffused or touched with violet, maroon or reddish pigments.

The ninety-odd species of elasipods are known from upper slope levels to depths of 10,000 metres or more. One of the best studied genera, *Elpidia*, is represented by species in all parts of the deep ocean down to nearly 10,000 metres (Fig. 85). The arctic species *E. glacialis*, like numerous other deep-sea invertebrates that are represented in polar regions, extends upwards into shelf waters (in the arctic). In the richer trenches, such as the Kermadec and the Kurile–Kamchatka, species of this genus are very abundant at depths well below 7,000 metres. Indeed in the Kurile–Kamchatka Trench, holothurians, consisting largely of *Elpidia*, made up as much as 80 per cent of the total weight of invertebrates at depths below 8,000 metres. The outstanding features of this genus are the ovoid body with four pairs of laterally set podia and the

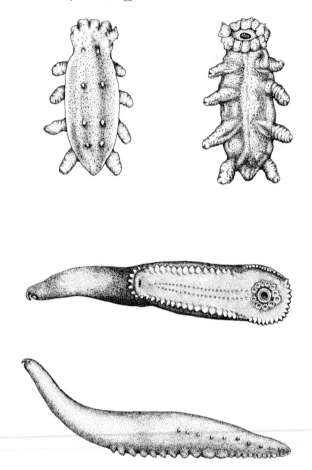

Fig. 85   Two elasipod sea-cucumbers. Top, *Elpidia glacialis*, viewed from above (left) and below (right); centre and bottom, *Psychropotes longicauda*, viewed from below and the side respectively.

few pairs of dorsal papillae, which in some species are long and slender. The related *Scotoplanes*, with a few extra pairs of tube feet to lurch over the sediments, has been photographed in herds on the continental slope off California and New England. In the San Diego Trough one such herd was estimated to move at a rate of nearly 40 cm per hour. One of the most striking elasipods seen in deep-sea photographs is *Psychropotes*, which has a tail-like appendage and may reach half a metre in length (Fig. 85).

'The characteristic tracks of most asteroids and ophiuroids are easily recognized: the footprints of the various arthropods are mostly too small to be seen in normal photographs and real walking fish are not known in the abyss. Thus, only the holothurians can be responsible for most of the large prominent footprints so commonly seen on the deep-sea floor.' The kind of trail, as Heezen and Hollister (1971) go on to reason, will depend on the arrangement of the tube feet and lateral papillae (if present). For instance, *Psychropotes* leaves a wide track consisting of four parallel rows of holes, and so do *Pelopatides* and *Euphronides*. In the two latter genera and others the body is framed by a thin row of fused papillae, which leave the outer rows of holes and is called a 'floating device'. Elasipods of this type were photographed in the Puerto Rico Trench and some appeared to be swimming. In a recent study of more than 80,000 sea-floor photographs, Pawson (1975) concluded that half the known species of elasipods '... may be capable of swimming for extended periods of time. Elasipodids usually indulge in selective detrital feeding but opportunistic scavenging may also comprise a significant part of their feeding activity.' Indeed, certain kinds have left the deep-sea floor entirely and become bathypelagic. Species of *Pelagothuria* have no trace of a skeleton, and the forepart of the small body bears a large swimming brim supported by long papillae. They float mouth uppermost '... now and then gently flapping the brim against the body'. Species of *Enypiastes* look even more like jellyfishes (Hansen and Madsen, 1956).

Epifaunal 'worms'

We have seen that acorn-worms, which are among the most expert bur-rowers in the sediments of the continental shelf, have almost become members of the epifauna in the deep sea (p. 154). Deep-sea sipunculid worms, most of which belong to the genus *Golfingia*, are also likely to be epifaunal. Species of *Golfingia* are slender worms of small to moder-ate size (10–60 mm) with one or more circlets of tentacles around the mouth (except in the smallest species) (Fig. 60). *Golfingia anderssoni*, *G. muricaudata* and *G. schuttei* live from slope levels to depths over 6,000 metres (Murina, 1974). These and other deep-sea species may well obtain organic material by ingesting surface sediments and by using their ten-tacles (if these have a ciliary-mucus feeding system).

Flatworms (Turbellaria) and nemertean worms also live on the deep-sea floor, but little is known of their biology. The most successful of the epifaunal worms are various kinds of polychaetes. The sea-mouse family (Aphroditidae) is well represented. Aphroditids are broad-bodied worms with relatively few segments that bear prominent parapodia. The deep-sea species, which may be tinged with violet or purple pigments, are of modest size (25–80 mm). Some, such as *Mallicephala mirabilis*, known from all three oceans and the polar basin at depths between 46 and 8,400

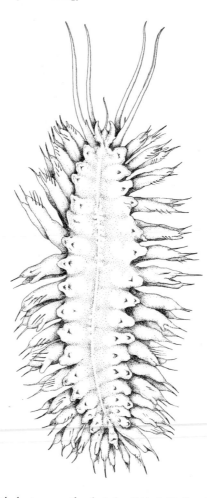

Fig. 86   A hadal polychaete worm (family Aphroditidae), *Mallicephala hadalis* (length
about 25 mm), from the Kermadec Trench (6,660–8,300 m). (Redrawn from
Kirkegaard, 1956)

metres, have wide geographic and bathymetric ranges. Species of *Mallicephala* and *Mallicephaloides*, each with a long proboscis armed with
several pointed teeth, have a predatory look (Fig. 86). They are certainly
active worms, able to swim easily over the bottom (Lemche et al., 1976).
A species of the second genus, which was taken in the Marianas Trench
at a depth between 10,630 and 10,710 metres, is the deepest known polychaete worm. The worms best known from deep-sea photographs are

eunicids of the genus *Hyalinoecia* and they live in a long calcareous tube. The cosmopolitan species *Hyalinoecia tubicola* (length to 15 cm) was one of the largest and commonest invertebrates seen in photographs taken at depths between 346 and 564 metres off the American coast between Georges Bank and Cape Hatteras. Most of the tubes were occupied and the worms that were emergent (from the broad end of their tube) seemed '... to be exploring the sediment surface, possibly searching for food, but not burrowing or digging deeply in the sediment' (Wigley and Emery, 1967). Grooves in the sediment appear to be made as the worms crawl along the bottom.

Epifaunal arthropods

PYCNOGONIDA

Though called sea-spiders, the pycnogonids are but distant relatives of the arachnids. The body of a sea-spider consists merely of a proboscis, a cephalothorax, which bears the appendages, and an abdomen formed of one segment. At most, as in certain nymphonids, there are nine pairs of appendages: palps, chelifores, ovigerous legs and six pairs of walking legs (most pycnogonids have four pairs). Palps and chelifores are absent in some species and ovigerous legs are frequently found only on the males. Most species are a few millimetres in length but the largest species, which live in the deep sea, have walking legs that may attain a length of 40 cm. Apart from the few species that gather detritus, pycnogonids use their proboscis to extract the living substance of certain invertebrates, particularly cnidarians and bryozoans. They live in all parts of the ocean from intertidal regions to trench levels near 7,000 metres (King, 1973).

Most of the deep-sea species of pycnogonids belong to the families Nymphonidae and Colossendeidae (Figs. 87 and 88). The former generally have a slender body, headed by a relatively short proboscis, and long limbs, which in some species are used in swimming as well as walking. Some fifteen species have been taken at depths beyond 2,000 metres, two (*Nymphon longitarse* and *N. tripectinatum*) penetrating to the deepest level known for pycnogonids (7,370 metres in the Japan Trench).

All species of *Colossendeis* have four pairs of extremely long and thin legs, which give some of the largest forms a span of at least 60 cm. Fage (1956) writes, '*C'est là un avantage incontestable pour des formes ayant à déambuler dans un milieu calme, sur la vase molle des grands fonds*'. ('There is, without a doubt, an advantage for those forms that must walk, in calm conditions, on the soft mud of the deep ocean.') Elsewhere, he recalls that the biologist Théodore Monod saw several sea-spiders slowly ambling over the ooze (when on a 1,400-metre dive in a bathyscaphe off Dakar). Most likely, the specific gravity of *Colossendeis* and other sea-spiders is so close to that of sea water that their long legs are not

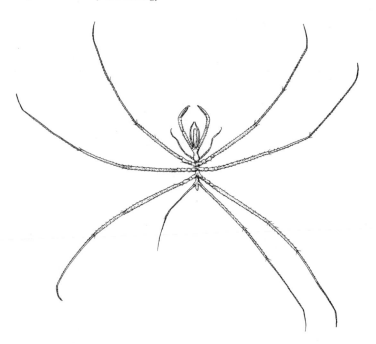

Fig. 87   An abyssal sea-spider (pycnogonid), *Nymphon femorale*, male (about 70 mm in width). (Redrawn from Fage, 1956)

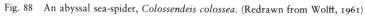

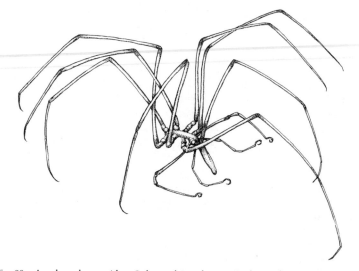

Fig. 88   An abyssal sea-spider, *Colossendeis colossea*. (Redrawn from Wolff, 1961)

so much to enable them to walk on soft oozes but rather to make slow giant strides in search of food or mates. Furthermore, Grassle et al. (1975) saw *Colossendeis* swim, using its legs like the opening and closing of an umbrella; alternately swimming up then sinking, it moved along the bottom. No doubt the deep-sea species of *Nymphon* are also able to swim and they too have more chances of suitable food organisms. One might even speculate that striding or swimming powers of exploration are behind the success of colossendeids and nymphonids in colonizing the deep sea. During the Michael Sars Expedition in the North Atlantic, deep-sea-spiders were nearly always caught together with hydroids, and often with black corals, sea-pens and sea-anemones (Murray and Hjort, 1912). If such cnidarians enable sea-spiders to make a living, one must also remember that they are patchily distributed over the deep-sea floor.

About twenty species of *Colossendeis* are known from depths beyond 2,000 metres. After giving a list of all such deep-sea pycnogonids, Fage (1956) remarks that all the genera involved are represented over wide depth ranges (eurybathic). Of the thirty-odd species of *Colossendeis*, sixteen have been found in waters south of the subtropical convergence (at about 40°S).

EPIFAUNAL CRUSTACEANS

In the entire ocean the benthic species of decapod crustaceans far outnumber those that are wholly pelagic. If we recall that there are over 4,000 species of true crabs (Brachyura) this faunal disparity between the two main marine environments does not seem surprising, but even in the active swimming group of decapods (Natantia) the ratio of benthic to pelagic species (out of a total of nearly 2,000) is about ten to one. In the deep-sea benthos some of the typical representatives of the Natantia are stoutly armoured crangonoid shrimps (e.g. *Sclerocrangon* and *Glyphocrangon*) or large and heavy-bodied kinds of caridean prawns (e.g. species of *Pasiphaea*, *Acanthephyra* and *Parapandalus*). Such decapods must be good bonanza strategists (p. 158), for they are usually the first animals to appear around baited deep-sea cameras, even at a depth of 5,850 metres in the northwestern Pacific Ocean (Hessler, 1972a).

Of the crawling kinds of decapod crustaceans (Reptantia) one lobster-like group, the Eryonidea, is confined to the deep sea between bathyal levels to a depth of 4,500 metres. In the Triassic and Jurassic periods, as Ekman (1953) says, the eryonids were littoral animals with well-developed eyes. Thus, the marked reduction of these organs in modern forms must indicate that their life in the deep sea is secondary. Eryonids, which are usually pink or reddish in colour, have a flattened carapace, long claw-bearing limbs (chelipeds), followed by four pairs of

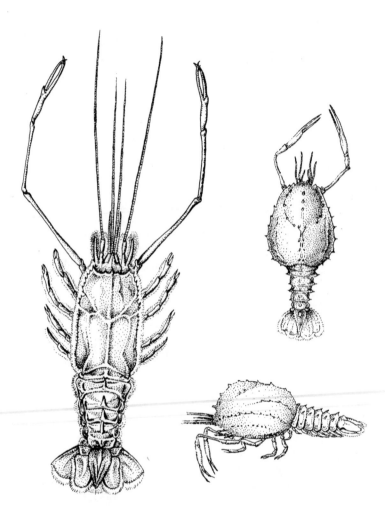

Fig. 89   Left, the eryonid decapod crustacean *Willemoësia*; right, two *Eryoneichus* larvae. (Redrawn from Wolff, 1961)

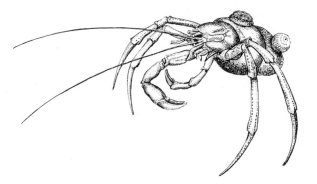

Fig. 90   A deep-sea hermit-crab, *Parapagurus*, bearing two sea-anemones.

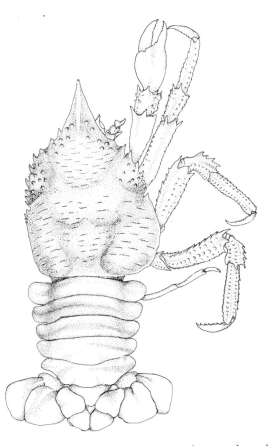

Fig. 91   A galatheid crab, *Munidopsis subsquamosa latimana* from the Kurile–
Kamchatka region (5,035–5,210 m). (Redrawn from Birstein and Zarenkov,
1970)

walking legs, each ending in a small claw. The 20–30 species of *Polycheles* live at bathyal levels down to about 2,000 metres, whereas *Willemoesia* is represented by some six species between depths of 2,500 and 4,400 metres (Fig. 89). Eryonids have feeble mouthparts and are evidently deposit feeders (Madsen, 1961b). Like certain other benthic deep-sea animals, they produce large and advanced larvae (*Eryoneichus*) that have a prolonged pelagic existence (pp. 465–7).

Most of the anomuran decapods are either hermit crabs (Paguridea) or squat lobsters (Galatheidea), and both groups have deep-sea representatives. Deep-sea hermit crabs, such as *Parapagurus*, may be overgrown with sea-anemones or zoantharians (Fig. 90). Some occupy tusk

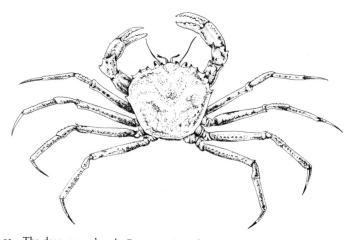

Fig. 92   The deep-sea red crab *Geryon quinquidens*. (Redrawn from Wigley, Theroux and Murray)

shells (*Dentalium*). Deep water hermit crabs and squat lobsters both range from the upper slope to depths near 5,000 metres. Between these bathymetric limits the most diverse deep-sea genus of galatheids, *Munidopsis*, is known from more than a hundred species, five penetrating to depths beyond 4,000 metres. Lengths range from a few to about 20 cm (Fig. 91).

The true crabs of the deep sea range from large robust types, such as *Geryon* (Fig. 92), to relatively small, fragile forms, such as *Ethusa* and *Ethusina*. Crabs of the latter kinds, some of which have very reduced eyes, may be found clinging to sponges, cnidarians and tunicates. *Ethusina* with eight species ranging between 500 and 4,300 metres, has colonized the deep sea better than any other crab genus. Wolff (1961) states that out of a total of 3,500 crab species only 125 or so are known from depths beyond 200 metres. The limited colonization of the deep-sea floor, not only by crabs, but by all decapod crustaceans, may well be due to their predatory and scavenging habits, which cannot be deployed in the food-poor depths of the abyssal ocean. If so, it is hardly surprising that large species, such as the red crab, *Geryon quinquidens*, which may reach a weight of 1 kilogram, are most abundant at upper slope levels (Schroeder, 1955). Even so, certain shallow-water hermit crabs and squat lobsters are, at least in part, deposit feeders. Perhaps this is also true of their deep-sea relatives.

Molluscs

All classes of molluscs are well represented in the deep-sea benthos. In the infauna there are diverse bivalves, particularly sediment-eating species, scaphopods (tusk-shells) and aplacophorans (worm molluscs) pp. 149–51). Certain epifaunal species, which are now placed in the class Monoplacophora, caused quite a stir when they were seen to be closely related to the Tryblidiacea, known only as fossils from Cambrian and Silurian times. The first species, *Neopilina galatheae*, which was taken off the west coast of Mexico at a depth of 3,570 metres, has been thoroughly described by Lemche and Wingstrand (1959) (Fig. 93). Since then other species of *Neopilina* have been discovered and all may be descended from a Cambro-Devonian genus, *Pilina* (Menzies et al., 1973). These molluscs, which look superficially like a limpet, have five or six pairs of gills and a broad flat foot. They are deposit-feeders and presumably feed as they move slowly over the ooze.

Chitons or mail-shells (Polyplacophora) are most diverse and abundant in warm inshore waters, where the largest species grow to a length of one third of a metre. A few species, such as certain of the genus *Lepidopleurus*, have invaded the deep-sea floor to depths of 4,000 metres or more. The inshore species browse on algae and encrusting diatoms etc., which are probably 'replaced' by organic material in the oozes at deep-sea levels.

Snails (Gastropoda) are well represented in the deep sea, where they extend to the deepest trenches. For instance, in samples taken (in an epibenthic sled net) between New England and Bermuda at depths between 478 and 4,862 metres, there were 93 species of prosobranchs and 30 species of opisthobranchs (Rex, 1976). Of the more familiar kinds there were representatives of the limpet, top-shell, pelican's foot, spire-shell, moon-shell, whelk, cone and turret-shell families. Rex found that all of the archaeogastropods (e.g. limpets and top-shells) were deposit-feeders, whereas all neogastropods (e.g. whelks, cone-shells and turret-shells) were predators. Species that live in the trenches may be without eyes or eye-stalks and form thin white shells with no sculpturing.

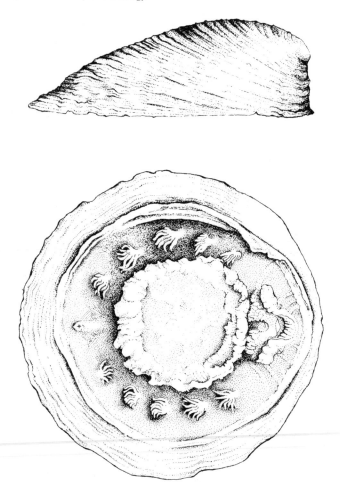

Fig. 93   The *Neopilina galatheae*, a monoplacophoran mollusc (order Tryblidiacea), length about 35 mm. Above, shell; below, animal seen from below, showing mouth, anus, foot and gills. (Redrawn from Lemche and Wingstrand, 1959)

The most remarkable of the epifaunal bivalves are surely those of the wood-boring genus *Xylophaga* (family Pholadidae). Seventeen species of this genus were collected by the *Galathea* at depths between 545 and 7,290 metres, but it is known also from inshore waters. These molluscs, which live in globular shells with a prominent beak (shell sizes range from 1 to 12 mm), have a long, proboscis-like siphon. Numerous individuals were found on water-logged trees or twigs or associated with plant debris. Species of *Xylophaga* certainly bore into trunks or twigs and, while their faeces contain wood fragments, their source of food is still uncertain. Perhaps bacteria digest the swallowed wood fragments and are, in turn, digested by their hosts (Knudsen, 1961).

Regarding the epifaunal, filter-feeding bivalves, we have seen already that with increasing depth they make up less and less of the bivalve fauna, which becomes dominated by the shallow-burrowing, deposit-feeding protobranchs (pp. 148–9). For instance, 'when it comes to feeding habits, the composition of the bivalve mollusc fauna in the abyssal zone of the north-western part of the Pacific Ocean would show that most of them belong to the group that gathers detritus from the surface of the sediment. The number of species belonging to the filter-feeders (feeding on detritus suspended in water), which are so common in the sublittoral zone, sharply decreases in the abyssal zone and as regards quantity is of secondary importance' (Filatova, 1959). Instances of filter-feeding families of bivalves with deep-sea representatives are mussels (Mytilidae), scallops (Pectinidae), file-shells (Limidae) and Noah's Ark shells (Arcidae).

While squid of various kinds, both neritic and oceanic, may spend part of their life near the ocean floor, they are essentially a pelagic group. The true kinds of benthic cephalopods are either octopods or sepioids (cuttlefishes etc.) both of which have pelagic representatives (pp. 119–21). Indeed, the cirrate octopods, as we have seen (pp. 141–2) are bentho-pelagic. They drift and swim near the deep-sea floor.

Octopods of the continental slope '...are muscular but soft and slightly gelatinous. They are pinkish or light red. Most of them have rather short arms; none possesses the long arms found in some littoral species' (Voss, 1967). Such octopuses may be attracted to baited cameras, and, like true bonanza strategists, they may appropriate their finds. 'In one motion picture sequence made at about 4,000 metres a small octopus squats on the bait, keeping the grenadiers at tentacles' length, and in an unusual sequence made with a still camera off Cedros Island in Lower California, two octopuses fend off a single grenadier from the bait' (Isaacs and Schwartzlose, 1975), who use the common name grenadier for macrourid fishes or rat-tails). Perhaps the most highly modified of the benthic octopods are species of *Opisthoteuthis*, which are so flattened that they have been called flapjack devilfish. When alive these octopods are dark reddish brown and they grow to a width of 20 cm or more. They are known from Pacific and Atlantic localities between depths of 125 and 2,250 metres.

Little is known of the biology of deep-sea sepioids. Species of *Sepiola* and *Rossia* have a short, squat body, broad fins and a reddish coloration. Voss (1967) says that most of the upper-slope-dwelling squids belong to the family Sepiolidae and have the ink sac associated with a light organ containing luminescent bacteria. The combined discharge of the two organs produces a luminous ink, which is best known in the midwater genus, *Heteroteuthis*.

Benthic fishes

Diverse deep-sea fishes, whether by virtue of a gas-filled swimbladder, a large buoyant liver or reduced tissue systems (especially muscular and skeletal), have become weightless in water, and are thus able to swim easily (and habitually) near the ocean floor. We have called such fishes benthopelagic (p. 135). The habit of benthic deep-sea fishes, which have no special weight-reducing systems and are negatively buoyant (but see below), is to rest on the bottom. Here they may be seen from deep submersibles or taken on film.

If a benthic fish swims over the bottom, part of its muscular effort is needed to keep it from sinking. Skates (Rajidae), which are classic types of benthic fishes, and the most diverse group of rays (Batoidei), are active predators, cruising over the bottom in search of fishes, crustaceans, molluscs, worms and so forth. Most species live in relatively productive waters over the continental shelf, especially in temperate regions. The few deep-sea species range down the slope. For instance, in the northeastern Atlantic, *Raja bathyphilus* is known from 670 to 2,180 metres, *Bathyraja richardsoni* from 2,000 to 2,540 metres and *B. pallida* from 2,410 to 2,950 metres. The two latter species, which have soft bodies and extremely large, oil-charged livers, are probably only just less than perfectly buoyant (Bone and Roberts, 1969). Clearly, when seeking their prey, these two kinds of rays will use very little energy to keep clear of the bottom, which must be advantageous in an environment where food organisms are less plentiful than those on the shelf.

In the fish faunas of coastal seas the classic types of 'sit and wait' predators have broad massive heads, large jaws and a relatively small trunk and tail. In the cold and temperate waters of the northern hemisphere there are bullheads (Cottidae and Cottunculidae) and scorpion fishes (Scorpaenidae). Their counterparts in the Southern Ocean are nototheniiform fishes, some of which, though but distantly related, look very like northern bullheads. These are all lurking, usually well-camouflaged fishes that are able to engulf a wide size range of food organisms, notably fishes and crustaceans. Certain species have spread, but not very far, into the deep sea. In the antarctic dragon fishes (Bathydraconidae), which are actually less bullhead-like than other nototheniiforms, have been taken down to nearly 2,600 metres (Andriashev, 1965). In the northern North Atlantic the cottunculid bullheads live at slope levels down to 1,750 metres. At upper slope levels over warm temperate to tropical parts of the ocean, the main 'sit and wait' predators are scorpion fishes.

Two other groups of benthic fishes, the eel-pouts (Zoarcidae) and sea snails (Liparidae), have their headquarters in arctic to temperate waters of the northern hemisphere. In these regions there are nearly 150 species of eel-pouts, whereas there are only about 10 in the Southern Ocean. In the northwestern part of the Pacific Ocean the species of the deep-sea genus *Lycenchelys* fall into two groups, bathyal (800–2,500 metres) and abyssal (3,000–4,000 metres). The latter species differ from the former in their smaller eyes, smaller lateral line pores on top of the head, and reduced extent of the dorsal fin. Species of *Lycenchelys* grow to a length of about 35 cm. Sea snails are also diverse in the northern Pacific, where about 100 species are known. One species, *Careproctus amblysto-mopsis*, which was taken in the Kurile–Kamchatka Trench at a depth of 7,230 metres, is one of the deepest recorded fishes.

The benthic fishes that are typical of the warm temperate to tropical parts of the ocean are well designed to rest on the deep-sea floor. The pelvic fins are inserted well forward of the fishes' centre of gravity so that the outer pelvic rays and the lower rays of the tail fin, which are specially strengthened, form an undercarriage. In the tripod-fishes (Bathypteroinae) these supporting rays have become very long, but in *Ipnops*, *Bathymicrops*, the green eyes (*Chlorophthalmus*) and the Bathysauridae, they are relatively short (Marshall, 1961).

Tripod-fishes, which have firm fusiform bodies and pleasing scale patterns, range in length from about 10 to 45 cm. They are best known in the Atlantic Ocean, where the deepest ranging species is *Bathypterois longipes* (2,500–5,600 metres). The shallowest living species, *Bathypterois bigelowi*, known from the Gulf of Mexico and Caribbean localities between depths of 475 and 1,200 metres, appears in magnificent photographs taken by Ron Church from *Deep Star* in the De Soto Canyon (Gulf of Mexico). Each fish, with its piebald body perched on dark pelvic and caudal fins, faces a weak current, the pectorals turned forward so that the long rays are outthrust like multiple antennae (Fig. 94).

Ipnops and *Bathymicrops* are slim-bodied and somewhat smaller than their relatives, the tripod-fishes (Fig. 95). Study of *Ipnops murrayi* recorded from about 1,500 to 4,000 metres from various Atlantic localities (including the Gulf of Mexico and the Caribbean) showed it to have a well-developed system of free-ending lateral line organs on the head, trunk and tail, but the olfactory organs are very small (Marshall and Staiger, 1975). The extraordinary eyes, consisting largely of flattened but functional retinae (p. 392), shine in a photograph taken of an *Ipnops* in the Virgin Island Basin (Roper and Brundage, 1972). *Bathmicrops* is known from all three oceans between depths of about 4,260 and 5,850 metres.

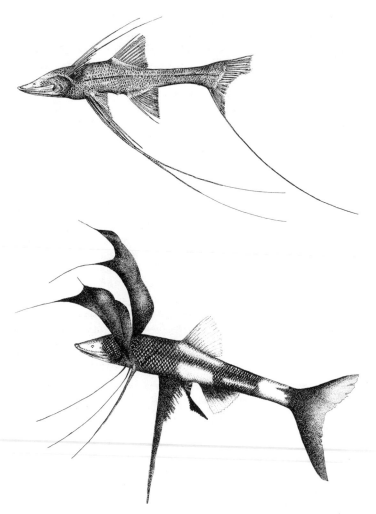

Fig. 94 Tripod-fishes (Bathypteroidae). Above, *Bathypterois grallator* (length about 270 mm); below, *Bathypterois bigelowi*. (Above, redrawn from Bruun, 1956; below, drawn from a photograph by R. Church)

The tripod-fishes and their relatives are all monoecious, and so far as is known, are synchronous hermaphrodites. Beside cross-fertilization '...the ability to self fertilize may be present, and may provide the "insurance" permitting reproduction in the absence of a mate—an absence possibly common in low density abyssal populations. It should be noted, however, that one such monoecious group, the Chlorophthalmidae, is

Fig. 95   The bathyteroid fish, *Ipnops*.

chiefly composed of species which are quite abundant on suitable continental shelf substratas' (Mead, Bertelsen and Cohen, 1964) (p. 460).
Indeed, species of *Chlorophthalmus* are common in the North Atlantic between depths of 200 and 750 metres. Lengths range between 15
and 30 cm.

Bathysaurus, which is most closely related to the lizard fishes (Syno-
dontidae), has strong jaws armed with bands of pointed teeth and a fusiform, well-muscled body. Like their relatives, they are well designed
to take off and seize their prey. Fully grown fish frequently reach lengths
of 60 cm or more. There are two species, one *B. mollis*, occurring at depths
beyond 4,000 metres in the Gulf of Mexico and the eastern North Atlantic.

SECTION II
DEEP-SEA FOOD WEBS

Introduction

Henry Bigelow wrote, 'When one picks up a fish one may be said, allegorically, to hold one of the knots in an endless web of netting, of which the countless other knots represent other facts, whether of marine chemistry, physics or geology, or other animals and plants. And just as one cannot make a fish-net until one has tied all the knots in their proper positions, so one cannot hope to comprehend the web until one can see its internodes in their true relationship.' Clearly, the more one knows of deep-sea food webs the closer one comes to comprehending the endless web of life. The prospect is daunting, for there are many cross connections between the knots, which are continually being tied and untied. There is immense variation between one generation of knots and the next, from a few hours for a bacterium to many years for a whale. In the deep ocean our knowledge of food webs is far from adequate. Indeed, we do not even have a proper inventory of the living knots: there are still many new species to be discovered and described. Even so, systematic and biological investigations are advanced enough to give some understanding of how the food web is made and maintained, though we may well agree that '... the single most important unresolved problem in the fields of marine biology and chemistry' concerns the '... mechanism by which organic matter is transported to depth and how the energy requirements of the deep seas are met' (Menzel, 1974).

Our preceding surveys of oceanic plants and the animals of the pelagic and benthic environments have been partly concerned with patterns of productivity and animal means of making a living. It seems clear that benthic animals are largely dependent, whether directly or indirectly, on organic materials that have fallen from, and through, the food webs of the pelagic zones to eventually reach the bottom. First then, we must look at trophic events in the pelagic sphere.

Pelagic food webs

6

Epipelagic food webs

Deep-sea animals are most diverse under the surface waters of the sub-tropical and tropical belts. Here are clear blue waters, where photosynthesis by the microscopic algae of the plankton is possible down to a depth of 100 metres or more. Apart from the divergent regions of the equatorial current systems, where upwelled water brings nutrient salts to the surface and leads to relatively high levels of productivity, these warm central oceanic waters are low in nutrients and standing crops of plankton (p. 64). Blue-green algae thrive in such surroundings and a dominant form, *Trichodesmium*, whose rust-coloured blooms may at times cover wide stretches of the warm ocean, is reported to use fixed molecular nitrogen for its growth (Dugdale, 1976). In other respects the composition of the phytoplankton of the tropical and subtropical belts is much like that of other oceanic regions. But the size range of the algae extends below that of floras in nutrient-rich waters, such as are found in the antarctic and in upwelling waters off Peru and elsewhere. For comparable forms the smaller the cell the less are its nutrient requirements, but the more its surface for nutrient absorption compared to its volume. Thus, it somehow seems right that much of the flora of nutrient-poor tropical waters should consist of mobile μ-flagellates, such as crypto-monads and chrysophytes.

To be but a few microns in size confers another advantage. In subtropical and tropical regions in particular, it seems that the multiplication of the phytoplankton barely keeps pace with its consumption by all manner of herbivores in the zooplankton. One might then expect selection pressure to be high for means of countering the grazing inroads of zoo-plankters. To be small enough to pass through the meshes of a filtering system, particularly of copepods, is advantageous, though there is less escape from the mucus nets of the pelagic tunicates (see p. 81 and below).* But in overall terms of nutrition μ-flagellates may form no more than a small part of the grazing of herbivores. For instance, during an expedition to the eastern tropical Pacific, counts of phytoplankton constituents were made (under an inverted microscope) from 200 samples covering 48 stations. When each count was expressed in terms of its carbon content, the main conclusion was that though flagellates (2–5 μ in diameter) formed about 80 per cent of the catches in numbers of individuals, they made only a minor contribution to the total content of phytoplankton carbon. In equatorial waters most of this carbon was contained in the diatoms, whereas dinoflagellates were the main carbon contributors north and south of the equatorial region (Zeitschel, 1971). No doubt the relative amounts of the major groups of algae change from time to time, and we must always remember that the fragile and elusive μ-flagellates are likely to be underestimated.

There is no off-season in the production of phytoplankton in warm oceanic waters: throughout the year diverse zooplankters live on thinly dispersed crops particularly in the subtropical regions bounded by the gyres. In temperate and cold regions there is rich grazing in the favourable season, during which animals form stores of fats and oils to help tide them over the winter months. Whatever the place, there are times when the phytoplankton crops are so meagre and thinly spread that one may wonder how the herbivores manage to exist. Indeed, such thoughts, stimulated by the quantitative studies of Hensen, Lohmann and Brandt towards measuring the productivity of the seas, led Pütter to investigate the nutritional requirements of marine animals (as measured by their rates of carbon dioxide production). After comparing the metabolic needs of diverse species with ambient amounts of algal food, Pütter concluded that marine animals must depend largely on dissolved organic matter, which they absorb through their integument. Krogh, who was drawn to this problem through his experiments to determine whether aquatic animals are able to absorb organic substances, first concentrated on an accurate method of measuring dissolved organic matter (DOM) in sea water. Though he found that there is much more DOM than particulate forms of organic matter in the ocean, he also observed that DOM is very largely resistant to bacterial decomposition. Krogh concluded that the food of aquatic animals is generally in the form of organisms and organic detritus.

Since the 1950s numerous experiments on diverse aquatic invertebrates have shown that most marine, but not freshwater, forms can take up glucose and amino-acids through their integument. But Krogh's con-

---

* The evolution of thick gelatinous sheaths is evidently another means of protection. Such algal species, though eaten by grazers, are frequently intact and viable after passage through their consumers.

clusion still stands. Even the Pogonophora, which are confined to sediments rich in organic matter (pp. 177–8), have a restricted rate of absorption when in their tubes (see the review by Barker Jørgensen, 1976).

Members of the micro-zooplankton (ciliated protozoans of the tintinnid and oligotrich groups, foraminiferans, radiolarians, naupliar and postnaupliar stages of crustaceans, etc.), certain copepods and euphausiids, shelled pteropods and pelagic tunicates form the main groups of oceanic herbivores. More than we realize, the lives of these diverse forms (and their predators) depend on the natural fact that the phytoplankton is not randomly dispersed. Statistical analyses of well-planned series of samples show that the constituents of phytoplankton are 'overdispersed', which implies the aggregation of 'clumping' of individuals but not necessarily the forming of physically observable clusters (Cassie, 1963; Steele, 1976). Studies of a water column down to 150 metres off Bermuda, which was so homogeneous that measures of temperature and salinity varied by less than 0·05 C° and 0·05 per cent respectively, showed that the concentrations of phytoplankton species differed by several orders of magnitude at different depths. There are discrete layers of phytoplankton (as shown by measurements of chlorophyll) and at times such concentrations lie near the surface, at others near or even below the depth of the euphotic layer (Strickland, 1968; Steele and Yentsch, 1960). Members of the micro-zooplankton aggregate in these layers (Beers and Stewart, 1969). In one way or another enough individuals of any herbivorous species seem to refute Pütter's conjecture, which is also true of the largest marine animals. For instance, fin whales were estimated to need zooplankton concentrations in excess of 1·5 gm per cubic metre, whereas the average standing stock of zooplankton in the Subarctic Pacific is rather less than one-tenth of this amount. During crossings of this region Barraclough et al. (1969) used a 200 kHz echo-sounder and recorded shallow scattering layers that net hauls showed were composed almost entirely of the copepod, *Calanus cristatus*. Such concentrations were considered to be more or less sufficient to satisfy the appetites of fin whales.

While there is no off-season in the production of life in subtropical and tropical waters, there are seasonal changes. Indeed, monsoon comes from the Arabian word *mausim*, meaning a time or season. During the northeast monsoon (November to March) in the Indian Ocean measurements of carbon assimilated by the phytoplankton yielded a mean figure of 0·15 gmC per square metre per day, which is less than one-third of the production during the southwest monsoon (Cushing, 1973b). The invigorating effect of the southwest monsoon is widespread, but in the warm Pacific Ocean seasonal changes are most pronounced in equatorial regions. Here from July to December (when the southeast trades are blowing and upwelling increases) there is greater productivity than during the rest of the year, when there are northeasterly or variable winds

(King and Hida, 1954). In the great subtropical gyres, where the (so-called) 'seasonal' thermocline is more or less permanent nearly everywhere (so barring the upward transfer of nutrient salts), there can be little seasonal change in the standing stocks of plankton.

Marked seasonal changes in primary production no doubt lead to changes in the biomass and composition of the zooplankton. Concerning the productive zones of the equatorial Atlantic in the (northern) summer, Bernikov et al. (1966) were impressed by the high biomass of plankton, the bulk of which consisted mainly of jelly-like animals, notably of pelagic tunicates (salps, doliolids and pyrosomes). In the autumn most of the zooplankton consisted of crustaceans.

Pelagic tunicates with their efficient plant-gathering nets of mucus and relatively little living tissue to maintain, seem well organized to respond quickly to an increase in primary productivity. Indeed, one salp (*Thalia democratica*) with an instantaneous rate (r) of population increase of o·4 to o·91 per day, can double its numbers in a single day. Swarms of this salp, which may be very widespread, are generated immediately after phytoplankton blooms, and thus are transient resources exploited to the full (Heron, 1972). Their small relatives, the appendicularians (pp. 94–6), most of which live in subtropical and tropical regions, are even livelier opportunists. In the Black Sea the turnover time (the interval between one generation and the next) of *Oikopleura* is about three days (Greze, 1970).

Appendicularians are among the commonest groups of zooplankton (often second only to the copepods) and, like their relatives, are able to feed on the smallest kinds of phytoplankton (p. 95). The mesh size of the filtering windows in their houses varies from species to species and when the filter becomes clogged with food, another house is quickly secreted (which may be several times in one day). In each discarded house is a concentrated package of phytoplankton, detritus and mucus, food for other planktonic animals and bacteria. Appendicularians themselves are prey to such animals as jellyfishes, siphonophores, arrow-worms and lantern-fishes (p. 244). By-products of zooplankters, particularly of appendicularians, salps and pteropods, form many of the flakes and larger organic aggregates of the 'marine snow' so often seen in the open ocean by divers and observers in submersibles. Alice Alldredge (1976) concludes, 'These large amorphous bits of mucus and detritus serve as surfaces on which many planktonic organisms can rest or feed. Bacteria and protozoans use them as a permanent habitat. Appendicularian houses and other particles of marine snow provide tiny solid substrates and introduce heterogeneity into an environment that is generally considered to be relatively homogeneous physically. Although these miniature habitats have not been completely explored, it is certain that they have influenced the adaptation and feeding strategies of many planktonic organisms.'

Pelagic tunicates are what theoretical ecologists call *r*-strategists, opportunistic animals (usually small) that grow and reproduce quickly when conditions become favourable. Indeed, nearly all gelatinous forms in the zooplankton, both herbivorous and carnivorous, seem to be *r*-strategists. Concerning the latter, medusae, siphonophores and ctenophores evidently have '...a versatile adaptation system, which facilitates flexible synchronisation of the population booms with abundance of planktonic forms which they use for food. This synchronisation includes the capacity to vary the number of eggs; timing, rates and types of reproduction; to withstand long periods of starvation; to produce resting stages, and to form aggregations. These properties (many of them characteristic of ephemeral species) provide passive pelagic predators with the capacity of "waiting" for a favourable biotic situation to occur' (Zelickman, 1972). At first sight, these gelatinous herbivores and carnivores seem to break the rule that *r*-strategists are small, but if large in bulk, as most species are, they are small in protoplasmic content. To find small kinds of *r*-strategists, one need look no further than the protozoans of the micro-zooplankton.

Compared to those of cold regions, zooplankton communities of the warm ocean have a low biomass, a high diversity with no very dominant species, a higher proportion of carnivorous species, more species with a short life span (several generations in a year) and numerous herbivorous species with a reduced body size (but there is no such diminution in carnivorous species) (Motoda et al., 1974). As Motoda and his colleagues say, one advantage of greater size in herbivorous zooplankters of cold waters is that they have extra space to store fats etc., to help them over the winter season. The smaller herbivorous species of the warm ocean live faster and breed faster but store little in the way of food reserves, for phytoplankton is always available, even if the supply may be meagre, as it most often is in the deep-blue gyral deserts of the subtropical belts. In the North Pacific gyre productivity is generally limited by the quantity of nitrogenous nutrient present in the water. Indeed, nitrates are insignificant in quantity; the main source of nitrogen for plant growth seems to be ammonia and urea produced as waste products by the zooplankton and nekton. The impoverished nature of subtropical waters, compared to equatorial and more northerly waters of the western North Pacific, may be seen in Fig. 96, which shows the geographical variations in phytoplankton crops and the biomass of zooplankton in the surface waters. (See also the general chart of primary productivity in the ocean, Fig. 16.)

In subtropical regions, where the upper waters are kept stable but impoverished by a marked thermocline, plant production barely keeps pace with herbivore consumption: there is, as Cushing (1975) calls it, a quasi-steady state production cycle. In the Indian Ocean, Timonin

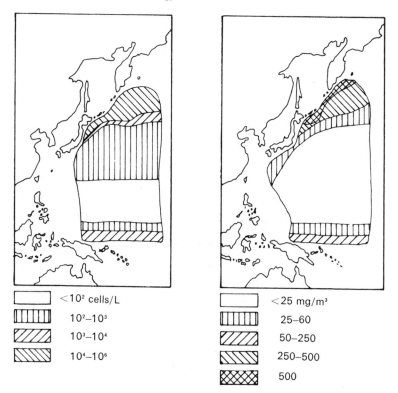

Fig. 96   Left, geographical variations in the standing crops (measured in number of cells per litre) of phytoplankton in the western North Pacific Ocean.
Right, geographical variations in the biomass of zooplankton (milligrams of wet weight per cubic metre) in the upper layers in the western North Pacific Ocean. (After Motoda, Taniguchi and Ikeda, 1974)

(1971), who analysed zooplankton communities, especially their cope-pod, euphausiid and chaetognath constituents, in specific and trophic structure,* contrasts regions of stable water stratification with those, as in the equatorial belt, where current divergences lead to upwelling and greater productivity. Stable regions support communities with a low bio-mass but high species and trophic diversity (carnivores predominating),

---

* The 86 species of copepods were divided into fine filter-feeders on phytoplankton (16 species), coarse filter-feeders (15 species), omnivores (30 species), seizing and masticating carnivores (7 species) and piercing and sucking carnivores (18 species). The 15 species of euphausiids were either coarse filter-feeding herbivores (9 species) or seizing and masticating carnivores (6 species). The 13 species of chaetognaths, as to be expected, are seizing and swallowing carnivores.

whereas communities in regions of intensive divergence have a higher biomass with lower species and trophic diversity (coarse filter-feeders predominating). Communities in regions of low divergence were more or less intermediate in the values of the above ecological indices.

After one organism becomes the food of another, matter and energy are lost in the processes that transform prey tissue into predator tissue. Any such transformation can be seen to have a certain *transfer efficiency*, which Ricker (1969) defines as '...the percentage of the prey's annual production that is incorporated into the body tissues of the consumer species. This, in turn, is compounded of two processes characterized by separate coefficients:

E, the *ecotrophic coefficient*—the fraction of a prey species' annual production that is consumed by predators (trophic referring to nutritive or food levels):

K, the *growth coefficient*—the predators' annual increment of weight divided by the quantity of food they have consumed.'

Ricker suggests that an overall average growth coefficient for herbivores could hardly exceed 15 per cent, though the figure may be around 20 per cent for fish and other consumers of animal food. Concerning the ecotrophic coefficient of (primary) plant consumers (including bacteria) a figure of 66 per cent seems reasonable, while for secondary consumers, such as lantern-fishes which feed on zooplankton, 75 per cent is proposed. Eventually, one may construct a simplified aquatic food pyramid such that the area of each rectangle representing a particular trophic level is proportional to the estimated production of that level (see Fig. 97). But the four levels shown in this figure, as Ricker points out, are given too low a trophic value. Successive steps in a food pyramid (trophic levels) need to be distinguished from ecological levels (groups of animals of generally similar size and habit). In order to assess an average trophic level one has to consider within-level predation ('cannibalism'), represented by the horizontal arrows in Fig. 98. The effect is to reduce the net biomass produced by any ecological level. For instance, the production of lantern-fishes, which feed on zooplankton, is reduced by members of the zooplankton, such as jellyfishes and arrow-worms, that feed on larval lantern-fishes. More realistic estimates of average trophic level are shown on the right-hand side of Fig. 98.

The overall transfer of production from one trophic level to the next is, of course, represented by the product KE, and Ricker estimates average values of about 10 per cent for the primary consumer stage and 15 per cent for subsequent stages. (Both figures include recycling adjustments.)

Relative productivities, as in Fig. 97, are thus: plants 100, herbivorous zooplankton 10, small pelagic fishes that feed on zooplankton (e.g. lantern-fishes) about 1·6, and so forth. The more the links in a food chain, the very much less is the terminal productivity. In some oceanic

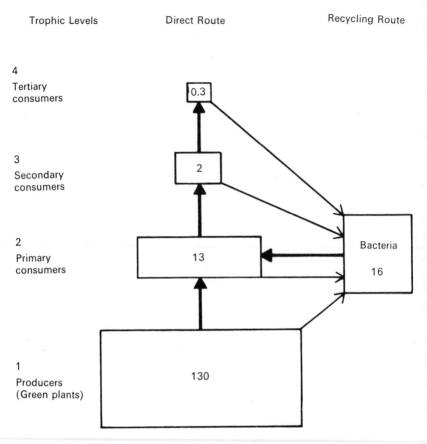

Fig. 97 Simplified aquatic food pyramid, illustrating direct and 'recycling' routes for conversion of plant material into animal tissue. Enclosed areas are proportional to the estimated production (*not* the standing crop) of material at each trophic level; production figures are in thousand millions of metric tons of organic matter per year. (From Ricker, 1969)

communities, at least, Ryther (1969) considers there may be five links in the food chain:

nanoplankton (small flagellates) — micro-zooplankton (herbivorous protozoa) — carnivorous crustacean zooplankton — larger carnivorous zooplankton (e.g. chaetognaths and euphausiids) — feeders on zooplankton (e.g. lantern-fish and saury) — fish eaters (e.g. squid and tuna).

But tuna, which feed, inter alia, on lantern-fishes, saury and euphausiids, may be put between trophic levels 5 and 6. Even so, such lengthy food

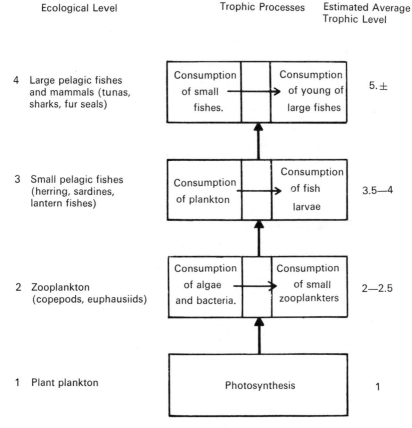

Ecological Level   Trophic Processes   Estimated Average Trophic Level

Fig. 98   Diagram illustrating the relation of ecological levels in the pelagic environment (not to scale). 'Cannibalism' (indicated by the horizontal arrow) within each ecological level increases the average distance of the members of that level from the primary plant food. For simplicity, the bacterial 'recycling' shown in Fig. 97 has been omitted. (From Ricker, 1969)

chains are much less productive than those in persistent upwelling areas, as off Peru and in antarctic seas. Peruvian anchovies feed on phytoplankton as well as small forms of zooplankton, while baleen whales in antarctic regions feed on krill (*Euphausia superba*), which of all euphausiids comes closest to being a complete herbivore. Anchovies may thus be given a trophic level of about 1·5: whales are near level 2. Ryther, who assumed that short food chains would be the more efficient, assigned *KE* values of 20 per cent to both upwelling areas, while the oceanic food chain was assessed at 10 per cent. He concluded that the annual primary production of oceanic areas was generally low and gave 50gmC/m²/year as an average figure. The corresponding figure for the Peruvian area was estimated as 300gmC/m²/year.

Such estimates are provisional and may well need revision. For instance, Cushing (1973b), after analysing data from the International Indian Ocean Expedition, concluded that the transfer of living material from primary to secondary trophic levels is three times as efficient in the open ocean as in upwelling areas (15 v. 5%). Perhaps there are less than five links in the oceanic food chain. If there are four rather than five, then the productivity of planktivorous fishes and piscivorous tuna will be ten times greater than estimated. There is still much uncertainty. As Rothschild (1972) argues, one could take a number of plausible alternative values for the trophic coefficients and arrive at very different conclusions.

Mesopelagic food webs

The open ocean, which extends over nine-tenths of the entire oceanic area, has a total productivity that is more or less equivalent to a plant assimilation of about 15 thousand million (15×10^9) tons of carbon per year. Investigations in the Sargasso Sea have led Riley (1970) to estimate that about 75 per cent of such production is consumed by zooplankton and heterotrophic organisms (particularly bacteria) in the uppermost part (ca. 300 metres) of the water column. Twenty per cent of the total production falls to the animals and heterotrophs of mesopelagic levels, the bathypelagic zone and the bottom. Mesopelagic metabolism down to and including the level of the oxygen minimum layer accounts for most of the deep-sea consumption of organic matter, which leaves less than 5 per cent to the bathypelagic and benthic organisms. Riley suggests that total amounts consumed in the bathypelagic and benthic zones are probably about equal. In the former, consumption is thinly (but unevenly) spread over a water column with a mean depth range of some 3,000 metres, whereas benthic consumption is concentrated into a narrow depth range in, on and just over the deep-sea floor. In temperate and subarctic regions of the Pacific, which are something like three times as productive (per unit area) as the waters of the Sargasso Sea, Bogdanov (1965) estimated that 10 per cent of the organic matter produced in the euphotic zone reaches depths below 3,000 metres, but in tropical and south temperate latitudes the corresponding figure is less than 5 per cent.

Clearly, there are certain outstanding questions. How is organic material transported from the euphotic zone to the depths? What are the metabolic requirements of deep-sea animals and how are they met? Concerning the first question, there is the gravitational settling of particulate matter (the 'perpetual snowfall') and the movements of animals, whether daily, seasonal or ontogenetic, up and down the water column.

Gravitational transport

When particulate material in sea water is collected on a membrane filter, or in a settling cylinder, microscopic scrutiny always reveals the presence of many particles of irregular shapes and sizes. These particles, known collectively as detritus, consist of organic and inorganic material. (Particulate material that is composed of detritus and minute planktonic organisms is called 'seston'.) Organic particles (the concentrations of which are expressed in amounts of particulate organic carbon—POC), are derived largely from metabolic waste (faecal materials etc.), the dissolution of pelagic organisms, and from dissolved organic matter (expressed as DOC) drawn out of solution to form organic aggregates. Concerning the first and third sources, Sieburth (1976), under a section entitled 'Fecal Fragments', writes:

'It is obvious that much of the amorphous and flake types of suspended debris occurring throughout the water column is the debris resulting from feeding. The increase in suspended organic debris following phytoplankton blooms led to the concept of organic aggregates in which large reserves of DOC formed flakes and amorphous particles through a foam-tower or bubble-scavenging mechanism or through film collapse. A re-evaluation of bubble-salvaging indicated that it could be a laboratory artifact. It has been shown that particle formation (organic aggregation) is mainly in the $< 4 \mu m$ range and is due to bacterial growth. This is in contrast to some 90% of the particles ($> 2 \mu m$), which are in the 8–44 μm size range.

'Not all feces or their fragments are consumed by coprophagy. Just as there is a threshold level for DOC utilization there is also a minimum threshold level for particulates below which filter-feeding organisms will not feed. This ensures a more or less steady state input of debris into the sea floor. This material as well as larger debris undergoes biodeterioration on the sea floor to form a secondary production for the development of benthic ecosystems.' (See also pp. 287–8).

The concentrations of both dissolved and particulate organic matter vary widely in the upper waters, but at depths below 200–500 metres there is relatively little change over most of the ocean. Deep DOC ranges from 0·5 to 0·8 mgC/litre, while the corresponding figure for particulate organic carbon is 3–10 μgC/litre (Menzel, 1974). Thus, DOC is hundreds of times the more plentiful, though nearly all of it seems to be unusable by living organisms. How nutritious, then, is particulate organic matter?

When, for instance, one looks in the house of an appendicularian, the gut of a copepod or the filtering-basket of a euphausiid, fragments of detritus are discernible. Deep-sea detritus, as histochemical tests reveal (see Gordon, 1970a), contains proteins and carbohydrates, some of which should be digestible. But laboratory tests have at best provided conflicting evidence that zooplankters, such as copepods, thrive on a detrital diet. This seems curious in view of Gordon's (1970b) finding that 20 to 25 per cent of deep organic detritus is hydrolysable by proteases. There is even more direct evidence from vertical profiles of the amounts of adenosine triphosphate (ATP) in the water column (Holm-Hansen and Booth, 1966; Holm-Hansen, 1970). Since ATP, the high-energy basis of metabolism, is maintained in fairly uniform concentrations in all living cells, but is soon destroyed after death, measurements of total ATP concentrations should indicate amounts of living substance. Indeed, analyses in the laboratory show that the quantity of ATP should be multiplied by 250 to give the cellular content of organic carbon (Holm-Hansen, 1970). Investigations in the eastern Pacific off California reveal that the ATP content (per litre), which is proportional to biomass, is very high in the euphotic zone but diminishes very rapidly between depths of 100 and 200 metres, below which there is a much more gradual decline to 3,000 metres. Analyses of samples taken between 500 and 1,000 metres showed that 3 per cent or less of the particulate matter was alive but above 100 metres the live fraction ranged from 17 to 79 per cent. After concurring with other investigators that most of the dissolved and detrital carbon in deep water is unusable by micro-organisms, Holm-Hansen (1970) suggests that a small fraction of this organic carbon turns over rapidly in support of microbial populations that may well be important as the first stage in oceanic food chains.

What are the first animal stages of mesopelagic and bathypelagic food chains? At such levels the micro-zooplankton seems to be represented largely by foraminiferans and radiolarians; the remnants of which may be common in the guts of copepods, mysids, euphausiids and so forth. Concerning the relatively large (> 2 mm) species of deep-sea zooplankton, the conclusions from Chindonova's (1959) analyses of stomach contents is that most of the 64 species examined, which included 15 out of 18 species of copepods, were carnivorous. But her results show also, together with Harding's (1974) more detailed studies, that some copepods feed at lower levels of the food chains. Harding examined deep-water copepods that were taken in closing nets fished in the Sargasso Sea (4,000–3,000 metres) and slope water off Nova Scotia (2,000–1,000 metres). The considerable range of food items that he found in about 80 species included mineral particles and detrital remains, phytoplankton (whole cells and remains of diatoms, dinoflagellates, coccoliths, pigmented cells and silicoflagellates), protozoans (ciliates, Radiolaria and Foraminifera), cysts and eggs, remains of cnidarians (nematocysts), copepods and euphausiids. He concludes that heterotrophic protists (phytoplankton and protozoans) are not only the main food of filter-feeding species of deep-sea copepods, but also form much of the diet of omnivorous and carnivorous species. In turn the heterotrophic forms '... must be supported by the dissolved and particulate organics, bacteria and possibly detrital remains of organisms present in deep waters'. Harding's idealized chart, which outlines the food relations of filter-feeding, omnivorous and carnivorous species of deep-sea copepods and possible energy pathways, is shown in Fig. 99. Further studies of this kind, coupled with investigations on the quantitative distribution and metabolic requirements of the main components in midwater food chains, should

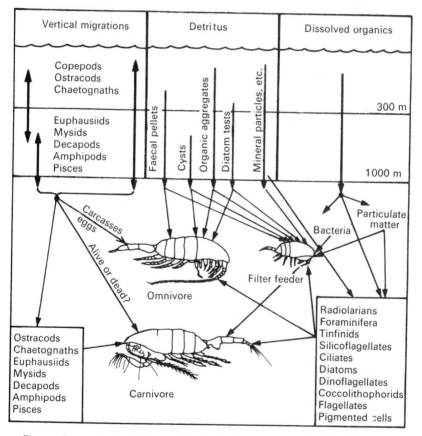

Fig. 99 Reconstruction of the possible and likely food sources of filter-feeding, omnivorous and carnivorous deep-sea copepods. Energy transfer from the productive surface waters is by way of vertical migrations, the fall of detrital material and the formation of usable organic substances in solution. (Redrawn from Harding, 1974)

resolve some of the present uncertainties. Concerning the metabolic aspect, measurements of the oxygen consumption of crustaceans and fishes from different midwater levels showed, inter alia, that species from about 1,000 metres respired at a tenth of the rate of shallow-water forms (Childress, 1971). Metabolic needs are evidently modest, but how are they met?

Active organismal transport

Organic material is actively transported to the depths during daily, seasonal and ontogenetic movements of pelagic animals (pp. 234–52). Though the daily vertical migrations of aquatic animals was first recorded over 150 years ago, it is only now that there is enough direct evidence to show that migrants climb to more productive parts of the water column in search of food. Towards nightfall, over all main regions of the ocean, myriads of planktonic and nektonic animals make their ascent, and, after spending some part of the night at higher grazing or hunting grounds, descend before daybreak to their deeper, daytime residence. In the open ocean these migrations are virtually confined to the epipelagic and mesopelagic zones. The deepest records concern daily migrations (of arrow-worms, mysid shrimps and prawns) that seem to reach a depth of 1,700 metres.

Every day, then, vast quantities of planktonic and nektonic forms descend from their feeding grounds, bearing means for their maintenance and growth. Vertical migrations '... certainly result in far greater mobilization of biological material than any other animal migration: using an average figure for plankton biomass of 25 gm/m^2 to a depth of 100 metres as a reasonable estimate for all oceans and assuming that diel migration adds 10 per cent to this at night as a minimal figure for an average daytime residence depth 250 metres deeper, then one arrives at vertical translocation of 25 tons/km^2/day over a distance of 250 metres'. Longhurst (1976) adds that 'such calculations are not to be taken very seriously', but his minimal figure is likely to be well exceeded in certain groups. For instance, Baker (1970) estimates that about two-thirds by weight of adolescent and adult euphausiid populations are involved in vertical migrations between 50 and 960 metres off the Canary Islands. In taxonomic terms, Omori (1974) states that daily vertical migration seems to occur in nearly all epipelagic and mesopelagic prawns. The same is true of mesopelagic lantern-fishes.

Most kinds of deep-sea migrators are cnidarians (especially siphonophores), crustaceans (copepods, ostracods, amphipods, euphausiids and prawns), arrow-worms, cephalopods and fishes. Relevant investigations on the vertical distribution of such migrators reveals that some species make extensive migration of several hundred metres, usually ending in the productive surface waters, while others are partial migrators. Members of a third group stay more or less at the same level throughout the day, but there are instances of reverse migrations. Such categories of behaviour may also occur in individual species. At certain times some of the adults migrate while others stay still and advanced larval stages and adolescents may be partial migrators. Concerning these and other matters, there is much to be found in the results of the *Discovery* Sond Cruise (1965) to an oceanic area off Fuerteventura, Canary Islands (Badcock, 1970; Baker, 1970; Foxton, 1970a & b; Roe, 1972a, b, c & d; Angel and Fasham, 1974; Pugh, 1974; and Thurston, 1976a & b; and also Marshall, 1960; the papers by Clarke and Lu—or Lu and Clarke; Omori, 1974; Roper and Young, 1975; and Badcock and Merrett, 1976).

During vertical migrations animals are bound to experience changes in hydrostatic pressure. Over much of the ocean, particularly in the subtropical and tropical belts, they are exposed to considerable changes in temperature and, if they live in an oxygen minimum layer, to wide ranges of oxygen tension. Moreover, in seeking biological changes, they experience others, notably the attacks of predators. But their commuting seems to be timed and regulated so as to keep the light around them at a constant intensity.

There is diverse evidence that light has a controlling and dominant influence on vertical migration. In the open ocean, daily migrations are virtually confined to the euphotic and twilight zones, and the ascent of the migrators, at least, is closely related to nightfall. Investigations of the daily migrations of deep scattering layers that are recorded by echosounders, and the main sound scatterers are fishes with a swimbladder, particularly lantern-fishes, supports earlier conclusions that migrators move so as to keep the ambient light field at a constant level: they are 'imprinted' on a particular isolume. During the daytime a given DSL will seek a higher level if the sky is darkened by clouds or a solar eclipse. Artificial lights will delay a dusk ascent or drive down a DSL by night. It would seem, then, that animals will only be able to shadow an isolume if their eyes are sensitive enough to perceive relatively small changes in light intensity. Mesopelagic fishes have very sensitive eyes, and so have cephalopods and decapod crustaceans. But what of migrators with much simpler eyes? Concerning these and other aspects of light and vertical migrations, Cushing (1951, 1973a), Clarke (1970), Kampa (1970) and Blaxter (1976) may be consulted.

The biological advantages of daily vertical migrations

Modes of pelagic life are other than those elsewhere. The contrasts have been illuminated by Elton (1966), who wrote '...plankton has no cover structure in which predator hunts for prey, though in most other communities it seems quite likely that without such cover the community would actually collapse before long. I believe that for this reason plankton is profoundly different in its organization from terrestrial and most aquatic communities. Two features can be noted, though the whole subject merits much deeper comparative attention. For one thing, plankton has rather few species of animals living together; for another, its survival seems to necessitate very intricate vertical and other movement.' Perhaps Elton was thinking of temperate or cold seas when writing about the 'rather few species' in the zooplankton. A qualification is needed to cover the much greater diversity of plankton (and nekton) in tropical regions, but this simply means that the 'whole subject' needs even 'deeper comparative attention'.

The biological significance of daily vertical migrations is not to be considered in isolation. An outstanding question is this: Why have some species taken to such vertical commuting seemingly to make a living, while other species, which may be near neighbours of the migrators, habitually stay in the depths? It would seem that one needs the concepts of ecological niche and competitive exclusion. A niche comprises the space and resources involved in a species' way of life. 'Niche breadth is simply the sum total of the variety of different resources exploited by an organismic unit' (Pianka, 1976). Competitive exclusion means that no two species can occupy the very same niche.

Investigations during the *Discovery* Sond Cruise (1965), which centred on the vertical distribution and migrations of deep-sea animals in the upper 950 metres of a restricted oceanic area, close to the island of Fuerteventura (Canary Islands), are again particularly relevant. Vertical patterns of euphausiids make a suggestive basis for comparison with other taxa, especially in view of Baker's (1970) conclusion that in principally non-migrant genera (e.g. *Stylocheiron*) there is a decided tendency for congeneric species to occupy separate strata in the water column, whereas migrants, such as *Euphausia* spp., have overlapping, or even coincident, vertical ranges (see Fig. 100). Recent trials in the Western North Atlantic with a multiple opening/closing net (see Wiebe et al., 1976) have yielded samples that give strong support to Baker's generalization (see also Youngbluth, 1975).

Now it is evident from Roger's (1973) work in the equatorial Pacific that migrating euphausiids feed principally and most actively by night (20.00 to 06.00 hrs). If the same is true of the Canaries region, then it seems natural that migrating species should occupy overlapping or coincident vertical ranges by day. They are hardly competing for food, but merely for non-trophic living space, and evidently the different species have evolved compatible arrangements. If they feed at discrete subsurface levels by night then interspecific competition will again be minimized.

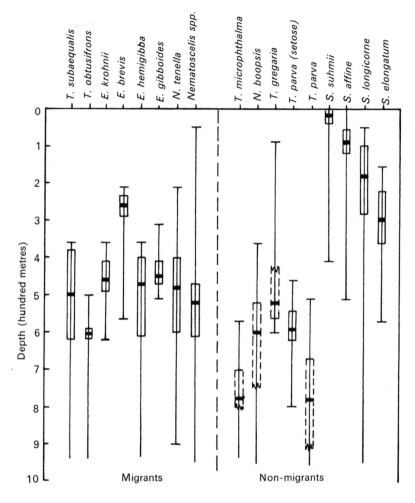

Fig. 100 The daytime distribution of migrant and non-migrant species of euphausiids (adults) off the Canary Islands (Fuerteventura). T, *Thysanopoda*; E, *Euphausia*; N, *Nematobrachion*; S, *Stylocheiron*. The depths above which 25% and 75% occurred are indicated by the top and bottom of the rectangle and 50% by the solid black bar. The vertical centre line indicates the overall range for each species. Where the lower limit of the range was not reached or for other reasons sampling was thought to be inadequate, the rectangle is pecked. (From Baker, 1970)

To obtain evidence of such closer-packed nocturnal segregation requires a carefully planned series of hauls, but there is some indication from Roger's investigations. For instance, in the genus *Thysanopoda*, two species, *tricuspidata* and *equalis*, migrate to levels above the thermocline (depth ca. 160 metres), while the populations of three others, *pectinata*, *monacantha* and *orientalis*, though also found above the thermocline, are concentrated mainly between 160 and 300 metres at night. But even if *tricuspidata* and *aequalis* feed at the very same levels, they will not be competing, for the first (mean length 20 mm) is largely a herbivore, while the second (mean length 15 mm) is a carnivore. The three other species reach much the same size (30–40 mm) and all are essentially carnivorous, and, as Brinton (1962) found, have nearly the same (wide) distributional ranges. It seems reasonable that two or more related species will avoid competition if differences are evolved in their size (see later), diet, feeding places (such as by diel migrations) and feeding periods. There is no evidence from Roger's (1973) study that the three above *Thysanopoda* have segregated feeding periods, but there is plenty of space, say between 100 and 300 metres, for the occupation of distinct feeding levels. Further investigation is needed.

In the Canaries region non-migrant euphausiids seem to be separated by a middle mesopelagic association of migrant species into an upper (epipelagic–upper mesopelagic) group of segregated species of *Stylocheiron* and a lower (mainly lower mesopelagic) group consisting of five species belonging to three genera (see Fig. 100). Concerning the last group, Baker (1970) states that *Thysanoessa gregaria* and the typical form of *T. parva* occupy different depth ranges, with the setose form of *T. parva* lying in between. There is also evidence from Rogers' (1973) observations that non-migrant euphausiids are less closely tied to nocturnal feeding periods than are migrants. Indeed, *Stylocheiron* spp., most of which range from epipelagic to upper mesopelagic levels, are daytime feeders (Roger and Grandperrin, 1976).

Migrating and non-migrating euphausiids differ trenchantly in structural plan (Brinton, 1967). The migrators, such as species of *Euphausia* and *Thysanopoda*, are generalized forms with spherical, undivided eyes and rather uniform, coarsely setose, comb-like thoracic limbs, designed to sweep or strain food from the sea. These species tend to be omnivorous, but at one end of the trophic scale there are mainly herbivorous types, at the other preferential carnivores. All the non-migrators belong to more specialized genera that have a conspicuously elongated, anterior pair of thoracic legs, behind which is a much-reduced thoracic basket. In the most diverse genus, *Stylocheiron*, each eye is formed of a distinct upper and lower lobe (see also pp. 400–2), and the elongated legs are more or less chelate. These euphausiids have definite carnivorous tendencies (no doubt catching their prey, such as copepods, with the aid of their long chelate limbs) and they are generally much less abundant than the predominantly herbivorous kinds of migrators.

Migrating and non-migrating euphausiids are also generalists and specialists in other respects. Consider a subtropical or tropical species that migrates between 400 and 50 metres. Besides the possession of an adaptive organization, particularly biochemical, to tolerate considerable changes in pressure (ca. 40 to 5 atmospheres), temperature ($10C°$ or even more) and oxygen tension (if the migrator lives in an oxygen minimum layer), the species must also have the right set of physiological and behavioural attributes to spend a divided life in two quite distinct environments. It is surely a generalist. The specialists that stay at one level display no such ranging adaptability.

As regards the population structure in the water column, the generalist migrators are certainly eurytopic, the specialist stenotopic.* Are there the same differences in the horizontal plane? If the two large genera, *Euphausia* and *Stylocheiron*, are compared, and both consist predominantly of epipelagic to upper mesopelagic species, it emerges that species of the former cover a much wider area. *Euphausia* is represented in the subantarctic, antarctic and subarctic zones and in the subtropical and tropical belts while *Stylocheiron* is essentially a warm-water genus (Brinton, 1962). Thus, evolutionary potentiality in the genus *Euphausia* has been such that species have emerged that are not only adapted to the wide range of physical and biological conditions associated with diel vertical migrations in the warm ocean, but also, inter alia, to the widely ranging seasonal conditions, especially biological, in colder regions. One is reminded of the recent work of Jackson (1974), who finds that the infaunal bivalves adapted to the rigours of life in depths of 1 metre or less have distinctly wider geographic ranges than other infaunal species (not deep-sea) restricted to depths greater than 1 metre. Again, the set of species with wider environmental tolerances has the greater geographic representation.

Sergestid prawns, some of which prey on euphausiids, are active migrators. Off Fuerteventura, Foxton (1970b) found that species of the subgenus *Sergestes*, which are semi-transparent and bear internal, ventral light organs, live at levels above 750 metres, whereas *Sergia* spp., uniformly red and with external light organs (if present), live below this level. Here, in a daytime descent from 400 metres and considering only the migrators, there are three pairs of species that each live together in concentrations between the same levels of the water column: *Sergestes sargassi* and *S. armatus* (400–650 metres); *S. vigilax* and *S. pectinatus* (600–700 metres) and *Sergia robustus* and *S. splendens* (750–950 metres). There is also *Sergestes corniculum* (650–800 metres). Partly or entirely below the last pair,

---

* Gr. *eury*, broad, *topos*, place: Gr. *steno*, narrow, *topos*, place.

there are two non-migrating bathypelagic species, *Sergia japonicus* and *S. tenuiremis*.

For investigations on both the vertical pattern and food of sergestids, we turn to the Ocean Acre project off Bermuda (Donaldson, 1975). During the day there is again a step-down occupation of the water column with groups of two or more migrant species concentrated between the same levels: *Sergestes cornutus* and *S. sargassi* (450–600 metres); *S. pectinatus*, *S. vigilax* and *S. armatus* (500–700 metres); *S. atlanticus*, *S. corniculum* and *S. grandis* (600–850 metres) and *Sergia robustus* and *S. splendens* (800–1,100 metres). Again there are the same bathypelagic species, *Sergia japonicus* and *S. tenuiremis*, but in the Bermuda area they appear to be partial migrators (see Fig. 101). During both day and night the only species that occurred together were *Sergestes sargassi*, *S. pectinatus* and *S. vigilax*, but dissimilar features of their third maxillipeds, a key pair of limbs involved in feeding, led Donaldson to suggest that they have different diets.

Donaldson found that most species took a variety of food organsims, but three species (*grandis*, *corniculum* and *splendens*) contained euphausiids more often than other kinds of prey. Fish were most frequently found in *Sergia robustus* and *grandis*. The bathypelagic species, *Sergia japonicus*, which has a flaccid body with reduced musculature, appears to be a detritus-feeder. In most species food items were found most often at night, but no species was exclusively a nocturnal feeder. On the other side of the Atlantic, however, Foxton and Roe (1974), who analysed the food of six sergestid and three oplophorid prawns, concluded that intensive night-time feeding seemed to be the rule in all nine species.

Once more, partitioning of resources and the water column by a set of migrating species seems to be such as to reduce interspecific competition. During the daytime in particular, groups of two or more species may coexist in relatively narrow strata, which is quite unlike the mutually exclusive stratification in the non-migrating euphausiid, *Stylocheiron*.

How far do copepods show the same trends? They are the prey of innumerable marine animals from cnidarians to whales (including euphausiids and sergestids) and at any one point in the warmer parts of the ocean, many species occupy the water column from top to bottom. In the Fuerteventura area 212 species of calanoid copepods were identified in hauls made between 960 and 40 metres (Roe, 1972a–d). The bulk of the population, which consisted largely of immature stages, was centred at 500 metres by both day and night. Of the adult forms, some were conspicuous migrators but most species showed restricted migratory tendencies or were non-migrators.

Considering first the non-migrators, it is evident that related species do not necessarily follow the discretely stratified (*Stylocheiron*) pattern.

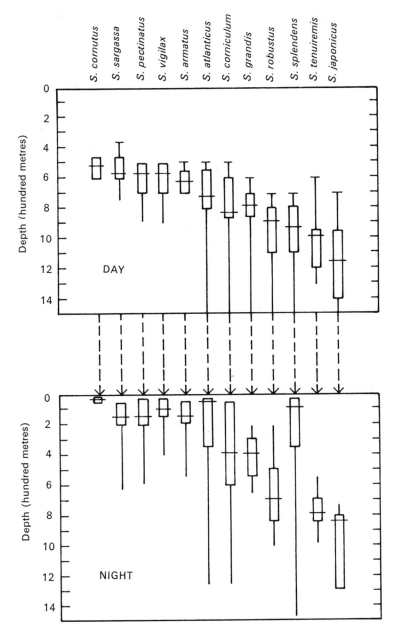

Fig. 101 Daily vertical migrations of species of the deep-sea prawn genus *Sergestes*. The vertical distribution by night is shown immediately below the daytime depth range. Each rectangle represents at least 75% of the catch. The horizontal lines indicate the depths above which at least 50% of the prawns were caught. The vertical lines show the entire depth range. (From Donaldson, 1975)

For instance, three pairs of species: *Calanus helgolandicus* (2·63–3·31) and *C. carinatus* (2·05–2·81); *Rhincalanus nasutus* (4·26–5·60) and *R. cornutus* (2·81–4·16), and *Spinocalanus magnus* (2·28–2·96) and *S. spinosus* (1·37–1·98) were concentrated between much the same depth ranges (400–700 metres for the first two pairs, 600–900 metres for the third), throughout the entire day. After the name of each species in parentheses is the length range of the females (in millimetres), and it will be seen that in each pair one species is decidedly the larger. The significance of size differences in reducing competition between coexisting, congeneric species of animals has been well demonstrated by Hutchinson (1959), who found that the ratios of mouthpart sizes in relevant species of insects, birds and mammals ranged from 1·2 to 1·4. As Wilson (1975) observes, larger predators are able to utilize food sizes that are not available to smaller predators, while the reverse is much less true. Within a species competition for food may be reduced by sexual differences in size (see also p. 261), but in most oceanic copepods the males are slightly the smaller sex. On the other hand, the adult males of some species take no food (Harding, 1974). There is every indication, then, that Hutchinson's rule also applies to certain coexisting, congeneric species of oceanic copepods.*

Species of *Undeuchaeta* and *Pleuromamma* are among the conspicuous migrators. By day *Undeuchaeta major* (4·64–5·28) and *U. plumosa* (3·34–4·41) were concentrated between 450 and 900 metres: by night the migrators had reached levels above 250 metres (Fig. 102). But if most of their food is taken at night there need be little competition, for once more, the size difference between the females (in parenthesis) is considerable. Two common species of *Pleuromamma*, *xiphias* (4·18–4·96) and *abdominalis* (2·81–3·36), had overlapping distributions by day, the first centred at 570 metres, the second at 400 metres. By night most of the extensive migrators were clustered around levels of 250 and 100 metres, but yet again, size differences are such as to reduce competition (Fig. 102). Moreover, in the uppermost waters Roe (1972c) has evidence to suggest that *xiphias* was most abundant between 50 and 25 metres, *abdominalis* between 10 and 0 metres. Lastly, three other species, *borealis* (1·67–1·90), *pisekei* (1·75–1·98) and *gracilis* (1·67–1·98), are centred at 350, 300 and 200 metres, respectively, during the day; by night the first is concentrated at 250 and 100 metres, the other two at 50 metres. Nocturnal competition between *pisekei* and *gracilis* is possible. They are of much the same size and all three species, as Roe (1972c) observes, are closely related. He adds also that samples from the uppermost 50 metres showed that *borealis*

---

* However, discrete stratification does occur in related non-migrating species of copepods, as in the four species of *Pareuchaeta* in the northern region of the North Pacific (Sekiguchi, 1975).

was scarce and *pisekei* more abundant than *gracilis*. Further investigation should be of interest.

The evolutionary interplay of mesopelagic life seems to have been such that potential competitors in the zooplankton and micronekton have come to be more or less segregated through mutually exclusive combinations of spatial, temporal and biological attributes. The same is true of the productive upper waters, as may be seen in Zalkina's (1970) paper on the vertical distribution of cyclopoid copepods in the equatorial western Pacific. During a 24-hour station a series of hauls was made between

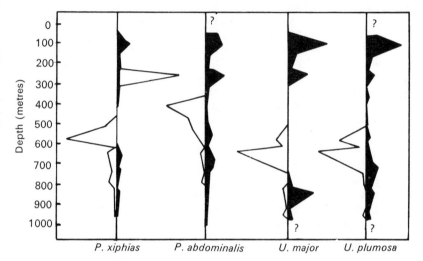

Fig. 102   Vertical distribution of four deep-sea copepods, *Pleuromamma xiphias*, *P. abdominalis*, *Undeuchaeta major* and *U. plumosa*, in the uppermost 1,000 metres off the Canary Islands (Fuerteventura). The distribution by day is shown in white, that by night in black. The numbers plotted on either side of the vertical lines represent the numbers of a species at each depth expressed as a percentage of the total number of that species in the water column. (Redrawn from Roe, 1972)

the surface and 300 metres. The catches showed that the species of Oithonidae, which are non-migrators and concentrated between 75 and 150 metres, have little overlap in the water column (except during part of the day) with those of the families Oncaeidae and Corycaeidae. It is interesting also to compare the changing vertical dispositions of the two dominant species, *Oithona plumifera* and *Oncaea media*. Again, there is a little overlap, except from 10.00 to 18.00 hours, when the second (migrant) species is keeping its daytime station. Zalkina found that both migrating and non-migrating species tended to maintain a step-like distribution. Some species differed in the times of their ascent and duration of stay at higher feeding levels, and Zalkina concluded, 'This peculiarity of disposition tends to lessen the intensity of food competition between species with similar nutritional habits.' She also quotes her colleagues, particularly M. E. Vinogradov (1968), who also found that dominant species of oceanic zooplankton with similar food requirements occupy discrete strata in the water column, and that such segregation is particularly marked' in regions of low productivity.

The Californian sardine (*Sardinops caerulea*) is an active fish, and during its virtually restless life, respiration consumes most (82·99 per cent) of the calorific content of its hard-won food in the zooplankton. Reproductive processes take a mere 1 per cent of assimilated calories (Lasker, 1970). The lives of mesopelagic fishes are not so strenuous, but whether they migrate or not, they are highly organized predators, needing the means to run elaborate tissue systems, particularly the substantial muscles of their jaws and motor complex. How do they share available resources?

Most kinds of mesopelagic fishes are lantern-fishes (Myctophidae, > 250 species), alepisauroids (> 100 spp.) or stomiatoids (ca. 300 spp.) (see figs. in Chapter 4). Of all the twelve families that form these groups, daily vertical migrations are by far the most prevalent in the lantern-fishes. Indeed, except for some of the deepest-living kinds, such as certain *Lampadena* species and those of the genus *Taaningichthys*, well-planned investigations reveal that lantern-fishes are inveterate migrators. Judged by their biomass, ubiquity and species diversity, they are also the most successful family of midwater fishes.

In relative length and other features of their jaws, which bear rows of small conical teeth, there are three main groups; small-jawed (e.g. *Myctophum*, *Diogenichthys*), medium-jawed (e.g. *Gonichthys*, *Ceratoscopelus*) and long-jawed (e.g. *Stenobrachius* and *Lampadena*). The genus *Diaphus* is unusual in having representatives in all three groups (Paxton, 1972). One species, *Diaphus taaningi*, a rather small lantern-fish with medium-sized jaws, is known from the Gulf of Mexico, the Caribbean Sea and off West Africa. In the Caribbean it lives in the Cariaco Basin, a depression in the continental shelf off Venezuela with a shallow northern sill that restricts the horizontal exchange of water between Basin and Sea. Due to this restricted circulation and other factors the Cariaco Basin is anoxic (without oxygen) below a depth of about 300 metres (Richards, 1975). But *D. taaningi* lives above the anoxic waters. Baird and his colleagues (1972) found that it migrates from a daytime depth of about 250 metres to appear in the upper 50 metres at night. Nearly all of their fish ranged from 38 to 45 mm in standard length, and all were feeding largely on herbivorous zooplankters, particularly copepods and larvaceans, which formed about 70 per cent of the total prey biomass. Analyses suggested that most of the prey (density about

150 items/m³) was taken in the early evening, leaving the rest of the night for digestion. Fish taken before the evening ascent were essentially empty. In their discussion the authors argue that *Diaphus taaningi* and certain other midwater fishes, such as the hatchet-fish, *Polyipnus*, are bound to near-neritic waters where there is an abundance of small zooplankton. 'As prey density and distribution at a given time are thought to determine the amount of food a predator can obtain as well as the cost to the species in energy of pursuit (Schoener, 1969; Brocksen et al., 1970) it would appear that small prey in high density in the night time depth zone may well be a niche requirement for *D. taaningi*.'

Two other lantern-fishes, *Stenobrachius leucopsarus* and *Triphoturus mexicanus*, which are long-jawed and larger than *D. taaningi* (both grow to more than 70 mm in length), were the dominant midwater fishes taken during a cruise off California. Collard (1970), who found that both preyed mainly on small and medium-sized crustaceans, particularly copepods and euphausiids, concluded that diets were diverse, with no sign of selection. But some mesopelagic fishes feed selectively, as Merrett and Roe (1974) found after analyses of fish caught by *Discovery* in a series of eleven hauls over a 24-hour period between 230 and 266 metres in the eastern North Atlantic (30°N, 23°W). By comparing the composition of the zooplankton (in the net) with fish prey, the authors were able to determine whether or not food items had been selected by some of the more abundant fishes. They looked also for periodicities in feeding. The small stomiatoid *Valenciennellus tripunctulatus*, virtually a non-migrator, fed mainly by day on selected calanoid copepods. Two partial migrators, the hatchet-fishes, *Argyropelecus hemigymnus* and *A. aculeatus*, which seemed to feed mainly at dusk, differed in diet. The first fed randomly on mixed copepods and ostracods, the second selectively on ostracods (*Conchoecia curta* adults). There were parallel results from two fully migrant lantern-fishes, *Lobianchia dofleini* (day range 300–500 metres: night range 25–200 metres) and *Lampanyctus cuprarius* (day range 700–900 metres: night range 25–200 metres). Both fed at night, the first randomly on copepods, the second on selected amphipods, particularly the males of *Phrosina semilunata*. The viper-fish *Chauliodus danae*, an avid predator, seemed to feed at any time and randomly (on euphausiids and fish). Merrett and Roe concluded that vertical migrations and feeding habits are closely connected, whether by the migration of prey species away from levels occupied by their predators (e.g. the species of the copepod *Pleuromamma* away from *Valenciennellus tripunctulatus*, so that the latter must feed by day) or by the upward migration of the predators to the headquarters of their prey (e.g. *Lampanyctus cuprarius* in search of *Phrosina semilunata*).

At the same station another series of nets was fished down to 2,000 metres, and from these catches Badcock and Merrett (1976) have presented an overall analysis of vertical patterns in the midwater fish fauna. Considering first the lantern-fishes, their results complement those of Badcock (1970) from the Fuerteventura area off the Canary Islands. All species, or at least all well-represented species, were diel migrators. By day the vertical ranges of silvery species (e.g. *Lobianchia*) largely overlap or are coincident between levels of 300 and 700 metres, while dark species (e.g. *Lampanyctus*) have similar range correspondences below 700 metres. Silvery lantern-fishes and transparent species of sergestid prawns have much the same depth range: so do dark lantern-fishes and red sergestid prawns (Foxton, 1970). Indeed, lantern-fishes and migrating species of sergestid prawns and euphausiids coexist at similar or widely overlapping depth ranges by day. They also tend to take most of their food by night. There is also evidence of dietary differences amongst them (see p. 245).

Over most regions of the ocean between the subarctic and subantarctic zones the most numerous non-migrators at mesopelagic levels are the more or less transparent species of *Cyclothone*, stomiatoid fishes known as bristle-mouths. Dark *Cyclothone* live deeper and like other bathypelagic animals, which are considered later (pp. 253–8), are also non-migrators. The transparency of mesopelagic bristle-mouths is interrupted by black markings, especially at internal levels, but in *Cyclothone pseudopallida* there is also an integumentary dusting of small melanophores over much of the body. When adult they have a gas-filled swimbladder, which regresses and becomes fat-invested in their dark relatives. Further differences between the two groups are considered elsewhere (pp. 447–9).

At mesopelagic levels in the Canary Islands region there are two 'transparent' *Cyclothone*, *braueri* and *pseudopallida* (Plate II). There is also a third species *pallida*, which ranges from lower mesopelagic to bathypelagic reaches. The first is concentrated between 400 and 600 metres, the second between 500 and 800 metres and the third between 600 and 800 metres. There is thus relatively little overlap between *braueri* and the other two species and *braueri* is not only the dominant species (catch numbers 6,479 v. 222 and 112) but also the smallest (maximum standard length 38 mm v. 45 mm and 63 mm). Competition between *braueri* and the deeper-living pair is evidently slight and is reduced also between mature males and females of *braueri*, to consider only the dominant species. As in other *Cyclothone*, the males are not only the smaller sex (maximum length 25 mm v. 33 mm) but nearly all live at higher levels (400–500 metres) than the maturing to ripe females (500–600 metres) (Badcock and Merrett, 1976). Apart from reducing intraspecific competition, the segregation of the sexes may well facilitate eventual sexual congress (pp. 404–7). Competition between mature individuals of *pseudopallida* and *pallida* is likely to be modulated by their difference in size (see above), but other information is needed, particularly on their diet and trophic periodicity. The concentrations of these two species also overlap with the upper part of the depth range of *Cyclothone microdon* (500–2,000 metres), but this is made largely of young individuals (<25 mm), which are not likely to compete much with mature fishes of the other two species.

Lastly it will be recalled that Badcock and Merrett's (1976) paper is based on sampling at 30°N, 23°W, about 600 miles west of Fuerteventura, Canary Islands. At this station *Cyclothone microdon* is a dominant species, but off Fuerteventura it was not recorded (Badcock, 1970). Faunal composition in the water column changes from place to place but there is still the interplay of non-migrating species to reduce competition. Off Fuerteventura the relations between *C. braueri*, *pseudopallida* and *pallida* are rather like those at the westerly station, though there is less overlap between the second and third species, and *microdon* is replaced as it were, by two other dark-coloured bathypelagic species, *C. acclinidens* and *livida* (see Badcock, 1970, Tables I and II).

Seven of the nine dominant (and migratory) species of lantern-fishes in the Bermuda area are small or diminutive (20–50 mm) (Gibbs et al., 1971), about the same size as the mesopelagic species of *Cyclothone*. Like *Diaphus taaningi* (p. 244), these lantern-fishes must be restricted by their relatively small gapes to small prey organisms. If any one of the seven is compared to a *Cyclothone* of the same length, the latter will be seen to have a considerably wider gape, for the jaws are about four-fifths the length of the head, which is about a fifth to a quarter of the fishes' standard length. The *Cyclothone* should be able to take a wider size range of prey items than can a diminutive lantern-fish. But the latter migrates up to epipelagic levels or just below, where there is likely to be a greater abundance of small prey, particularly of herbivorous copepods, both young and adult. To make a living at deep-sea levels, where, with increasing depth, there is a more or less exponential decline in the biomass of suitable food, it would seem that a non-migratory fish should be adapted to secure the widest possible size range of food organisms. Indeed, investigations off southern California showed that *Cyclothone acclinidens*, which lives below the transparent species, *C. signata*, at depths greater than 400 metres, takes a wider size range of prey, from copepods to arrow-worms and amphipods. This is understandable, considering that the deeper-living species bears the larger jaws, due both to its greater size (maximum standard length 50 mm v. 33 mm) and to its relatively larger head (DeWitt, 1972; DeWitt and Cailliet, 1972). Similarly, Ebeling and Cailliet (1974) found that the deepest-living and non-migratory fishes called big scales (Melamphaidae) not only have larger jaws than shallower-living species, but also develop a gill raker system fine enough to retain and concentrate food items as small as 1 mm or so.

The larger a fish, the less energy it needs per unit of body weight. Even so, a large fish cannot exist in a food-poor environment: there are no deep-sea tunas and billfishes. Some of the largest mesopelagic fishes are alepisauroids, but unlike their relatives the lantern-fishes, they are nearly all non-migratory. The lancet-fish *Alepisaurus ferox* grows to about 2 metres, but it is not a heavy fish. A five-foot individual weighed about 4 pounds, which is not only indicative of its long compressed body but especially of its lightly ossified skeleton and sparely developed motor muscles (Marshall, 1955). It is above all due to the reduction of skeletal and muscular systems that deep-sea midwater fishes nearly attain neutral buoyancy. Thus, as well as being economically made, they need little energy to keep to a preferred level in the water column (Denton and Marshall, 1958). The barracudinas (Paralepididae), which range in length from a few to over 50 cm, have both the look and economical build of their relatives, the lancet-fish: so have other alepisauroids.

These fishes have also the jaws, swallow and maw to take very large meals. An individual of one species, *Anotopterus pharao*, which was swallowed by a whale captured in the Ross Sea, Antarctica, and measured nearly 75 cm, contained two barracudinas (*Notolepis coatsi*) of 18 and 27 cm in length. The barracudinas were bloated with krill, *Euphausia superba* (Marshall, 1955). More recently, Wassersug and Johnson (1976) have shown that alepisauroids have a sac-like stomach that is both large and greatly distensible. The intestine is simple and straight. Species of all six families swallow intact prey (e.g. lantern-fishes) that are at least one-third to one-quarter of their length. As the two authors say, in energetic terms midwater predators in the deep sea must not be much larger than their prey. But given the above and other adaptations (see Marshall, 1955), alepisauroids can afford, as it were, to be non-migrators.

But there are non-migratory, tiny-mouthed mesopelagic fishes (e.g. *Opisthoproctus* and *Rhynchohyalus*). Such argentinoids have large upturned tubular eyes with parallel optical axes, which, together may well provide extra sensitivity. Even more important, such eyes presumably afford much-needed accuracy in pin-pointing small prey that is to be sieved in tiny jaws (see also p. 388). Hatchet-fishes of the genus *Argyropelecus*, which at most are partial migrators, have similar eyes but larger jaws. Even so, accurate prey taking is still needed, particularly when the diet is selected, as in *A. aculeatus* (p. 245). Where prey of a suitable size (and kind) may be sparse, chances of nourishment should not be missed. And there is a further aspect. The hatchet-fish *Argyropelecus affinis* and the alepisauroid *Scopelarchus analis*, a voracious fish, also with tubular eyes, have yellow lenses. The function of such lenses may well be to enable their owners to 'see through' the bioluminescent camouflage of various midwater animals (Somiya, 1976; Muntz, 1976; and p. 388). Meals (and enemies) may be seen that might be missed by predators without special filters.

More evidence could be adduced to support the idea that diverse mesopelagic fishes have each evolved a unique and special organization to
enable them to stay in the depths and make a living after their metamorphosis to adult form. The same is true of non-migratory euphausiids (pp.
238–9). Migrators are generalists, adaptable to wide ranges of physical
and biological conditions not only vertically but also in the horizontal
plane (p. 239). Is this also true of fishes? For instance, one can compare
the distribution of the migratory lantern-fishes with their non-migratory relatives, the alepisauroids. To take but one criterion of extensive distribution, about 15 species of lantern-fishes are known from the antarctic
zone but only three or four alepisauroids (see Andriashev, 1965).

Daily vertical migrations, evidently in search of food, are thus ingrained in many kinds of epipelagic and mesopelagic animals. One advantage is that during their resting, non-feeding periods, potentially competing species can reside together between much the same depth levels
(pp. 236–44). Indeed, the entire nexus of vertical migrations would seem
to be fundamental to species packing and community structure in the
water column. Both these aspects are well illustrated by fishes, particularly as they are evidently the outstanding components of the deep
scattering layers (DSLs) (sonic scattering layers) recorded on ships' echo-
sounders (see Backus and Craddock, 1977, and Hersey and Backus,
1962). The prominence of fishes as acoustic scatterers is due largely to
their gas-filled swimbladder, which is present in most lantern-fishes and
other mesopelagic forms, notably in the smaller stomiatoid fishes
(Marshall, 1951, 1960 and 1970). Some scattering layers undergo daily
vertical migrations, but others are more or less stationary. In the present
context a ranging survey of DSLs in the North Atlantic by Haigh (1970)
will be considered. Records of a 10 kHz echo-sounder from 1963 to 1967
and between latitudes 10° and 68°N revealed five mesopelagic layers
during the day (given the letters A B C D E). The first three (ranges of
maximum mid-layer depths; A 135–170 fms, B 140–245 fms, C 215–280
fms) were purely migratory and found in latitudes south of 49°N. Layer
D, which was partially migratory and attained a maximum depth of
about 330 fms, was the most widespread, being recorded at all latitudes.
Layer E was mainly migratory and recorded only in restricted areas in latitudes south of Iceland (the maximum depth ranged from 300 fms near
Iceland to nearly 500 fms in more southerly waters). Thus, north of
49°N there was one main layer D joined locally by layer E. South of this
latitude there were up to five layers and over wide areas three occurred
together. The conjecture is that the number of DSLs is a reflection of
species diversity, which is much greater under the armer waters south of
ca. 40°N than in the colder waters to the north. Where there are more
species there is need for more statified species-packing in the water
column. Indeed, considering, for instance, that there are over 200 species

of mid-water fishes off Bermuda (most with a swimbladder) it is remarkable that there are so few deep scattering layers. That there are so few would seem to be related to the phenomenon of daily vertical migrations, whereby sets of potentially competing species can live together between the same depth ranges when they are resting by day. During the night, all but the non-migratory and partially migratory layers are in the upper 150 metres, where broad and prominent sound scattering is recorded.

The biomass and number of species in a DSL may be greater than those outside (see Dunlap, 1970) but do the sound scatterers and associated species form communities? Off Santa Barbara, southern California, factor analyses of the catches of deep midwater animals revealed a complex of four resident communities, which interact with a number of transitory, presumably less stable, groups (Ebeling et al., 1970). On a larger scale there are the investigations of Backus and his colleagues (1970) in the Atlantic. Daylight hauls were generally made at the depth of sound scattering maxima, while night hauls were set (by echo-sounder and bathythermograph) to sample concentrations of migrating fishes at depths above 300 metres. Though largely concerned to sample layers of strong sound scattering, the investigators are readily able to refer the distributions of the mesopelagic fishes in their catches to six or more pelagic regions, each with a '. . . more or less unique fish fauna with its characteristic assemblage of species in characteristic proportion, its characteristic diversity and so on'. This definition could also be applied to the communities that presumably contribute largely to the individual characteristics of a pelagic region. In turn, the more or less stable structure of a community is integrated by the many relationships between its constituent species (particularly predator–prey interactions and interspecific competition) and the environment (see MacFadyen, 1963). With such community criteria in mind, it will be interesting to see how far they apply to the animals in sonic scattering layers. There is much to do. As Kinzer (1970) says, we know little about the predator–prey interactions of organisms within the diurnal depth range of such layers.

If daily vertical migrations are feeding sorties, the metabolic benefits must exceed the energetic costs of ascent and descent. No doubt the benefits will vary from day to day, but even so, the energy needed for vertical migrations by small crustaceans (and *Chaoborus* larvae) is but a few per cent of their daily metabolic rate (see Swift, 1976). For planktonic crustaceans ranging from 0.5 to 30 mm in length, the amount of energy expended during vertical migrations was calculated (by estimating the work required to overcome hydrodynamic resistance and lift their net body weight) to be no more than 20–40 per cent of their resting metabolism (Klyashtorin and Yarzhombek, 1973). Comparable figures for small fishes may well be of the same order.

   The adaptive value of daily vertical migrations in energetic terms has been considered by McLaren (1963). In a study of migrating zooplankton McLaren followed the implications of a mathematical model based on von Bertalanffy's growth equations. The model predicts that a migration from cool to warmer waters can be advantageous: an animal that is mainly active when feeding at upper levels and then descends to rest in cooler waters, will use its food to good effect. For instance, growth and fecundity should be enhanced, for during its resting periods an animal should eventually grow larger than it would in warmer surroundings. Though subsequent tests of this hypothesis (see Lock and McLaren, 1970, and Swift, 1976) have not been encouraging, it should not yet be abandoned.

   Besides the gain in food, the lessening of competition and the partitioning of resources, there are other likely adaptive aspects of vertical migrations. For planktonic animals, at least, Hardy (1956), after pointing out that during their migrations currents may carry them considerable distances in the horizontal plane, suggests that, '... the development of a habit of regular periodic vertical migration has been evolved to enable the animal continually to sample new environments—new feeding grounds—and so to have some power of choice, of varying its behaviour within limits'. Following this line of thought, Isaacs, Tout and Wick (1974) have shown how animals (in sound scattering layers) may be transported into (or maintained within) areas of high standing crops of phytoplankton.

   Besides gaining food, vertical migrators may escape predators. For instance, in the tropical Pacific yellow fin tuna and albacore evidently feed during the day (mainly on non-migrating euphausiids of the genus *Stylocheiron*), when lantern-fishes and other likely prey are residing at lower levels (Roger and Grandperrin, 1976). Though one cannot imagine lantern-fishes with their sensitive nocturnal eyes in the surface waters by day,* it seems also that they avoid lancet-fishes (*Alepisaurus*), voracious mesopelagic predators. In both the North Atlantic and the southeastern Pacific the fish prey of *Alepisaurus* consists largely of hatchet-fishes (*Sternoptyx*), barracudinas (paralepidids) and others of their kind (Haedrich and Nielsen, 1966). All such prey are non-migrators; lantern-fishes, nearly all of which migrate, are rarely taken. Clarke (1973) also observes that studies of the feeding habits of tuna, dolphin-fishes (*Coryphaena*) and *Alepisaurus*, indicate that they consume very few vertically migrating forms such as lantern-fishes and gonostomatid fishes (see also Matthews et al., 1977). Thinking of epipelagic predators, particularly yellow fin

---

* But *Benthosema panamense* (and other lantern-fishes) have been seen near the surface by day off Central America, when, significantly, they are fed on by yellow fin and skipjack tuna.

tuna and albacore, Roger and Grandperrin (1976) argue that if such predators do not feed at night, they cannot retrieve the energy won by the migrating mesopelagic fauna. They see the whole process as an 'energy valve' permitting the downward transfer of energy but reducing the upward one. They conclude that if this process exists in other parts of the food chain '. . . then one can suggest that the mechanism is fundamental, as it explains how the deeper layers of the ocean are able to support a fairly large biomass'. Their ideas seem well worth pursuing.

Lastly, ontogenetic and seasonal migrations also lead to transfers of organic matter between the euphotic zone and midwater levels. Concerning the former, we shall see (pp. 429–55) that in many kinds of mesopelagic and bathypelagic animals the spawning place and buoyant properties of the eggs are such that the larval life history takes place in the productive surface waters. During and after metamorphosis there is a descent to the adult living space. There must be heavy mortality as the eggs and early larvae ascend to their euphotic nursery ground and during their stay in this layer. The descending metamorphosis stages will also fall to predators, but the young are now large and active enough to avoid some of their enemies.

Seasonal movements of zooplankton between epipelagic and lower levels seem to be largely confined to temperate and cold waters and upwelling regions (see Longhurst, 1976). Conditions in antarctic waters are considered by Foxton (1964), whose investigations (between the surface and 1000 metres) showed that the bulk of the standing crop of zooplankton was concentrated in the upper 100 metres during the summer months of November to March. In the underlying waters there were also seasonal variations such that maximum quantities (between 500 and 1,000 metres) were reached from July to September, when the standing crop in the upper 100 metres was at a minimum. But over the whole water column from 0 to 1,000 metres there were no significant monthly variations in the mean total volume of plankton. Foxton continues; 'This perhaps unexpected result can be explained by what is known of the life histories of some of the Antarctic zooplankton. It has been shown that one of the most marked features of the plankton cycle in Antarctic waters is that some of the commoner species, including the copepods *Rhincalanus gigas* and *Calanoides acutus*, and the chaetognaths *Eukrohnia hamata* (Mackintosh, 1937) and *Sagitta gazellae* (David, 1955) migrate seasonally so that the bulk of the adult population during the summer is concentrated (by day and night) in the surface layer and during the winter in the waters of the warm deep current' (see p. 103). Indeed, Mackintosh (1937) proposed that such migrations are of biological significance in that species populate the northerly flowing surface water in the summer but return south in the counter-flowing, warm deep current in the winter, thereby staying within their proper geographic range.

Bathypelagic food webs

The bathypelagic zone, the largest environment on earth, is cold and dark (apart from fitful sparks of living light) and the most deserted life zone in the ocean, both in numbers of organisms and of species. Apart from their ontogenic migrations, the deep-set ways of bathypelagic animals would seem to give them little or no direct contact with the more productive waters above their habitat. There may well be overlap between populations of lower mesopelagic and upper bathypelagic animals. Thus, dark-coloured lantern-fishes (*Lampanyctus* spp.), most of which live at lower mesopelagic levels and undertake daily vertical migrations, are sometimes eaten by female angler-fishes (Bertelsen, 1951). But virtually nothing is known of the exchange of organic material in the overlapping zones. Certainly the zooplankton biomass near 1,000 metres is considerably greater than that at lower levels, but the main factor may simply be depth of residence: those highest in the bathypelagic water column are the first recipients of organic crumbs from above and thus have the means of highest productivity.

Copepods, which must be the main metazoan users of organic particles, are well represented at bathypelagic levels (especially by species of the families Spinocalanidae and Scolecithricidae). At all depths, copepods, including their copepodid stages, make up the bulk of the zooplankton. Indeed, off Bermuda, Deevey and Brooks (1977) found that the relative proportions of copepods in the total zooplankton catch increased with depth (from 73.5 per cent in the upper waters to 91 per cent between 1,500 and 2,000 metres). The decrease in numbers and quantities of zooplankton is greatest between the upper waters and a depth of about 500 metres. Below 500 metres the vertical gradient for volumes or numbers of zooplankton falls exponentially, the rate of decline being greater between 500 and 2,000 metres than that below this depth range. For the biomass of total zooplankton, as measured by displacement volumes, Deevey and Brooks found that the slope of the gradient between 500 and 2,000 metres was -5.66×10^4 log units per metre, which they compare with Vinogradov's figure of -3 to -2×10^4 log units per metre for the gradient between 2,000 and 4,000 metres in the Indian Ocean.

There is also a decline with depth in the numbers of species. For copepods, Grice and Hulsemann (1967) recorded 153 bathypelagic species in the western Indian Ocean, of which 122 occurred between 1,000 and 2,000 metres, 73 between 2,000 and 3,000 metres and 13 between 3,000 and 4,000 metres. Of the 326 species they identified from catches made between the surface and 2,000 metres off Bermuda, Deevey and Brooks recorded 128 species between 0 and 500 metres, 204 between 500 and

1,000 metres, 172 between 1,000 and 1,500 metres and 119 between 1,500 and 2,000 metres.

At upper levels (1,000–2,000 metres) and in an area of high productivity (deep slope water off Nova Scotia), it may well be, as Harding (1974) found, that protists (diatoms, flagellates and protozoans) are an important food source for bathypelagic copepods. But in the poorly productive Sargasso Sea at deeper levels (3,000–4,000 metres) his analyses show that the most frequent gut contents of copepods were mineral particles, detrital remains and detrital balls. Deevey and Brooks (1977) conclude that phytoplankton production in the Sargasso Sea is insufficient to support resident populations of filter-feeding copepods. Throughout the water column, they suggest that most copepods are omnivorous opportunists, consuming whatever particulate material, micro-organisms and nauplius larvae are available. They contrast their findings with those of Vinogradov (1970) from the Kurile–Kamchatka region, where carnivorous species increase in importance between 1,500 and 3,000 metres while omnivores prevail at lower levels. But, as they say, such trophic stratification is probably typical of regions of high primary productivity.

Clearly, there is much we do not know of bathypelagic food pyramids, but recent investigations are suggestive. Light scattering measurements (at nephelometer stations) of the standing crops of suspended particles at bathypelagic levels in the Atlantic Ocean reveal a quantitative distribution that matches the pattern of surface productivity (Biscaye and Eiltrim, 1977), which is also reflected in the standing crops of bathypelagic and benthic organisms.

Even more significant, we shall see later (pp. 285–8) that the faecal products of copepods and other zooplankters, especially from the epipelagic zone, are a major source of food for benthic, deep-sea animals. Many such waste packages must be intercepted and used by bathypelagic zooplankton. The greater the surface productivity the more will be the 'snowfall' of organic particles and the 'hail' of faecal materials, and both together, particularly the latter, may well go far to determine levels of productivity in bathypelagic and benthic environments. At mesopelagic levels, as we have seen, there is also the powerful influence of vertical migrations.

The copepods, which are so predominant numerically at bathypelagic levels, must go far to nourish resident populations of arrow-worms, euphausiids (e.g. 'giant' species of *Thysanopoda*), prawns (e.g. *Hymenodora glacialis* and *H. gracilipes*) and diverse fishes (especially black *Cyclothone*). Apart from the small *Cyclothone pygmaea* in the Mediterranean and *C. livida* off Western Africa, black species of *Cyclothone* range over the three oceans from lower mesopelagic to bathypelagic levels. The open ocean species are not only larger (lengths ca. 50–75 mm) than the transparent mesopelagic species, but they have a relatively larger head and mouth (see also p. 247). Their prey ranges widely in size from other *Cyclothone* through euphausiids, amphipods and arrow-worms to ostracods and copepods. As stressed already, where food is scarce it is fitting to have the means for securing such a ranging regimen.

Thinking particularly of this desideratum, we turn to the deep-sea angler-fishes (Ceratioidea), which are the most diverse group of bathype-lagic fishes: indeed, the 100-odd species make up at least two-thirds of the bathypelagic fish fauna. Between adolescence and when they are ready to seek their partners the free-living dwarf males (length < 50 mm) have relatively small jaws and a more or less fusiform body. Unlike the females, they have no angling system. In some families (e.g. Ceratiidae and Linophrynidae) the males, which eventually become parasitic on the females, are not known to take food after metamorphosis. Males of other families (e.g. Melanocetidae and Himantolophidae) have jaws that are not only suitable for taking prey but also, presumably, for hanging on to the skin of a female during the breeding season. Such males grow after metamorphosis and those of *Melanocetus* at least, feed on zooplankton (Bertelsen, 1951; Pietsch, 1976).

The larger females, which in most species tend to a globular shape, bear large jaws armed with recurved, depressible teeth. The length of the jaws is usually more than half the length of the head (Bertelsen, 1951). Except in *Neoceratias*, all females bear a rod (illicium) and attached bait (esca), containing in nearly all species a luminous gland. The illicium projects from the head between or just before the eyes and is articulated to a horizontal basal bone set in a groove on the cranium. There are muscles from the basal bone to the base of the illicium such that the latter can be moved between an upright and a horizontal position. Moreover, in some angler-fishes (ceratiids and some oneirodids) the basal bone is so long and movable that the rod and bait can be extended well in front of the mouth and then retracted (Bertelsen, 1951). Williams (1966) rightly urges critical circumspection before deciding that features have an adapt-ive significance and he writes, 'A frequently helpful but not infallible rule is to recognize adaptation in organic systems that show a clear analogy wi'h human implements.' The angling system of ceratioids is surely an outstanding example.

Female angler-fishes feed largely on crustaceans (copepods, ostracods, amphipods, euphausiids and prawns) and fishes. They also take arrow-worms and squid. As in *Cyclothone*, an individual may contain organ-isms of a wide size range. Thus, in the stomach of an adolescent *Himanto-lophus* (length 123 mm) there were small copepods, two amphipods, five euphausiids (15–45 mm), four *Cyclothone* (15–30 mm), one hatchet-fish (*Sternoptyx*, 19 mm), one lantern-fish (10 mm), one melamphaid fish (25 mm), one 'sudid' fish (22 mm) and one squid beak. Moreover, the tissues of the stomach and abdominal wall are distensible enough to accommodate very large meals, which may consist largely or entirely of one prey species. In two *Melanocetus johnsoni* (80–95 mm) were found a single large black lantern-fish (*Lampanyctus crocodilus*), each twice to three times the length of the swallower. A *Linophryne quinqueramosus* (length 89 mm) contained a large deep-sea eel (*Serrivomer*, length 330 mm), a hatchet-fish (*Sternoptyx*, 25 mm), two *Cyclothone* (26 and 60 mm) and five 'shrimps' (12–30 mm). After presenting these and other findings, Bertelsen (1951) remarks that though female angler-fishes have diverse kinds of angling devices, there is no evidence of differences in their choice of food. Indeed, as he and Pietsch (1974) have shown, the luminous complex and dermal decoration of the bait (esca) differs from species to species (Fig. 103). Acting on conjectures that different kinds of esca might attract different kinds of prey, Pietsch analysed the stomach contents of *Oneirodes* species but found no signs of prey selection. Again, it seems clear that there is no choice but the widest possible range of diet in food-poor surroundings.

The exponential decline in the food available for production between succeeding links in a food chain is believed to limit the number of possible links to five or six. If this is so, very productive environments should support more links than do impoverished ones. There are certainly five or six links in the euphotic zone of oceanic regions, but how many are there at bathypelagic levels? If copepods feeding mainly on faecal material, detritus and micro-organisms, correspond to herbivorous epipelagic species, then one might say that first-level carnivores, in that they may well be largely dependent on the 'herbivores', are other predatory copepods, amphipods, euphausiids and arrow-worms. Second-order carnivores would be prawns and fishes, but are the fishes, for instance, largely sustained by first-order carnivores? We have seen that they feed on each other and at all levels of the food chain. Trophic rules seem to be overruled in bathypelagic deserts.*

The predatory design of bathypelagic fishes is very impressive, but how efficient are they as predators? When on RRS *Discovery* I unwittingly staged a striking demonstration of the predatory powers of female angler-fish, which are described elsewhere (Marshall, 1974).

Ceratioid angler-fishes '… may be kept alive for some time after capture, particularly in cool surroundings. I wanted to test the escape responses of ceratioids, for my guess is that the crossed pair of Mauthnerian neurones and fibres, a system essential in the quick escape movements of fishes, might be better developed in the males'.†

---

* Trophic rules, in so far as very productive environments might be expected to have long food chains, are broken in antarctic waters. There are two links in the food chain: phytoplankton–krill–large whalebone whales.

† My recent observations of the fine structure of the brain and spinal cord of a male ceratioid do not support this guess.

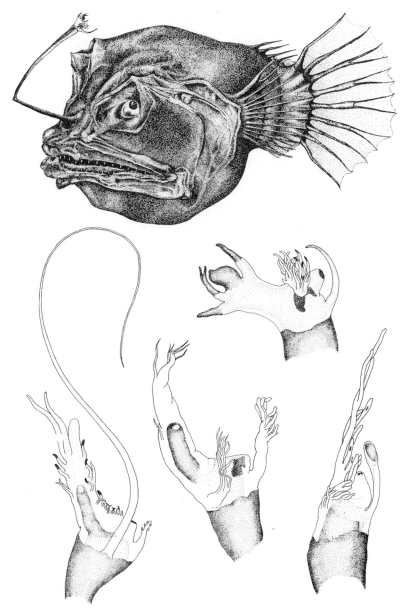

Fig. 103   Patterns of esca structure in some species of the ceratioid angler-fish genus
Oneirodes. Above, a female *Oneirodes eschrichtii* and immediately below the
esca of a 118 mm long specimen. Below, left to right, the escae of female
Oneirodes anisacanthus (length of fish 78 mm), *O. macrosteus* (124 mm) and
O. basili (115 mm). (Redrawn from Pietsch, 1974)

'When I gently held the tail-fin of a male ceratioid in a pair of forceps he wriggled hard as though to escape. Using thumb and forefinger, I did the same to several females but they promptly turned round and tried to bite me. In discussion, Mr Peter David* said that the quick turn-round of female ceratioids reminded him of Mr Nubar Gulbenkian's taxicab. Mr Gulbenkian once remarked that his taxi could turn round on a six-pence, "whatever that may be". I thus came to call the negative escape response of female deep-sea angler-fishes the "Gulbenkian reflex". Indeed, the females of most species have relatively short luminous lures, a globular or deep-bodied form and fins nicely related to the centre of gravity, features that are admirably fitted to a quick rounding on their prey. After all, a female cannot always expect prey to swim both towards her flashing lure and straight for her large and gently smiling jaws. Prey may approach from behind, where, incidentally, the body is studded with free-ending lateral line organs, able to detect nearby disturbances in the water.'

Studies of the organization of bathypelagic fishes and the 'Gulbenkian reflex' of female ceratioids led me to Sir D'Arcy Thompson's views on deep-sea fishes. In his book *Growth and Form*, Sir D'Arcy remarks, 'The great depths of the sea differ from habitations of the living, not least in their eternal quietude. The fishes which dwell therein are quaint and strange; their large heads, prodigious jaws and long tails and tentacles are, as it were, gross exaggerations of the common and conventional forms. We look in vain for any purposeful cause or physiological explanation of these enormities; and are left under a vague impression that life has been going on in the security of all but perfect equilibrium, and that the resulting forms, liberated from many ordinary constraints, have grown with unusual freedom.'

Contrary to the thought behind Sir D'Arcy's last sentence, the more one looks the more one comes to appreciate how cunningly and closely angler-fishes (and other midwater forms) are designed for life in their deep and difficult environment.† It is true that bathypelagic animals do not face the constraints imposed by vertical migrations. They live in virtu-ally changeless physical surroundings, and they are thus freed from de-veloping physiological and biochemical means to cope with fluctuating conditions. But they are not liberated from one ordinary constraint, the continual need to make a living in impoverished circumstances. Indeed, their life in the most deserted part of the ocean, as will now be argued, is ingrained in their whole organization.

---

* Head of the Biological Section, Institute of Oceanographic Sciences.
† But my admiration for his great work remains.

Organization of midwater life

The dominant theme of this closing section is that life is simplest at bathy-pelagic levels. Simplest is more or less synonymous with least complex, but how is relative complexity assessed? One is led to a fundamental concept in biology, that of organization. More complex biological systems may have more levels of organization (integrative levels) than simpler ones, and at each particular level the former systems are more complex than the latter. A set of levels of organization ('a level is an assembly of things of a definite kind, i.e. a collection of systems characterized by a definite set of properties and laws', Bunge, 1973) forms a hierarchy such that each level emerges from the one immediately below. In the present context, the hierarchy is: ecosystems–organisms–organ system–cells–cytoplasm–nucleus. Thus, if we say that the bathypelagic life zone is simpler than its mesopelagic counterpart, we mean that it has simpler ecosystems, organisms, organ systems and so forth.

An ecosystem may be regarded as a community in its physical-environmental setting (Emlen, 1973). Integration in an oceanic community is predominantly through the food chain of primary producers (phytoplankton)–herbivores–omnivores/carnivores, and sooner or later all fall to decomposers (bacteria etc.), which produce, inter alia, nutrients that may contribute to renewing the phytoplankton. Mesopelagic and bathypelagic ecosystems are incomplete in lacking primary producers. Through vertical migrations mesopelagic zooplankters may graze on the phytoplankton but, as we have seen, the 'herbivores' of the bathypelagic zone must depend on whatever micro-organisms and waste materials (faecal–detrital) they can obtain. The intricate ecosystems of mesopelagic levels are undoubtedly sustained through vertical migrations but in the present context we need only note that they are more complex than their bathypelagic counterparts in containing more species and individuals. To return to Deevey and Brooks's (1977) investigations off Bermuda, between depths of 500 and 1,000 metres, copepods were represented by 204 species at a concentration of 19 individuals per cubic metre: between 1,500 and 2,000 metres the figures were 119 species and 4·5 individuals per cubic metre. More than anything else, these and other such contrasts are surely due to the constraints imposed by the very low quantities (and poor quality) of food available below the mesopelagic zone. In turn, these contrasts are seen along the food chain: catches of bathypelagic fishes are not only much smaller than those from mesopelagic levels, but consist of smaller numbers of species. Off Bermuda no more than a tenth of the 200 or more midwater species live in the bathypelagic zone. Naturally, in thinking of numbers of species, one must think also of their ecological niches. But in living so far below the zone of primary production, bathypelagic animals are evidently denied the diversity of niches made available through vertical migrations.

The range of complexity in organisms is a product of evolution. Ayala (1977) writes, 'The earliest organisms living on earth were no more complex than some bacteria and blue-green algae. Three billion years later their descendants include the orchids, the bees, the dolphins and man.' In many-celled organisms the number of organ systems has increased during evolutionary time. At the cellular level of organization one may think in microcosm of differentiated cell types as species and the complement of such cells in a tissue as a number of individuals. The number of kinds of cell 'species' give a striking measure of complexity. Thus, Stebbins (1977) estimated that there are 47–52 kinds of differentiated cells in flowering plants, about 66 in an earthworm, 100 to 150 in an insect and 200 to 250 in a mammal.

If the relatively simple organization of bathypelagic animals is a response to poor levels of subsistence, then one might predict that an overall response could be a reduction in size: the smaller the individual the fewer the cells it has to support. In freshwater fishes, for instance, stunted individuals are produced under adverse conditions (see Fryer and Iles, 1972).

In copepods the mean size increases with depth to reach a maximum at levels between 500 and 2,000 metres but below 2,000 metres there is an appreciable decrease. In the Indian Ocean between 1,000 and 2,000 metres the mean size of adults studied by Grice and Hulsemann (1967) was 2·34 mm; between 2,000 and 3,000 metres the mean was 2·19 mm, a decrease of 8 per cent. Similar results were obtained in the Atlantic and we should recall that females are somewhat the larger sex. This decrease in size may well reflect the decrease in available food, but the correlation is likely to be partial. The size attained by organisms is perhaps always a compromise between conflicting selection pressures, and Parsons (1976) has reviewed certain of these constraints in the zooplankton. During the phase of active growth (and the data considered comes from studies on the metabolism, growth, food intake and growth efficiency of copepod-sized zooplankters), the smaller the animal, the more food it needs for metabolism as opposed to that used in growth.* (In cold bathypelagic waters both processes are presumably slow. But this disadvantage may be partly offset in that smaller animals are reported to have a greater reproductive capacity: thus the allocation of energy to reproductive tissue is much greater per unit of body weight in small crustaceans as compared to larger species, and deep midwater copepods produce large eggs (Sekiguchi, 1974).) There is also a relationship

---

* Growth efficiency ranges from 17 per cent for an animal of wet weight 0·005 mg to 57 per cent for an animal of 5·0 mg.

between size and feeding efficiency. In the present context, it may well be that omnivorous bathypelagic copepods need to be large enough to collect and handle a range of faecal and detrital fragments, or even catch nauplii of their own kind. Thus, economy in the number of cells to support can be attained by a reduction in size, which may have reproductive advantages, while growth efficiency and food-procuring ability are enhanced by an increase in size. Two factors seem to conflict with the other two, and there may well be more.

Euphausiid shrimps, the 'copepods' of the micronekton, get larger with depth. The giant bathypelagic species of *Thysanopoda* (*cornuta, egregia* and *spinicaudata*), which are 60–70 mm in length at maturity and may eventually exceed 100 mm, are much larger than their shallower-living congeners. The young of these species are also bathypelagic, and at comparable stages they are also much larger than the larvae of their relatives (Brinton, 1962). Presumably, they produce much larger eggs, which, if fecundity is to be maintained, require the extra space provided by a giant female for their development. Moreover, if, as seems likely, these species are omnivorous, then both sexes could be large and quick enough to prey on their neighbours in the zooplankton and have a large thoracic basket for the collection and holding of waste food materials. Larger forms, as seen already, have a lower metabolism and greater growth efficiency compared to smaller ones. Here then, in contrast to the copepods, all three factors call for increased size. Most likely the limit is set by the principle that eggs should not be put into too few baskets, particularly where means for making them are in very short supply. In other words, the overall fecundity needed to maintain a population of bathypelagic euphausiids should not be contained in a few exceedingly large females, for where the sexes are equal in size (and presumably in numbers of individuals) they may fail to meet if each is poorly represented (see also pp. 433–4).

In bathypelagic fishes the classic instance of reduced size concerns the dwarf males of ceratioid angler-fishes. Apart from the economy in the total amount of living tissue needed to maintain any one species, the two sexes, except during their early life in the euphotic zone, will hardly compete for food. Moreover, we may recall that potential parasitic males do not appear to feed after metamorphosis. The males of *Cyclothone* are also smaller than the females (by about two-thirds), but here there is unlikely to be an overall metabolic economy: the males have much more red muscle and gill surface than their partners (Marshall, 1971). But intersexual competition for food may be somewhat reduced (see p. 246) and in any event, the males may live at higher levels than the females for part of their life (Badcock and Merrett, 1976). These two workers also found evidence for sex reversal in the bathypelagic species *Cyclothone microdon*, but there were no such signs in the mesopelagic *C. braueri*.

This recalls the earlier work of Kawaguchi and Marumo (1967), who found in *Gonostoma gracile* that the males are not only smaller than the females (males < 60 mm; male or female 70–90 mm; females > 90 mm), but change through a hermaphroditic phase into females. Thus, in both species the males are not discarded after use but are saved to become females, which is both economical and a means of boosting fecundity.

Apart from male angler-fishes, the bathypelagic fish fauna is not composed of dwarfed individuals. Gulper-eels reach a length of 1 metre or more and several of the whale-fishes (Cetunculi) grow to well beyond 100 mm. Moreover, as stated already, the bathypelagic species of *Cyclothone* are both larger than their mesopelagic relatives and bear relatively larger heads and jaws. It is significant that the one exception occurs in a most impoverished sea, the Mediterranean, where the bathypelagic dwarf form (*pygmaea*) of *Cyclothone* is smaller than the mesopelagic *C. braueri*. In the related genus *Gonostoma*, the bathypelagic species *bathyphilum* grows to much the same size as its mesopelagic congeners, but again it has relatively longer jaws (Marshall, 1960 and 1971). The advantages of a wider gape at deeper levels have already been stressed. What is now relevant is that *Gonostoma bathyphilum*—and other bathypelagic species—are not only sizeable fishes but have lower levels of organization than those in mesopelagic fishes.

An apt model for such contrasts is got by comparing *Gonostoma denudatum*, an upper mesopelagic species, with its bathypelagic relative *G. bathyphilum* (Fig. 104). In the North Atlantic both grow to a length of 200 mm or more. The former is silvery-sided, has well-developed eyes and light organs and undertakes daily vertical migrations: the latter is black, has regressed eyes and light organs and stays in the depths. Concerning organ systems, *G. denudatum* has one more—a capacious gas-filled swimbladder—than its relative. By losing the swimbladder, *bathyphilum* has lost the need, as it were, to develop and maintain the following tissues; mesenteries to suspend the swimbladder in the body cavity; tissues of the swimbladder wall, notably the connective tissues* (tunica externa and the submucosa), the smooth muscle layer and the inner epithelium, locally differentiated to form the gas gland; the blood system, particularly the complex rete mirabile to supply the gas gland, and the nerve supply. The presence of the swimbladder in *G. denudatum* enables it to carry, inter alia, a well-ossified skeleton and firm muscles at neutral buoyancy: the skeleton of *bathyphilum* is poorly ossified and its lateral muscles are not only smaller in bulk, but also rather soft and watery. A third species, *Gonostoma elongatum*, black and from lower mesopelagic levels, has a regressed swimbladder (presumably functional

---

* Although these need no maintenance after their formation.

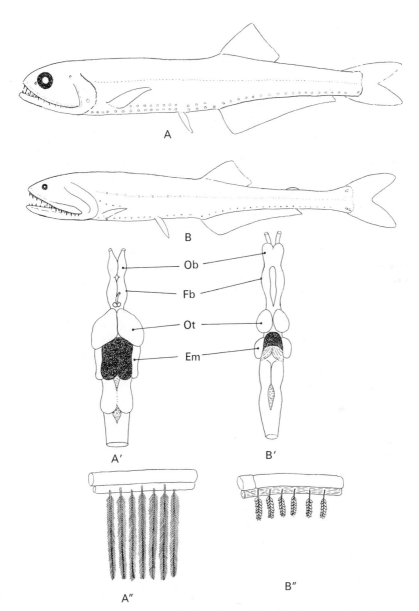

Fig. 104 A comparison of two related midwater fishes. A, *Gonostoma denudatum* from the mesopelagic zone; B, *Gonostoma bathyphilum*, from the bathy-pelagic zone. Note the much smaller light organs and eyes of the latter. Below are shown (drawn to the same scale) their brains and a few of the longest filaments on the first gill arch; *denudatum* (A[1] and A[11]); *bathyphilum* (B[1] and B[11]). In the brain the corpus cerebellum, the size of which is directly related to the amount of lateral muscle, is shown in black. Ob, olfactory bulb; Fb, forebrain; Ot, optic tectum; Em, eminentia granularis. (Redrawn from Marshall, 1971)

in the young) when fully grown. This congener is also a vertical migrator with well-developed eyes and light organs but, due largely to the reduction of its skeletal and muscular systems, which are intermediate in development between those of *denudatum* and *bathyphilum*, it comes close to attaining neutral buoyancy (Denton and Marshall, 1958; Marshall, 1960, 1971). Presumably *bathyphilum* is virtually weightless in water, and so may stay at a chosen level with very little effort. This alone is a considerable saving in energy (Denton and Marshall, 1958; Alexander, 1966). Naturally, the paring down of its skeletal and muscular systems and the slimness of its trunk and tail reduces its powers of locomotion, but parallel changes also occur in its potential competitors in the bathypelagic fish fauna.

Extending these comparisons throughout the organism, one eventually finds that the relative levels of organization are depressed in nearly all the organ systems of *Gonostoma bathyphilum*. The exceptions are the olfactory system, which is highly developed in the (smaller) males, but regressed in the females (see also pp. 404–8), and the acoustico-lateralis system. Concerning the latter, the ears, which seem never to regress in fishes, are needed in balance (through gravity receptors) to detect angular accelerations of the body and sounds, and to maintain and regulate muscle tone (Lowenstein, 1957). Nearby disturbances in the water, whether caused by partners, prey or predators, are detected and located by the lateral line organs (Dijkgraaf, 1962). All systems being considered, *G. bathyphilum* seems very well fitted to make a living in the most deserted life zone of the ocean.

There remains the cellular level of organization, and here there is Post's (1974) extraordinary discovery that *Gonostoma bathyphilum* has six chromosomes, the smallest known number in fishes, whereas there are 24 in *G. elongatum*. The usual number in teleosts is 46 to 52, but ranges from six (eight in the killifish, *Notobranchius rackowi*) to over 100 (Ohno, 1970). Apart from the microchromosomes of birds and certain reptiles, fishes have the smallest chromosomes of all vertebrates.

Naturally, we need an estimate of the DNA content of *G. bathyphilum* and one or more of its mesopelagic relatives. The DNA content of teleost nuclei as a percentage of that in placental mammals, varies from 14 per cent in the puffer fish *Spheroides maculatus*, to 125 per cent in the armoured catfish, *Corydoras aeneus* (from 8×10^8 nucleotide pairs in another puffer fish, *Tetraodon fluviatils*, to 8.8×10^9 in *Corydoras aeneus*). In presenting these and another data, Ohno (1970) remarks that '. . . the smallest genome to be found in vertebrates is possessed by remotely related teleost fish belonging to diverse orders. The genome of the surf smelt, the tetrazona barb, the swordtail, the hornyhead turbot, the fantail sole, the seahorse and the puffer is slightly larger or about the same as that of the amphioxus, which is 17% of the genome size of placental mammals. Their close relatives, however, may be endowed with a genome of considerably larger size.' If the genome size of *G. bathyphilum* proves to be much less than that of its mesopelagic relatives, one will be tempted either to correlate this with its much simpler organization or to wonder whether a further economy is to dispense with redundant DNA. Meanwhile, there is Ohno's Olympian conjecture that the '. . . apparent chaos which prevails with regard to genome size in fish is merely a reflection of nature's great experiment with gene duplication'.

The *Gonostoma* model may be used as a guide for a general survey of levels of organization in mesopelagic and bathypelagic fishes. This is elaborated elsewhere (Marshall, 1960, 1971) and is summarized in the later publication thus:

'Considered broadly, there are three groups of mesopelagic fishes: small-jawed plankton eaters, most of which have a swimbladder; predators with a swimbladder; and predators without a swimbladder.'

'The most diverse members of the first group are small gonostomatids, hatchet-fishes, argentinoids, lantern-fishes and melamphaids. Such species are compared in the table below with bathypelagic fishes: Cyclothone spp., ceratioid angler fishes, gulper eels and so on.'

'Mesopelagic predators with a swimbladder include species of *Astronesthes*, trichiuroids, and the chiasmodontids. The contrasts between the organization of these fishes and that of bathypelagic species are also very marked. For a paradigm, consider the systems of *Astronesthes niger* and a female angler fish of the species *Neoceratias spinifer*. Every tissue system of the former is elaborated to a far greater extent, much as we saw in *Gonostoma denudatum* by comparison with *G. bathyphilum*. For instance, the lateral muscles of *Astronesthes* are larger, more closely knit and contain many more outer red fibres than do those of a *Neoceratias* of the same size.* This difference is reflected in the great contrast between the size of the corpus cerebellum in these two species (Fig. 105). In fact, the corpus cerebellum of the *Astronesthes* is nearly equal in volume to the entire brain of the *Neoceratias*, every part of which, except for the acoustico-lateralis centres, mirrors the reduction of associated organs.

'More generally, the main contrasts between the organization of mesopelagic predators with a swimbladder and that of bathypelagic species are just as trenchant as those given in the above table for the latter fauna and mesopelagic plankton-eaters. If the mesopelagic predators were put in this table in place of the plankton eaters, the main descriptive changes needed could be those correlated with their predatory life: black skin, large jaws and teeth, and so forth.

---

* Waterman (1948) wrote as follows about another female ceratioid: 'Like the skeletal elements, many of the muscles in *Gigantactis* are reduced in size and some commonly present in the less highly specialized actinopterygians are lacking altogether.'

Organization of mesopelagic and bathypelagic fishes

Features	Mesopelagic, plankton-consuming species	Bathypelagic species
Colour	Many with silvery sides	Black
Photophores	Numerous and well developed in most species	Small or regressed in gonostomatids; a single luminous lure on the female of most ceratioids
Jaws	Relatively short	Relatively long
Eyes	Fairly large to very large with relatively large dioptric parts and sensitive pure-rod retinae	Small or regressed, except in the males of some anglerfishes
Olfactory organs	Moderately developed in both sexes of most species	Regressed in females but large in males of *Cyclothone* spp, and ceratioids (most species)
Central nervous system	Well developed in all parts	Weakly developed, except for the acoustico-lateralis centres and the forebrain of macrosmatic males
Myotomes	Well developed	Weakly developed
Skeleton	Well ossified, including scales	Weakly ossified; scales usually absent
Swimbladder	Usually present, highly developed	Absent or regressed
Gill system	Gill filaments numerous bearing very many lamellae	Gill filaments relatively few with a reduced lamellar surface
Kidneys	Relatively large with numerous tubules	Relatively small, with few tubules
Heart	Large	Small

'Mesopelagic predators without a swimbladder belong mainly to the stomiatoids (Melanostomiatidae, Stomiatidae, Chauliodontidae, Malacosteidae) and alepisauroids (Scopelarchidae, Evermannellidae, Omosodidae, Alepisauridae, Anotopteridae and Paralepididae). Here a good guide to the contrasting organization of these and bathypelagic fishes is given by comparing *Gonostoma elongatum*, a representative of the first group and *G. bathyphilum*. The first species, which grows to a standard length of 250 mm or more, feeds on fishes (for example, *Cyclothone*), prawns, euphausiids and so on. If the reader will refer to the earlier comparison of *G. denudatum* and *G. bathyphilum* the main descriptive

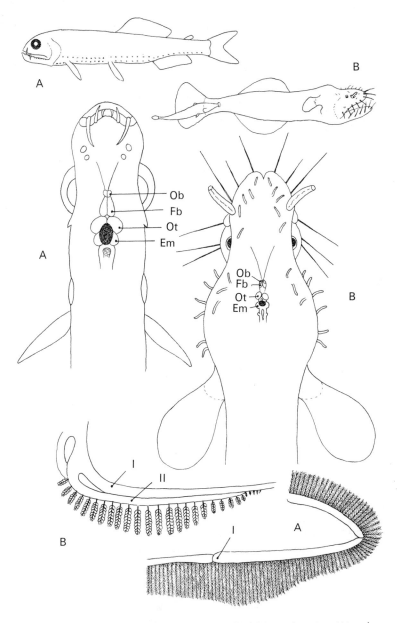

Fig. 105   Comparison of a common mesopelagic fish *Astronesthes niger* (A) and a female angler-fish *Neoceratias spinifer* (B) with attached male. All figures are drawn to the same scale. The head and brain structure are shown below. Ob, olfactory bulb; Fb, forebrain; Ot, optic tectum; Em, eminentia granularis. Note the much smaller brain of *Neoceratias*. At the bottom, the first gill arch (I) of *Astronesthes* is compared to the two first gill arches (I, II) of *Neoceratias*. Note that the first gill arch of the latter has only a few minute terminal lamellae. (Redrawn from Marshall, 1971)

changes needed to put *elongatum* in place of *denudatum* would be these: (1) *elongatum* is centred at lower, rather than upper, mesopelagic levels; (2) the colouration is black and brassy; (3) the eyes and optic centres of the brain are moderately developed; (4) the lateral muscles and skeleton are not so well formed as those of *denudatum*; (5) the swimbladder is regressed and invested by adipose tissue; (6) the gills are not so well developed.' *

Moreover, mesopelagic fishes without a swimbladder, such as *Gonostoma elongatum*, *Bathylagus*, *Chauliodus*, *Photostomias* and *Melanostomias* have soft bodies with a very high water content (88·95%), low haematocrits (5–9%), small hearts and large lymph ducts (Blaxter, Wardle and Roberts, 1971). The lymph spaces in the bathypelagic species *Eurypharnx pelecanoides* and *G. bathyphilum* are even more enlarged. Such fishes can be both relatively large and simple, for their reduced organ systems are disposed and held in a large volume of body fluids. Size is maintained so that jaws may be wide, and as we saw, there is also the acquisition of an almost buoyant body, both advantageous in a poor environment.

There are parallel changes in the larger crustaceans. For instance, the bathypelagic prawn, *Sergestes japonicus*, has softer and more watery tissues than those of its mesopelagic relatives. Donaldson (1976) found that the protein content of *japonicus* was 46·8 per cent of its dry weight compared to a value of 65·7 per cent in the mesopelagic *S. splendens*. This is not altogether surprising, but Donaldson, who also analysed the chemical composition of other species, points out that *S. splendens*, which undertakes the most extensive vertical migrations, has the highest protein and lowest lipid content: *S. japonicus* is at most a partial migrator. Omori (1974) contrasts the 65 per cent protein content of epipelagic and mesopelagic prawns (0–400 metres) with 39·3 per cent in bathypelagic species, and the latter also lose weight in reduced calcification of the carapace. Like their fish counterparts, they become more buoyant with the loss of skeleton and muscle: they also store considerable quantities of relatively light wax esters.

Besides these advantages in the buoyancy balance sheet, reduction in the muscle system decreases the proportion of actively metabolizing tissue and thus saves more energy. Indeed, Childress (1971) found that the mysid *Gnathophausia gracilis* and the ostracod *Gigantocypris agassizii*, which were taken between depths of 900 and 1,300 metres (off California), had a mean respiratory rate (expressed as mg O_2/kg dry weight/min) of 1·2, about a tenth of that measured in crustaceans (e.g. *Sergestes*

---

* On the first gill arch of a 120 mm *G. elongatum* I counted about 200 gill filaments, the longest measuring 2 mm and there are about 30–35 gill lamellae per mm of length. Comparable figures for a 130 mm *G. denudatum* are 210 gill filaments, the longest 2·5 mm with about 50 lamellae per mm.

similis and *Euphausia pacifica*) taken between 0 to 400 metres. Thus, as Macdonald (1975) argues, if an animal's rate of respiration provides a measure of its food requirements, the food needed by the deeper-living crustaceans should be about one-tenth that of the upper-water species. This order of reduction in metabolism leads him to suggest that animals such as the angler-fish *Ceratias holboelli* '... may only require prey equal to 10 per cent of its own weight at three month intervals or longer'.

In summary, from ecosystems to organ systems, even perhaps to cellular systems, bathypelagic life is simpler than that at mesopelagic levels. Indeed, where relevant comparisons can be made, one finds a fall in levels of organization from upper to lower mesopelagic levels before the more trenchant reduction at greater depths. The evolutionary aspects are intriguing. Evolutionary theory, as it should be, is very adaptable, but we have yet to comprehend the increase in levels of organization in evolutionary series—why it is, as Medawar wrote in 1971 '... that organisms seem, as it were to propound increasingly difficult and complicated solutions to problems of remaining alive in a hostile environment'. A related aspect concerns adaptive radiations in extremely hostile environments. As Waddington (1975) says, '... the deep-sea fauna is one of those which present in dramatic form one of the great problems of evolution. How did it come about that any animals ever went into such an uninviting and difficult environment?' Thinking along these lines, Cairns-Smith (1971) remarks '... many organisms seem to have gone to great lengths to survive and thrive in the most unpromising environments'. Concerning bathypelagic animals, it would be surely right to say that they have gone to great and simple lengths to flourish in their unchanging but impoverished environment. Entry into the largest life zone on earth has demanded a far-reaching loss of complexity.

Food webs near the deep-sea floor

7

Animals that live near the deep-sea floor form a benthopelagic fauna, which has already been under review (pp. 134–42). Fishes, especially rat-tails, are prominent members of this fauna, both in numbers of individuals and species. If one looks closely at a series of fishes over a complete range of depths—from the upper continental slope to abyssal levels of 5,000 metres or more—one soon becomes aware of two striking contrasts between the benthopelagic and midwater faunas. Fishes of the latter, as we saw, become much simpler at bathypelagic levels below 1,000 metres; in the benthopelagic fauna there is no such change. Moreover, benthopelagic fishes are more highly organized than their bathypelagic counterparts.

To pursue the last statement, let us compare *Nematonurus armatus*, one of the deepest-dwelling rat-tails, and bathypelagic gulper-eels of the genus *Saccopharynx*. The rat-tail, which grows to a length of a metre or more, has a world-wide distribution between depths of about 280 and 4,700 metres (and is most common between 2,000 and 4,000 metres). The gulper-eels, also widely distributed, mostly between depths of 800 and 2,500 metres, reach lengths of well over a metre (up to 183.6 cm) (Bertin, 1934). In both kinds of fishes a long tapering tail comprises more than two-thirds of their entire length (Figs. 49 and 50).

Though it may live well below the gulper-eels, *Nematonurus* is a much more complex kind of fish. It is many times heavier than a gulper-eel of the same length and thus has many times the total number of cells. Furthermore, it has two organ systems not found in *Saccopharynx*: a capacious gas-filled swimbladder with a highly developed gas-secreting complex and a full investment of well-ossified scales. Where systems may be compared, the rat-tail has the higher level of organization. In the brain, for instance, any one major nervous centre of the rat-tail has a greater volume than the entire brain of a gulper-eel. The one outstanding development in the gulper-eels resides in their greatly elongated but slender

270

jaws—wide enough to engulf large fishes as well as smaller prey (see also p. 131).

More depth to this comparison, which is between distantly related fishes of somewhat similar form and fin pattern, may be got by looking at the midwater species of rat-tails. These must have evolved from benthopelagic ancestors. One species, *Odontomacrurus murrayi*, grows to a length of at least half a metre and lives from lower mesopelagic to bathypelagic levels in parts of all three oceans. It has evolved away from its ancestry and towards a midwater predator in these outstanding respects: it is dark-coloured; the jaws are terminal and armed with sharply pointed teeth (long and few in number in the lower jaw); the stomach has a long blind-sac, able to accommodate sizeable crustaceans and fishes; the swimbladder is regressed in the adult; the skeleton and scales are lightly ossified and the muscles are rather soft and watery. In these and other features it is more like a deep midwater fish than a benthopelagic rat-tail. Whatever the causes, the change of habitat from near-bottom to deep midwater levels has necessitated a far-reaching loss of complexity (see also Marshall, 1964, 1965, 1967 and 1973).

Thus, conditions of life near the deep-sea floor must be such that relatively complex fishes are not only able to increase and maintain their complexity but be disposed in populations dense enough to ensure the continued existence of their kind. Conditions are most favourable over the upper continental slope (150–1,000 metres), and, as we shall see, these are due largely to the relative nearness of the sea floor both to the surface productive zone and to the efflux of nutrient materials from the land. In both the benthopelagic and benthic faunas of the deep sea, the centres of numerical abundance of most of the larger species are over or on the upper continental slope. If one takes a bathymetric chart and traces the course of the 1,000-metre contour, it is soon plain that such species are supported under the very margins of the deep ocean. In relative terms the depth interval between 200 and 1,000 metres extends over an area of 4.3 per cent of the ocean. Thus, near or on about one-twentieth of the entire interface between sea and land live most of the larger deep-sea animals (both in numbers of individuals or species) that belong to the benthopelagic and benthic faunas. If we add to this the percentage area of 4.2 between the depth interval 1,000–2,000 metres then less than one-tenth of the above interface 'sees' or bears at least three-quarters (as a conservative guess) of the above two faunas. For instance, out of 40 relatively well-known species of rat-tails in the western North Atlantic, including the Gulf of Mexico and the Caribbean Sea, 24 have centres of abundance between 200 and 1,000 metres, 11 between 1,000 and 2,000 metres and five at depths beyond 2,000 metres (Marshall, 1973). More will be said of the biomass at a later stage.

The benthopelagic fauna

Knowledge of the benthopelagic fauna is not easily gained. Until suitable nets are designed, the small members of this fauna must be sought almost entirely inside the kinds that are large enough to appear in bottom photography. Observations from deep submersibles are costly but more revealing: better still, we saw how the *Alvin* was used to sample the copepods that live close to the deep-sea floor (p. 141). Turning back to the earlier review (pp. 134–42), what are the outstanding components and features of the benthopelagic fauna?

Beside the inveterate members, there are temporary inhabitants derived from the pelagic and benthic faunas. Considering first the former, representatives of all major groups of pelagic animals are likely to appear near the bottom, especially over the upper parts of the continental slope. During extensive underwater photography off southern California, many camera lowerings, largely at depths below 1,000 metres in certain of the basins, took trachyline medusae, jellyfishes (Scyphozoa) and siphonophores, which tended to occur near the bottom. Hartman and Emery (1956) even suggest that these cnidarians may take benthic food. Over the northwest African continental slope in the Cap Blanc area, a sound scattering layer, which was recorded about 30 metres above the bottom, was evidently caused by a mass of large planktonic animals. Some of these, such as coronate jellyfishes, ctenophores and pyrosomas, were found in benthopelagic fishes (alepocephalids) trawled at depths of 1,000 metres or more. These fishes also ate deep-sea angler-fishes and gulper-eels, which were also found in trawls fished below 1,200 metres. Indeed, Golovan (1974) found that over both the upper slope (500–1,000 metres) and lower slope (> 1,000 metres) the predominant benthopelagic fishes (rat-rails, alepocephalids, etc.) fed mainly on mesopelagic or bathypelagic organisms: benthic animals were rarely found in their stomachs. Their main sources of food were shrimps (euphausiids), jellyfishes, pyrosomas, angler-fishes, gulper-eels and squid.

In a more recent paper, Parin and Golovan (1976) record that otter trawls fished over the African continental slope (mainly off Mauritania, Spanish Sahara, Namibia and South Africa) at depths between 220 and 1,700 metres took 117 fish species belonging to 35 families usually considered as midwater forms. But, as they say, some families, such as lantern-fishes, hatchet-fishes and gonostomatids, include both midwater species and some that live partly or entirely at benthopelagic levels. For instance, there are the interesting '... observations of J. E. Craddock (personal communication) on *Alvin* Dive 325 in the western North Atlantic. While the submersible was close to the bottom in about 1,025 metres and moving at 1 knot, two species of lantern-fish (*Lampadena* sp. and *Lampanyctus macdonaldi*) all ca. 6 inches in length, were seen in roughly equal numbers over a period of several hours, distributed singly about 3 m apart and less than 2 m off the bottom' (Marshall and Merrett, 1977). Off northwest Africa, these two authors found that lantern-fishes were the preferred prey of benthopelagic squaloid sharks.

Off the northwest African coast the surface waters are enriched by periodic upwellings of nutrients. Primary production is high and through downward transfer of organic matter contributes, as Golovan (1974) says, to the abundant development of zooplankton at near bottom levels. Conditions may well be similar in other upwelling areas, as off Peru, Panama, California and southwest Africa. Where productivity is lower the pelagic contribution to the benthopelagic fauna is presumably also lower. Concerning pelagic sources of food for benthopelagic fishes, Marshall and Merrett (1977) conclude that the extent of this contribution may vary widely. 'Clearly conditions are better in areas of considerable productivity, especially upwelling areas. For instance, Geistdoerfer (1975) found that *Nezumia sclerorhynchus* (a rat-tail) took much less pelagic food (copepods etc.) in the relatively unproductive western Mediterranean than off northwest Africa.'

Movements of benthic animals off the bottom are likely to be in search of food or mates, or due to disturbance. After a baited automatic camera is in position on the deep-sea floor, these animals, which are mainly crustaceans, are much in evidence. Beside the rat-tails and other benthopelagic fishes, large prawns and swarms of amphipods, some quite large, gather round the bait. It may take some time for them to arrive, especially at great depths, where food is very scarce (Isaacs and Swartzlose, 1975). Presumably the crustaceans use good olfactory powers to eventually find the bait. Concerning the large lyssianassid amphipods that were taken in a baited trap placed at 5,720 metres on an impoverished bottom in the central North Pacific, Shulenberger and Hessler (1974) suggest that their slicing mouthparts, very distensible gut and lack of any gut contents other than bait denote a scavenging habit as a normal mode of existence. Mallicephalid worms of the polychaete family Polynoidae, probably also scavengers, have been photographed swimming over the bottom at depths of more than 8,000 metres (Lemeche et al., 1976).

After stating that many neotanaid crustaceans are very widely distributed Gardiner (1975) concludes that transport in near-bottom currents has probably led to such dispersal. He believes that the older animals are almost certainly able to swim effectively, if only over short distances: they may also drift passively with the current. Like most other benthic animals, neotanaids have no pelagic larval stage (pp. 463–5). The young hatch at an advanced stage and some may well be caught up in currents before they settle on the bottom.

The inveterate members of the benthopelagic fauna include copepods, especially spinocalanids, siphonophores (*Stephalia*), cirrate octopods and diverse fishes. As these have already been introduced (pp. 134–42), we may turn to the trophic strategies of the fishes, which are best known in the rat-tails.

Food resources of benthopelagic fishes

It is not altogether surprising that benthopelagic fishes are more highly organized than their relatives at bathypelagic levels. A bathypelagic fish must angle or hunt for food in an impoverished three-dimensional world. If it is versatile enough, a benthopelagic fish may not only forage near or on the bottom, which is a two-dimensional feeding ground and the recipient of all that falls from above, but also range upwards to feed on midwater prey. Considering first the last source, a number of large rat-tails are known when mature to feed at midwater levels. The best known is *Coryphaenoides rupestris* of North Atlantic slope waters, which is trawled by the Russians. 'On the slope off Labrador, Pechenik and Troyanovskii (1971) found that *Coryphaenoides rupestris* were feeding on migrant zooplankton and they correlated the fluctuations in catch rate with the diurnal migrations of zooplankton and the feeding rhythm of the fish. At night, as the forage organisms migrated towards the surface, *C. rupestris* concentrated near the bottom and the catches increased. In the morning the shoals dispersed, the fish beginning to feed actively on the descending zooplankton up to 100–200 m from the bottom, and the catches dropped. Haedrich and Henderson (1974) proposed extensive migrations to explain food items such as the midwater fish *Chauliodus* in the stomachs of abyssal *Coryphaenoides armatus* and Haedrich (1974) reported the pelagic capture of *C. rupestris* up to 1440 m from the sea-bed. Similarly, among deep-sea squaloid sharks, Forster (1971) reported the capture of *Centrophorus squamosus* (13) and *Etmopterus princeps* (1) on vertical lines between 229–1,900 m above soundings of 1,537–3,000 m. Thus, there is a body of data to indicate that fishes hitherto considered to be bottom dwellers do make excursions well off the bottom and that these and other benthopelagic species rely to varying degrees on pelagic prey' (Marshall and Merrett, 1977).

It may well be, though, that such excursions are restricted to the largest members of the benthopelagic fish fauna. Turning now to the deep-sea floor, there are three sources of food; benthopelagic, epifaunal and infaunal. Concerning the last, a fish might make a living if it skimmed off the top layer of the oozes and digested the indwelling nematode worms, harpacticoid copepods, foraminiferans, bivalves and so forth. But only one ooze-eater is known, *Lipogenys gilli*, a relative of the spiny-eels (Notacanthidae) (Fig. 52). This fish, which lives in parts of the western North Atlantic between depths of about 600 and 1,600 metres, has an inferior sucker-like mouth with toothless jaws. Its stomach is 'U'-shaped and very distensible: its intestine is very long and complexly folded (McDowell, 1973). In the stomachs of five individuals, McDowell found a '... seemingly formless mass of pulpy material, sometimes with scattered sponge spicules, plant fibres, crustacean parts and sand grains'. As no fish had an empty stomach, he suggests that '... *Lipogenys* may feed almost continuously on a diet of low nutritional value, rather than selectively on items of high nutritional value'. He believes also that the mouth is modified as a vacuum-cleaner to suck up unconsolidated material from the bottom.

A fish might also swallow the ooze and blow it through its gill slits, leaving the infaunal organisms on its screen of gill rakers. The macrourine rat-tails have a restricted first gill slit, and in many species the gill rakers are short and stumpy and bear small projecting denticles. Such species have very protrusible inferior jaws, which seem capable, on eversion, of sucking in the top layer of sediment. As the mouth cavity of a fish acts as a pressure pump during respiration, the swallowed ooze-laden water could then be gradually forced through the gill slits and expelled through the opercular valves (Marshall, 1973). In sequences taken with their automatic camera, Isaacs and Schwartzlose (1973) have seen rat-tails presumably feeding in this way '... with a sudden explosive thrust into the sediment, throwing a cloud of sediment through the gills'. This way of feeding is used by certain lake-dwelling cichlid fishes (*Lethrinops*) of Malawi (Fryer and Iles, 1972). The fish hovers over the bottom at an angle and then plunges its snout in the sand. Having filled its mouth with sand and water, the fish moves off the bottom and on blowing the contents over its gills, midge larvae are retained by its rows of stumpy gill rakers, which look like those of rat-tails. Each thrust leaves a pit and elevation in the sediment. Similar depressions may be seen in the deep-sea floor and some, as Heezen and Hollister (1971) say, look as though they have been caused by '... a rather short blunt probe into the bottom' (perhaps by rat-tails). Though the gill rakers of rat-tails are not fine enough to retain the meiofauna, the members of the miniature infauna (e.g. bivalves, polychaete worms and tanaids) are large enough to be screened.

Concerning the epifauna, one outstanding fact is that the attached forms are eaten by relatively few kinds of benthopelagic fishes. The exceptions are species of spiny-eels (*Notacanthus*) with a cutting dentition (Fig. 52). *Notacanthus chemnitzi* feeds principally on sea-anemones, and off Genoa, *N. bonapartei* contained many pieces of bryozoans and hydrazoans, but also copepods and amphipods. Species of *Polyacanthonotus* also feed on cnidarians and mobile prey. McDowell (1973) concludes, 'That the premaxillary teeth of *Notacanthus* should be modified to form a cutting edge is in keeping with the food habits, where sessile bryozoans must be sliced from the substrate and portions of anemones cut from the attached animal form conspicuous parts of the diet'. In *Polyacanthonotus*, which is without such teeth, the little known about the diet '... suggests that the genus feeds on whole organisms that do not require any cutting or scraping action for their capture and engulfment'.

Most species of benthopelagic fishes are versatile enough to feed on a mixed diet of mobile prey, both benthopelagic and benthic, but some depend largely on the former source. The classic instances are bathygadine rat-tails, which take such prey as copepods, amphipods, mysids, euphausiids, small natant decapods and arrow-worms (Geistdoerfer, 1975; Marshall and Merrett, 1977). Most likely the amphipods and decapods are taken during their excursions off the bottom: slowly moving benthic forms, such as bivalves and echinoderms, are unknown in bathygadine rat-tails. Indeed, these fishes seem well adapted to take prey moving over the bottom: the wide jaws are terminal and armed with rows of small holding teeth; the gill rakers are long and closely set, so forming a screen to retain small prey such as copepods, and the long dorsal fin is higher than the opposed anal fin. The fin pattern feature is hydrodynamically significant in that undulations of the tail and its fins will tend to lift the head and depress the tail, for the downward component of force exerted by the dorsal fin will be greater than the opposing upward force executed by the anal fin (Marshall and Bourne, 1964).

These features, but not so much the last, are most closely approached in the macrourine rat-tails by the genus *Hymenocephalus* (Fig. 50), the species of which have wide, almost terminal jaws with narrow bands of small pointed teeth and more numerous and longer gill rakers than have their relatives in other genera. Thus, it is not surprising that one species at least, *Hymenocephalus italicus*, exists on much the same diet as a bathygadine rat-tail (Marshall and Merrett, 1977). One would expect the other species to have similar feeding habits.

Lastly, some of the largest kinds of benthopelagic fishes, notably the squaloid sharks, synaphobranchid eels and the slickheads (*Alepocephalus* spp.), also seem to depend largely or entirely on (sizeable) pelagic prey taken near the bottom. Species of both the first and second groups have been found to contain the remains of fishes (e.g. myctophids) and cephalopods (Marshall and Merrett, 1977). The kinds of pelagic food that *Alepocephalus* spp. may take in productive areas have already been covered (p. 272).

Returning to the macrourine rat-tails, it is of more than taxonomic interest to consider how they differ from their bathygadine relative. Outstanding contrasts in their character complex are: the jaws are inferior and protrusible; the snout is reinforced, if only at the tip and side angles, by large bony scales; the anal fin has longer rays than the second dorsal fin and the first gill slit is restricted. When a macrourine rat-tail swims, the action of the opposed dorsal and anal fins will be the reverse of that in bathygadines (see p. 276). Indeed, photographs and movie sequences often show them nose-down and tail-up over the sediments: in contrast, the only bathygadine photograph known to us showed the fish swimming more or less parallel with the deep-sea floor (Marshall and Bourne, 1964). As it forages over the bottom, a macrourine rat-tail might feed on prey moving just above the sediments or it might also protrude its jaws to pick up epifaunal prey. And we have already seen how they may screen the deposits for infaunal food (p. 137).

Comprising more than 50 species, *Coelorhynchus* is the most diverse genus of rat-tails. Their triangular snout is well armed with scute-like scales and the small, inferior jaws are markedly protrusible. In the very productive waters off southwest Africa, where there are profitable fisheries for sardine and hake, a common species is *Coelorhynchus fasciatus*. It feeds on prey moving over the bottom (euphausiids, prawns, lantern-fishes and *Maurolicus*—a small gonostomatid fish), and also on benthic forms (mantis-shrimps, hermit crabs, spider crabs and echinoderms). In Japanese waters *Coelorhynchus smithi* and *C. japonicus* have a rather similar mixed diet. Off northwest Africa, the diet of *Coelorhynchus coelorhynchus* was also mixed, but benthic organisms were predominant. Species of *Coryphaenoides*, *Nezumia* and *Macrourus* are likewise predators on benthopelagic and benthic animals (see Marshall, 1973; Geistdoerfer, 1975; and Marshall and Merrett, 1977).

But some rat-tails, notably of the genus *Ventrifossa*, may depend much more on pelagic than on benthic prey. Species of *Ventrifossa* have relatively long, subterminal jaws, each with an outer row of enlarged pointed teeth. Individuals of *Ventrifossa occidentalis* fished off northwest Africa between depths of 185 and 440 metres were feeding predominantly (in terms of weight) on natant decapods (55·6%) and fishes (35·1%), but also took a few crabs (6·2%). Geistdoerfer (1975), who gives these figures, contrasts this predacious species, bearing large, subterminal and slightly protrusible jaws, with *Coelorhynchus coelorhynchus*, which takes, as we saw, much benthic food and has small, inferior and very protrusible jaws.

Lastly, halosaurs, which are relatives of the spiny-eels, have converged with macrourine rat-tails in certain respects. The snout projects beyond small, inferior jaws and the long tapering tail bears a many-rayed fin along its entire extent. Again, undulations of tail and fin enable their possessor to swim nose-down tail-up over the bottom (Marshall and Bourne, 1964). Halosaurs feed on a mixed diet of pelagic species (e.g. copepods) and benthic species (e.g. polychaete worms and bivalves) (McDowell, 1973; Marshall and Merrett, 1977). Perhaps on soft bottoms they root in the sediments rather in the manner of rat-tails. They are known to eat tanaids, some of which live in burrows.

Benthopelagic fishes thus have varied ways of making a living, and those with the most versatile ways are the most successful. The macrourine rat-tails, which form the outstanding group, are represented by about 250 species from upper slope to abyssal levels (to ca. 6,000 metres) in all parts of the ocean. The more specialized bathygadines are represented by about 25 species with centres of abundance above 2,000 metres in restricted regions. Since they are dependent on pelagic food, they may tend to be centred, where they can gain living space, in the more productive slope waters. But further investigation is needed.

Investigation of space and food partitioning by the dominant species in particular regions is also desirable. Off northwest Africa, the depth range and centres of abundance of common slope dwellers may coincide, overlap or be well separated. For instance, the rat-tails *Nezumia aequalis* and *N. micronychodon* seem to be coincident in these distributional features. They are both mixed feeders, but how different are their detailed distributions and diets? Another rat-tail, *Hymenocephalus italicus*, has much the same distribution and concentrations over the upper slope as *Hoplostethus mediterraneus*, a berycoid fish. But if these two live side by side they will hardly compete for food: the first, as seen already, lives on small pelagic forms, whereas the other takes larger pelagic species (e.g. natant decapods). Clearly there are interesting problems, but what is needed in a particular area is statistical assessment of species associations, together with careful study of each species' diet.

Energetics and locomotion

Earlier we saw how much more highly organized is an abyssal rat-tail (*Nematonurus armatus*) than a bathypelagic gulper-eel (*Saccopharynx*). The contrasts are great but what is their expression in metabolic terms? The metabolism of a gulper-eel has yet to be measured, but that of one rat-tail is known. To achieve this, Smith and Hessler (1974) used a remote under-water manipulator (RUM) in the San Diego Trough off California at a depth of 1,230 metres: here the bottom temperature was 3·5°C and the dissolved oxygen measured 0·71 ml/litre. The RUM is a vehicle with a mechanical arm that can be lowered to the sea-bed by a conducting cable, and when in position can be monitored and controlled through television cameras. The vehicle carried a fish trap respirometer, and after a rat-tail had entered this trap it was closed and watched. After the fish had been in the trap for over 3 hours, the oxyegen it had used during this time was measured. It was a *Coryphaenoides acrolepis* of length 68 cm and weight 1·8 kg. The rate of respiration per kilogram of its wet weight proved to be 2·4 ml oxygen per hour. The comparable figure for cod (*Gadus morhua*) at a temperature of 3°C was over 20 times greater (actually 55·6 ml O_2/hr). One might guess that this measure of the resting metabolism of the rat-tail will be at least ten times greater than that of a gulper-eel. Thus, weight for weight, the food requirements of a cod are likely to be about twenty times those of a rat-tail and at least two hundred times those of a gulper-eel. If one remembers that the gill surface per unit of weight is correlated with the level of activity in fishes, these contrasts are by no means excessive.

A resting rat-tail's use of oxygen seems modest, but how much more is needed when it swims. Clearly, a fish with a long tapering tail is not an active kind. The part played by a laterally moving segment of a fish in its total propulsive effort depends partly on its depth (by d_2): thus, the suceeding segments of a tapering tail will contribute less and less thrust for a given velocity and degree of lateral swing. Indeed, movie sequences show that even when a halosaur or a rat-tail is fully undulating its tail, it is moving unhurriedly over the bottom. Most often, though, photographs show these fishes simply undulating the rear half of the tail (Marshall and Bourne, 1967). Evidently such means, presumably aided by the pectoral fins and outwelling water from the gill chambers, are enough to keep them in slow and steady motion. Moreover, as relatively little muscle is in play, it must be a very economical way of locomotion, needing relatively little oxygen more than that used in resting metabolism. One must also remember that benthopelagic fishes, mostly through a gas-filled swimbladder, are neutrally buoyant and so need little or no energy to stay at a given level (pp. 338–9). Thus, a gently cruising rat-tail is likely to consume much less than one-twentieth of the amount of oxygen (per unit weight) than does a cod (also neutrally buoyant) in its quicker cruising motion.

Recently, the swimming of the deep-sea cod, *Antimora rostrata*, has been observed. It grows to over 65 cm in total length and lives over much of the ocean, mostly between middle slope depths of 800 to 1,800 metres. The deep-sea cod family (Moridae) is predominantly benthopelagic and consists of about 70 species: *Antimora rostrata* is one of the commonest (Iwamoto, 1975) (Fig. 51).

During a dive of the deep submersible, *Alvin*, near the Hudson Canyon at a depth of about 2,400 metres, an *Antimora* was photographed as it swam ahead of the vessel (for 4·2 minutes over a distance of 99 metres at 0·76 knots). It was probably swimming near its maximum speed, and Daniel Cohen (1977), who has analysed this record, points out that its rate of 1·45 body lengths per second (estimated body length 27 cm) agrees well with the speeds of 1 to 3 body lengths per second that have been obtained for endurance swimming in shallow-water cod-like fishes. At this observing speed, he calculates that the deep-sea cod has much the same swimming powers as rainbow trout (*Salmo gairdneri*) of comparable size (though at its maximum speed a 27 cm trout will swim at about

10 lengths per second, a cod at 4–5 lengths per second). Moreover, if *Antimora* respires at the same rate as the rat-tail *Coryphaenoides acrolepis*, then its power should be developed with extra efficiency. Certainly the body form and fin pattern of *Antimora* (see Fig. 51) ought to give it the better swimming powers. It may well be that it is the more active fish, which should be indicated by comparing its gill surface (per unit weight) with that of the rat-tail. Perhaps the maximum speed of the rat-tail is no more than half a knot.

Thus, the rat-tail kind of fish seems well fitted for benthopelagic deep-sea life. At first sight, the rat-tails, halosaurs and spiny-eels seem excessively well endowed in numbers of vertebrae and fin rays. But all these parts are perhaps infrequently used at once: a small, low cost, rear tail section not only provides the power for slow foraging, but also keeps the fish in the right position for feeding near or on the bottom. Eels, which are relatives of the halosaurs and spiney-eels, also have the right form and long fin pattern, etc., for a bottom-dwelling existence, particularly species of the family Synphobranchidae. Little is known of the food of brotulid fishes, most of which are benthopelagic. Evidently they are much less common around baited automatic cameras than are rat-tails. But they are rat-tail like in their long, many-rayed dorsal and anal fins and tapering tail, though in most species the jaws are terminal. Brotulids are almost as diverse and widespread as the rat-tails, but most species seem to be thinner on the ground. It will be interesting to learn more of their means of making a living.

Standing stocks

Perhaps of all marine animals, the fishes, which are complex and muscle-packed, but adaptable creatures, reflect most clearly the biological structure of the ocean. The more productive stretches of the epipelagic zone support large and fast-moving predators, notably the mako-sharks (Isuridae) and tunas (Thunnidae), the only warm-blooded and probably the most highly organized of all fishes. By comparison, nearly all mesopelagic fishes are Lilliputians. Even so, a minnow-sized lantern-fish is also highly organized, especially in its luminous, visual, locomotor and prey-taking systems, parts of an elaborate organism that in most species is carried at neutral buoyancy by a well-developed swimbladder. Life in the mesopelagic world, as we saw, is maintained and integrated by daily upward migrations to feed in more productive water levels. Though bathypelagic fishes must be too far below the productive upper waters to undertake such trophic commuting, they are cunningly fitted to food-poor surroundings by much simpler levels of organisation (pp. 259–69). Benthopelagic fishes are much more complex. But how much fish flesh is supported in the major pelagic strata of the deep sea? Concerning the epipelagic zone, we need only remember that there are profitable fisheries for tuna.

Russian investigations in the western tropical Pacific Ocean and the seas of the Indo-Australian Archipelago produced a wealth of mesopelagic fishes. Besides their taxonomic and biogeographic studies, Parin and his colleagues (1977) measured the biomass of these fishes in the entire water column to 1,000 metres or more. Their results are expressed as biomass per cubic metre and are as follows: Kuroshio zone 5·0 to 6·0 mg/ m^3; equatorial waters, 3·1 to 8·1 mg/m^3; central water mass 0·6 to 2·0 mg/ m^3. Their investigations barely extended to the bathypelagic fish fauna, but in the water column, say from 1,000 to 4,000 metres, the comparable biomass is likely to fall to a tenth (or less) of the mesopelagic value (see Vinogradov's 1968 figures for zooplankton biomass).

Concerning bottom-dwelling fishes, most of which are benthopelagic, Haedrich and Rowe (1977) have provided valuable estimates of biomass from investigations over the continental slope, rise and abyssal levels in the western North Atlantic (south of New England). Beside the fish taken in 116 trawl hauls, which were identified and weighed, estimates of absolute abundance were obtained from series of exposures in a pair of cameras mounted on DSRV *Alvin*. The cameras fired at a rate of one frame per 10 seconds as the submersible cruised along a measured transect, and over 3,500 frames were analysed from eight dives.

The density of fish, expressed in weight per square metre of bottom over areas between 497 to 2,780 metres, varied from 0·63 to 5·78 gm/ m^2, and was particularly remarkable in one respect. Though the numbers of fish (per 1,000 m^2) declined with depth, the mean fish weight suddenly and markedly increased at the beginning of the continental rise at 2,000 metres. By means of samples taken by box cores they also estimated the density of benthic invertebrates. The biomass of both the miniature infauna (their macrofauna!) and larger forms (megafauna) was much the same and could be fitted on a regression line that fell from a $\log_{10}$ biomass of about 1·4 gm/m^2 at 400 metres to 0·6 gm/m^2 at 2,800 metres. Indeed, at depths beyond 2,000 metres the biomass of either invertebrate fauna was not much more than the biomass of bottom-dwelling fishes. Thus, if the infauna is to support the large invertebrates and fishes, its turnover rate through successive life cycles must be rapid but there is no such evidence. Haedrich and Rowe suggest that the dramatic increase in fish size over the continental rise endows individuals with extra powers of foraging for bonanzas falling from above in the form of bodies of decapod crustaceans, squid, fishes and whales. They point out that the large rattail, *Coryphaenoides armatus*, which accounts for 80 per cent of the fish biomass on the continental rise, depends mainly on pelagic food.

In a paper presented to the Joint Oceanographic Assembly at

Edinburgh on 15th September 1976, J. A. Musick, who reported on trawl hauls, made off the middle Atlantic coast of the USA down to 3,000 metres, also found that though the numerical abundance of bottom-dwelling fishes fell with depth, there was a great increase in the mean weight per individual at depths beyond 1,500 metres. As he says, large size not only endows an individual with extra foraging powers, to smell out large food packages, but also increases the size range of prey that can be taken. It is also of interest that the deepest ranging rat-tail of the genus *Coelorhynchus* in the North Atlantic (*C. occa*), is also the largest, growing to a length of 50 cm or more. On the other hand, *Lionurus carapinus*, a relatively common rat-tail at depths beyond 2,000 metres, is of moderate (35 cm) size (Marshall, 1973). Evidently there is more to be discovered. At all events, it is abundantly clear that life over and on the deep-sea floor supports much more fish flesh (and, of course, invertebrate flesh) than is carried at midwater levels. As we have seen, the biomass of mesopelagic fishes is a few milligrams per cubic metre, which probably falls to a tenth or less at bathypelagic levels. But the weight of deep-sea bottom-dwelling fishes is a few grammes per square metre, about a thousand times as much as that in a cubic metre of the mesopelagic zone. We might also recall that weight for weight the food requirements of a cod are likely to be twenty times those of a rat-tail. Compared to a cod, how much greater are the metabolic needs of a blue-fin tuna? Clearly, from tuna to gulper-eel there is an extraordinary range of metabolic and organizational levels in fishes.

Food webs of the   8
deep-sea floor

Except in stagnant basins, communities of animals exist over the entire
deep-sea floor. The main topographic and sedimentary features of this
vast living space have been outlined already (pp. 11–12), as have the
functional design and modes of life of the inhabitants (pp. 143–215). The
concern now is to appreciate, so far as we can, how all these animals
manage to make a living and how they form food webs. Our aim is
towards an integrated understanding of trophic life on the deep-sea floor,
but we must remember that biological studies of the smallest kinds of
animals are not more than 25 years old. First, though, what is the organic
basis of the food pyramids?

Organic food sources

Metabolism on the deep-sea floor is sustained by organic matter derived
from both sea and land. The main supply, which is related finally to the
productivity of the phytoplankton, falls to the bottom in the form of
detrital particles, faecal material and animal carcasses—from copepods
to whales. Fragments of large marine plants, notably of Sargassum weed
and eel-grasses, also reach the deep-sea floor, sometimes in appreciable
quantities.

In neritic waters, where most of the benthic invertebrates produce
planktonic larvae that consume small planktonic organisms, potential
food and organic matter reaches the bottom as young stages settle after
metamorphosis. This kind of supply is virtually absent in the deep sea,
though it may well be that many invertebrates produce advanced benthopelagic larvae that drift and feed over the bottom before they settle (see
pp. 463–5). Lastly, the organic matter in ships' garbage must contribute
considerably to benthic life. Research vessels, which may need to stay
in one place for extended periods, tend to avoid busy shipping lanes,

but it is still surprising how often human artifacts come up in trawls. Human waste will surely have to be considered when we come to measure the productivity of deep-sea life on the bottom. How much, for instance, do ocean weather ships contribute to local productivity?

Organic matter and nutrients reach the ocean through run-off from the land. Apart from bearing high concentrations of terrigenous detritus in suspension, continental shelf waters owe some of their relatively high productivity to nutrient salts formed on land. Deposits on the shelf are thus rich in organic materials, some of which are transported to the continental slope. The means and extent of this shelf to slope transfer are not well understood. The most rapid and far-reaching transport is by turbidity currents to abyssal plains at the foot of the continental rise. All such plains are evidently connected to land-based sources of sediment and organic material through canyons and other channels down the slopes (p. 13).

The remains of terrestrial vegetation also play a part in benthic deep-sea life. Heezen and Hollister (1971) write, 'Logs float out to sea from most of the major rivers of the world and some are eventually cast up on the shore. But much of this floating debris, particularly the less resistant types, must sink to the bottom. Camera explorations along the south wall of the Puerto Rico Trench have revealed logs, cane and palm fronds, littering the sea floor; and in dives to the floor of this trench, dead black leaves, remains of reeds and branches of trees, have been observed.' It is thus not entirely surprising to find wood-boring molluscs in the deep sea (p. 210). Sea-urchins and amphipod crustaceans must also benefit.

All the above forms of organic material are made from solar energy through photosynthesis. It now seems that energy derived from the earth is the basis of rich food chains at certain places along the rifts between tectonic plates of oceanic crust. Besides the eruption of lava at these places, sea water enters newly formed fissures, becomes hot and is chemically changed. This heated sea water is ejected upward through hydrothermal vents and it contains, inter alia, hydrogen sulphide derived from the breakdown of sulphates. Chemosynthetic bacteria that use hydrogen sulphide as a metabolic source of energy are thus able to flourish, thereby, it is presumed, forming food for suspension feeding invertebrates such as clams, mussels and polychaete worms. These and other invertebrates, notably beard-worms, are not only abundant, but very large (see Corliss and Ballard, 1977). These authors, who took part in investigations along the Galapagos Rift, regard these prolific places as 'oases' in a cold sunless desert (see also pp. 306–7).

The fall of organic material

Most of the small organic particles come from decaying life in the euphotic zone, and they are prominent scatterers of light. When the amount of light scattered by suspended particles is measured by lowering a nephelometer* from surface to near-bottom levels, the readings decrease to a minimum in midwater and then increase towards the bottom. The minimum in light scattering is called the 'clear water minimum', and at this level, which ranges from about 1,000 to 3,000 metres, large-scale variations in the concentrations of suspended particles have been investigated over much of the Atlantic Ocean (Biscaye and Eittrim, 1977). The pattern of these variations is strikingly similar to that of the surface productivity (see Figs. 106 and 16). The lowest and most widespread concentration levels underlie the expanse of the North and South Atlantic gyres, where productivity is least. But there are higher concentrations under the more productive equatorial, temperate and near-continental regions.

Organic particles are just heavier than sea water and they settle at a rate of about 1 metre per day. At 4,000 metres, the mean depth of the ocean, particles from the surface waters will take over ten years to reach the bottom. How many nutritious particles pass through the food webs of midwater detritus feeders and reach the bottom before they are consumed by bacteria? Clearly, the longer the fall, the more the decay. It is understandable that the red clay deserts not only extend over the deepest parts of the ocean basins, but also underlie the gyral regions of least productivity. The richer deposits on the slopes are at modest depths and under more productive waters. In the deeper regions, at least, one might well presume that the deposit of small organic particles is too thin and impoverished to play more than a minor role in the support of benthic life.

Means of gauging the fall of sedimentary components are clearly desirable. Recently, a sedimentation trap designed for use just above the deep-sea floor was positioned at a depth of 2,050 metres in the Tongue of the Ocean off the Bahamas Islands (Wiebe, Boyd and Winget, 1976). Sixty-three days after its descent to the bottom, the trap was closed and recovered by DSRV *Alvin* (though it can be set and retrieved as a free-fall vehicle). The rich coating of brown particulate deposit on the filters was analysed for total carbon content, and after elimination of the inorganic fraction, the daily accumulation of organic carbon (from 18 subsamples) was estimated to be 5.7 mgC per square metre. Faecal pellets were the most easily recognized source of organic material, and their

---

* A device that '... consists of a white-light source, a calibrated attenuator and a camera which records water depth and both the attenuated direct light beam ... and the light scattered in a forward direction ... by particles in the water' (Biscaye and Eittrim, 1977).

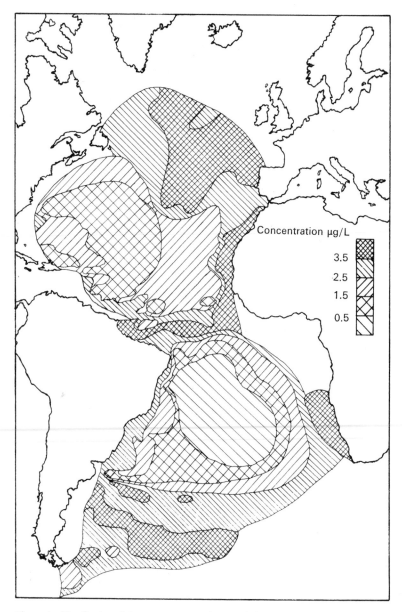

Fig. 106   Distribution of the concentration of suspended particulate matter at the clear
water minimum level in the Atlantic Ocean. (From Biscaye and Eittrim,
1977)

mean fall per day was about 660 per square metre. In laboratory tests, these pellets, which had an average length and breadth of 241 and 109 μm respectively, sank at rates varying from 50 to 941 metres per day. After analysing six *in situ* bell-jar measurements of oxygen uptake at the bottom (made near the trap by K. L. Smith), the investigators estimated the total energy requirements of the benthos to be 38 to 42 milligrams of organic carbon per square metre per day. Their conclusion is that rapidly sinking organic material could supply about 14 per cent of this requirement. But, considering the loss of sediment in recovering the trap and their exclusion of amorphous faecal material from their reckoning, the authors suggest that this estimate could well be about ten times too small.

At their station in the Tongue of the Ocean, the tests of Wiebe and his colleagues indicate that most of the faecal pellets took between 9 and 40 days to reach the bottom at 2,000 metres. Most of such pellets come from copepods, and those packed with the siliceous remains of eaten diatoms must be particularly heavy. Their rapid and direct descent to the bottom in faecal pellets explains why the abundance and distribution of diatom remains in sediments reflects clearly the patterns of primary productivity, phosphate content and annual production of silica in the surface waters (Maynard, 1976). Concerning the entire fall of organic material, recent work by Bishop and his colleagues (1977) suggests strongly that much more comes from amorphous faecal material than from pellets.

At a station in South Atlantic Central water off Liberia, these investigators made close studies of the chemistry, biology and vertical flux of particulate matter in the upper 400 metres of the water column. Light microscopy and scanning electron microscopy were used to determine the sizes of particles and organisms. In the surface mixed layer (0–40 metres) the maximum concentration of the former coincided with that of the latter, which at sizes above 53 μm consisted largely of diatoms, foraminiferans, radiolarians and acantharians. Most of the large particles over this size comprised the appendages and carapace material of crustaceans, mucoid material, hyaline or sheet-like material, faecal pellets and faecal matter. Both kinds of faecal waste, which contained the remains of the organisms cited above, became increasingly important with depth, but pellets were much rarer than amorphous faecal material.

The chemical studies of Bishop and his colleagues suggested that at least 87 per cent of the organic carbon fixed by photosynthesis is recycled in the ecosystems of the uppermost 400 metres. If the remaining carbon reaches the deep-sea floor and is consumed by benthic organisms, the resultant oxygen consumption would be 2·4 ml at STP per square metre per hour, which compares well with *in situ* measurements of benthic consumption (p. 310). At their station the investigators conclude that the organic flux through a level of 388 metres is enough to meet the respiratory requirements of bottom-dwelling organisms. Moreover, virtually all of this flux consists of faecal waste, especially in amorphous form, which they estimate would take 10 to 15 days to fall to a level of 4,000 metres. If the same is true in other parts of the ocean, there is good support for the contention by Wiebe and his colleagues that their estimate of these requirements (based on faecal pellets alone) was about an order of magnitude too small. It thus looks as though benthic deep-sea life is energized much more by 'hails' of faecal waste than by gently settling 'snowfalls' of organic flakes and motes (see also Sieburth, 1976). No doubt the latter kind of fall plays most part over the shallower reaches of the continental slope, where the waters, as we saw (p. 285) are relatively rich in suspended material. Lastly, these recent discoveries add even more to our appreciation of the key role played by copepods in the economy of oceanic life—and we should not forget the euphausiids, the 'copepods' of the micronekton.

The size spectrum of organisms in benthic food webs

Bacteria live in the sediments at all levels of the ocean. When a nutrient medium is inoculated with a small subsample of sediment, whether from the North Sea or the Philippine Trench (at 10,000 metres), bacterial growth is soon evident. The first laboratory experiments on deep-sea bacteria, which were carried out some thirty years ago by Frank H. H. Johnson and Claude E. Zobell, showed, inter alia, that increased pressures reduced the metabolism of the bacteria. But tolerance to pressure varied considerably from culture to culture, and it even looked as though certain (barophilic) forms grow best at pressures (and temperatures) comparable to those in the depths (see also pp. 289–90). Indeed, bacteria, which should be the key organisms at the base of food pyramids on the deep-sea floor, form enzymes for converting diverse kinds of organic material, including recalcitrant kinds, such as chitin, into their own substance.

Deep-sea bacteria are not easily studied. Since the kinds of bacteria in deep-sea sediments include some that are common on land, there is the problem of identification; which forms are truly oceanic? When bacteria are cultured at deep-sea pressures and temperatures many grow actively, but what is the effect of decompression as they are brought to the surface? How active are they in their natural surroundings?

Concerning the last question, the incident of the abandoned lunch on DSRV *Alvin* was first perplexing, then stimulating. In 1968, when about 100 miles south of Nantucket Island, the *Alvin* accidentally sank to a depth of 1,540 metres, mercifully after the pilot and two scientists had made their escape. When the *Alvin* was recovered 11 months later, their packed lunch, which consisted of bouillon, a bologna sandwich with mayonnaise and a dessert of apples, was still very well preserved. But after three weeks refrigeration at an appropriate temperature of 3 °C, the food was spoiled.

As Jannasch and Wirsen (1977) relate, the incident stimulated them to lower packages of various organic nutrients to the deep-sea floor (for periods of two to six months at depths of 5,300 metres). For each submerged package, there was a duplicate in the laboratory that was incubated at bottom temperature (2–3.5 °C) and at atmospheric pressure. When the samples were recovered and analysed, it eventually became clear that the metabolic rate of deep-sea bacteria *in situ* was much lower than that of those cultured in the laboratory: in many instances it was 100 times lower. These experiments led to the design of deep-sea samplers that after collecting bacteria keep them at *in situ* temperature and pressure on their return to the laboratory, where respiration and nutrient uptake can be measured daily. One sampler works to 200 atmospheres (2,000 metres), the other to 600 atmospheres (6,000 metres). Results gained in this way were much the same as the apparent findings of the earlier *in situ* incubation tests: undecompressed deep-sea bacteria are much less active than those cultured at the same temperature but at atmospheric pressure. In one experiment a population kept under deep-sea pressure was 64 times slower in carbon uptake and 11 times slower in respiration than one kept at atmospheric pressure. As for the riddle of the *Alvin*'s sandwiches, they were untouched because of their packing. The last words belong to Jannasch and Wirsen, 'In retrospect it appears that if the *Alvin* sandwich had not been properly kept in a lunch box, fishes and crustaceans would have eaten it and microorganisms would not have had a chance. Nor for that matter, would microbiologists.'

The standing crop of bacteria in a gram or millilitre of deep-sea sediment (ca. 4,000–10,000 metres) seems to be about a million (Thiel, 1975). In sediment samples from the central Pacific at depths from 4,600 to 5,800 metres, Sorokin made direct counts of from 4 to 84 million bacteria per cubic centimetre of deposit, which gave an estimated biomass of 3.3 to 39.2 mg/1,000 cm^3. After testing the bacterial assimilation of CO_2 at atmospheric pressure and 13 °C, he reduced his results to a deep-sea temperature of 2 °C by a (Q_{10}) value of one third. This gave a production of 0.04 to 4.8 mg per 1,000 cm^3 per day or ca. 0.15 to 17.5 gm per square metre per year. But if bacterial activity at deep-sea pressures is but a few per cent of that at atmospheric pressure (Jannasch and Wirsen, 1977), Sorokin's estimate of productivity is still too high by an order of magnitude. If this is taken into account the revised estimate is still comparable with a suggested productivity of 0.05 to 0.1 gm/m^2/year for the meiofauna of the central Pacific (see Thiel, 1975).

Bacteria are the smallest but not the only micro-organisms at the base of the food pyramid. Between the bacteria and the meiofauna there are organisms ranging from 2 to 50μm in size, which include bacterial clumps, yeasts, fungi, amoebae and ciliates. Again in the central North Pacific, Burnett counted from 17,300 to 27,600 individuals of such organisms (the microbiota) in a square centimetre of sediment. Apparently the microbiota live in a thin layer of surface sediment and Burnett (1977) suggests they could well be food for many of the small deposit-feeding invertebrates in the so-called macrofauna. They could even be more important as food for the meiofauna. Presumably the microfloral organisms of the microbiota live on organic particles and dissolved organic matter, and are in turn consumed by the microfaunal protozoans, which may also subsist on bacteria.

Meiofaunal organisms, which range in size from about 0.05 to 1 mm, live in the surface sediment at all depths of the ocean. In sublittoral silty sands the meiofaunal biomass is generally a few per cent of that subsumed by the larger infaunal invertebrates (Gerlach, 1971), but in descending to abyssal levels the relative decrease of the former is much the lesser. For instance, analyses of samples taken at stations from the upper slope to 3,000–5,000 metres in the Atlantic showed the maximum decrease of meiofaunal biomass to range between 16 and 83 per cent (mean 50%) (Thiel, 1975). The comparable decrease in biomass of the larger invertebrates is likely to be at least ten times as much. As intimated already, (pp. 143–6), the dominant meiofauna members are free-living nematode worms, foraminiferans, harpacticoid copepods and ostracods. Less robust members, such as small cnidarians, flatworms (Turbellaria), gastrotrichs and archiannelids, many of which are destroyed on attempted preservation, are known to live at depths down to 3,800 metres or more (Thiel, 1975). There are also rotifers, mites and tardigrades.

In numbers and biomass, and no doubt in activity, the key members of the meiofauna are nematodes and foraminifera. Excluding the latter, nematodes form 39 to 96 per cent (mean 85%) of the meiofauna in 165 samples taken from 11 areas at bathyal to abyssal levels. Moreover, they maintain their dominance with increasing depth or distance from land (Wolff, 1977). Recently, Tietjen (1976) has investigated the distribution and species diversity of deep-sea nematodes off North Carolina, USA. In samples taken either with a Pierce box dredge or a gravity corer, he identified 209 species, of which 106 were restricted to one of four habitats: quartz-algal sands (50–100 metres, 35 species); foraminiferan sands (250–500 metres, 17 species); sandy-silts (500–800 metres, 5 species) and clayey-silts (800–2,500 metres, 49 species). In the shallower sandier habitats, where water percolates briskly through the upper sediment, the most abundant nematodes had long bodies and setae, means to 'hang on' in an active environment. In the deeper, finer and less disturbed sediments the dominant forms had shorter bodies and setae. In the present context the finer sediments were evidently dominated by deposit-feeding nematodes, toothless species that presumably feed by sucking in small particles and organisms. Coarser sediments can be dominated by 'epistrate' feeding species (those with small teeth that may well scrape off fine organic particles and organisms from the grains) and 'predator-omnivores' (species with large teeth), but deposit-feeders are sometimes present as well.

It looks as though nematodes have an intimate and key role in the economy and productivity of benthic life. In the deep sea they must surely have many predators, such as foraminifera, Xenophyophoria, harpacticoid copepods, ostracods and small carnivorous types of infaunal invertebrates, especially polychaetes and bivalves. Virtually nothing is known, but it is tempting to speculate.

The prevalence of foraminiferans also increases with depth. Calcareous forms predominate at depths above 2,000 metres: beyond 3,000 to 4,000 metres virtually all are large agglutinating species. Tests of all kinds are common in sediment samples, but as Hessler (1972b) cautions, even after histological staining one still cannot be sure how many contained protoplasm. Even so, '...the importance of Foraminifera in the community cannot be doubted. They display a diversity of 20 to 25 species in 0·25 m² samples.'

In the deep impoverished sediments of the ocean basins the dominant foraminiferans are members of the Komokiacea, a new superfamily recently described by Tendal and Hessler (1977) (see also pp. 159–62). They are fragile agglutinating forms with a test composed of a cluster of fine branching tubes, the entire complex spanning a few millimetres (Fig. 64). On a few individuals there appears to be the remnants of a fine reticular system of pseudopodia. Over the abyssal floor of the basins box core samples show that the volume of Komokiaceans well exceeds that of the entire assemblage of many-celled animals. The same can be true in the trenches. Their complete depth range is from 400 to 9,600 metres and covers all three main oceans. Like other benthic foraminiferans, they presumably trap food in the pseudopodial system. They are very diverse and most species live in the top centimetre of sediment. But the presence of substantial numbers at sediment depths from 1 to 4 cm makes one realize that their mode of life has yet to be discovered. Again, it is not easy to distinguish living from dead individuals; thus an estimate of biomass is hazardous. Even so, they are clearly a key group in the trophic organization of the deep-sea floor, particularly at abyssal and hadal levels (Tendal and Hessler, 1977).

The allied, but much larger Xenophyophoria, are also important (Fig. 63). Their radiating pseudopodial system, as interpreted from bottom photographs, may extend over a diameter of 6 to 12 cm (Lemche et al., 1976). In some parts of the southwest Pacific trenches these authors estimate that 30 to 50 per cent of the bottom is constantly covered by their pseudopodia. If one thinks of the amoebae in the microbiota, the realization grows that the sarcodine protozoans flourish exceedingly in deep-sea sediments. Flowing from Foraminifera and Xenophyophoria there are pseudopodial systems from millimetres to centimetres in span, cytoplasmic nets to entangle organic particles and organisms ranging in size from bacteria through microbiota to small invertebrates. This is conjecture, but by no means far-fetched.

The nets, covers and screens of American deep-sea biologists (p. 148) have revealed and concentrated attention on a 'macrofauna' of small invertebrates, mostly from 1 to 10 mm in size. Earlier this was referred to as a miniature fauna (p. 148): could it not be called a minifauna? The minifauna consists largely of polychaete worms, peracarid crustaceans, bivalve molluscs and small brittle-stars, most of which are either permanent or part-time members of the infauna (pp. 148–9). In two relatively well-investigated areas, the western North Atlantic and the central North Pacific (in the first area anchor dredge samples were taken between New England and Bermuda at depths from 200 to 5,000 metres, in the second from box cores lowered to about 5,600 metres), most of this fauna consisted of polychaete worms, bivalves and peracarid crustaceans, especially tanaids. Compared to the total number of benthic individuals (some epibenthic), the percentage composition of these three groups in the order just given is: western North Atlantic, $< 4,000$ metres, $70.4 + 13.0 + 6.7$; $> 4,000$ metres, $55.6 + 4.3 + 33.0$; central North Pacific, 5,600 metres, $54.4 + 7.0 + 24$ (see Hessler, 1972). Much the same faunal composition is evident in core samples in the Californian basins (1,130–1,230 metres) and in the Arctic Sea (1,060–2,350 metres). In comparable epibenthic sled samples from the western North Atlantic the same three groups are still important, but epifaunal species, notably small brittle-stars, make up more of the catch (see Wolff's 1977 summary).

The functional design and feeding of deep-sea bivalves were considered earlier (pp. 148–51), but little said of polychaete worms. The forms that are so dominant in soft sediments at all deep-sea levels are mostly but a few millimetres in length, and they are very diverse. For instance, between New England and Bermuda, and in other Atlantic areas, 12 samples from the slope at depths less than 2,000 metres contained 244 species: even at abyssal levels below 5,000 metres, 9 samples contained 89 species (Hartmann and Fauchald, 1971). Some burrow actively, but many species, both errant and sedentary, form tubes, often coated with silt or ooze, that may be several times the length of the occupant. Besides being small, the body tends to be linear and plain, and the relatively few segments bear parapodia reduced to small, papillar elevations (Fig. 107). In many abyssal polychaetes, Hartmann and Fauchauld found also that the body is flaccid and thin-walled, due to the marked reduction of the

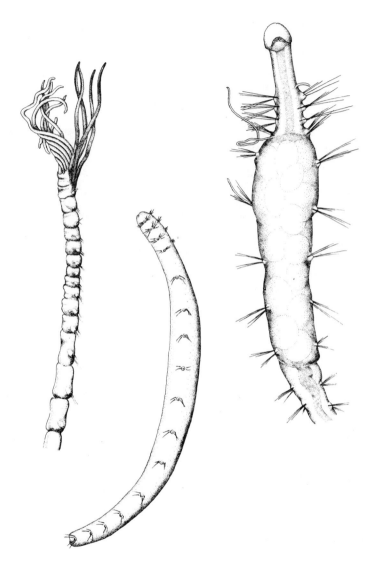

Fig. 107 Polychaete worms from the minifauna of the abyssal Atlantic Ocean. From left to right, *Euthelepus atlanticus*, *Fauveliopsis brevis* and *Tharyx nigrorostrum*. (Redrawn from Hartman and Fauchald, 1971)

muscular system. The strongest muscles, which are in the anterior or masticatory region and the posterior region, are involved in movements through tubes or burrows and the expulsion of waste products. In the 56 families they studied, some species had a muscular pharynx with jaw pieces suggestive of a predatory, carnivorous existence; some had a ciliary apparatus round the mouth like that of filter-feeders, while a third large group had an unarmed pharyngeal pouch characteristic of deposit-feeders. In their analyses of abyssal communities from box cores taken in the central North Pacific (5,500–5,800 metres), Hessler and Jumars (1974) found that more or less actively burrowing or creeping, deposit-feeding polychaetes, such as cirratulids and capitellids, are overwhelmingly predominant. This mode of life prevails also in many peracarid crustaceans (see Barnard, 1961, 1962; Menzies, 1962; Wolff, 1962, and Gardiner, 1975).

Some of the peracarids appear to subsist, at least partly, on the remnants of land plants. In his Galathea Report on gammarid amphipods, Barnard (1961) mentions that a few species contained woody debris. Both amphipods and isopods are presumed to feed on the remains of sea grasses (Wolff, 1976) and we have already seen how boring bivalve muluscs (Xylophaga) depend on large pieces of wood for their livelihood (see p. 210).

The largest invertebrates (the megafauna) are those sampled by large trawls. Many are epifaunal and may be seen in bottom photographs. Octopuses and fishes are also part of the megafauna. As in the meiofauna and the minifauna, there are deposit-feeders, suspension-feeders, carnivores and scavengers. As implied already (p. 19), the suspension-feeders are bound to be underestimated in the ecological organization and economy of the ocean. Volcanic activity, which has above all thrust up the mountain-chains and foothills of the mid-oceanic ridges, sea mounts and islands, has provided attachment sites for countless sponges, hydroids, octocorals, sea-anemones, bryozoans, barnacles, sea-squirts, sea-lilies and so forth. But for bottom photographs and observations from deep-sea submersibles, our knowledge of this fauna would be very sketchy. For a suspension-feeder, conditions of existence are likely to be more favourable on the higher reaches of these rocks, which are nearer, of course, to upper-water sources of life and falling organic materials. Even on sediment-covered expanses of the deep-sea floor, forms attached on flexible stalks are liable to bend under the footings of a trawl and be missed. To uproot and sample such organisms more adequately one needs a heavy dredge with wide biting edges.

Crinoids, classic types of suspension-feeders, have been photographed to depths of about 8,000 metres. The feather-stars (comatulids), as we saw (p. 188), flourish on the slopes down to 3,000 metres or more. They crawl or swim to perches on rocks or attached invertebrates, there raised above the substratum and in positions favourable for intercepting their food from currents. There they live, as Meyer and Macurda (1977) say, as functional stalked crinoids. They can also change their living site: they are less at the mercy of their surroundings than the immovable, stalked sea-lilies. On a dive off the north coast of Jamaica to 3,000 metres, Macurda and Meyer saw how isocrinid sea-lilies could form a filtration-fan (p. 188) by recurving their arms into the current, even at speeds of more than half a knot. They suggest that deep-sea lilies live where there are currents to bring them food. Gardens of sea-lilies grow in coral reefs and elsewhere (p. 189), but in the deep sea they are thinly distributed. Heezen and Hollister's (1971) conclusions from their photographic surveys are that '... they have a spotty distribution for, like other attached forms, they require at least a moderate current and are not found in the more tranquil environments'. Indeed, their photographs show feather-stars attached to other large suspension-feeders, notably sponges and sea-fans, while sea-lilies and sponges may also be seen on one exposure. Sea-fans and sea-pens may also live side by side (p. 182). But are there currents in the trenches where sea-lilies live? There are certainly ripple marks in the sediments of the Puerto Rico trench at 7,000 metres or more, caused by the flow of Antarctic Bottom Water (Heezen and Hollister, 1971). Sea-lilies (*Rhizocrinus*) have been photographed on the floor of the Palau Trench (8,021–8,042 metres) and the South New Hebrides Trench (6,758–6,776 metres) (Lemche et al., 1976). In the latter they were quite common and occurred singly or in groups of three to six individuals, especially on exposed regions. Grouped sea-lilies usually held their expanded arm-crowns, the axes of which were oblique or even at right angles to the bottom, '... in the same direction, as if facing a current'.

Thus, large attached suspension-feeders are rheophiles (current-lovers), but the waters must bear enough food. Off the east coast of America (Carolinas to Georgia), sponges, sea-pens (*Umbellula*), brachiopods (*Pelagodiscus*) and sea-lilies (*Bathycrinus*) live on the lower abyssal red clay zone (5,070–5,340 metres) (Menzies et al., 1973). In this region, especially at shallower levels, western boundary currents along the bottom bear very high concentrations of resuspended particles (Biscaye and Eittrim, 1977). But in a comparable descent off the west coast of America one finds impoverished red clay deposits with virtually no large suspension-feeders. Hessler (1972) says that sedentary animals are so sparse that they are rarely collected. But there is no western boundary current, of course, and the fall of suspended organic material, judging from the rate of sediment, must be slight.

Sea-squirts (ascidians) are also classic types of suspension-feeders, but some of the deep-sea forms have evolved other modes of life (p. 173). Inshore species feed essentially on suspended particles, but deep-sea species, according to the Monniots (1975), fall into four main trophic types. Those with a fine-meshed and ciliated branchial basket like that of inshore forms are much smaller than the latter and more or less spherical in shape (e.g. certain pyrurids and molgulids). The largest forms are rarely over 15 mm in diameter. Others, such as *Culeolus* (and allied stolidobranchiates), have a very large, wide-meshed branchial sac devoid of cilia. In *Culeolus* the body is hydrodynamically shaped (and buoyant enough) to stream at the end of its stalk so that the wide intake siphon, which leads backward into an outlet siphon, faces the current (Fig. 70). In this way, by means not understood, these animals collect suspended organic particles. According to Vinogradova (1972), they filter detritus stirred up from the sediment, and in rare instances the intestine may even contain minute tanaids or amphipods. A third group, called 'traps', and exemplified by octanemids (p. 173), evidently have a mixed diet, shown by the presence in the digestive tract of a twisted cord made of fine organic particles and small crustaceans. In *Octacnemus* the oral siphon is a small opening in the middle of a crown formed by eight large membranous lobes with muscular means of movement. Small invertebrates, particularly crustaceans, are caught by the lobes and conveyed to the mouth. Lastly, there are 'macrophagous prey-hunters' (Sorberacea), which are small, with no branchial sac, and the oral siphon modified into a prehensile device. The large stomach receives prey seized by six muscular oral lobes. In the stomach of an 8 mm *Gasterascidia lyra*, taken in the Bay of Biscay, were more than a hundred foraminiferans, about a hundred nematode worms, many setae from polychaete worms, 11 copepods, 8 ostracods, 4 isopods and the ossicles of brittle-stars. Conditions of existence in the deep sea have clearly evoked incredible adaptive powers in ascidian organization. But the Monniots conclude that the trophic diversity of abyssal ascidians is not entirely related to the scarcity of food.

Attached suspension-feeders are bound to be the dominant forms of life on the exposed slopes of the vast complex of volcanic mountains and sea mounts of the deep-sea landscape. Where sediment can accumulate and is firm enough, or there are other more solid means of attachment, sessile suspension-feeders, as we have seen may also make a living. But deposit-feeders are necessarily limited to the sediments, which between the continental slopes and the mid-ocean ridges cover most of the deep-sea floor (see also pp. 13–15).

Deposit feeders seem to be the dominant invertebrates in soft sediments. In the megafauna there are sea-cucumbers, irregular sea-urchins, certain sea-stars and brittle-stars, gastropods and various 'worms'; sipunculid, echiuroid, hemichordate and polychaete (e.g. terebellids), which for our present purpose have been adequately considered in Chapter 5. This purpose is to try to assess the degree of dominance of deposit-feeders on and in the sediments. Deep-sea trenches and hydrothermal 'oases' will be considered separately at a later stage.

We have seen already that the sediments near the continents and under the more productive oceanic waters are likely to be richest in organic matter. In a descent to abyssal levels the sediments change from hemipelagic muds on the continental slope and rise through pelagic oozes, mainly calcareous and siliceous, until the red clays of the central ocean basins are reached. The muds accumulate at rates between 5 and 100 cm per 1,000 years, the calcareous oozes (Globigerina) at 1–3 cm/1,000 yrs, the siliceous oozes (diatomaceous and radiolarian) at 1–2 cm/1,000 yrs, and the red clays at a mere 0·1–1 mm/1,000 yrs (Heezen and Hollister, 1971). It is thus hardly surprising that the biomass of benthic organisms also declines markedly at greater depths. As Rowe has shown, the benthic biomass is directly proportional to the primary productivity in the overlying surface waters and inversely proportional (exponentially) to the depth. In the transition from shelf to slope, exceptions to this rule occur in rich upwelling regions (Nichols and Rowe, 1977). In the western North Atlantic south of New England the decline in biomass of small infaunal invertebrates between 400 and 4,400 metres is from 15 to 1 gm per square metre of the bottom. There is a parallel decline in the megafauna over much the same range of biomass (Haedrich and Rowe, 1977). In a comparable descent off California towards the impoverished sediments in the North Pacific Basin, the biomass decreases from high nearshore values, due particularly to local enrichment from upwelling, to very low values at 5,000 metres or more. For instance, off California the biomass of large invertebrates declined from 36–40 gm/m² at 200 metres (50 miles from the coast) to 0·2–0·5 gm/m² at 4,350 metres (500–600 miles from the coast). At a station of the same latter depth, but considerably nearer the coast (100–200 miles), the biomass was much higher (3·5 gm/m²) due presumably to the site's greater reception of nutrient material from the land and greater overlying productivity in the euphotic zone (see Thiel, 1975). The ratio of biomass to numerical abundance of animals also decreases with depth, at least in impoverished basins, which means that the animals are diminishing in size (Nichols and Rowe, 1977; see also p. 311).

Trophic conditions of existence on the deep-sea floor are the subject of an illuminating series of investigations by Sokolova, which she has recently synthesized (Sokolova, 1976). Correlating her quantitative studies with surveys by her colleagues on oceanic productivity and sedimentation, she divides the deep-sea floor into eutrophic and oligotrophic regions. In eutrophic (good feeding) regions rates of sedimentation are relatively higher, and the organic matter, all of which may not be decomposed at the surface but part buried in a usable form, is 10 to 15 times that in oligotrophic (scant-feeding) regions. The sedimentation is so slight in oligotrophic regions that virtually all the biologically usable matter is consumed at the surface: the part buried is refractory. Manganese nodules and the teeth of Tertiary sharks are prominent in oligotrophic sediments. As will be evident already, eutrophic conditions obtain on the continental slope, in temperate to polar regions, both north and south, and in much of the equatorial belt. Oligotrophic sediments prevail in the deeper parts of the ocean basins in subtropical latitudes under the oceanic gyres. In the Atlantic and Pacific oceans there are oligotrophic regions both north and south of the equatorial belt, but in the more latitudinally restricted Indian Ocean, there is a main southerly region. In the northern Indian Ocean eutrophic sediments cover much of the Arabian Sea and Bay of Bengal. The change from eutrophic to oligotrophic conditions is not necessarily abrupt but is especially gradual in transitional regions.

Considering first the oligotrophic sediments of the deep ocean basins, Sokolova (1972) concludes that conditions favour suspension-feeders. The biologically useful surface layer of sediment is too thin and impoverished, at least for large deposit-feeders. 'For bottom invertebrates living under oligotrophic conditions, the most effective mode of feeding is on suspended food particles and, to a lesser extent, on particles settled on the very bottom surface. Feeding from subsurface layers is here, as a rule, most unfavourable, as the organic matter buried in a greatly transformed state possesses but little nutritive value. Consequently, in the bottom populations of oligotrophic regions, suspension-feeders are usually dominant and, among deposit-feeders, forms feeding on the surface of the sediments prevail.' The suspension-feeders include small sponges, barnacles, serpulid polychaete worms and brachiopods (*Pelagodiscus*) (Fig. 108).

American investigations on oligotrophic sediments under the Sargasso Sea and the North Pacific gyre indicate that deposit-feeders are predominant, at least in numbers of individuals. Polychaete worms, bivalves and tanaids, the dominant groups, are largely or entirely represented by deposit-feeding species (see also p. 300). Hessler and Jumars (1974) base this conclusion on o·25 m² samples taken by a box corer from a red clay bottom at 5,500 to 5,800 metres in the central North Pacific (ca. 28°N, 155°W). Their analyses of these samples show that 93 per cent are deposit-feeders and 7 per cent potential suspension-feeders. But, as they say, Sokolova's conclusions apply only to the large benthic animals (megafauna) that are taken in trawls. Such large, coarse-meshed nets are clearly not suitable for sampling the small kinds of animals (minifauna) taken in a box core. Conversely, the gear used in American investigations (box corer, anchor dredge and epibenthic sled) will virtually miss the sparsely distributed megafauna. But comparable gear (bottom grab 'Okean', also taking a sample of o·25 m²) has been used by the Russians

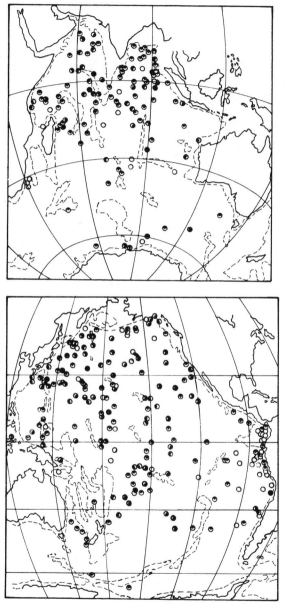

Sponges
Barnacles
Polychaetes
Absence of all
three in trawls

Sponges
Barnacles
Polychaetes
Absence of all
three in trawls

Fig. 108 Distribution of suspension-feeding sponges (orders Hexactinellida and Tetraxonida), barnacles (family Scalpellidae) and serpulid polychaete worms on the floor of the Indian Ocean (above) and Pacific Ocean. (From Sokolova, 1972) (see also Fig. 16)

in the Pacific. The screening of such samples yielded invertebrates in the size range 0·5 to 5 mm, which largely covers the minifauna of American investigations (see also p. 292). In the poorest oligotrophic region in the South Pacific, Sokolova estimated that sponges formed 41 per cent of the total biomass, followed mainly by deposit-feeding polychaete worms (27%) and tanaids (21%). If one adds the megafaunal suspension-feeders taken in trawls (see above), this would still seem to leave this trophic class predominant. One would like to know more of the sponges taken by grab: are they dwarf species adapted to impoverished conditions? There can be no doubt that minifaunal deposit-feeders are numerically prevalent in oligotrophic oozes and that large deposit-feeders, such as holothurians, are poorly represented. It is clear also that investigations of trophic structure in the deep-sea benthos require types and sizes of gear to cover the size spectrum of the fauna. Lastly, there seems to be hardly any place for large, slow-moving carnivores (e.g. sea-stars) in the benthos of the most deserted sediments. But we have seen already that active crustaceans and fishes must be able to smell out sites where relatively large carcasses have fallen.

Eutrophic sediments not only bear detritus- and suspension-feeders but carnivores and scavengers may also be prominent (Figs. 109 and 110). Distributions of all three trophic types are shown in Sokolova's (1972) paper, which was cited earlier (p. 292). Concerning depths between 3,000 and 6,000 metres, Russian stations in the Indian and Pacific Oceans show that deposit-feeding echinoderms (e.g. porcellanasterid sea-stars, pourtalesiid and urechinid sea-urchins and various sea-cucumbers) are largely confined to eutrophic sediments in equatorial and off-continental regions (mostly between 3,000 and 4,000 metres). The same is true of carnivores, such as the predatory sea-stars and ophiurid brittle-stars. Suspension-feeders, such as hexactinellid and tetraxonid sponges, serpulid polychaetes and barnacles, also flourish in these eutrophic regions but extend also to oligotrophic sediments.

Deposit-feeders, suspension-feeders, carnivores and scavengers are most diverse in numbers of individuals and species on the slopes of the continents, especially at upper levels. If it were not for their populations in the most productive trenches, this would apply also to the beard-worms (Pogonophora). They are not abyssal animals: nearly all kinds live with their tubes largely buried in rich soft sediments. The hemipelagic muds of the slopes must here and there contain enough of the right kinds of organic substances to be absorbed by, and meet the needs of beard-worms (see also pp.175–8). The sediments of most slopes are muds and silts, derived from continental erosion. The eroded material bears organic waste, and there is a further addition from primary productivity in the overlying waters, which is enhanced by nutrient run-off from the land. In updwelling areas the slopes are still more productive.

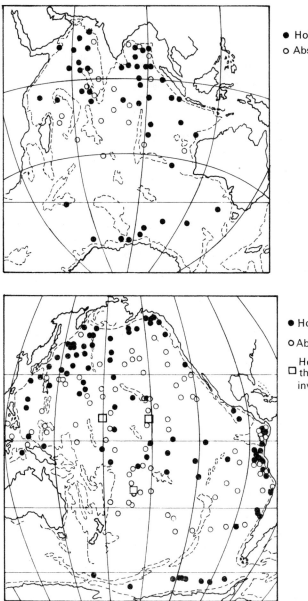

Fig. 109   Distribution of deposit-feeding sea-cucumbers (Holothuroidea; families Mol-
padiidae, Synaptidae, Gephyrothuriidae, Synallactidae, Elpidiidae, Deimati-
dae and Laetmogonidae) on the floor of the Indian Ocean (above) and the
Pacific Ocean. (From Sokolova, 1972) (See also Fig. 16)

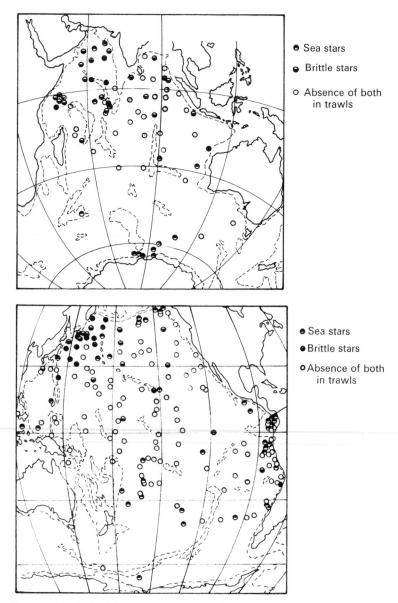

Fig. 110 Distribution of carnivorous sea-stars (of the families Brisingidae, Pterasteridae, Astropectinidae, Benthopectinidae and Goniasteridae) and brittle-stars
(of the family Ophiuridae) on the floor of the Indian Ocean (above) and the
Pacific Ocean. (From Sokolova, 1972) (See also Fig. 16)

Concerning the slope-dwelling species of large invertebrates, Le Danois (1948) has reconstructed the sequence of dominant forms off European coasts. The zone of deep-sea coral (*Lophohelia* etc.) between 400 and 1,000 metres bears other carnivores (hydroids and sea-stars) together with many suspension-feeders (sponges, brachiopods and bivalves). These trophic types are also found at lower levels (1,000 to 4,000 metres) along with deposit-feeders, notably elasipod sea-cucumbers. Actually, all three trophic types may be found at all slope levels, but conditions vary from place to place. For instance, over steep areas where little sediment can settle, suspension-feeders are likely to be dominant. But where sediment can accumulate, deposit-feeders are prevalent (see Sokolova, 1972, and Dayton and Hessler, 1972).

Conditions of existence on the upper slope must have facilitated the invasion of the deep-sea floor by diverse deposit and suspension-feeders, many of which subsequently evolved away from their shallow-water ancestors. Such evolutionary divergence is least evident in a deposit-feeding group, the sipunculid worms (Sipuncula), represented by numerous species that extend from the littoral zone to bathyal and abyssal levels (of 4,000 metres or more) (Cutler, 1977). The suspension-feeding (eulamellibranch) group of bivalves flourish on the upper continental slope, but towards abyssal levels they are replaced more and more by the carnivorous septibranchs and the deposit-feeding protobranchs, particularly the latter (p. 148). Of the passive suspension-feeders, most of the deep-sea corals (pp. 179–86), which probably depend more on live than dead food, have ranges that are much narrower than the eutrophic zones surrounding the continents (Pasternak, 1977).

This is true also of many mobile carnivores of the megafauna. On the continental shelf, gastropods are prominent among the slow-moving carnivores, and the same seems to obtain in the deep sea. In the Woods' Hole transect from New England to Bermuda, samples taken by epibenthic sled between 478 and 4,862 metres contained 93 species of prosobranchs and 30 species of opisthobranchs (Rex, 1976). More than half of the entire gastropod fauna are predators, probably feeding on polychaete worms and bivalves. On the continental slope they are more diverse (in numbers of individuals and species combined) at lower than upper levels and least diverse at abyssal levels. Unlike the deep-sea corals they do not tend to be centred in the richer, nearer-shore belt of eutrophic areas surrounding the continents. But most species are small, and because of this and their low density on the sediment, Rex (1977) suggests that their predatory impact is relatively small. He also suggests that faunal zonation in the deep sea, as shown by depth-correlated changes in species composition, is much more pronounced in epifaunal groups dominated by 'croppers' and predators, than in infaunal deposit-feeders. The vertical ranges of the former higher trophic types may be more compressed by interspecific competition than by that among the latter.

Croppers, as defined by Dayton and Hessler (1972), are animals that ingest living particles '... whether exclusively or in combination with dead or inorganic materials'. Croppers include not only deposit-feeders, such as holothurians, which are bound to swallow small animals in the sediment, but also sea-stars, brittle-stars and those cephalopods, decapods and fishes that prey on small deposit-feeders.

All kinds of sea-stars are thus slow-moving croppers. In the eutrophic northeastern Pacific samples taken off central and northern Oregon from sublittoral to abyssal levels (46 to 4,260 metres) contained 491 sea-stars represented by 29 species (Carey, 1972). Seven species are predators (they always contained much small prey and fragments of larger animals); 13 are omnivores (they contained sediment and considerable quantities of animal remains); 6 are deposit-feeders (they consistently contained sediment with meiofaunal and other animal remains) and 3 were indeterminate. On the shelf and upper slope levels of this transect about two-thirds of the sea-stars were predators, but deeper down this proportion declined until predatory species were absent from the catches at abyssal depths. From the outer shelf to abyssal levels the proportion of omnivores steadily increased from one-fifth to seven-tenths of the sampled fauna. Changes in the proportions of detritus-feeders were not so pronounced. Carey suggests that food supplies are relatively poor at the deeper levels, which will favour omnivorous and detritus-feeding sea-stars; they take what they can get. Like most deep-sea corals, the predatory sea-stars are confined to the richer and higher reaches of the near-continental eutrophic zone (see also p. 300).

The last kind of zonation is typical of certain deep-sea crabs. Most crabs are omnivorous, but they prefer animal food, particularly that seized by themselves (Warner, 1977). As already stated (p 209), large active crabs, such as *Geryon*, are virtually confined to the upper slope. Those, like *Ethusina*, that range to deeper levels (ca. 4,000 metres) are smaller and much less robust than *Geryon*. Crayfish-like kinds of reptant decapods that live below 3,000 metres are represented by species of the eryonid genus *Willemoesia* (Gordon, 1955) (Fig. 89). Perhaps the most successful of the deep-sea Reptantia are squat-lobsters (Galatheidae) and hermit crabs (Paguridea), members of the Anomura (Figs. 90 and 91). There are over a hundred species in the squat-lobsters genus *Munidopsis*: most live beyond 800 metres, close on 20 to depths between 3,000 and 5,000 metres (Gordon, 1955; Birstein and Zarenkov, 1970). Most of the deep-sea hermit crabs live on the upper slope, but some, such as species of *Parapagurus* and *Tylaspis*, range to abyssal reaches beyond 3,000 metres (Wolff, 1961a; Menzies et al., 1973). Unlike the true crabs (Brachyura), the hermit crabs are primarily deposit-feeders. 'Although capable of scavenging, and even of minor predation, their most frequent method of feeding consists of scooping up muddy or sandy deposits with the slightly spooned minor chela and sorting it with the mouthparts' (Warner, 1977). Squat-lobsters feed in rather similar fashion (Gardiner, 1972). This mode of making a living surely goes far to explain their success in the deep sea. But both groups are more or less confined to eutrophic regions.

Benthic fishes and their mode of life have been considered (pp. 212–15). As they rest on the bottom, they presumably tend to be 'sit and wait' predators, which would seem to put them at a disadvantage compared to the neutrally buoyant ranging kinds of benthopelagic fishes, such as rat-tails and brotulids, able to sniff out bonanzas (p. 273). But if a benthic fish can raise itself above the bottom, it may gain an advantage. For instance, Robilliard and Dayton (1969) describe how an antarctic ice-fish, *Pagetopsis macropterus*, will perch on a sponge, where it has more chance of snapping-up pelagic animals. In the deep sea, the tripod-fishes (*Bathypterois*), as we saw (p. 213), are raised on long specially strengthened rays of their pelvic and caudal fins (Fig. 94). Their food consists primarily of very small planktonic crustaceans, especially copepods (Marshall and Merrett, 1977; Sulak, 1977).

Tripod-fishes are rather closely tied to continental and major insular water masses: there are few records from the deep ocean basins, though half the 18 species of *Bathypterois* live to depths beyond 2,000 metres and at least 3 beyond 5,000 metres (Sulak, 1977). In numbers of individuals they form a constant and considerable part of the benthic fish fauna at depths from 250 to 6,000 metres. But at any place they are likely to be outnumbered and outweighed by benthopelagic fishes, such as synaphobranchid eels, rat-tails, brotulids and alepocephalids. Both *Bathypterois* and the related *Ipnops* are common in the deep basins off the Bahamas, where a recent trawling survey showed that species of these genera were third in numerical abundance (15%) after synaphobranchids and brotulids. But in biomass they formed only 3 per cent of the total fish catch (Sulak, 1977). Tripod-fishes are also common in the Panama Basin. During dives in DSV *Turtle*, Heezen and Rawson (1977) saw thousands between depths of 1,000 and 2,000 metres (... they invariably stand on their fins and point directly into the current'). They also saw some of their rat-tail competitors. Rat-tails swam with their mouths near the bottom, the body held at an angle of 30° to 40° to the substrate. ('They respond to any current by swimming directly into it but if the current stops, they immediately swim in random patterns, quickly rearranging themselves into the current when it resumes.') These are invaluable observations of two dominant forms of bottom-dwelling fishes, the first among benthic species, the second among benthopelagic.

Hydrothermal 'oases'

Conditions of existence under hydrothermal vents were introduced earlier (p. 284). Such vents may well emerge in very many places along the 65,000-kilometre-long system of oceanic spreading centres between tectonic plates. Our present knowledge comes from investigations in the Galapagos Rift and East Pacific Rise west of Ecuador at depths from 2,500 too 3,000 metres (Lonsdale, 1977; Corliss and Ballard, 1977). Photographs showed that close to a vent the joints and fissures in the basalt are encrusted by bright yellow and white accumulations of chemical precipitates: within 15 metres of the fissure there can be a community of large and abundant benthic organisms. While surface productivity is high in this part of the equatorial eastern Pacific, Lonsdale suggests that the plumes of warm water draw in adjacent water rich in suspended materials and nourishment for large suspension-feeding bivalves attached to the lava. At the Eastern Pacific Rise photographs show large sea-anemones and sea-pens. There are local pockets of sediment burrowed and tracked by deposit-feeding animals, but clearly the geologically recent fields of pillowed basalt and rich feeding conditions favour the development of sessile suspension-feeders.

As earlier implied (p. 284), Corliss and Ballard have evidence that hydrogen sulphide emerging from the vents energizes the metabolism of chemosynthetic bacteria, which may well go far to meet the needs of suspension-feeders. Whatever the basic sources of food, it is clear that hydrothermal oases are richly endowed enough to support local populations of large beard-worms. In the first of the series of fascinating photographs in the article by Corliss and Ballard, there appear clusters of 46 cm long beard-worms, several cancroid crabs, a reddish brotulid fish and lava pillows encrusted with limpets and sponges, all living at temperatures near 17°C. Presumably the local breakdown of abundant organic matter yields the dissolved food substances to nourish giant beard-worms. Perhaps the limpets browse on bacterial films on the lava, and certainly, in another photograph, the crabs can be seen feeding on serpulid worms. Beside serpulids, other suspension-feeders include mussels, giant clams (vesicomyids) and octocorals. There are also large (3.8 cm wide) organisms 'resembling dandelions gone to seed', which may be unknown forms of protozoa. They are attached by filaments to the rocks and from the body emerge long radiating translucent threads (pseudopodia?). Beside the crabs (there are also squat-lobsters, *Munidopsis*) and brotulid fish, there are other carnivores in the shape of an octopus (hunting crabs?), a skate and a rat-tailed fish.

The oases thus attract abundant populations of those Putterian animals, the beard-worms, suspension-feeders and roving carnivores. Corliss and Ballard cite the founder principle '... which designates the establishment of a new population by a few original members ... which carry only a small fraction of the total genetic variation of the parental population' (Mayr, 1963). First arrivals at a vent may come to dominate an oasis. This, they suggest, may explain why giant beard-worms prevail in a site called the Garden of Eden, whereas bivalves are dominant at Clambake I. Vesicomyid clams taken by the *Alvin* and photographed *in situ*, measured 30 to 40 cm in length: those taken by nets in the poor abyss average less than a centimetre. Clearly, of whatever genes the founders bring, there seem to be none that truncate their growth. Perhaps the dwarfing in at least some deep-sea invertebrates is simply a phenotypic response to oligotrophic conditions.

Trench faunas

Deep-sea trenches were introduced in Chapter 2. Recent research in plate tectonics suggests that they are found at the shear zones between sinking oceanic plates and overriding plates: along these subductions trench-like depressions appear in the deep-sea floor, and become floored with sediment, terrigenous and oceanic. Though not so extensive as the system of mid-oceanic ridges, there is a long broken chain of trenches down the western side of the Pacific Ocean. Southward from the Aleutian Trench, this western chain, beginning with the Kurile–Kamchatka Trench, runs past Japan and divides round the Philippine Plate into Ryukyu–Philippine and Mariana sections. Continuing from the Philippines in an easterly direction along the Pacific Plate there are the New Britain Trench, the Vityaz and New Hebrides Trenches and then the southerly running Tonga and Kermadec Trenches. Close to the continent on the eastern side of the Pacific stretch the Middle America Trench and the Peru–Chile Trench. There are two island-arc trenches in the Atlantic, the Puerto Rico Trench round the eastern side of the Caribbean Plate and the South Sandwich Trench marking the eastern boundary of the Scotia Sea. In the eastern Indian Ocean there is the long Java Trench. Depths in the trenches range from about 6,000 to 10,000 metres.

Thanks to Russian investigations, much is known of life in the Kurile–Kamchatka Trench. Like most other trenches, the Kurile–Kamchatka Trench, as implied by its name, lies off large land masses. Besides receiving sediment and organic matter from the land, this trench lies under waters of high productivity. Belayaev (1972) records a biomass of 1.65 gm/m^2 from 120 large invertebrates at 6,150 metres and 3.44 gm/m^2 from 200 at 6,938 metres. Recent Russian investigations in the South Sandwich Trench, which also underlies productive waters, gave a biomass value of 8.8 gm/m^2 at 6,875 metres (Vinogradova et al., 1974). Trenches far from land masses and in poorly productive areas, such as the Mariana Trench and Tonga Trench, carry a low standing crop of benthic animals. Three samples from these trenches yielded an average biomass of 0·008 gm/m². In another productive area, the Aleutian Trench, samples at 6,460 and 7,286 metres gave 35 and 100 megafaunal invertebrates/m² respectively (Beleyaev, 1972). At 7,298 metres in this trench, Jumars and Hessler (1976) found 1,272 minifauna representatives/m². Lastly, in the Peru–Chile Trench, both near the land and rich upwelling waters, Frankenburg and Menzies (1968) estimated that there were over 60 megafaunal invertebrates/m². Though 6,000 to 10,000 metres below the euphotic zone, the most productive trenches, which support much higher standing crops than exist on oligotrophic sediments between 4,000 and 6,000 metres (< 0·1 gm/m²), the inverse relation between depth and standing crop (p. 297) is evidently more than covered by local productivity, both terrigenous and oceanic.

The richness of the Kurile–Kamchatka Trench is well indicated by the presence of beard-worms (Pogonophora), which in places may be very abundant (Ivanov, 1963 and pp. 175–8). They also live in the South Sandwich Trench, where the studies of Vinogradova and her colleagues (1974) reminded them forcibly of earlier work in the Kurile–Kamchatka Trench. The similarity in biomass and trophic structure of the benthos in both trenches at the '... same latitude of the Southern and Northern Hemispheres is proving the fact that the biological structure of the Ocean is apparent even at its greatest depths'. In both there is an abundance of deposit-feeding sea-cucumbers, mainly of the genus *Elpidia*. In a general survey of the megafauna of trenches between depths of 6,200 and 10,000 metres 54 per cent of the total individuals trawled were sea-cucumbers, followed by bivalves (19%), polychaete worms (7%), peracarid crustaceans (5%), brittle-stars (2%) and other (11% ; 'mainly pogonophorans due to a mass occurrence in one sample from the Kurile–Kamchatka Trench') (Wolff, 1977).

Concerning the minifauna, Jumars and Hessler (1976) have analysed a 0·25 m² box core from the Aleutian Trench (50° 58′N, 171° 37·5′W). The high density of individuals already cited consisted largely of polychaete worms (49%) and bivalves (11·5%), but what surprised them was the relatively high proportion of aplacophoran molluscs (10%), acorn-worms (8%) and echiurid worms (3%). In the meiofauna, foraminiferans (allogromiinid), nematodes and harpacticoid copepods were the dominant groups.

The dominance of polychaetes is typical of deep-sea sediments, but in terms of areal species richness their diversity in the Aleutian Trench is low compared to that found in the San Diego Trough and central North Pacific. Jumars and Hessler conclude that; 'sediment instability caused by rapid sedimentation and frequent seismic activity in turn appears to maintain low species diversity and to prevent sessile polychaetes from comprising a large portion of the fauna.' As they say, trench walls are steep, which coupled with copious sedimentation (as at their station) and the frequent redistribution of sediment in island-arc trench systems following intense seismic activity (Menzies et al., 1973), the fauna is bound to be periodically disturbed, but not necessarily extinguished *en masse*: physical features, as shown by temperature, salinity and dissolved oxygen values, are highly stable. Local falls and disturbances of sediment may not only hinder the establishment of sedentary suspension-feeders, but place '... a premium on the capacity to move'. For instance, elasipod sea-cucumbers are probably sufficiently large, buoyant and mobile to extricate themselves from minor disturbances. But even if a sponge is not covered by a sedimentary avalanche, the ensuing and persisting cloud of silt may clog its water channels and eventually kill it. The latter danger is not confined to trenches, and, as Koltun (1972) writes, 'It is not surprising, therefore, that almost all deep water sponges possess a long stalk or an elongated body, as if striving to bring the vital openings of their body as far as possible from the deadly vicinity of the silty bottom.' Water may flow through a sponge in a current without the aid of its flagellated cells, which fits observations that the total flow through the sponge varies directly with the ambient water velocity and that the height of the oscular chimney in the same species is higher in slowly moving than in rapidly flowing water (Vogel and Bretz, 1972; Vogel 1974). Thus, a stalked or tower-like deep-sea sponge may not only be raised above sedimentary smothering, but have the height for life in slow-moving currents subject to frictional retardation by the bottom.

Though unstable, trenches may still support a considerable diversity of megafaunal invertebrates. This is well shown in the photographic survey of southwest Pacific trenches already cited (Lemche et al., 1976). Deposit-feeders are represented especially by Elasipoda (*Elpidia*, *Peniagone* and *Scotoplanes*) peracarid crustaceans and hemichordates, possibly of an undescribed group (Fig. 1). There are also a surprising number of suspension-feeders: sponges (*Cladorhiza*?), octocorals, sabellid worms, brachiopods, crinoids and solitary stalked ascidians (Fig. 1). There are also diverse carnivores, including the giant hydroid *Branchiocerianthus*, actiniarians, polynoid polychaetes and most remarkably, certain brisingid, pterasterid, goniasterid and asteriid sea-stars. The sea-stars are not only well beyond their usual depth range on more level bottoms, but also attest to the richness of life in the Palau, New Britain, North Solomon and South New Hebrides Trenches at depths between 6,758 and 8,662 metres. There is also foothold and sufficient stability for suspension-feeders, which of all trophic types are most likely to be missed by trawls and grabs. At all events, estimates of benthic biomass in the deep sea are likely to be low, especially in trenches.

The rate and mode of life

Life on the deep-sea floor proceeds very slowly. The metabolism of deep-sea bacteria may be 10 to 100 times less than those kept at the same temperature but at surface pressure (p. 289). As first clearly shown in Smith and Teal's (1973) investigations, the respiration of benthic communities is also relatively low. Bell jars, each covering an area of 48 cm² and containing an oxygen electrode, were placed by deep submersibles on the deep-sea floor (at 1,850 metres) south of Cape Cod, Massachusetts. Here the temperature is between 3·7 and 4·5 °C and there is a slight bottom current. Measurements of oxygen uptake in the jars, which ended after 48 to 72 hours when formalin was released, were between 0·39 and 0·62 ml per square metre per hour. Comparable measurements in shallow-water habitats, as in Buzzard's Bay and Woods Hole outfall (at 22 °C), were over one hundred times greater. Most of the oxygen uptake by inshore benthos is due to bacterial activity, which may also be true at deep-sea levels. Smith and Teal suggest that community respiration may be still lower at levels beyond their stations, and they conclude that the 'metabolic activity of deep-sea benthic communities, which occupy more than 76 per cent of the world's ocean bottom is low'. Later measurements by Smith, already cited (p. 287) support this conclusion.

Large individuals (8·4 mm) of the deep-sea bivalve *Tindaria callistiformis* have been estimated (radioactively) to be at least 100 years old (Turekian et al., 1975), which means that annual increments of shell growth are barely measurable. On the other hand, wood-boring xylophagine bivalves, given a supply of wood at 1,830 metres for 104 days, proved to have all the qualities of opportunistic species (rapid growth, high population density, early maturity and high rate of production) (Turner, 1973). We should recall the gigantism of dwellers in hydrothermal oases (p. 307). But these are surely exceptions to the general slowness of activities in the deep-sea benthos, which is also well shown by recent observations of the slow recolonization of local sediment deprived of its life (Grassle, 1977). In September 1972 the research submersible *Alvin* placed boxes (50 × 50 × 10 cm) of azoic sediment at a depth of 1,760 metres at the Permanent Bottom Station (39° 46′N, 70° 40′W) of the Woods Hole Oceanographic Institution. After 26 months one box was retrieved and the other cored *in situ* by eight 35 cm² tubular cores, which were also used in 1972 and 1973 to obtain 25 control cores of sediment near the boxes. A third box retrieved after 2 months was compared with a similar azoic box left for the same time at 10 metres in Buzzard's Bay. Recolonization of the boxes after 26 months was such that they contained 10 times fewer individuals and species than the samples from the surrounding sediment. Moreover, after 2 months the Buzzard's Bay box contained thousands of individuals ranging from postlarvae to adults of at least

47 species; the deep-sea box held 41 larval and postlarval individuals and 2 adults representing 14 species. Grassle concludes, 'Life processes as reflected in rates of colonisation and growth are remarkably slow in the deep sea. Thus, small scale spatial and temporal mosaics resulting from disturbances on the scale of individual organisms may have a major role in structuring highly diverse deep-sea communities.'

Some of the largest disturbers of the sediment are sea-cucumbers. While largely confined to eutrophic sediments, the slow rate of recovery after their cropping activities presumably forces individuals to seek new grounds from time to time. And many, as we saw (p. 201), are considerably mobile. Even so, the species of the elasipod genus *Elpidia*, which of all deep-sea holothurians have the lowest content of organic matter in relation to their bulk, are largely restricted to rich sediments in highly productive regions near the continents, including the trenches (Hansen, 1975). Acorn-worms (hemichordates) are also restricted to rich sediments. Though less mobile than sea-cucumbers they have efficient ways of cropping the sediment in tightly wound spiral paths (see also Seilacher, 1967). Concerning suspension-feeders, they become smaller and sparser in a descent to impoverished red clay sediments.

Indeed, there is a general decrease in size as the sediments become deeper and poorer: more precisely, due largely to the persistence of meiofaunal and minifaunal components, the relative decline in the number of individuals is less than that of the biomass (p. 297). The size structure of the deep-sea benthos has been reviewed by Thiel (1975), who argues that the fall in mean organismal dimensions with depth is correlated with the decrease of available food and the need to maintain populations of adequate densities for reproductive purpose. There is another factor: organisms may not only be smaller but also simpler, which is certainly true of many infaunal polychaete worms (pp. 292–4). Discussing the low diversity of gastropods at abyssal depths, Rex (1976) suggests that some species are unable to adapt, by reducing their body size and the complexity of their organ systems, to very low levels of trophic resources.[*] When there are no density-dependent limitations to body size in the individuals of growing populations, Fenchel (1974) argues that the ratio of energy used for maintenance to that used for production increases with increasing body size. Thus, small organisms have a productive advantage, especially if they are simply organized. Such adaptations to conditions of existence in the deep sea end present considerations, but they are surely relevant to life-history tactics, which are considered in a later chapter.

---

[*] It will be recalled that fishes in the poorest (bathypelagic) midwater zone maintain or even increase their size in comparison with their mesopelagic relatives, but are still much the simpler (p. 262).

LIFE FROM DAY TO DAY:
PHYSIOLOGICAL ASPECTS

Against gravity

In oceanic biology one always comes back to the 'water column', which everywhere holds an uppermost layer of planktonic plants and animals that give life to the underlying deep-sea animals. Indeed, '... down to the deep-sea floor exist thousands of pelagic organisms, poised, drifting or actively moving at many different levels. Each species lives within a certain range of depth, there to feed, grow and multiply; and, as far as can be seen, the individuals of a species must keep to their particular environment in order to flourish. Study of the means by which plants and animals maintain their level in the ocean is one way of appreciating the latter aspect of living activities' (Marshall, 1954).

During evolution pelagic animals have spread into different deep-sea layers of the ocean: upper mesopelagic (ca. 200–600 metres), lower mesopelagic (600–1,000 metres), upper bathypelagic (1,000–2,000 metres), lower bathypelagic (> 2,000 metres) and benthopelagic. There is overlap between these vertical zones, but as we have seen elsewhere, each is occupied by a recognizable community of organisms that exists in stratified but dynamic stability.

Pelagic organisms, which have diverse means of keeping their level in the water column, are alike in one negatively buoyant respect: their living substance, with a range of about 1·03 to 1·11 in specific gravity, is heavier than oceanic sea water (s.g. 1·02–1·028). The proteins that form much of their tissues have a specific gravity of about 1·33. Skeletal materials, as one might expect, are also negatively buoyant. Thus, skeletons containing calcium salts may have a density of about 3. On the positive side, fats (s.g. ca. 0·9), certain body fluids and gas-filled cavities provide the organism with lift. Clearly, in a neutrally buoyant organism, which can stay at one level with virtually no effort, there must be equipoise in the negative and positive factors of buoyancy. The buoyancy balance sheet, to use E. J. Denton's concept, must remain balanced.

Oceanic plants

For the phytoplankton, light strong enough to energize photosynthesis exists only in the uppermost 150 metres. The continued existence of oceanic life depends on the means by which microscopic plants are suspended in the productive euphotic zone.

Diatoms, whose glassy cell walls make up 20 to 75 per cent of their dry weight, would seem to need appreciable means of positive buoyancy. Most likely, all species are slightly denser than sea water. Their growth and reproduction is not only offset by the grazing of herbivorous zoo-plankters, but also by the fallout below the euphotic zone of living as well as dying cells. Where there are heavy crops of diatoms, as in the Southern Ocean, the Subarctic Pacific and in upwelling regions, samples of diatom ooze may be green and contain living cells.

Diatoms may contain a positively buoyant cell sap, produced by the selective and partial exclusion of heavy sulphate ions, as in large 'bladder-like' species with thin siliceous cell walls. There may also be changes in buoyancy related to nutritional factors and the effect of light (pp. 59–60). Colony formation by linkages between individual cells of the same species increases density, for smaller colonies sink more slowly than larger ones (Smayda, 1971). The larger the colony the greater the 'need' presumably, for the gelatinous threads, spines, setae and other kinds of projections that increase the surface area and retard the rate of sinking. Moreover, diatoms of the open ocean may not only be smaller but also have lighter siliceous skeletons than their relatives of the neritic province.

Most of the other members of the phytoplankton have active means of movement in the form of one or more flagella. The naked micro-flagel-lates should easily maintain their level, which must require more effort by the armoured dinoflagellates and calcareous flagellates. The buoyancy and behaviour of these forms is discussed elsewhere (p. 60 and Marshall, 1954).

The buoyancy of planktonic animals

Beside the coccoliths of calcareous flagellates the calcareous oozes of the ocean are formed of the shells of pelagic Foraminifera, while Radiolaria as well as diatoms form siliceous oozes. Little is known of buoyancy in these two groups of pelagic protozoans. The great surface area of their reticular pseudopodial system, and such area in some species is enhanced by projections of the skeleton, must do much to retard their rate of sinking. The latter, according to figures quoted by Parsons and Takahashi (1973), is 30 to 4,800 metres per day in Foraminifera and about 350 metres per day in Radiolaria. If so, how do members of a population remain long enough at one level to complete their life cycle? Actually foraminifera with spiny tests are generally epipelagic, whereas the non-spinose species, which range also into the depths, may have much thickened shells (Bé, 1966). The latter must have ways of countering their calcareous ballast.

More is known of flotation of radiolarians in the genus *Thalassicola*, which may reach up to 5 mm in diameter. *Thalassicola pelagica* has a central capsule encircled by a froth of large cytoplasmic 'bubbles', each filled with a watery solution (see also Fig. 111). The central capsule, even

Fig. 111   A radiolarian, *Thalassophysa pelagica*.

though it contains droplets of oil, sinks in sea water: evidently, it is the light fluid contents of the 'bubbles' that provide enough uplift (Jacobs, 1935). Observations have shown that the number (and perhaps the contents) of the 'bubbles' can be so regulated that the protozoan can rise or sink in the sea. It is said to descend during stormy weather.

Of the acantharians, species of *Acanthometra* have other means of flotation. Radiating from the central capsule are a number of symmetrically arranged spicules of strontium sulphate provided distally with contractile threads (myonemes), which are inserted also on the gelatinous extracapsular cytoplasm. When the myonemes contract the gelatinous matrix is greatly expanded and the animal eventually rises in the water:

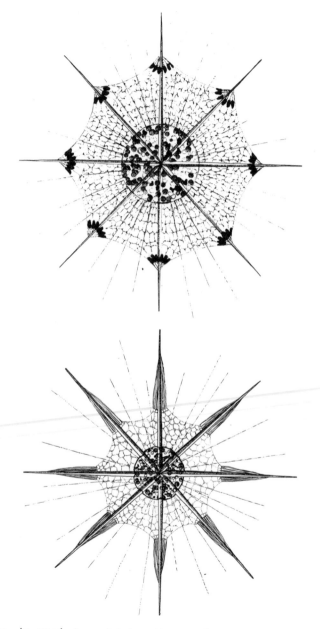

Fig. 112   An acantharian radiolarian. Above, with outer cytoplasm expanded (myonemes contracted); below, with cytoplasm withdrawn (myonemes relaxed). (After Jacobs)

on relaxation of the myonemes the matrix shrinks and is followed by sinking (Jacobs, 1935) (Fig. 112). The expansion is presumably linked to inner changes that provide uplift.

The many larger gelatinous forms of the zooplankton, whether herbivorous pelagic tunicates or carnivorous cnidarians and ctenophores, have many common adaptations, including their means of flotation, to a pelagic mode of life (p. 81). Their very watery tissues, which contain very small quantities of sinking materials, require just the uplift provided by body fluids that are isotonic with sea water, but in which the heavy sulphate ions have been replaced in part by lighter chloride ions. This insight came from the investigations of Denton and Shaw (1961) on the jellyfish *Pelagia noctiluca*, the ctenophores *Beroe* and *Cestus*, the tunicates *Salpa maxima* and *Thalia democratica* and the heteropod mollusc *Pterotrachea coronata*. Except for *Beroe*, which was somewhat over-buoyant, these animals were close to neutral buoyancy. Their body fluids were not only less dense than sea water, because of the partial exclusion of sulphate ions, but also the uplift (mg/m) varied directly with the degree of such exclusion.

The attainment of neutral buoyancy by single planktonic animals is fitting, but the flotation of siphonophores evidently depends largely on over-buoyant members of the colony. When the bracts (hydrophyllia) of a colony are detached from the central axis they float to the surface. Moreover, in physonects there is also uplift from the gas-filled float (pneumatophore) at the top of the colony. The float supports the upper part containing the swimming-bells (nectophores), but the heavy feeding and reproductive members that form most of the colony are buoyed by the gelatinous bracts (Jacobs, 1937) (Fig. 113).

Siphonophores without a pneumatophore (Calycophorae) also consist in part of buoyant members. In the diphyids the feeding and reproducing members are propelled and buoyed by two gelatinous swimming-bells. One may contain an oily cavity (somatocyst), but in *Sulculeolaria quadrivalvis* at least, the oil can have little buoyant effect, for the bell containing it floats underneath the other; the upper swimming-bell together with gelatinous bracts on the stem are the main supports of the colony. In *Hippopodius hippopus*, a representative of another family, there is a cluster of gelatinous swimming-bells, all lighter than sea water (Fig. 114). There are also means for altering the specific gravity in *Hippopodius* and *Diphyes*: in a half-hour to an hour these forms can become heavier or lighter than sea water and so vertical movements may be facilitated (Jacobs, 1937). Presumably the changes of specific gravity are effected by the sulphate-excluding mechanism in the buoyant members of the colony.

Arrow-worms (Chaetognatha), which are also dominant forms of the zooplankton, have much more muscle relative to their size than has a

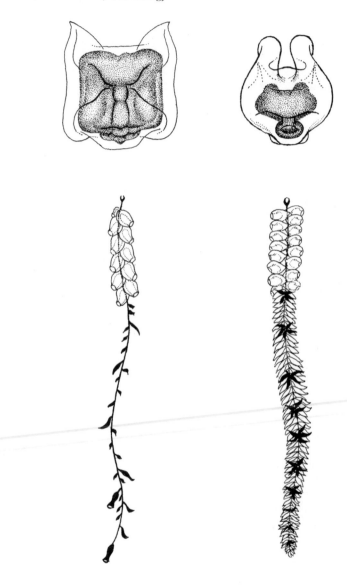

Fig. 113   Two siphonophores. Left, *Nanomia bijuga* with relatively few small bracts
(see figure above) and a gas chamber with a pore; right, *Agalma elegans*
with many gelatinous bracts (see figure above) and a closed gas chamber.
(After Jacobs)

siphonophore, a jellyfish, a comb-jelly, or a pelagic tunicate. Indeed, *Sagitta setosa* (2·1–2·2 cm in length) from the Black Sea have been timed to swim at speeds of about 0·5 to 0·6 metres per second, which is striking proof of their propulsive muscles. Such individuals had a mean density (gm/c.c.) of 1·02, whereas 1·01 was the density of the water in which they were caught (Aleyev, 1977). Their slightly negative buoyancy and tendency to sink can be countered by propulsive movements and, as Aleyev says, the flattened body with its horizontal lateral fins will improve its properties as a lifting surface. Aleyev's figure of 1·02 for

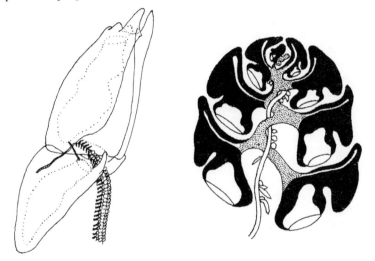

Fig. 114   Two siphonophores. Left, *Sulculeolaria quadrivalis*, lower bell with somato-cyst, lighter than upper bell; right, swimming-bells of *Hippopodius*, show-ing (in black) the great development of gelatinous tissue. (After Jacobs)

the density of *Sagitta setosa* is intriguing in that an arrow-worm of this density would be neutrally buoyant in oceanic waters. But some species are more muscular and thus carry more sinking material than *S. setosa*. Clearly, further investigations are needed.

There is also much uncertainty regarding the buoyancy of copepods and ostracods. When the copepod *Calanus finmarchicus* is anaesthetized it sinks in sea water. Individuals 2 mm in length took 3 minutes to sink through a 25 cm column of water (of temperature 18·5 °C and salinity to 35·01‰): *Calanus* of twice this length, thus with less surface area rela-tive to their bulk, took only half a minute to sink the same distance (Gar-diner, 1933). It would seem that *Calanus* would need to keep swimming upwards to stay at one level. But underwater observations do not bear this out, for, as Bainbridge (1952) found, 'An outstanding feature of all the populations observed was that, at any one time, as high a proportion as 50 per cent would be quite motionless, most of these remaining so for long periods and many so delicately balanced as not to be even sinking in the water. On some occasions, especially when the sky was overcast, the whole population would be hanging motionless or drifting passively.' When at rest *Calanus* hang from their outstretched antennae and plumose setae, from which one might conjecture that their antennal posture and level is kept by downward compensatory kicks of their swimming legs.

Some of the copepods of the open ocean, such as species of *Calocalanus*, *Augaptilus*, *Euchirella* and *Oithona*, and there are others, have most intricate feathery or fan-like setae on their antennae and elsewhere. When these forms spread their plumes they must, like *Calanus finmarchicus*, hang in the sea. But there are many copepods in the sea without such excessively feathered setae. The relative buoyancy of two such species, *Bathycalanus princeps* and *Gaussia princeps*, has been studied by Childress and Nygaard (1974). They measured the relative buoyancies of various crustaceans by weighing living individuals in sea water (5·5°C, 34·3‰, 1 atm.) and in air. Their findings, which are expressed as ratios of the weight in sea water (mg) to the weight in air (gm), thus represent the weight per 1 gram of wet weight that an animal would have to support to maintain its position in the water column.

Both copepods were negatively buoyant, *Gaussia*, with a relative buoyancy of rather less than 30 mg per gm wet weight was appreciably denser than *Bathycalanus* (10 mg/gm wet weight). In these two species and other crustaceans, as Childress and Nygaard show, the relative buoyancy depends largely on the lipid content (assumed specific gravity 0·91 gm/ml) and the protein content (assumed s.g. 1·33 gm/ml) of the animal. Thus, one can see why *Gaussia* with a lipid content of about 3 per cent of the wet weight and a protein content of about 50 per cent of the wet weight is more negatively buoyant than *Bathycalanus*, in which these two figures are about 6 per cent and 40 per cent.

By contrast, the mesopelagic ostracod *Gigantocypris agassizii* gave quite different results. It was neutrally buoyant, and combined with an extremely high water content (95·6%) it had very low lipid and protein contents (0·1% and 0·6% of the wet weight). As Childress and Nygaard observe, its chemical composition is more like that of a jellyfish than a crustacean. Moreover, according to Macdonald (1975) *Gigantocypris mulleri* has body fluids that are isosmotic but lighter than sea water due to the presence of ammonium ions (see also pp. 324–5).

Evidently, *Gigantocypris* can float at one level, but Angel (1971/72) has described how individuals of another deep-sea ostracod, *Conchoecia spinirostris*, sank head first at about 1 cm/sec, with the valves gaping and the limbs moving, after swimming movements. Several times individuals hung, as though neutrally buoyant, in midwater with the valve gape closed and only occasional bursts of activity by the vibratory plates.

Buoyancy of the micronekton

Most members of the micronekton are crustaceans, and we will recall that they are muscular animals with complex skeletons. Euphausiids, which were called the 'copepods' of the micronekton, concern us first. When observed in an aquarium they sink when they stop swimming. Indeed, the common epipelagic species, *Euphausia pacifica*, proved to be the most negatively buoyant of the midwater crustaceans studied by Childress and Nygaard (1974). Its relative buoyancy of 40 mg/gm wet weight may be set beside its low lipid and high protein contents, which were about 2 per cent and 60 per cent of the wet weight. Even more than arrowworms, euphausiids are muscular animals without special means of uplift. Evidently they have muscles and legs enough to keep their position in the water column.

Sergestid prawns are also appreciably heavier than sea water. Thus, the relative buoyancy, lipid and protein contents of two mesopelagic species, *Sergestes phorcus* and *S. similis* are: 32 and 24 mg/gm wet weight, 4.5 and 3.3% and 63 and 53% wet weight (Childress and Nygaard, 1974). Like most sergestids these are active, muscular animals that undertake daily vertical migrations. But species from lower mesopelagic to upper bathypelagic levels, whose vertical migrations, if any, are small in extent, have reduced muscles. Two such species are *Sergestes robustus* and *S. japonicus*, and Donaldson (1976) concludes that they have adapted to the low food supply of their living spaces by reducing their muscular tissues and building up their lipid reserves. Even so, their chemical composition suggests that they are still somewhat negatively buoyant.

By contrast, three of the four oplophorid prawns that Childress and Nygaard (1974) studied were neutrally buoyant. The negatively buoyant exception, *Acanthephyra curtirostris*, had a relative buoyancy of nearly 20 mg/gm wet weight and lipid/protein contents of 7.5/35 per cent of the wet weight. Two of the just buoyant species, *Hymenodora frontalis* and *Systellaspis cristata*, had lipid/protein contents of 19/30 and 12/35 per cent of the wet weight, which would suggest that the former is more buoyant, but it had a lower water content (63.8 per cent) than the other (72.8 per cent). In the third just buoyant prawn, a bathypelagic species of *Notostomus*, the water content was very high (91.3 per cent) and the lipid protein contents (1.5/22 per cent) low. But buoyancy in *Notostomus*, with its 'inflated' cephalothorax, largely resides in ammoniacal body fluids (Herring, 1973), which, as we shall see, is a common means of buoyancy in deep-sea squids.

Buoyancy of the nekton

We begin with the cephalopods, and must note first that active forms of epipelagic squid with much muscle in the mantle have no special buoyancy devices. They are heavier than their environment. For instance, the hooked squid, *Onychoteuthis banksi* and *Ommastrephes pteropus*, which have a density (gm/c.c.) of 1·05, came from sea water with a density of 1·03 gm/c.c. (Aleyev, 1977). Mesopelagic squids, such as *Pterygioteuthis*, *Abraliopsis* and *Lycoteuthis*, with no obvious buoyancy mechanisms, are likely to be negatively buoyant. But Clarke (1977b) estimates that 53 to 78 per cent of the midwater squid that are consumed by sperm whales belong to species with light ammoniacal body fluids.

At least five species of cranchiid squid, which can hang almost motionless in the sea, have a large, fluid-filled buoyancy chamber. The fluid has a relative density of about 1·010 (sea water is about 1·026) and its principal ions are ammonium, sodium and chloride (in concentrations of about 470, 83 and 630 m mol/l respectively). It is isotonic with sea water but nearly all of the cations have been replaced by ammonium ions. Without its buoyancy chamber, the relative density of the animal is 1·046 and thus a large volume of (moderately buoyant) ammoniacal fluid is required to make it neutrally buoyant (Denton, 1971; Denton and Gilpin-Brown, 1973). Moreover, such high concentrations of ammonium, which stop the conduction of squid nerves, must clearly be segregated from vital tissues (Denton, 1971).

The buoyancy chamber of cranchiid squid, which takes up so much of the body volume (about two-thirds), hardly leaves space for the development of a capacious mantle cavity. Indeed, such forms, though capable of rather quick escape reactions, are small-finned, unhurried swimmers. In larger and more active kinds of squid there is no single buoyancy chamber: instead there are many small vacuolar chambers over the body, particularly in the arms and mantle, that are filled with ammoniacal fluids very similar in composition to those of cranchiid squids. In *Histioteuthis* these chambers are so disposed over the body that the animal may assume any attitude with very little effort. The arms are the most buoyant parts of the giant squid (*Architeuthis*), *Mastigoteuthis* and enoploteuthids, which may thus hang vertically in the sea, suspended by them (Denton and Gilpin-Brown, 1973). Representatives of at least 11 of the 25 families of squid are buoyed by ammoniacal body fluids. As in species of the families Chiroteuthidae, Octopoteuthidae and Histioteuthidae, they are approximately weightless in water. Their buoyancy balance sheet has been analysed by Denton (1971), who writes, 'Now if the only components denser than sea water in a marine animal were protein and amino-acids, and the only component giving lift a body fluid isosmotic with sea water but in which sodium ions were replaced by ammonium ions, then the animal would be neutrally buoyant if the ratio of ammonium nitrogen to total protein and amino-acid nitrogen was about two to three. We can, therefore, by analysing whole animals determine whether or not the accumulation of ammonium is a principal buoyancy mechanism. For a specimen of *Octopoteuthis danae* this ratio was found to be 0·69 while for a specimen of *Calliteuthis reversa* it was 1·3. Clearly animals like *Calliteuthis* and *Octopoteuthis* derive the lift giving them neutral buoyancy mostly from the ammonium ions which they contain.'

Lastly, at least one squid (*Bathyteuthis*), like numerous sharks, has a buoyant liver. In this squid, which swims slowly and often hangs motionless in the water, the liver consists of two parts, the forward part consisting of two large chambers filled with reddish-orange oil (C. P. E. Roper, quoted by Denton and Gilpin-Brown, 1973).

Ammoniacal devices are not known in midwater octopods, most of which, like *Vampyroteuthis*, have the look and consistency of a jellyfish. Indeed, at least one species, *Japetella diaphana*, has jellyfish ways of providing uplift. This octopod, which weighs only 0·12 gm in sea water, excludes sulphate ions to the extent that it is half way towards neutral buoyancy (Denton and Gilpin-Brown, 1973).

A gas-filled buoyancy chamber, which is so common among teleost fishes, is found in but one deep-sea cephalopod, the sepioid, *Spirula spirula*. Like cuttlefishes (*Sepia*) and *Nautilus*, *Spirula* has means of pumping liquid out of chambers that become filled with gas (Fig. 115).

Spirula is a mesopelagic cephalopod, which off the Canary Islands is found at depths of 600–700 metres by day, and after vertical migrations, between 300 and 100 metres by night (Clarke, 1966). Living individuals float gently upward in a vertical position with the buoyant, shell-containing part of the body uppermost, the head and tentacles pointing downwards (Fig. 115). In adults the spiral shell is formed of thirty or more calcareous chambers through which runs a tubular siphuncle, differentiated in part as a fluid-pumping device. Whatever the depth of capture, the pressure of gas in the shell chambers is less than atmospheric (0·8 atm.). Under pressure the shell implodes at pressures corresponding to a depth range of 1,300 to 2,300 metres, well below the centre of concentration of *Spirula* (Denton, 1971). As *Spirula* grows new chambers are added to the shell. Both the newest and oldest chambers contain liquid. A newly forming chamber is first full of liquid isosmotic with sea water, then after some removal of solutes followed by withdrawal of liquid, the chamber contains a sizeable gas bubble, the liquid around it now being very

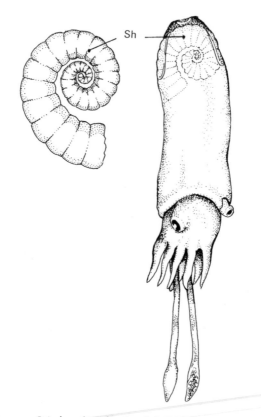

Fig. 115   *Spirula* and its chambered shell (Sh). (From Marshall, 1971)

hyposmotic to sea water. By this time a still newer chamber is partly formed, and when the penultimate one is nearly full of gas the last chamber is not only sealed but also contains a fluid isosmotic with sea water (secreted into it by surrounding tissue) and now ready for evacuation. The liquid is removed by a special epithelial section of the siphuncular tube, but how this pump works against high hydrostatic pressures has yet to be determined (Denton and Gilpin-Brown, 1973).

The buoyancy of midwater fishes

There are some 1,000 species of midwater fishes, and those containing a gas-filled swimbladder live at mesopelagic levels. These species, which are mainly lantern-fishes and small stomiatoids, form about 40 per cent of the mesopelagic fauna (ca. 850 spp.). In numbers of individuals, as judged by the catches of large midwater nets, well over 75 per cent of mesopelagic populations consist of fishes that derive most or some of their buoyancy from a gas-containing swimbladder. Most of these individuals (those of *Cyclothone* spp. are outstanding exceptions) undertake daily vertical migrations. More and more evidence indicates that the swimbladders of such fishes are predominantly the cause of the sonic scattering layers that are recorded on ships' echo-sounders in the open ocean (Marshall, 1951, 1960, 1970; Brooks, 1977).

Many of the mesopelagic fishes lacking a swimbladder are predatory forms. Some such as *Chauliodus*, *Stomias* and various melanostomiatids, are migrators: others, like the alepisauroids and giganturids, are non-migrators. There are also non-migrators with small jaws (bathylagids) and moderately long jaws (searsiids). Even though lacking a swimbladder, such fishes, as we shall see, are only slightly heavier than sea water.

Bathypelagic fishes either have no trace of a swimbladder, as in ceratioid angler-fishes and gulper-eels, or a regressed swimbladder (*Cyclothone* spp.). More than mesopelagic fishes that lack a swimbladder, bathypelagic species approach a neutrally buoyant condition.

To make its user weightless in water, the swimbladder of a marine fish should have a volume equal to about 5 per cent of the animal's total volume (Jones and Marshall, 1953). In mesopelagic fishes this percentage is more likely to range between 4 and 1, but species with the smallest swimbladders may derive buoyancy from reserves of lipid substances. A marine teleost also gets uplift from dilute body fluids that provide about one-sixth the lift exerted by a 5 per cent swimbladder. The main heavier-than-sea-water components are proteins, especially in its lateral muscles, and skeletal materials (Denton and Marshall, 1958).

In deep-sea fishes, as in other teleosts with a closed swimbladder, gas is secreted into the swimbladder by a gland that receives blood through closely parallel systems of arterial and venous capillaries (retia mirabilia). In lantern-fishes an artery–vein pair that enters the fore part of the swimbladder subdivides to form three retia mirabilia, each made of hundreds of contiguous arterial and venous capillaries that run in parallel to a corresponding lobe of the gas gland* (Fig. 116). Blood flows to the gland through the arterial capillaries, and after circulating among the gland cells in capillary loops, returns to the venous capillaries of the rete (Fig. 117). There is a single posterior rete mirabile in the swimbladder of stomiatoid fishes but the capillaries do not pass directly into the gas gland:

---

* Gas glands, which are rich in glycogen, are specialized parts of the inner epithelial lining of the swimbladder. Between this epithelium and the tough outer fibrous coat of the swimbladder wall is a loose gelatinous connective tissue layer (submucosa) that may hold a layer of smooth muscles.

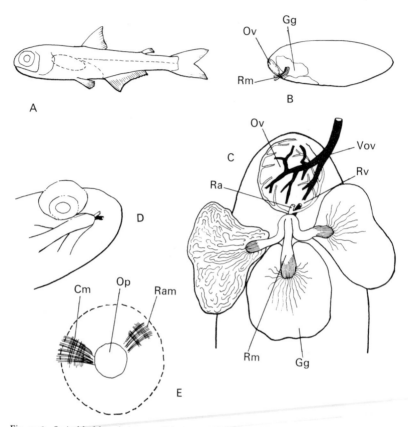

Fig. 116   Swimbladder of a lantern-fish, *Myctophum punctatum*; A, in position in body
cavity of fish; B, laterally; C, ventrally (anterior part of organ). D and E,
the gas-resorbing part (oval) of the swimbladder of another lantern-fish,
Diaphus rafinesquei. Cm, circular muscle of oval; Ram, radial muscle of oval;
Gg, gas gland; Op, opening from oval into swimbladder chamber; Ov, oval;
Ra, retial artery; Rv, retial vein; Rm, rete mirabile; Vov, vein to oval. Veins
shown black, arteries white. (From Marshall, 1960)

instead they reunite to form larger vessels that run into the gland and
then form the looping circulation.

The countercurrent system of capillaries in a rete mirabile serves both
to keep gas within the swimbladder and to build the requisite gas tensions
for secretion through the gas gland. The first function is by way of gas
exchange between the blood in the two sets of capillaries. Blood leaving
the swimbladder will enter the venous capillaries bearing dissolved gases
at a pressure equal to that in the swimbladder (which in atmospheres
at deep-sea levels is about one-tenth of the depth in metres). As there

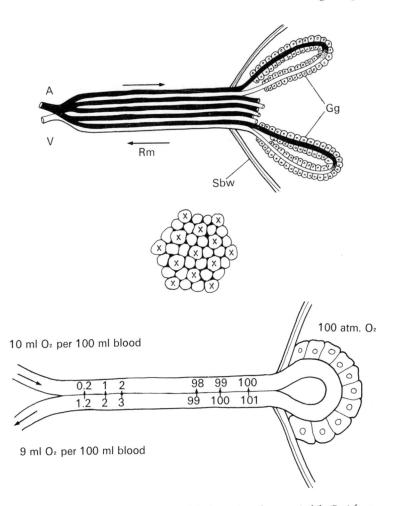

Fig. 117 Top, schematic representation of the formation of a rete mirabile (Rm) from an artery (A) and a vein (V). The arterial capillaries are shown in black. Within the swimbladder (Sbw is the swimbladder wall) are shown two capillary loops and associated gas gland cells (Gg).

Centre, a few of the capillaries of a rete mirabile as seen in cross-section. (The larger venous capillaries are labelled X.)

Bottom, schematic representation of the countercurrent multiplier system of the teleost fish swimbladder. The numerous capillaries of the rete mirabile are represented by a single loop. As the gas gland produces lactic acid, the oxygen tension increases (above that in the swimbladder) in the venous capillary (below), and gas diffuses across to the arterial capillary (above), thus remaining in the loop. As it leaves the rete, the venous blood contains less oxygen than the inflowing arterial blood. (After Schmidt-Nielsen, 1975)

are merely two thin capillary walls between the venous blood and the blood in the arterial capillaries, gas will diffuse freely into the latter and continue to do so as the blood flows down the venous capillaries. When the blood is about to leave the rete it flows by ingoing arterial blood that contains gas tensions (notably oxygen at about 0·2 atm.) acquired during circulation through the gills. Thus the venous blood in the rete can lose gases until it is in diffusion equilibrium with the inflowing arterial blood (Fig. 117).

While gaseous exchange is facilitated by the short diffusion distance ($< 1 \mu m$) between the blood in the two sets of capillaries, the efficiency of the rete as a gas retainer depends largely on the total capillary surface available for countercurrent exchange, which in turn depends on the number and length of the retial capillaries. Indeed, these two dimensions in the rete of a deep-sea fish ensure inter alia that it is a very effective system for maintaining high gas pressures (even in excess of 4,000 atm.) in the swimbladder (Scholander, 1954).

How a rete mirabile can build gas tensions through countercurrent multiplication was first clearly shown by Kuhn and his colleagues (1963). As they assumed, and later study has revealed, the gas gland produces lactic acid, which on entering the circulation, has two effects on the venous blood that flows into the retial capillaries; it releases oxygen from haemoglobin (Root effect) and decreases (by a 'salting-out' process) the solubility of all gases in the blood. The gas tension generated by both processes causes the released gases to diffuse into the blood of the ingoing arterial capillaries and gradually accumulate at the gas-gland end of the rete: and here, as the blood continues to circulate, the pressure will steadily increase (Fig. 117).

Considering the salting-out process alone, the multiplying-gas-generating power of a rete depends directly on its length and the degree of permeability between the two blood systems. The inverse factors are the speed of flow in the capillaries, the bore of the capillaries and the solubility coefficients of the gases. In theoretical equations based on these relationships, Kuhn and his colleagues have shown that a rete 20 mm in length can easily build the gas tensions that exist in the swimbladders of deep-sea fish. For oxygen itself—the main gas in the swimbladder of a deep-sea fish—a small part (about 1 to 2 mm) of the retial capillary length is enough to generate, by both the salting-out and Root effects, pressures between 10 and 100 atmospheres. At pressures above 100 atmospheres further increase in the tension of oxygen is solely by way of the salting-out effect. High tensions of gas may thus be presented to the gas gland, but just how gas is secreted is still unclear.

The organization of the swimbladder in deep-sea fishes may now be set against these physiological sketches of the maintenance and generation of gas tensions. After survey of the mesopelagic fauna it will be

(Fig. 116), *M. affine*, *Diaphus garmani* and *Lobianchia garmani*, and hatchet-fishes of the genus *Argyropelecus*), the length of the rete is about 0·75 mm in the stomiatoid fishes *Maurolicus*, *Valenciennellus* and *Vinciguerria* (Fig. 118). The retia in the latter three genera are unusual also in having narrower capillaries (arterial 2–4 μm; venous 5–10 μm) than those of the majority (7–18 μm). Moreover, to fit the narrow-bore capillaries the red cells (erythrocytes) of the blood are unusually small and lack nuclei. (*Maurolicus muelleri* has rod-shaped erythrocytes 5 × 2.5 μm; *Valenciennellus* has globular erythrocytes 7 × 7 μm). The erythrocytes of other deep-sea teleosts (*Bathylagus*, various stomiatoids, lantern-fishes, rat-tails, tripod fishes, etc.) are larger nucleated and flattened, the largest dimension ranging from 7 to 20 μm (Hansen and Wingstrand, 1960). These authors suggest that the functional significance of small, non-nucleated erythrocytes is in their high surface to volume ratio, which will ensure rapid exchange of gases, and in their increased oxygen-carrying capacity through a high content of haemoglobin, owing to loss of nuclear ballast. Moreover, despite their shortness the small bore of the retial capillaries still provides, relative to the contained volume of blood, a large surface area for countercurrent functions (Marshall, 1972). There is also a high number of capillaries in the rete, which in *Vinciguerria* is about 5,000, a figure that may be compared to the 2,000 odd larger-sized capillaries in the longer retia (by 2·5) of the lantern-fish, *Myctophum punctatum* (Marshall, 1960).

In lantern-fishes from lower mesopelagic levels instances of length of the retia mirabilia are: *Lepidophanes guentheri* 3·0 mm, *Hygophum benoiti* 3·0 mm, *Lampanyctus pusillus* 3·0 mm, *L. alatus* 4–5 mm, *Lampadena chavesi* 6·0 mm, *L. speculigera* 7·0 mm (Fig. 118), *Taanigichthys bathyphilus* 7·0 mm. The greater retial dimensions of the last three species, whose depth ranges extend into the upper bathypelagic zone, should be noted.

The depth ranges of benthopelagic fishes with a gas-filled swimbladder extend from about 200 to 7,000 metres. Such species, as we will recall, are mainly rat-tails (Macrouridae), deep-sea cods (Moridae), brotulids, halosaurs, notacanths and synaphobranchid eels. The number of retia mirabilia, length of the retial capillaries and depth ranges of representatives from these families is shown in the table (Marshall, 1972).

It will be seen that the first four species, whose commonest depth ranges most closely correspond with those of upper mesopelagic fishes, have retia mirabilia of 4–6 mm in length (cf. mesopelagic species 1–2 mm). Species such as the rat-tails *Coryphaenoides rupestris* and *Coelorhynchus occa*, with commonest depth ranges at lower mesopelagic levels, have retia of 8–10 mm (cf. mesopelagic species 3–7). These contrasts in retial dimensions '... are most probably related to differences in the vertical spread of mesopelagic and benthopelagic fishes. The former, as echo-

Depth ranges of benthopelagic deep-sea fishes and lengths of swimbladder retia mirabilia

(Initial letter of family name to the left of each specific name: S. Synaphobranchidae; H. Halosauridae; Mo. Moridae; M. Macrouridae; B. Brotulidae; T. Trachichthyidae.)

Species	No. of retia	Length of retial capillaries (mm)	Depth range (depth range where most common) (m)	
M. Malacocephalus laevis	2	3·4	150–1315	(150–600)
M. Ventrifossa occidentalis	2	5–6	150–585	(300–500)
M. Coelorhynchus carminatus	4	6	90–850	(300–600)
M. Macrourus berglax	4	6	100–1350	(200–600)
M. Hymenocephalus italicus	2	6	200–2080	(300–800)
T. Hoplostethus mediterraneus	2	5	235–780	
M. Coryphaenoides rupestris	4	8	160–2260	(500–1400)
M. Coelorhynchus occa	4	8–10	460–2220	(700–1100)
M. Nezumia aequalis	2	11	200–2320	(400–1000)
Mo. Antimora rostrata	4	8–9	400–29000	(800–1800)
S. Synaphobranchus kaupi	2	10	200–3600	(800–2000)
M. Bathygadus melanobranchus	2	10	640–1590	(800–1400)
M. Gadomus longifilis	4	12	630–2170	(800–1500)
M. Cetonurus globiceps	2	15	860–4260	(1100–1830)
M. Coryphaenoides guntheri	4	20	830–2600	(1000–2000)
B. Basssogigas profundissimus *	1	15	5610–7160	
B. Bassozetus compressus	1	17·5	1920–2750	
H. Halosaurus ovenii	2	c. 20	440–1620	(800–1300)
H. H. parvipennis	2	c. 20	780–1610	
H. Aldrovandia affinis	2	c. 20	840–2615	(1000–2500)
B. Bassozetus taenia	2	25 +	4570–5610	
M. Nematonurus armatus	5	25 +	280–5700	(2600–3600)
M. Lionurus carapinus	6	25 +	1210–5300	(mainly below 2000 m)

* Data from Nielsen and Munk (1964).

sounder records suggest, and as proper surveys of vertical distribution are revealing (see Badcock, 1970), have relatively narrow depth ranges. Such vertical disposition is presumably related to the coactive place of each species in a complex ecosystem, and is clearly of essential biological value in species that undertake daily vertical migrations. Benthopelagic fishes tend to range downward well beyond the lowest depth of their most common occurrence (see the table). Hence, it is suggested, their built-in safety factor in the shape of comparatively long retia mirabilia' (Marshall, 1972).

The deepest living benthopelagic fishes, centred at levels below 1,500

metres, have swimbladder retia from 15 to 25 mm in length. Thus in a
series of mesopelagic and benthopelagic fishes ranging between depths
of 200 metres and 7,000 metres there is a direct relationship between
depth range and the length of the retia mirabilia in the swimbladder (Fig.

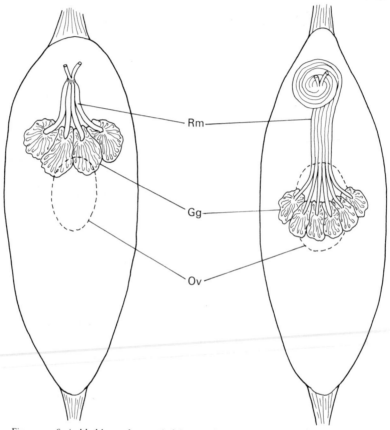

Fig. 119   Swimbladders of rat-tail fishes. Left, *Coelorhynchus coelorhynchus*, an
upper slope-dwelling species; right, *Lionurus carapinus*, an abyssal species.
Note the much longer retia mirabilia (Rm) of the latter. Gg, gas gland; Ov,
outline of oval. (Redrawn from Marshall, 1965)

119). To recapitulate, the efficiency of the swimbladder, whether in maintaining or generating gas tension, depends directly on the length and
interchange surface area of the retial capillaries. Both functions become
harder with increasing depth and pressure; hence, the graded increase
in retial capillary length in a series of species that range from shallow
to deep reaches of the ocean. Lastly, we should note how well the

theoretical maximum length (20 mm) of a rete (Kuhn et al., 1963) fits the lengths (25 mm) found in the deepest living benthopelagic fishes.

Returning now to the mesopelagic fish fauna, some species when adult contain a relatively small swimbladder. In adult species of *Cyclothone* (*braueri, alba, pseudopallida*) the swimbladder has much the same position and relative size that it once had in the postlarval stages. It still contains gas: indeed, the length of the rete (3-4 mm) is comparable to that in fishes with well-developed swimbladders from the same levels. The *Cyclothone* are neutrally buoyant and the swimbladder in concert with a high lipid content and dilute body fluids evidently provides enough uplift to counter the relatively slight negative buoyancy of the muscles and skeleton (which are much less well developed than those in relatives of comparable size, e.g. *Maurolicus*, with a capacious swimbladder).

In other mesopelagic fishes, including various myctophids, melamphaids, stomiatids and hatchet-fish (*Polyipnus*), the swimbladder regresses during the life history. In the adult it is invested with fat and ranges in structure from a small thick-walled organ containing a small amount of gas to a barely recognizable rudiment of the gas-filled organ of earlier stages in the life history.

In his analysis of buoyancy relations in some of the lantern-fishes Bone (1973) divides them into certain functional types. Species of the genera *Myctophum, Hygophum, Protomyctophum* and *Symbolophorus* which have a capacious swimbladder, have a low lipid content (< 5%) and are neutrally buoyant, as probably are species (e.g. *Diaphus rafinesquei* and *Ceratoscopelus maderensis*) with a higher lipid content (8.2-8.9%). Species of the genus *Stenobrachius* and certain species of *Lampanyctus* (e.g. *ritteri*), which have a high lipid content (14-19%), made up largely of low density wax esters, have little or no gas in the regressed adult swimbladder, but they are still neutrally buoyant. Lastly, species (e.g. *Notoscopelus kroyeri* and *Tarletonbeania crenularis*) also with little or no gas in the adult swimbladder have a low lipid content (3-5%) and, not surprisingly, are negatively buoyant.

Vertical migration and buoyancy

In numbers of individuals most of the fishes that commute daily between midwater and higher levels have some form of swimbladder. In both numbers of species and individuals the majority of the migrators are lantern-fishes, which are now considered.

If a neutrally buoyant species with a capacious gas-filled swimbladder (say, 4 per cent of the body volume) is to maintain its buoyancy at all levels the volume of the swimbladder must be kept more or less constant. When the lantern-fish descends after feeding in the surface layer the swimbladder will be compressed by the increasing pressure. Thus, during a descent from 100 to 500 metres the volume of the swimbladder will decrease by about a fifth (if no gas is secreted). During an upward migration from 500 to 100 metres the swimbladder will undergo a five-fold increase in volume. Clearly, gas must be lost if the fish is not to become dangerously over-buoyant.

In lantern-fishes gas is lost through a thin-walled fore part of the swimbladder wall known as the oval. When fully exposed through the contraction of its radial muscles the highly developed capillary circulation of the oval comes very close to the gases in the swimbladder. These will diffuse into the oval at a rate depending on the area of the capillary bed, the rate of its circulation, the pressure (the difference in tension between the swimbladder gases and those in the blood) and the temperature. Apart from the first two factors, there is no way of increasing the rate of gas resorption; in fact, the area of the capillary bed is not especially large compared to the volume of the swimbladder (in a lantern-fish the ratio between these two dimensions is about twice that in a freshwater perch, *Perca*), but judging by the size of the venous vessels that drain the oval the capillary circulation must be very well developed (Marshall, 1960) (Fig. 166C).

While deep-sea fishes that make extensive migrations have both a very well-developed gas-producing and gas-resorbing system, it seems unlikely that they can secrete and reabsorb gas quickly enough to stay neutrally buoyant at all depths. In most sound scattering layers Hersey and Backus (1962) found that the peak frequency of reverberation increases with depth during the course of a diurnal migration, and they continue, 'The relationship that would obtain were the swim bladder and its gas passively responding to changes in ambient pressure $(F/F_0 = (P/P_1)^{5/6})$ has been observed during both a sunrise and sunset migration. This seems to imply that the retention of the swimbladder is worthwhile even though the fish is at neutral buoyancy only at the top of its depth range.' Certainly lantern-fishes when netted at the surface are neutrally buoyant (Kanwisher and Ebeling, 1957; Bone, 1973).

Taking as a model a free spherical bubble rising at a rate such that its expansion due to decrease of pressure is exactly opposed by the decrease in volume due to gas resorption, D'Aoust (1970) has shown that for a spherical bubble of a size equal to or less than the volume of swimbladder normally found in mesopelagic fishes, the times theoretically needed to rise 200 and 300 metres are up to an order of magnitude less than those actually got from diurnal records of sound scattering layers. Thus, once secreted there need be no stress in removing gas at rates relevant to observed rates of vertical migration. The more difficult physiological task, as perhaps reflected by the reduction in size or regression of the swimbladder in numerous species, may well be the secretion of gas during and after descent. Moreover, even though the deepest living lantern-fishes (e.g. *Taaningichthys* spp.) have the longest retia mirabilia in the swimbladder, they are non-migrators.

If the swimbladder of a migrating deep-sea fish falls short of its ideal hydrostatic function, a better alternative would be to replace gaseous lift by lipid lift. As implied already, certain lantern-fishes with a regressed swimbladder but with a lipid content high enough to give them neutral buoyancy should retain this advantage at all depths. This seems an ideal arrangement and makes one wonder why so many species retain a swimbladder. Evidently the provision of high quantities of wax esters is not for all species.

In any event, it is likely that most mesopelagic fishes without a gas-filled swimbladder are somewhat negatively buoyant. Concerning two common species, *Gonostoma elongatum* is often within 0·5% and *Xenodermichthys copei* within 1·2% of neutral buoyancy. Their fat content (about 3% of the wet weight) is not high, but this in concert with the greater lift given by their dilute body fluids, comes close to matching the relatively small negative buoyancy of their muscular and skeletal system (Denton and Marshall, 1958) (see also Fig. 120).

Bathypelagic fishes, as already stated, lack a gas-filled swimbladder.

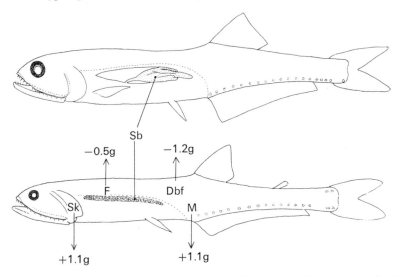

Fig. 120   Buoyancy balance sheets of two species of *Gonostoma. G. denudatum* (above) has a large gas-filled swimbladder (Sb) and is neutrally buoyant. In *G. elongatum* (below) the swimbladder is regressed but the uplift given by fats (F), and dilute body fluids (DBF) almost balances the two main heavier than seawater components, comprised by the skeleton (Sk) and proteins (M). (After Denton and Marshall, 1958; and redrawn from Marshall, 1971)

During their postlarval life in the surface layer *Cyclothone* spp. have a small gas-filled organ, but after metamorphosis and descent to the adult habitat the swimbladder regresses and is entirely without gas (Marshall, 1960). In a catch that includes both mesopelagic and bathypelagic species of *Cyclothone* many individuals of the former float whereas the latter sink to the bottom of the container. As stated elsewhere (Marshall, 1960), a fish with a capacious gas-filled swimbladder can carry a well-developed muscular system and a firm skeleton at neutral buoyancy. The simple organization of bathypelagic fishes, which fits so well with their food-poor living space, has necessarily involved the loss of the swimbladder. Again the reduction of the muscular and skeleton system means that relatively little lift will bring them close to neutral buoyancy. For instance, while a female angler-fish (*Melanocetus murrayi*) was somewhat heavier than water one of comparable size in another species (*Ceratias holboelli*) was very close to neutral buoyancy (personal observation).

The buoyancy of benthopelagic animals

To recapitulate; while at midwater levels a gas-filled swimbladder occurs only in mesopelagic fishes, such an organ is present, and well developed, in benthopelagic species that range from slope to abyssal levels. In an individual of the brotulid fish *Bassogigas profundissimus*, which was trawled by the *Galathea* in the Sunda Trench at a depth of 7,160 metres, Nielsen and Munk (1964) found a rather small swimbladder containing a rete, gas gland and resorbent area that showed no signs of regression. If, as in other deep-sea fishes, the swimbladder gas is mainly oxygen, its specific gravity at 7,000 metres will be about 0·7 and thus still provide considerable buoyancy (Alexander, 1966).

A second contrast between midwater and benthopelagic fishes is that whereas a gas-filled swimbladder is found in about 40 per cent of the former species, the comparable percentage in the latter is over 90 and comprised largely of macrourids, morids, brotulids, halosaurs, notacanths and various eels. Moreover, deep-sea squaloid sharks and chimaeroid fishes, which have no swimbladder, have virtually attained neutral buoyancy through the development of a large and oily liver (p. 135). In this organ the hydrocarbon squalene (specific gravity 0·86) is stored in such quantities that the uplift provided almost eliminates the fishes' weight in water (Corner, Denton and Forster, 1969). For instance, the shark *Centroscymnus coelolepis*, which weighed 5,260 gm in air, contained a liver of 1,550 gm, most of which (82%) consisted of oily squalene. The buoyancy balance sheet, as calculated by Corner and his colleagues, showed that at the depth of capture (1,500 metres) the shark is likely to weigh a mere 9·1 grams. Lastly, benthopelagic teleost fishes lacking a swimbladder, notably alepocephalids and ateleopids, must also be close to neutral buoyancy, judging by their rather fragile skeletons and watery, loose-textured muscles.

It is clear that neutral buoyancy, or a close approach to such conditions, is biologically valuable in fishes that live near the deep-sea floor. Concerning the species with gaseous uplift, it is not simply that their ancestors endowed them with a swimbladder, but rather that they live in surroundings that provide sustenance enough to support the heavier tissues, notably muscle and bone, that can be carried at neutral buoyancy by a gas-filled float. The best support for this conclusion involves mid-water macrourids (e.g. *Cynomacrurus* and *Odontomacrurus*) and ophidioid fishes (some of the aphyonids p. 458), that have evolved from ben-thopelagic ancestors. Both kinds, like many of the primary midwater fishes, have a regressed (or absent) swimbladder and reduced muscular and skeletal systems (Marshall, 1960, 1964).

Neutral buoyancy is apparent also in benthopelagic cephalopods. After study of the literature and a fine series of deep-sea photographs, Roper and Brundage (1972) concluded that *Cirroteuthis* spp. and probably some of the other cirrate octopods as well are not only benthopelagic in habit but also close to being weightless in water (see also p. 141). Perhaps in parallel with the fishes, the midwater *Cirrothauma murrayi* is more reduced than its presumed benthopelagic ancestors.

Buoyancy relations of benthic animals

Nearly all of the larger invertebrates that move over the continental shelf are plainly heavier than their environment. But the same is by no means true of some of their relatives on the deep-sea floor. As we saw elsewhere (p. 201) many of the elasipod sea-cucumbers are not much heavier than sea water. Thus, little effort is needed for them to rise above their resting place and be carried or move elsewhere. There is a parallel difference between shelf-dwelling and deep-sea rays. The latter, which have a large oily liver, are probably near to neutral buoyancy (Bone and Roberts, 1969, and p. 212). Even *Bathysaurus*, which from observations and photographs seems a classic type of benthic teleost, may not be much heavier than sea water. In a *Bathysaurus agassizi* (length 495 mm) the larger liver, which weighed about one-fifth of the total weight, consisted largely (70% by weight) of lipids with a specific gravity of 0.92 (R. J. Morris, quoted by Marshall and Merrett, 1977). One's impression is that diverse members of the deep-sea benthos have acquired the means to rise easily above the bottom and thus move on to new feeding grounds.

The biological significance of neutral buoyancy

Our survey of buoyancy relations in deep-sea animals poses an outstand-ing question: what is the biological advantage of neutral buoyancy or its near attainment 'sought' not only by simply organized planktonic forms but also by muscular cephalopods and fishes? Indeed, there is evi-dence that the sperm whale, which spends part of its life in deep-sea feed-ing grounds, has means of buoyancy control that involve the assumption of weightlessness (Clarke, 1978).

Even so, we must first recall that many midwater representatives of successful groups of animals, notably copepods, euphausiids, sergestid prawns and arrow-worms, are without special buoyancy devices. Evi-dently, they are not so negatively buoyant that the effort needed to main-tain height in the water column is a substantial drain on their total energy budget. The biological advantages of adequate buoyancy in the phyto-plankton have been considered elsewhere (p. 316). Until more is known of their buoyancy relations and life cycles little can be said of the midwater protozoans. Indeed, not enough is known of the slender energy budget of neutrally buoyant gelatinous forms, whether cnidarian or tunicates, to consider how much energy is saved by weightlessness in water. Perhaps the latter condition enables them to feed efficiently. Certainly, the proper setting of the deadly drift lines of siphonophores depends on appropriate assemblage of the over-buoyant members of the colony.

It is easier to consider the advantage of neutral (or near-neutral) buoyancy in nektonic animals. In a midwater fish one may consider the energy needed to maintain a level if the fish is deprived of its swimbladder. A marine fish with a well-developed gas-filled swimbladder will have a reduced weight in water of about 5 per cent of its weight in air. If it lacked such a swimbladder 'to stay at a constant level it would need to exert a downward force equal to its reduced weight (and unlike terrestrial animals, pelagic creatures can only exert forces on the surrounding medium by making movements). When it is remembered that even an active pelagic fish seldom exerts a force of more than 25–50% of its weight in air for more than a very brief period (Gray, 1953), the force necessary for a fish to maintain its level appears to be considerable' (Den-ton and Shaw, quoted by Denton and Marshall, 1958).

During swimming a just buoyant nektonic animal saves energy in that virtually all the power developed by its locomotor muscles is con-centrated in its forward motion. If an animal is heavier than water a con-siderable part of its propulsive power must go to maintaining its level. In most sharks, for instance, lift during swimming is generated mainly by the pectoral fins in concert with the asymmetrical tail fin. But in buoyant deep-sea sharks the pectoral fins no longer look like hydroplanes and the tail fin approaches a symmetrical form (Marshall, 1971).

In midwater fishes and squids the energy needed to secrete gases or ammoniacal fluids is evidently much less than that saved through the uplift derived by either of these processes. Even though a gas-filled swimbladder may provide neutral buoyancy only during their main feeding period, it is surely striking that in numbers of individuals the majority of migrating midwater fishes contain such a float. These migrators, notably lantern-fishes, may often have to work hard to obtain enough planktonic food. As they search for prey, neutral buoyancy will not only enable them to concentrate propulsive energy in forward motion, but give them easy manœuvrability in the seizure of prey. Turning to numbers of species, it is equally striking that the majority have dispensed with a gas-filled swimbladder. But the reduction of muscle and skeleton associated with near neutral buoyancy is fitting also in surroundings that contain limited supplies of food. Such reduction, as we saw, is most marked in bathypelagic fishes. These fishes, like many mesopelagic species that also lack a swimbladder, do not undertake daily vertical migrations.

Midwater cephalopods, which are largely confined to mesopelagic levels, differ in their buoyancy relations. The octopods, unlike so many squids, have not developed ammoniacal means of uplift. They are rather fragile, gelatinous creatures with relatively little muscle, and one at least (*Japetella*) has ionic means of approaching neutral buoyancy (Denton and Gilpin Brown, 1973). In their more active relatives the ammoniacal squids, one may argue again that neutral buoyancy not only saves energy as they hover and move but also gives them fluid manœuvrability in the capture of prey.

Near the deep-sea floor neutral buoyancy is more or less obligatory among the fishes. Those with a gas-filled swimbladder are predominantly elongated or rat-tail-like fishes with long, many-rayed median fins. They have slow deliberate ways, and as films also show, they are neutrally buoyant. At low and moderate speeds, according to Alexander's (1966) estimations, the energy saved through such buoyancy is considerable. Yet again, one thinks also of the postural versatility afforded by weightlessness in water—a versatility put to best use by the many species that feed on both benthic and free-swimming prey. After watching the unhurried ease and varied motions of rat-tails on films taken by baited cameras (see Isaacs and Schwartzlose, 1975), it is easy to forget that neutral buoyancy is at the centre of their existence.

Bioluminescence: 10
light from life

Life, and light from life, probably arose in the ocean. Today the most prevalent light-producers among the simple forms of marine life are bacteria and dinoflagellates. Bacteria, which are prokaryotic organisms, were no doubt prominent in the populations of monerans that inhabited the earth for the first three-quarters of its life history. The first unicellular organisms with a discrete nucleus, the protistan eukaryotes, which include the dinoflagellates, began much later in the Precambrian period. But the dinoflagellates did not flourish until lower Jurassic times, relatively late in geological history. As far as we know, the first luminescent marine organisms were bacteria.

When, during the vastness of pre-eukaryotic times, photosynthetic organisms, such as blue-green algae, began to produce oxygen—toxic, of course, for anaerobic bacteria—survival of the latter must have led to metabolic changes to offset the toxicity. One biochemical adaptation to remove oxygen, as postulated by McElroy and Seliger (1962), involved the production of light. (The biochemical sequence is from oxidative phosphorylation to final stages involving a luciferin—a complex of flavin mononucleotide, FMN, and an aldehyde, RCHO—which is oxidized in the presence of an enzyme, luciferase, to yield FMN, RCHO, water and light.) Thus the bioluminescence of present-day bacteria may well be a vestige of the past, for the elimination of oxygen is no longer necessary. Indeed, mutant forms of bioluminescent bacteria grow and metabolize well, even if one or another of the light-producing components is absent. Today, luminescent bacteria live at all depths of the ocean. Furthermore, certain squid and many fishes, both deep-sea and coastal, have evolved organs that contain cultures of luminescent bacteria as a source of light. The symbiotic light-givers belong to the genus *Photobacterium* (Herring, 1977b).

Dinoflagellates are prevalent enough to account for the ubiquity of luminescence near the surface of the ocean. Their light is most intense in the euphotic zone, but persists at deeper levels. The flash of a dinoflagellate is brighter than the glow of a bacterium, or even of a radiolarian. If the light-givers (some species are not luminescent) inherited their powers from prokaryotic ancestors, subsequent improvements on their inheritance have been considerable. Beside the ingrained circadian rhythms behind their nocturnal activity, there need only be a slight change in ambient light to stop them flashing. These attributes, together with other luminescent features, such as the structure and action of their scintillons, all indicate genetic backing that has evolved through natural selection. If so, what is the survival value of their light? Conjectures are that their flashing reduces the inroads of browsing animals, whose feeding is inhibited by their light, or whose movements stir up the light of their fodder, thus exposing browsers to their predators. Certainly the multiplication of phytoplankton often barely keeps pace with the cropping of herbivores. But the survival value of dinoflagellate flashing (with a wavelength of 480 nm, near that of light that penetrates farthest in sea water) remains an intriguing question. Comparative studies of luminescent and non-luminescent species might be revealing (see also Tett and Kelley, 1973).

Luminescence and living space

In *Crowds and Power*, Elias Canetti (1962) contrasts the symbolism of fire and sea. Though the sea '... lacks the mystery and suddenness of fire ... there is, nevertheless, mystery in it, a mystery lying not in suddenness, but in what it contains and covers. The life with which it teems is as much a part of it as its enduring openness. Its sublimity is enhanced by the thought of what it contains, the multitudes of plants and animals hidden within it.'

There is mystery also when the hidden life of the sea is revealed, sometimes suddenly, in the 'cold fires' of bioluminescence. For, as I wrote elsewhere, 'the connection between a certain type of luminescent display and the organism responsible is not easily traced. But it seems that whenever the light from the sea looks uniform from a distance, and is without large sparks of light, the luminous organisms are the small peridinians of the plant plankton. Professor Newton Harvey writes of a salt lake near Nassau in the Bahamas which is connected with the ocean and is full of luminescent peridinians: "Every fish that made the slightest movement was outlined with fire and every wave looked as if it were aflame." The winking, sparkling, points of light in the sea very likely arise from luminescing, pulsating jellyfish (*Pelagia noctiluca*) and tunicates (pyrosomas and salps). The light from pyrosomas has been likened to red glowing iron and is said to have been so intense as to illuminate the sails of ships at night and to make objects in the cabins easily visible' (Marshall, 1954).

Near and on the deep-sea floor there are also many forms of luminous animals. But voyagers to these regions in deep submersibles have not come back with glowing accounts in more than one sense. Less is known of the luminescence of bottom-dwelling animals than of those from pelagic levels.

'But the naturalists of a number of research ships have seen most beautiful displays of light from the alcyonarians, gorgonians and pennatulids (sea-pens) of the coelenterate fauna. When he sailed in the *Porcupine* Wyville Thomson saw that the sea-pens and gorgonians gave out a lambent white light so brilliant that the hands of his watch showed quite distinctly. During the voyage of the *Challenger* Moseley recorded that all the alcyonarians brought up in the trawl were brilliantly phosphorescent, while one haul of the trawl so delighted Wyville Thomson that he set down this glowing account:

' "The trawl seemed to have gone over a regular field of a delicate, simple Gorgonid ... the stems, which were from eighteen inches to two feet in length, were coiled in great hanks around the beam-trawl and engaged in masses in the net; and as they showed a most vivid phosphorescence of a pale lilac colour, their immense number suggested a wonderful state of things beneath—animated cornfields waving gently in a slow tidal current and glowing with a soft diffused light, scintillating and sparkling on the slightest touch, and now and again breaking into long avenues of vivid light indicating the paths of fishes or other wandering denizens of their enchanted region." '

'Among these wandering denizens are the rat-tailed fishes (Macrouridae), some of which have a long gland along the belly, a form of striplighting with millions of luminous bacteria providing the glow' (Marshall, 1954).

First, though, we consider the luminescence of midwater animals, especially of those with complex light organs.

Light and life at midwater levels

When a sensitive photomultiplier tube is lowered into the ocean, discrete flashes from luminous organisms can be detected to a depth of at least 3,750 metres. G. L. Clarke and his colleagues found that the number of flashes recorded in a minute varies from one to about 160. In the deep sea the most luminous depths are in the mesopelagic zone. When the photomultiplier descends below 1,000 metres the light flashes become more and more sporadic, which seems reasonable in view of the much smaller standing stocks of zooplankton at bathypelagic levels. At the brightest mesopelagic levels the background of twilight may almost disappear behind the bright and swift sequence of luminescent sparks (see Clarke and Backus, 1956).

Even so, the lights of marine animals are dim, rarely equalling moonshine. The intensity of luminescence is usually expressed in units of power (microwatts,μW) falling on a square centimetre (cm^2) of a receptive surface at a distance of one metre from the source of light. As a reference level, the intensity of moonlight may be taken as $10^{-1} \mu W/cm^2$ at 1 metre, which is thousands of times more intense than measurements of irradiance from luminous animals. Nicol (1971) lists the following:

Radiolaria   ca. $10^{-9} \mu W/cm^2$ at 1 metre
Jellyfish, Siphonophores 10^{-7} etc.
Ctenophores (comb-jellies) 10^{-5}
Copepods 10^{-8} to 10^{-4}
Euphausiids 2×10^{-7}
Lantern-fishes 0.5×10^{-7}

The irradiance of the brightest midwater flashes recorded by G. L. Clarke and his colleagues was greater than $10^{-3} \mu W/cm^2$ at 1 metre.

The colour of bioluminescence is blue, the wavelength ranging from 470 to 490 mm, which includes the wavelengths that travel farthest in sea water. Concerning the detection of light at a distance, Nicol (1971) has written, 'Under ideal conditions a very sensitive eye could just see a bright emitter such as a comb jelly at a distance of 40 to 120 m; in practice, weak luminescent lights probably become indistinguishable at much shorter distances than this, at 10 m or so. Luminescence is usually emitted in short flashes which causes them to be detected more readily, because the eye is usually very sensitive to small flashing lights.'

Light in the zooplankton

Bioluminescence comes from a diversity of animals in the plankton (Boden and Kampa, 1964). Of the main taxa, only the Chaetognatha (arrow-worms) and pteropod molluscs seem to be without luminescent members. Among pelagic cnidarians, members of at least seven genera of siphonopores are known to be luminous. Luminescence in siphonophores is not inhibited by light, but is induced by mechanical stimulation. Nicol (1958) found also that the light from species of *Vogtia*, *Rosacea* and *Hippopodius* appears as an intracellular bluish glow that lasts from 1 to 11 seconds depending on the strength of the stimulus. Beside *Pelagia* luminescent medusae include two deep-sea coronate jellyfishes, *Atolla* and *Periphylla*. Both luminesce when stimulated mechanically. The blue flash from *Atolla wyvillei* lasts up to 2 seconds. During exploratory probing the light of *Atolla*, which appeared in a thin streak just inside the outer edge of thick coronal muscle, was visible above but not below the jellyfish. Evidently, downward transmission is barred by the opaque and white coronal muscle, but it may reflect the light upward. Histological preparations indicated that the light may well come from tall, columnar, glandular-looking cells in the floor of the rhopalar canal (Nicol, 1958).

All pelagic ctenophores are likely to be luminescent. Their activity in surface waters is well known, but there are deep-sea species, which together with submerged members of epipelagic species, may account for some of the brighter flashes recorded at mesopelagic levels. The luminescence, which can be elicited mechanically or electrically, is intracellular and comes from parts of the meridional and interconnecting canals of the animal. Repeated stimulation leads to a fall in light intensity. Exposure to light dowses the luminescence.

Many pelagic tunicates are luminescent. As well as in pyrosomids, luminescence is known in most of the salps and some of the doliolids. Of the appendicularians, *Oikopleura* is certainly luminous, and so probably are other genera. In *Salpa* and *Cyclosalpa* the luminous bodies are housed between muscle bands in the body wall (Tett and Kelly, 1973). *Pyrosoma* luminesces only when excited and mechanical, electrical, photic and chemical stimuli are effective. 'The light appears in a punctate pattern over the whole surface; at each locus two closely spaced points of light can be resolved, which correspond to the pair of light organs at the entrance to the branchial sac of each zooid ... In addition, there is a diffuse glow, the result of scattering and internal reflexion within the colony' (Nicol, 1958). Thread-like inclusions in the photogenic cells have been interpreted as symbiotic bacteria, but after various histological tests, Nicol was unable to confirm that these inclusions might be luminous. During development the photogenic organs are formed from test cells, which are transmitted from one generation to the next. The follicle cells around the eggs of *Pyrosoma* are also luminous. The supposed symbiotic bacteria are thought to be transmitted from parental ascidiozooids to be luminous organs that develop in the ascidiozooids of the larval colony.

The luminescence of these gelatinous cnidarians, ctenophores and tunicates, like that of dinoflagellates, is surely not to be explained away simply as a vestige of the past, or as a by-product of metabolism. They have special light cells and light organs that are under some kind of control. They all have their predators (p. 83), but many are so transparent that they must be difficult to see. Why, then, do they luminesce? If the cnidarians and ctenophores luminesce to advertise, as it were, their noxious nature to potential predators, then it is curious that the quality of their light is much like that of other planktonic animals, such as the harmless tunicates. Aposematic animals usually display distinctive features. Luminescence, as we shall see, may be a means of attracting prey, but are the flashes of cnidarians and ctenophores repeated often enough to attract their food? There is certainly no such possibility in tunicates, which feed on phytoplankton (p. 98). Perhaps the explanation is simpler. Siphonophores, medusae, ctenophores and pelagic tunicates* are large, drifting animals, which because of their transparency will not readily be seen by active nektonic animals. If there is a 'collision', the luminescence induced in the gelatinous drifter may enable the blundering prawn, squid or fish to disengage itself (see Nicol, 1962). Perhaps the sparks struck may startle and deter a predator. At all events, here is enough conjecture to stimulate further observations and suggest experiments.

Pelagic polychaete worms are also transparent, and the tomopterids, at least, are luminescent. The light, which is said to be yellowish, is probably produced in yellow glandular structures on the parapodia (Fig. 34). In the better studied syllid worms of shallow waters luminescence is known only during sexual congress (Boden and Kampa, 1964).

Luminescence is widespread among pelagic crustaceans of all main groups. In the zooplankton diverse copepods and ostracods are light-producers and there are no doubt more to be discovered. Of the calanoid copepods, there are records of bright luminescence in representatives of the families Aetideidae, Augaptilidae, Heterorhabdidae, Lucicutiidae and Metridiidae, which contain numerous midwater species. *Gaussia princeps*, a mesopelagic species of the last family, has been studied recently by Barnes and Case (1972). Light glands on the head, mandibular palps, urosome, furca and underlying at least 14 body pores contribute to the display which, as in other luminous copepods, consists of a sudden discharge to the exterior of luminous blue material. In *Gaussia princeps* (Fig. 121) the display occurs in both sexes, and after discharge the luminous material glows at peak intensity for 1 to 3 seconds before fading as 8 (or less) discrete, low-intensity points of light over a period of 80 seconds. Displays induced by electrical and mechanical stimuli, which were recorded on photomultiplier and TV image-intensifier equipment, showed that responses may be facilitated, vary in duration of emission and be repetitive. The light glands need not act in concert nor do all necessarily participate in a given response.

In their discussion, Barnes and Case support Harvey's (1952) conjecture that the sudden release of luminous material may hold a predator's attention while the light-producer escapes. They quote David and

---

* Epipelagic appendicularians are small (p. 94). *Oikopleura dioica* flashes spontaneously and on being disturbed. The large gelatinous house also flashes, even when unoccupied, but only after mechanical stimulation (Galt, 1978). Perhaps the brilliant display of surface-dwelling forms sometimes deters or diverts their fish predators. The large deep-sea *Bathychordaeus* must have a very large house, which could well be damaged on contact with an actively swimming fish or squid. But is the house luminescent?

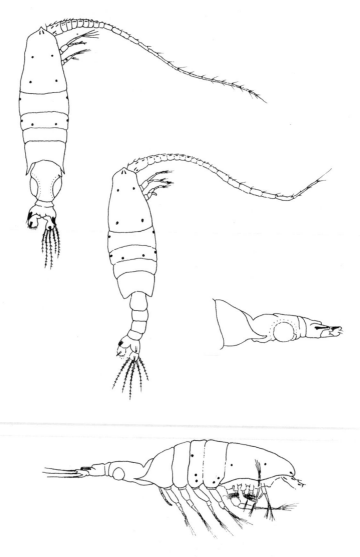

Fig. 121 Distribution of fluorescent glands, luminescent glands and luminous pores (shown as black dots) of a deep-sea copepod, *Gaussia princeps*. From top to bottom, adult female, adult male, lateral view of 5th thoracic segment and abdomen of female, lateral view of adult female. (Redrawn from Barnes and Case, 1972)

Conover's (1961) observations on the copepod *Metridia lucens* in the presence of a predatory euphausiid, *Meganyctiphanes norvegica*. The number of copepods seized by *Meganyctiphanes* was less than the number of flashes produced by the prey over the periods of observation. Evidently, the copepods were not only luminescing when confronted by their predator, but also escaping from many of such encounters. Copepods, as we saw (p. 78), have a very quick escape response, and in *Gaussia* it is significant that luminescence is usually coupled with the sudden initiation of rapid swimming movements. Barnes and Case conclude, 'The initial brightness of the light burst produced by *G. princeps* would seem likely to evoke a "startle" response in a dark adapted predator, and the fixed pattern of discrete points, glowing long after the animal has rapidly departed, seems ideally suited as an attention-holding decoy in an environment characterized by extremely low light levels.'

All species of ostracods known to be light-producers also extrude luminous material. Like copepods, they have many predators. *Gigantocypris* is said to be luminescent, and so are numerous species of *Conchoecia*. In the latter genus, Angel (1971/72) observed that 'sites below the rostrum and on the posterior margins of the carapace are concerned with release of luminescent secretion into the inhalent or exhalent water currents through the carapace'. But the extrusion of luminous clouds is not confined to copepods and ostracods. As we shall now see, the phenomenon is known also in certain mysids, decapods, squids and fishes.

Light in the micronekton and nekton

Many of the animals now to be considered have complex light organs of precise optical design and performance. But before turning to such aspects of bioluminescence, we continue the survey of forms that extrude luminous clouds, which come from relatively simple kinds of light organs.

In the large scarlet shrimps of the mysid genus *Gnathophausia* luminous secretions are expelled through a projection on the second maxilla. More is known about the pelagic prawns, among which the production of luminous clouds is confined almost entirely to the carideans (to oplophorids generally and to some pandalids, pasiphaeids and thalassocarids) (Herring, 1976). The luminous material comes from part of the liver. In actively swimming species of *Oplophorus* the secretion seems to be squirted forwards and downwards, and at the same time, but not always, there is a strong backward flip of the abdomen. The light issues from a region at or close to the mouth. After extrusion the intensity decays rather rapidly, but on repeated stimulation the animal is still able to produce a series of luminescent squirts. Herring's dissections led him to think it probable that '... the site of production and storage of the components of the luminescent reaction is the liver and that they are discharged from the mouth via the stomach'. The sudden discharge of luminous clouds '. . . is undoubtedly an alarm or escape response to sudden stimulus, be it mechanical, chemical or light'.

Many of the deep-sea squids have elaborate luminous organs, but luminous secretions are known only in certain sepiolids and loliginids. In sepiolid genera such as *Rossia* and *Sepiola* symbiotic luminous bacteria are cultured in paired organs lying against the ink sac near the anus. Light appears when the bacterial contents are secreted into the mantle cavity and expelled to the sea. Some sepiolids are bottom-dwellers. In *Heteroteuthis*, which is pelagic, the source of luminescence is unknown (Herring, 1977).

In midwater fishes the extrusion of luminous material is confined largely to members of one family, the Searsiidae. Whether or not they have discrete light organs on the body, searsiids have a black, sac-like shoulder organ just below the lateral line. The organ opens through a retrorse papilla. A living *Searsia koefoedi* '. . . was seen to discharge a bright luminous cloud into the water when handled. The light appeared as multitudinous bright points, blue-green in colour' (Nicol, 1958). The luminous glow lasted for about 4 seconds, and at a distance of 1 metre (in air) the light intensity was about $2 \times 10^{-6} \mu W/cm^2$ receptor surface.

The walls of the shoulder organ have an inner lining of small glandular cells, which stain violet in trichrome preparations. In the lumen of the organ are large spherical 'red cells'. Evidently 'violet cells' pass through transitional phases to become free 'red cells'. The red cells are packed with acidophilic granules, which appear to be the photogenic bodies. Nicol's interpretation is that on extrusion to the sea the red cells break up, so releasing and activating their luminescent contents.

The different means of producing and expelling luminescent material by remotely related midwater animals—copepods, ostracods, mysids, caridean prawns, squid and fishes, is a wide-ranging example of convergent evolution. The convergers have 'learned' that chances of staying alive in open surroundings (lacking solid cover) can be improved if there are not only means to make luminous decoys but also means of rapid escape. The latter are certainly well developed, particularly as they involve giant nerve fibres that swiftly excite the muscles used in escape, which are: the muscles that move the thoracic limbs of copepods; the powerful abdominal muscles that produce the tail-flips of prawns, so shooting the animal backwards and in the direction opposite to the squirt path of its luminous cloud; the mantle muscle of squids; and the lateral muscles of fishes. But luminous secretions are produced by female deep-sea angler-fishes of the genera *Ceratias* and *Cryptosaras*, which are without means of rapid escape (the lateral muscles are relatively small and there are no Mauthner fibres in the spinal cord). The light glands (caruncles) are carried on two or three curiously modified fin-rays set just before the soft dorsal fin. There are three caruncles in females of *Cryptosaras couesi*, and on pressing the large median organ in a living individual, Bertelsen (1951) saw a whitish-yellow secretion issue into the water and this spread into numerous points of light '... which could be observed scattered over the basin used for about a minute. The caruncles contain luminous bacteria' (Hansen and Herring, 1976). In both genera the caruncles, which are well developed after metamorphosis, may have a particularly important life-preserving function during juvenile exist-ence. But the caruncles are not simply a youthful feature. In *Cryptosaras* at least, the largest known female (length 440 mm) has very large caruncles (see Bertelsen, 1951). Perhaps these two kinds of female ceratioids are good escapers in angler-fish terms: when adult they both have a very large caudal fin and a better-shaped body than most of their relatives.

CAMOUFLAGE IN THE TWILIGHT ZONE

The setting of luminescent decoys is one way of surviving in open sur-roundings: another, which has led to close convergence in midwater fishes, squids, prawns and euphausiids, involves the evolution of complex photophores designed and set to cast a light field that tends to match the background field of twilight. If the match is exact the animal will disappear from the view of a predator.

In these animals the vanishing light field comes from large ventrally placed luminescent organs that converge in the following structural re-spects: the organs are partly screened by pigment cells and backed by a reflecting layer, which may have constructive interference properties; direct and reflected light from the photogenic body is directed through some kind of lens and in some fishes, squids and prawns, the light passes through filters before reaching the lens.

Luminescent camouflage is best known in fishes, thanks largely to the fascinating studies by E. J. Denton and his colleagues. The dominant light-producing mesopelagic fishes, both in numbers of individuals and species, are stomiatoids and lantern-fishes (Myctophidae). Certain of the argentinoid fishes are luminescent and so are a few alepisauroids, the large group of predatory fishes related to the myctophids. Other taxa of predatory fishes, such as gempylids, trichiurids and chiasmodontids, appear to be without luminous members, which is true also of eels and melamphaids. But at least two-thirds of the mesopelagic fish fauna (ca, 1,000 spp.) consists of luminous species.

In profile, each of the three hundred odd species of lantern-fishes may be recognized by the individual constellation of lights over the sides of the head and body. Most of the lights are set below the level of the lateral line. The pattern of lights seen from below is much more uniform. For instance, nearly all lantern-fishes have a row of three lights on each gill membrane. The twin rows of lights between the head and pelvic fins and then between pelvics and anal fin, each consist of four or five paired lights in most species. Then follow smaller and generally more numerous lights in twin series flanking the base of the anal fin and the lower midline of caudal peduncle (see Fig. 122).

Lantern-fishes are more or less fusiform, whereas stomiatoids range from short, laterally compressed forms through fusiform to anguilliform types. Many have more variable and complex patterns of lights than occur on lantern-fishes. But whatever their shape and photophore complement, nearly all species have a double row of lights along each side of the mid-ventral line between the head and pelvic fins, and even beyond (Fig. 122). These double rows are succeeded by a single row down the rear part of the fish. (In certain species there is a single row down the entire body.) Lastly, nearly all stomiatoids have one or more lights that shine directly into the eyes.

The large ventral lights of stomiatoid fishes look like series of inlaid jewels, which in some species are attractively coloured. In hatchet fishes each light has an inset filter holding a magenta pigment with a 'window' in the wavelength range (460–480 nm) of deep-sea twilight (Denton et al., 1970). In other, perhaps all other, stomiatoids the filter pigment has a deep red fluorescence (Herring, 1977d). After passage through the filter, the light focussed into the sea, as already implied, matches the blue background remnant of solar light.

Concerning the light field around animals with ventral photophores, Denton and his colleagues (1972) refer to Fraser (1962) and Clarke (1963) who '... suggested that by generating a downwardly directed light these animals might obliterate the shadows which they would otherwise cast and so make themselves difficult to see from below (Lord Blackett tells us that during the Second World War aeroplanes were camouflaged in a similar way by placing rows of lights shining forwards and downwards underneath their wings). This is a method of camouflage which could not be used effectively by animals living in bright daylight near the surface of the sea, for even the brightest photophores could not produce sufficiently intense lights.

'If deep-sea animals did camouflage themselves by producing light, they could make themselves completely invisible from all directions if they exactly replaced the light lost by absorption on their upper surfaces, and if the emitted light was of the right intensity, the right spectral distribution and the right angular distribution.' By placing luminescing hatchet-fish (*Argyropelecus affinis*) and viper-fish (*Chauliodus sloani*) on a rotating device so that their light could be measured from all directions in two vertical axes, one through the long axis of the fish, the other at

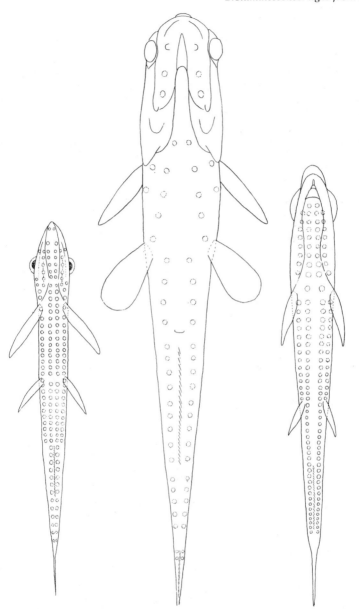

Fig. 122 Ventral light organs of mesopelagic fishes. Left, *Vinciguerria*; centre, a lantern-fish, *Myctophum nitidulum*; right, *Ichthyococcus*. (Redrawn from Marshall, 1971)

right angles to this axis, Denton and his colleagues found that the angular distribution of luminescence did match the external twilight field over a wide range of viewing angles (Figs. 123 and 124). Earlier, Denton, (1970) had analysed how the half-silvered outer surface over the ventral photophores of *Argyropelecus* could produce, by multiple reflection, the proper angular distribution of emitted light. More recently Herring (1977), by measuring the angular distribution of light from selected ventral photophores of *Argyropelecus olfersi* before and after peeling off the half-silvered reflector, has confirmed Denton's analysis. When this reflector

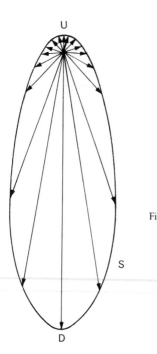

Fig. 123 Diagram showing the type of angular distribution of radiance of daylight found at mesopelagic levels where *Argyropelecus* and *Chautiodus* live in the daytime. From the centre from which the arrows radiate, the distribution in three dimensions is given by the surface formed by revolving the solid curve S around the axis UD. The relative radiance in a given direction is given by the length of the arrow in that direction. (From Denton, Gilpin-Brown and Wright, 1972)

is removed, the light is no longer diffused and cast downwards in the proper angular ways.

Even if they are daily vertical migrators, midwater fishes tend to stay with a particular level of light intensity (p. 235). But how is the intensity of their luminescence adjusted to match that of the background twilight? Concerning the orbital photophores of stomiatoids, which shine directly into the eyes, Nicol (1967) has asked, 'Are these small point-sources, permitting their possessors to compare luminous output with the light? Can the fish thereby adjust the intensity of its lights with reference to ambient light levels?' Herring (1977d) has provided affirmative answers, if only

partial, from his observations on hatchet-fishes, which show that the light intensity of the orbital photophores waxes and wanes with that of the ventral organs.

When do fish switch on their camouflage? There is a remarkable photograph taken from the deep-submersible *Alvin* of lantern-fishes (*Ceratoscopelus maderensis*) that are the source of comet traces in sonic scattering layers (Backus et al., 1968). The fishes are hanging and poised with their body axes at all angles between horizontal and vertical orientations. If luminescent camouflage is to be effective, a fish's body axis should be more or less in the horizontal plane. Most likely, the least energetic way of escaping predators is simply to hang in the sea doing nothing in particular, which is most feasible for animals that, like *C. maderensis*, are neutrally buoyant. Moreover, the nearer the poised inclination to the vertical, the less the ventral silhouette. Certainly fishes such as snipe-eels and barracudinas are known to adopt this posture. An active fish is much more likely to attract the attention of predators, but with its body in the horizontal plane, luminescent camouflage is now possible. All this is conjecture, though Boden and Kampa's (1964) observations on the luminescence of sonic scattering layers are suggestive. The number of light flashes recorded in particular sonic layers was greatest around sunset and sunrise, the times of upward and downward migrations of luminescent fishes, euphausiids and prawns.

There are, of course, other aspects. For instance, how much advance warning of a predator's approach does a fish gain through its acoustico-lateralis system? If the predator is large the warning might be enough for the fish to switch on and disappear from view.

The ventral camouflage of lantern-fishes must come from the small pearl-button lights that stud their underparts. Their other kinds of luminescent organs are considered elsewhere. The lens of each pearl-button light is formed by an aptly placed thickening of an overlying scale. The organ, which is flying-saucer-shaped, is backed by pigment and reflecting layers and within is a small photogenic body held in gelatinous connective tissue. Nicol (1958) observes that the lens and aperture, being much wider than the photogenic tissue, will enlarge the apparent source of light. The lens will also collimate the light rays. The spectral composition of luminescence in the lantern-fish *Myctophum punctatum* ranges from 410 to 600 nm but is most intense at about 470 nm, the wavelength of deep-sea twilight (Nicol, 1960). There are no light-filters, though in species of *Diaphus* each button-like photophore has a sky-blue mirror. In *Diaphus rafinesquei* Denton and Herring (1976) found that these mirrors selectively reflect light in a narrow band of blue. Evidently, the mirrors have the same function as the colour filters in the photophores of other animals.

Compared to stomiatoids, little is known of luminescent camouflage in lantern-fishes. In *Tarletonbeania crenularis* Lawry (1974) found that individuals would luminesce when illuminated with blue light from above, but not in darkness. This lantern-fish has a photophore above the eye so placed that its light and downwelling light from above can pass through the ventral aphakic aperture and the lens of the eye to illuminate adjacent areas of the retina. The fish thus has a means of comparing the intensity of its own luminescence with that of the ambient light.

Certain of the tubular-eyed argentinoid fishes develop a large rectal light gland that cultures luminous bacteria. In *Opisthoproctus* (p. 127) the light from this gland, after passing through a lens, is spread and reflected downward by an extraordinary light-guide chamber just above the flattened underparts of the fish. By this means the ventral 'sole' of *Opisthoproctus*, which must be conspicuous from below, may be illuminated and camouflaged (Herring, 1977b). If the retinal diverticula of the eyes (p. 388) receive both downwelling background light and light from the underparts, there may be a means of adjusting the level of luminescence.

Lastly, but not completing a full survey of luminous midwater fishes, certain scopelarchids, evermannellids and paralepidids are among the few alepisauroid fishes known to be light-producers. In an evermannellid, *Coccorella atrata* (and probably also in *C. atlantica*), light issues ventrally from a forwardly directed pyloric caecum under the head and from three regions of the intestine (Herring, 1976). The ventral organs in the paralepidid genera *Lestidium* and *Lestrolepis* consist of one or two longitudinal ducts that extend from the head to the pelvic fins. The ducts contain luminous bacteria and are said to emit a brilliantly yellowish-white glow (Rofen, 1966). The ventral lights of alepisauroids may well provide some camouflage, but this is probably not their primary function. Perhaps the lights of paralepidids help them to find one another; the luminescent species are virtually transparent but have an iridescent skin. Luminescent scopelarchids (*Benthalbella*) are also transparent fishes.

Euphausiid shrimps, as shown by measurements of their luminescence, are among the brightest midwater animals (p. 345). Luminescing swarms of *Euphausia superba* may be seen from a distance of 100 metres or more, but, considering the ubiquity of euphausiids, there have been few observations of their lights at sea. Most species bear ten light organs, one in each eye-stalk, one pair at the bases of the second and seventh thoracic limbs, and a single organ set between the first four pairs of swimming legs on the abdomen.

The luminescence and design of euphausiid photophores have been studied by Herring and Locket (1978), who have concentrated on the abdominal organs of *Euphausia gibboides*. Apart from the light aperture, each abdominal photophore of this euphausiid has an external coat of red pigment and just beneath is a reflecting layer. The light comes from the 'lantern', a small refractile body held centrally in the organ just above the lens. The lantern, which contains stacks of lamellae that are attached to cytoplasmic processes of special (B) cells and bathed in a blood sinus, glows spontaneously when isolated from the photophore. Between this blood sinus around the lantern and the reflecting layer are large (A) cells with prominent nuclei. There are also C cells, which receive many nerve endings and are found where blood capillaries join a sinus. Herring and Locket's reconstruction is that the B cells produce photoproteins that pass to the lamellae of the lantern, and as the latter are bathed in blood, vascular control could regulate the luminescence. Evidently this control is through the nerve fibres to the C cells, which hold what seem to be contractile filaments that could constrict the capillaries. The A cells may simply be supporting structures or producers of some component of the luminescent reaction (Fig. 125).

Light from the lantern passes through a highly refractile biconvex lens, which transmits and condenses the light outwards and downwards to the sea. Moreover, the lens is girdled by a lamellar ring so placed that rays impinging on it at angles away from the normal will be deflected towards the light axis of the photophore and reinforce the downward transmission of luminescence. The spectral quality of the light, with a maximal intensity at a wavelength of 463 nm (close to that of background twilight), comes from selective return by the reflecting layer of the photophore. More precisely, the reflector is formed of alternate laminae of high and low refractive index that have about the right thickness, according to Herring and Locket, to act as a constructive interference mirror. They act as twilight-blue reflectors.

Furthermore, much as in fishes, the angular distribution of luminescence in euphausiids (*Nematoscelis tenella*) closely matches that of ambient twilight at depth (Herring and Locket, 1978). Thus euphausiids, which are the prey of so many animals, have the means to disappear from view. By rotation of their photophores, they are also able to change the angular distribution of light in the sagittal plane. As Michael Hardy (1962) found, the thoracic and abdominal photophores of *Meganyctiphanes norvegica* can be turned through nearly 180° to face forwards through downwards to backwards; the sagittal arc of rotation in the eye-stalk photophores extends over more than a right angle. Moreover, the movements of eyes and photophores are coordinated (the abdominal organs are linked to two ventral longitudinal ligaments). All the photophores, including those on the eye-stalks, usually face downwards (or downwards and backwards), when viewed by red light. When the light is shaded the photophores flick forwards, and then backwards when the shade is removed. Thus, in its natural surroundings, a euphausiid has the means to keep its camouflage glowing downwards when, as must

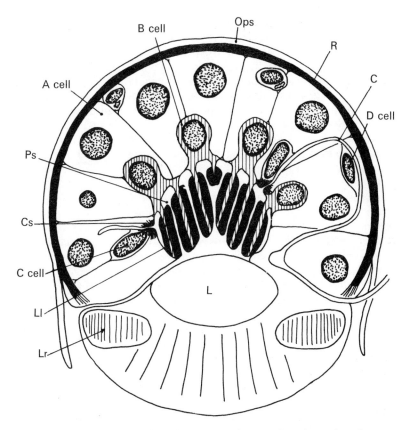

Fig. 125   General structure of an abdominal or thoracic photophore of *Euphausia gibboides* as seen in longitudinal section. Just beneath the outer pigment sheath (Ops) is the reflector (R), which invests the posterior cell mass. This mass contains four types of cells: A cells; B cells, are applied to peripheral sinus and send processes across sinus to contribute to the lantern; C cells, are present where capillaries join the peripheral sinus (Ps); D cells, are present along the course of the capillaries (C), which near the peripheral sinus have sphincters (Cs). Between the cell mass and the lens (L) is the lantern (Ll, lantern lamellae), encircled by a lamellar ring (Lr). (From Herring and Locket, 1978)

often happen, the long axis of its body deviates considerably from the horizontal plane.

Beside the organs that produce luminous secretions, caridean prawns may have definite dermal photophores (Fig. 36), which are known also in certain penaeids. These photophores are set in the cuticle and there may be 150 or more in particular species. The light may pass through a screen of cuticular pigment and in certain species of *Systellaspis* and *Sergia*, through a cuticular lens (Herring, 1976). More generally, Foxton (1970b) points out that his division of prawns into upper and lower meso-pelagic species, with the dividing line at about 700 metres off the Canary Islands, is well exemplified by the penaeid genera *Sergestes* and *Sergia*. Species of *Sergestes*, mostly from upper mesopelagic levels, are semi-transparent 'half-red' forms and possess internal ventral photophores (the hepatic Organs of Pesta). In contrast, species of *Sergia* from lower mesopelagic levels are red all over and photophores, where present, are dermal. Foxton concludes that upper mesopelagic species with semi-transparent bodies and ventrally directed lights, as in *Sergestes*, may well be fitted for ventral camouflage. Moreover, in certain species of *Sergestes*, the luminous organs (of Pesta) are rotated so as to stay facing downwards when the animal deviates substantially from a horizontal posture (Omori, 1974). This would seem to be the prawn parallel to photophore rotation in euphausiids. Lastly, in the caridean *Systellaspis debilis* Herring (1976) found a good agreement between the angular distribution of its light and that of the hatchet-fish, *Argyropelecus olfersi*. The light of *Oplophorus spinosus* had a greater ventral component. Clearly there is more to be discovered and present ideas seem both apt and well worth pursuing.

Bioluminescence in cephalopods is probably restricted to members of the squid group (Teuthoidea), the cuttlefish group (Sepioidea) and to *Vampyroteuthis*. There are no definite records of luminescence in octo-pods. Light organs below the eye form the most characteristic constella-tion in cephalopods. Some, such as *Pterygioteuthis*, have relatively few but complex light organs; others, like *Abraliopsis* and *Vampyroteuthis*, have a few large lights combined with many small lights; in a third group, exemplified by *Histioteuthis* and *Mastigoteuthis*, there are many, rela-tively complex organs. But whatever the pattern, all or most of the photo-phores are placed on the underside of the animal. The complex organs of cephalopods are essentially like those of fishes. The main difference is in the structure of the reflecting layer, which is probably formed of chitinous elements in cephalopods, whereas in fishes it nearly always con-sists of minute crystals of guanine (Herring, 1977c).

There is experimental evidence that midwater cephalopods adjust the intensity of their luminescence to match the intensity of downwelling light. Above an aquarium in a research vessel, Young and Roper (1977) placed a light variable over a fixed series of intensity steps (by factors 1, 2, 6.7, 20, 60, 200 and 300). Five species that off Hawaii may live as high as a level of 450 metres by day (*Albralia trigonura, Abraliopsis* sp., *Pterygioteuthis microlampas, Pyroteuthis addolux* and *Heteroteuthis hawaiiensis* (see Fig. 41), were able to match the intensity of their lumine-scence to that of the background light over steps of intensity 1 to 4.

Another species, *Octopoteuthis nielseni*, which has not been taken off Hawaii above 650 metres by day, could not quite match the intensity of overhead illumination at step 1 but was able to do so at intensities of 0·17 and 0·067.

As Young and Roper argue, ventral camouflage through luminescence can be effective only between certain depth levels. They estimate that the squid *Abralia trigonura*, with a ventral silhouette of about 4 cm^2 and a wet weight of 2·5 gm, would need at a depth of 400 metres off Hawaii to produce a light flux in the range of 0·9 × 10^{16} quanta per hour in order to countershade. Assuming that the efficiency of this squid's luminescence is much like that of fireflies, they calculate that 0·0009 calorie is necessary to produce the required quanta of luminescent camouflage at 400 metres, which is about 0·3 per cent of the energy that is likely to be consumed by the resting animal during the day. Since downwelling light in the sea off Hawaii changes by a factor of about 30 per 100 metres, the energetic cost of ventral camouflage at 350, 300 and 200 metres is 1·6, 9 and 270 per cent, respectively, of the resting metabolism. Thus, they suggest that concealment through luminescent countershading is limited to depths greater than 350 to 400 metres. In certain species of midwater squids, adjustments of the intensity of luminescence to match that of background twilight may be through extra-ocular light receptors, which may well be able to monitor both kinds of light (Young, 1977).

Through evolutionary processes, mesopelagic fishes, euphausiids, prawns and cephalopods have 'learned how to acquire' invisible cloaks of bioluminescence. Have they 'learned' through their eyes? Certainly, there are no midwater users of luminescent camouflage with poorly developed eyes. Perhaps the users 'learned' through their eyes two outstanding features of submarine light; that as the depth increases light becomes more monochromatic (towards blue) and that the main axis of the light field moves towards the vertical. Vertically migrating animals might be the best 'learners', but some users of ventral camouflage are non-migrators. And what were the steps in the evolution of the necessary photophores? Perhaps simple ventral lights were first evolved to enable animals to see one another, especially during sexual congress. Later, as such lights provided some ventral camouflage, the way may have been open for evolutionary steps towards more effective camouflage. Eventually, prey disappeared from hunters' eyes, despite their keen sensitivity to blue luminescent light and twilight. Even so, certain fishes and squid, which have yellow lenses in their eyes, may 'see through' luminescent camouflage (p. 388). Deep-sea red camouflage has also been exposed by certain predatory fishes (p. 363). There are no perfect adaptations.

At what depths does ventral camouflage become unnecessary? Clearly there is no need where the background light is very dim, which in most parts of the ocean is somewhere between 700 and 1,000 metres. Here live black species of lantern-fishes of the genera *Taaningichthys* and *Lampadena*, which have much smaller ventral photophores than species of the same size from upper mesopelagic levels. Even so, the small lights may be correlated simply with the low intensities of luminescence needed for camouflage in dim surroundings. The point is forcibly made by Denton and his colleagues (1972) when they consider the relative sizes of ventral photophores in *Argyropelecus affinis* and *Chauliodus sloani*; both of which, as they showed, produce light fields suitable for camouflage (see p. 354). For its size *Chauliodus* has much the smaller light organs, but during daytime it lives deeper than *Argyropelecus* and thus luminescence of lesser intensity suffices for camouflage.

In a descent to bathypelagic levels by way of the fishes *Cyclothone braueri*, *C. microdon* and *C. obscura* there are marked changes in the development of the ventral photophores. The first species, which lives at upper mesopelagic levels, has, compared to its ventral silhouette, appreciably larger photophores than *microdon*, centred between lower mesopelagic and upper bathypelagic regions. In *obscura*, a true bathypelagic species, the photophores are regressed. The bathypelagic species of euphausiids either have no photophores (*Bentheuphausia*) or very small ones (giant species of *Thysanopoda*). As there is no sunlight at bathypelagic levels, the regression or loss of ventral photophores is understandable. But comparison of light sizes is no sure guide as to where at lower mesopelagic levels luminescent camouflage is no longer needed. Off Hawaii Young and Roper (1977) suggest a lower limit of 775 metres, which is supported by their observations on *Octopoteuthis* (p. 361). Here, until further investigations are available, the question must rest.

CAMOUFLAGE BY COLOUR

Colourless transparency, which is more or less attained by a diversity of pelagic animals, seems the simplest way of being unseen. Here we are concerned with colours that hide their owners against backgrounds of blue light, both solar and luminescent.

An overall or partly red coloration is common among midwater invertebrates but rare in fish (see p. 131). Apart from the silvery-sided species of upper mesopelagic levels, many midwater fishes are dark brown or black in colour, as in diverse stomiatoids and nearly all ceratioid angler-fishes. Red or dark-coloured animals presumably reflect very little background light at depth and thus should be hard to see. Just how much light they reflect was determined by Nicol (1958). Measurements were made of reflexion from the integument of a red prawn (*Acanthephyra multispina*) and the dark skin of three fishes, *Chauliodus*, *Cyclothone* and *Xenodermichthys*. The fishes reflected very little light (2 to 4 per cent) over the visual range (400–700 nm). But the curve for the prawn showed that reflexion was minimal in the blue part of the spectrum (below 500 nm) and maximal in the red (above 650 nm). Thus, dark fishes and red prawns reflect very little blue light and must be very hard to see in the depths.

But certain predatory kinds of stomiatoid fishes bear large cheek photophores that produce red light. *Malacosteus*, sometimes called the rattrap fish, has a large light organ just below the eye that emits red light (Fig. 126). In *Pachystomias*, a genus of another stomiatoid family

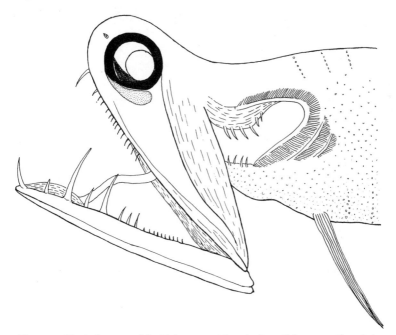

Fig. 126   Head of rat-trap fish, *Malacosteus*. Note the large light organ (dotted) just below the eye.

(Melanostomiatidae), Denton and his colleagues (1970) also observed red light from a large cheek photophore. Moreover, they showed that the rods in the retina of *Pachystomias* must be highly sensitive to red light (they contain a visual pigment with maximum absorption at 575 nm). More recently, another rat-trap fish, *Aristostomias scintillans*, was studied by O'Day and Fernandez (1974). This species has two large light organs near the eye: the suborbital organ emits red light, the postorbital organ green light. Extracts of retinal pigment gave two absorbance maxima, one at 551 nm, sensitive to red light, and the other at 526 nm, sensitive to green light.

Thus, these predatory fishes both emit and sensitively perceive red light. They are able to see red-coloured prey, which no doubt consists largely of crustaceans. Moreover, as Locket (1977) says, 'Not only can such a fish thus illuminate and see a red crustacean, but it is possible that the crustacean will be unaware that it is being illuminated, since its visual pigment may be relatively insensitive to red light.' O'Day and Fernandez suggest that red luminescence could be a good means of communication between members of a species. Such communication '... would require a relatively small repertory of signals compared with transmission via the common wavelengths of bioluminescence and, above all, messages sent in red light would be safe from interception by potential predators'. The biological significance of the production and perception of green light is something for further investigation.

LIGHT LURES

There are about a hundred species of deep-sea angler-fishes and in all but a few the female has a rod (illicium) that carries a light lure (esca) of individual design. The exceptions are species of the genera *Caulophryne* and *Neoceratias*, whose females have no light lure but instead, so to speak, have a highly developed lateral line system (p. 418). The individual shape, luminescent complex and dermal dressing of the lure do not seem to be related to attraction of individual food organisms (p. 256): in each species certain distinct features are presumably recognized by the males, which have well-developed eyes (p. 390).

Besides the short illicium and large esca, females of the genus *Linophryne* bear a chin barbel with luminescent tasselling. Recently, Hansen and Herring (1976) have shown that this genus has dual luminescent systems: the esca contains luminous bacteria whereas the beaded light organs of the barbel have their own system in the form of paracrystalline photogenic granules. The luminescent branches of the barbel bear series of small tubercles, each with a lens-shaped 'eye' encircled by a vascular coat. In the photogenic layer there are two elements, large glandular cells (photocytes) and capillary loops, formed from arterioles of the vascular system. The glandular cells contain a large number of mitochondria and paracrystalline granules. When stimulated by hydrogen peroxide, the barbel organs glow, and their light, as Hansen and Herring conclude, comes most likely from the paracrystalline bodies.

In *Linophryne arborifera* the esca is a simple ovoid bulb with a long serrated appendage (Fig. 137). Within the bulb are glandular tubules enclosed in a cup-shaped reflector. The tubules open into a collecting duct leading to a central cavity. Masses of bacteria may be seen in the spaces between adjacent glandular cells, which, according to Hansen and Herring, produce substances needed for bacterial light production. The light organ also has a ring-shaped smooth muscle system skirting the edge of the reflector. Beside the vascular supply, a pair of nerves enters the bulb.

The luminescent barbel of *Linophryne*, unlike that of various stomiatoid fishes, has no nerve supply. The light is evidently controlled through the vascular system, whereas nerves are also involved in escal light regulation. As Hansen and Herring say, the dual light system in *Linophryne* is apparently unique. Luminescence from both esca and barbel presumably attracts prey. The escal light of ceratioids is yellowish-green or bluish and is emitted in a series of flashes (Bertelsen, 1951).

Angler-fishes may not be the only users of light lures in the bathypelagic zone. Near the tip of the long tapering tail of gulper-eels (Lyomeri) is a light gland, which could well be dangled near the capacious mouth and draw prey. In the mesopelagic zone the main users are most likely to be among the predacious members of the stomiatoids, which also have a luminous barbel (Fig. 44).

'August Brauer, who made a close study of the luminescent organs of the fishes collected by the *Valdivia*, concluded that the luminous chin barbels of certain stomiatoid fishes were used in much the same way as the light lures of deep-sea angler-fishes. When the fantastic diversity of form in these barbels is considered, some tassel-shaped, some long and whip-like with a terminal luminous organ, others carrying a luminous bulb and streamers, yet others with a bulb and an intricate luminous tracery, a strong impression remains that these organs must play an important part in the day-to-day lives of these fishes. And it should be remembered that luminous barbels are present in many species, notably those of the melanostomiatid, stomiatid and astronesthid groups' (Marshall, 1954). Moreover, the barbels may not only attract prey but seem to be provided with sensory means for detecting its approach. Beebe and Crane (1939) have written this concerning *Echiostoma tanneri*, '... even a slight stirring of the water near the barbel would arouse the fish to the utmost so that it threshed about and snapped, striving to reach and bite the source of irritation. Again and again we proved the astonishing sensitiveness of this organ.'

Stomiatoid fishes, which seem to play the most variations on luminescent themes, are represented also by species that have photophores set within the mouth. These lights could well attract prey (Marshall, 1954). The oral light organs of the hatchet-fish *Sternoptyx diaphana* were recently studied by Herring (1977d), who writes, 'They appear as two elliptical patches covering quite a wide area of the roof of the mouth. The organs have no pigment surrounding them, no reflectors and no coloured "filters", each of which are found in other light organs. In fresh animals the organs are sometimes spontaneously luminous, whether or not the ventral series of organs are also luminescing, or they may

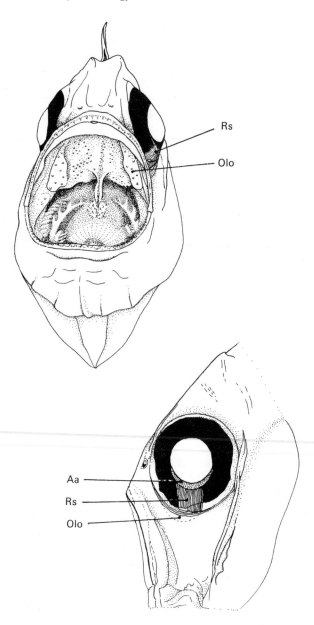

Fig. 127   Oral light organs (Olo) of the hatchet-fish *Sternoptyx diaphana*. Rs, reflective striae on iris; Aa, aphakic aperture. (Redrawn from Herring, 1977d)

luminesce if mechanically stimulated. In either event each organ glows steadily for periods of up to 30 minutes, fading gradually' (Fig. 127). Large, long-glowing oral lights seem made for attracting prey. If this is their main function, they might also, as Herring argues, serve as a reference standard for comparison with ambient daylight (see also pp. 354–5). The oral organs are so placed that some of their light must reach a tract of reflective striae over the ventral part of the iris and perhaps be guided into the eye.

Of all mesopelagic forms, viper-fishes (*Chauliodus*) come nearest to 'copying' the angler-fishes. In this genus the second ray of the dorsal fin is long, whip-like and tipped by luminous tissue. Elsewhere I wrote, 'During the daytime Dr Pérès has watched *Chauliodus* from a bathyscaphe and in a letter dated 17 November 1958 has written of his observations. These fishes hover in the water with the long axis of the body at an acute angle to the horizontal plane, the head being above the tail. At the same time the long second dorsal ray, which is tipped with luminescent tissue (Brauer, 1908) is curved forward over the head so that the extremity of the ray lies in front of the mouth. Clearly, the fish is behaving very much like a female ceratioid and is angling for prey' (Marshall, 1960).

SEXUALLY DIMORPHIC LUMINESCENT SYSTEMS

In deep-sea fishes sexual dimorphism is most highly developed in the ceratioids. Indeed the luminescent contrast between the sexes is extreme: the males appear to lack any kind of luminous system. Sexual dimorphism is also deeply engrained in the genus *Cyclothone* but does not extend to the photophores. The luminescent organization of deep-sea angler-fishes is covered elsewhere (pp. 364–5). Here we turn first to the stomiatoid fishes.

In these fishes sexual dimorphism is extreme in the genus *Idiacanthus* and the luminous system is strongly involved. 'The females are black and serpentine with a chin barbel and reach a length of at least a foot. The males are pale-coloured, have no teeth or pelvic fins or barbel and grow to little more than an inch and a half in length. Both sexes have two rows of light organs running along the underside of the body and the male carries a relatively large cheek light just behind the eyes: in the female this organ is poorly developed' (Marshall, 1954). Attraction between the sexes during the breeding season is no doubt facilitated by flashes from the males' powerful cheek lights.

Sexual dimorphism in other stomiatoid fishes also involves the cheek lights but the sexes are of much the same size. The males of the rat-trap fishes *Photostomias* and *Aristostomias* have much larger cheek lights than the females, which is true also of certain melanostomiatids and astronesthids. In *Astronesthes cyaneus* the luminous chin barbel is short in females but is about as long as the head in males (Goodyear and Gibbs, 1969).

More is known of luminescent organization in the sexes of lantern-fishes. A common sexual difference involves light organs that are housed on the caudal peduncle. In many species the male has one or more prominent luminous glands on the upper side of the peduncle: in females these glands, which may be less well developed, are on the lower side or absent

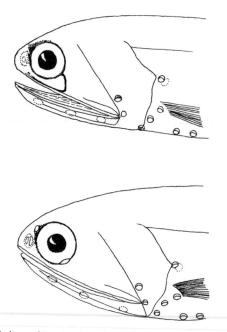

Fig. 128   Sexual dimorphism in the orbital photophore system in the lantern-fish *Diaphus diadematus*. Above, male, with a very large suborbital light; below, female.

entirely (e.g. *Protomyctophum, Electrona, Myctophum*). Both sexes of *Lampanyctus, Lampadena* and *Taaningichthys* have luminous glands on both the upper and lower surfaces of the caudal peduncle, which in the first genus consist of several scale-like glands and in the other two of single pearly-white organs. In the diverse genus *Diaphus* luminous caudal glands are not developed and sexual differences, where present, involve photophores around the eyes (Fig. 128). Certain organs before or below the eyes are much larger in male fishes (Nafpaktitis, 1968). Lastly, in *Bolinichthys* males and females differ in the pattern of luminous glands on the upper surface of the head between the eyes (Wisner, 1974).

The caudal glands of lantern-fishes emit brilliant flashes of luminescence. By chance, William Beebe found that an individual of one species (*Myctophum affine*) in an aquarium replied to the luminous dial of his watch. 'I happened to lift my wrist watch with its dully luminous dial close to the fish and it reacted at once, giving out two strong discharges of the caudal glands. Concealing the watch and later displaying the light of its face resulted in an instant reaction. This happened eight times. I then flashed on my much stronger flashlight with no result. For five minutes I altered the two artificial sources of illumination with identical results, the fish reacting vigorously to the wrist-watch, but paying no attention to the electric torch' (Beebe and Van der Pyl, 1944). Barnes and Case (1974), using mechanical and electrical stimulation, produced brilliant, rapid and transient displays from the caudal glands of several species of lantern-fishes. 'Caudal organ photogenesis was highly phasic, consisting of intense, short-duration pulses of light spatially arranged not as a uniform emission from the entire organ, but rather as a series of discrete subunits.' The pearl-button photophores over the body emitted a relatively steady glow of low intensity.

Both the photophores and luminous glands of lantern-fishes consist of 'fundamentally similar stacks of photogenic lamellae'. The caudal glands and other luminous patches simply contain much more of the photogenic tissue than do the primary photophores (Edwards and Herring, 1977). Both kinds of light organs are innervated. It is reasonable to suppose that bright flashes from the caudal glands play a part in sexual congress. In some species each sex has its own light signs, but in others (see above) there are no structural differences between the light organs of the two sexes. Where there is no sexual dimorphism in photophore pattern, it would certainly be wrong to conclude that light displays have no part in attraction between the sexes. Moreover, the caudal glands of lantern-fishes may have another function: their bright flashes could well deter predators.

Sexual dimorphism in luminescent organization seems to be rare among deep-sea invertebrates. In certain species of the euphausiid genus *Nematoscelis* the males have saddle-shaped thickenings of chitin on the dorsal surface of certain abdominal segments, which are associated with enlarged photophores on the segments immediately behind those with thickenings (James, 1973; see also Herring and Locket, 1978). In most euphausiids the photophore pattern is the same in both sexes and, as we saw, there is good evidence that it is involved in ventral camouflage (p. 358). That this may be not the only involvement of euphausiid lights is suggested by observations in the laboratory. In certain species luminescent response to stimulation is most readily elicited during the breeding season (Tett and Kelly, 1973).

Lastly, it seems odd that sexual dimorphism is unknown in the photophore patterns of deep-sea cephalopods. But as Herring (1977c) says, this does not automatically rule out the use of light organs for sexual displays. His conclusion that deep-sea cephalopods '... almost certainly make far more extensive and varied use of their impressive luminescent abilities than we can envisage', is surely apt.

DEIMATIC DISPLAYS OF LIGHT

Deimatic displays, known also as startle or fright displays, occur in diverse terrestrial animals. In the sea they have been observed in certain molluscs. For instance, if a cuttlefish (*Sepia*) is touched, two black spots appear suddenly on the back (Edmunds, 1974). The rapidly acting colour cells of cephalopods seem well suited to produce a sudden startle pattern. Perhaps photophores, which can be switched on and off rapidly, are also used in deimatic displays.

Such displays are certainly possible in stomiatoid fishes, as shown by O'Day's (1973) observations on *Stomias atriventer*, *Idiacanthus antrostomus* and *Chauliodus macouni*. Like other relatives, these fishes have very many minute photophores in the skin of the body and fins. Each photophore consists of an aggregation of secretory cells and is innervated. When stimulated by adrenalin, the whole fish became silhouetted and enlarged in a bright luminescent glow. Such display could well be deimatic and deter a predator, but as O'Day says, there are other possibilities, as in aids to mating or aggregation.

LUMINESCENCE AND ORIENTATION IN SPACE

Periodic luminescent displays of their form might space out predators in hunting territories. Nicol (1960b) considered this role of luminescence after study of the large subocular light organs of predatory stomiatoids. Each species (in the families Stomiatidae, Chauliodontidae, Astronesthidae, Melanostomiatidae and Idiacanthidae) has a large light organ below and behind the eye, which generally consists of a more or less spherical or elongate mass of photocytes backed by an inner tunic and a black pigment sheath. The tunic may contain reflecting material. A long band of muscle, attached to the hyomandibular bone, runs down and then under the light organ to an insertion on the ventro-lateral or external face of the organ. When the muscle contracts the light-emitting face of the organ will be pulled downwards and occluded by the ventral pigmented layer. The subocular organ is under nervous control and the light may be emitted in flashes. Nicol conjectured, 'The large subocular organs may be used as torches, for illuminating the surrounding water. The serially arranged photophores may permit the animals to recognize one another. The result may be mutual repulsion, thus keeping the fish spread out in hunting territories delimited by the intensity of the light and the distance at which it can be seen; or mutual attraction when the animals differ in sex.'

The closely coordinated behaviour within fish schools arises, as it were, from the continual and precise adjustment of attraction and repulsion between individuals of much the same size. The individual constellation of lights on each kind of lantern-fish could be used as recognition signs between members of a school, enabling them also to keep their distance. As observers have seen and catches often indicate, mesopelagic fishes may often be associated in groups, but whether these have the regimentation of schools is uncertain. But even if they are not schools, the interchange of light signals seems a good way of maintaining some kind of contact and cohesion between the members of a group. The same may be true in swarms of euphausiids.

Bioluminescence near and on the deep-sea floor
The benthopelagic fauna

Over the upper continental slope the waters are no doubt lit from time to time by the luminescent displays of mesopelagic fishes, euphausiids, prawns and squids. In permanent benthopelagic fauna (pp. 135–40) light organs are known only in fishes of the gadiform and beryciform orders. In the former group the main light producers are rat-tails (Macrouridae) and the deep-sea cods (Moridae). Both develop an elaborate light organ that opens near the anus or into the rectum and contains luminous bacteria, probably of the species *Photobacterium phosphoreum*.

In rat-tails, the light organ, which ranges from bulbous to tubular form, is confined to the slope-dwelling species that are centred above a level of 2,000 metres. Curiously enough, the deeper abyssal species (of the genera *Chalinura*, *Lionurus* and *Nematonurus*) have neither a light organ nor a drumming swimbladder (in male fishes) (see p. 426). Moreover, half of the 250-odd slope-dwelling species have some kind of light organ, which is housed along the mid-ventral line of the abdominal walls (Marshall, 1965, 1973). The symbiotic luminous bacteria are cultured in a glandular part of the organ.

The structure of the light organ is best known in the species *Malacocephalus laevis* (Hickling, 1925, 1926; Haneda, 1938). The visible parts of the system are two black scaleless 'windows' between the bases of the pelvic fins. There is a lens between each 'window' and the sac-like light gland, which opens through a duct just before the anus. An upper reflecting layer is placed so as to direct the light from the gland downward to the lenses. The entire organ is enveloped by a layer of black colour

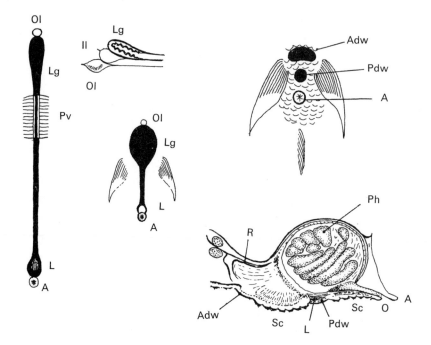

Fig. 129 Light organs of rat-tail fishes (Macrouridae). Left, the long tubular ventral light gland (Lg) of *Hymenocephalus italicus*, which extends from a position before the pelvic fins (Pv, symphysis of pelvic girdle) to just before the anus (A); L, lens; Ol, outer lens. Above centre, section of the light organ showing the relationship between the outer lens (Ol), the inner lens (Il) and the light gland Lg). Below centre, the relatively large light gland of a young *Hymenocephalus*. Above right, the anterior dermal window (Adw) and posterior dermal window (Pdw) associated with the light organ in *Malacocephalus laevis*. Below right, longitudinal section through the light organ of *M. laevis*; L, lens; Sc, scales; R, reflecting layer; O, opening of light gland near anus (A); Ph, photogenic folds of light gland, which hold capsules of luminous bacteria. (Above centre and below right after Haneda)

cells, which in the lens regions are of the dispersible type (Fig. 129). The dispersible chromatophores are presumably controlled so that when their pigment contracts light may pass through the scaleless 'windows' to the sea.

The light of *Malacocephalus laevis* was certainly well known to fishermen of Cezimbra, Portugal, who used to take a piece of dogfish flesh and rub it over the rat-tail's belly. 'Near the anus of the latter opens a gland full of luminous bacteria and from the opening discharges a viscid yellow fluid which shines with a sky-blue light. Thus the bait was smeared with a light which lasts for several hours and attracts fish to the hooks'

(Marshall, 1954). A bulbous light organ much like that of *Malacoce-phalus* is found also in species of *Nezumia* and *Ventrifossa*, which have a single dermal 'window'. In *Coelorhynchus* the form of the light ranges from a bulb to a long tube. Species such as *C. coelorhynchus* and *C. caribbaeus* have a dermal 'window' on the belly; in others there is no outer evidence of a light organ, which, as Haneda (1951) found in young *C. parallelus*, shines through the skin and scales. Each species of *Hymenocephalus* has a long, tubular light organ (Fig. 129). At the anterior end, before the bases of the pelvic fins, are two lenses, an inner one close to the gland and an outer, visible one, through which light is presumably transmitted. There is a third lens just before the anal opening of the organ (Marshall, 1965, 1973).

Light organs much like those of rat-tails are known in various species of deep-sea cod (of the genera *Physiculus, Brosmiculus, Gadella, Trip-terophycis* and perhaps *Antimora* and *Lotella*) (Marshall and Cohen, 1973). *Physiculus japonicus* has a bulbous light organ and from the lumen of the gland is a canal that opens into the rectum near the anus. There is a reflector above the gland and between the latter and a clear scaleless area of skin before the anus is a single lens. Haneda (1951) saw that light was emitted through the scaleless area.

Unlike its hake relatives, *Steindachneria argentea*, which lives in the Gulf of Mexico at depths between 200 and 400 metres, has a luminescent system (Cohen, 1964). The light gland surrounds the rectum just before the anus. Between the forwardly placed vent and the genital opening is a transparent gelatinous region, which apparently transmits the light backwards. Before and around the gland and gelatinous region the hypaxial muscles of the body wall are translucent and overlaid by skin with a striated colour pattern, formed of parallel lines of alternating dark and silver pigment. When fresh *Steindachneria* were injected with adrenalin, all these striated areas, which are presumably light guides, became luminescent (Cohen, 1964). Such indirect lighting systems are known in various shallow-water fishes. Besides *Steindachneria*, bentho-pelagic deep-sea fishes with a ventral light organ and silver-black stria-tions over the underparts include rat-tails (*Hymenocephalus* spp., *Lepido-rhynchus*) certain deep-sea cod and slime heads (Trachichthyidae), now to be considered.

These are beryciform fishes with a dark, laterally compressed body (3–15 cm) and they are commonest along the continental slope between depths of 120 and 600 metres. There are five genera and two, *Paratra-chichthys* and *Hoplostethus*, are represented by light-producing species (Woods and Sonoda, 1973). In *Paratrachichthys prosthemius* the light organ is visible externally as a black scaleless depression just before the anus. Internally the luminous gland, which encircles the vent, is formed of tubes that surround numerous luminous bacteria. These tubes drain into ducts with outer openings near the anal papilla. Below and forward of the gland is a hyaline lens that overlies the scaleless depression. Before the gland and extending forwards between the rami of the lower jaw is a translucent keel muscle that acts as a light guide. There is also a transparent pair of light-guiding muscles (the filiform body) that extends backwards from the gland to the rear of the anal fin. The skin in the region of the light guides has a silvery-black striated appearance (Haneda, 1957). In certain luminescent species of *Paratrachichthys* the striated areas are absent. When he looked at the underparts of a *P. prosthemius* in the dark, Haneda saw light shining from the keel muscles and the filiform body, but no light was seen from the side. The extent and intensity of the luminescence could be controlled by the fish.

There are thus deep-seated convergences between the ventral light organs of *Steindachneria*, numerous macrourids, morids and certain trachichthyid fishes. Do the convergences indicate a common biological function? The structure and position of the light organs are similar in both sexes, which need not imply that light signals play no part in sexual activities. In the slope-dwelling species of rat-tails, the periodic exposure of the ventral light by both sexes, combined with sound making by the males, could well facilitate sexual congress. In the singing midshipman (*Porichthys*), from shallow waters, Crane (1965) found that a male produced both luminescent flashes and made grunting sounds during courtship with a female luminescing after artificial stimulation. But as already mentioned, the absence of the means for making light and sound signals in abyssal rat-tails is curious. One would have thought that these ways of communication would be particularly apt in their sparsely inhabited surroundings. In other words, there is much we do not know.

The benthic fauna

In benthic deep-sea fishes, light organs are known only in the genus *Chlorophthalmus*, represented by species that live over the upper continental slope. Somiya (1977) found a small perianal gland in *C. albatrossis* and *C. nigromarginatus*. The activity of this organ became visible as a small spot of blue-green luminescence) only after 15 minutes of dark adaptation by the observer. Between the bases of finger-like structures in the perianal gland Somiya found colonies of luminous bacteria. He suggests that the light could be used to maintain contact between members of a group. If some of the latter are in spawning condition light signals could facilitate their mating.

Diverse kinds of invertebrates that live on the deep-sea floor are luminescent. Large oplophorid prawns, such as *Heterocarpus*, that rest on the bottom no doubt use their luminescent decoys from time to time. Sepiolid squid may do the same (p. 350).

Of the slow-moving forms, most is known of the echinoderms. Herring (1974b) observed luminescence in two species of brittle-stars (Ophiuroidea), six species of sea-stars (Asteroidea), nine species of sea-cucumbers (Holothuroidea) and two species of sea-lilies (Crinoidea), and more recently has made further discoveries (Figs. 130 and 131). Luminescence was induced mechanically, chemically (by 5% hydrogen peroxide in sea water) and by placing the animal in fresh water. The most intense light came from the elasipod sea-cucumber *Laetmogone violacea*, which emitted blue-green light from the tips of all the dorsal papillae. 'A multitude of smaller brightly luminous points appeared over the rest of the dorsal and ventral body surface, and there were two fainter lines of light laterally, in among the general flashes of the lateral body wall.' The measured intensity at 1 metre was $2,600 \mu W/cm^2 \times 10^{-9}$. The feather-star *Thaumatocrinus jungerseni* emitted points of light on each segment of the arms and cirri, the whole producing $654 \mu W/cm^2 \times 10^{-9}$ at 1 metre. Brittle-stars (*Ophiacantha aculeata*) glowed with a bright blue-green light when brought inboard. The luminescent cells of these echinoderms were not identified with certainty, but Herring (1974) concluded that control of luminescence is probably neuronal, as indicated by responses to acetyl choline and eserine.

In midwater animals one may make reasonable conjectures as to the function of their luminescence, but the echinoderms are baffling. If their light is to advertise their noxious qualities, it is curious that luminescence has not been observed in sea-urchins. Echinothuroid urchins, for instance, are not pleasant to handle. Comparison of nearly related species may be suggestive, as between the brittle-stars *Amphiura chiajei* and *A. filiformis*. The former, which is not luminescent, feeds with its arms on the surface of the sediment. The filter-feeding *filiformis* is luminescent and holds its arms in the sea to gather food. If one arm is removed the other arms flash and are quickly withdrawn under the sediment (Tett and Kelly, 1973). A predator might thus be denied a full meal.

Luminescence among the other slow-moving invertebrates is reported in the sea-spiders (Pycnogonida). The largest species *Colossendeis colossea* (p. 203) is said to be luminescent and there is a definite record of light emission (perhaps from the end segments of the first pair of legs) in *Nymphon gracile* (King, 1973). Of the larger polychaete worms luminescence may be at least suspected in the polynoids, which have light-producing relatives in shallow waters.

Besides the sea-lilies, the main light producers among the attached animals are alcyonarian (octocoral) cnidarians (pp. 179–83). In sea-pens Pennatulacea) luminescence has been best studied in the sea-pansy (*Renilla*). Investigations of the biochemistry of light in the sea-pansy have done much to clarify and enlarge present concepts (Herring, 1977b). Work on deep-sea pennatulids and related forms is in progress. During

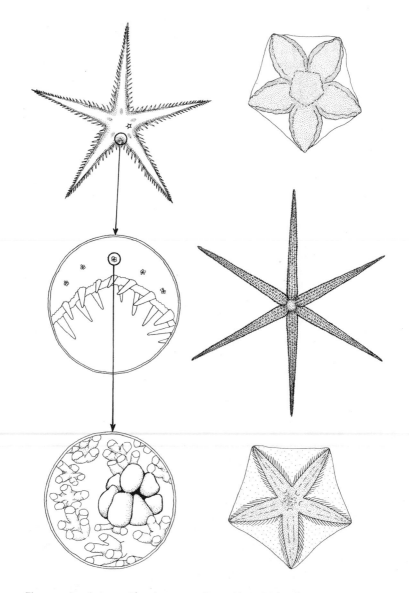

Fig. 130 Luminescent (deep-)sea-stars (Asteroidea). Right, from top to bottom,
Cryptasterias personatus, *Hydrasterias ophidion* and *Hymenaster* sp. Top
left, *Pectinaster forcipulatus* and, below, luminous points. Luminescent areas
stippled. (Redrawn from Herring, 1974b)

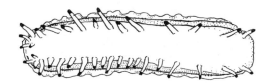

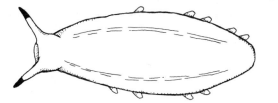

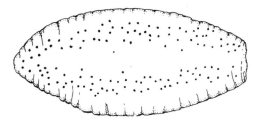

Fig. 131   Luminescent (deep)-sea-cucumbers (Holothuroidea). From top to bottom, *Laetmogone violacea*, *Peniagone théeli*, *Kolga hyalina* and *Paroriza* sp. n. The luminescent parts are shown in black. The body surface may also be luminescent. (Redrawn from Herring, 1974b)

a recent cruise (77) of RRS *Discovery*, J. G. Morin (1976) found luminescence in all the sea-pens trawled and in many of the sea-fans (gorgonians). Zoanthids were also luminescent, but no light could be elicited from sea-anemones and cup-corals. The luminescent animals were examined with a fluorescence microscope to determine if possible the appearance and distribution of the luminescent cells.

Nicol (1962) has suggested that as the luminescence of sea-pens and other attached forms may be elicited as actively swimming animals brush against them, the display of light could facilitate the disengagement of the intruders. Perhaps cnidarian lights may sometimes attract copepods and other small kinds of prey.

The sensory systems
of deep-sea animals

Survey of deep-sea bioluminescence leads naturally to consideration of visual systems. Midwater eyes, as we saw, are very sensitive to the blue wavelengths of luminescence, which match those of deep-sea twilight. But even in the clearest oceanic waters the twilight extends little below a level of 1,000 metres. In the midwaters the twilight zone is both the main visual and light-producing zone: in the sunless depths, where light displays are sporadic, eyes tend to regress. Even so, where biolumine-scence is all, eyes may be well developed if what they have to see plays a leading part in the continued existence of their users (p. 390).

The other senses are 'oceanic'. Scents, sounds, nearby disturbances and the contacts needed for taste and touch are possible anywhere. Moreover, most luminescent displays are visible only a few metres from the light-producer, whereas scents and sounds, especially the latter, may be per-ceived far from their source.

When a baited automatic camera reaches the deep-sea floor, the first animals to be filmed are usually brittle-stars, prawns, amphipods and perhaps a fish or two. Thereafter the number of fish around the bait gradually increases, reaching a maximum after a few hours. The fish leave after most of the bait has been eaten, and now crabs, sea-urchins, snails and such like move slowly towards the remnants (Isaacs and Schwartzlose, 1975).

As Isaacs and Schwartzlose continue, 'Sometimes the direction of the near bottom currents has been determined and it can be seen that these late comers to the banquet plod upstream toward the bait—an indication that they are probably following a scent. The fishes also probably depend on scent for close in detection of the bait; on some occasions we could relate the number of fish gathered round the bait to the strength of the current, which suggests that an increased current had carried the scent more widely. Like wide-ranging scavengers on land, however, fish that are far away are probably led to the area of the bait by other ones. They may sense the successive collapse of loosely established territories as the scavengers that held them move closer to the bait, each invading the area once held by an absent neighbour. The western prairie wolf, vultures and other terrestrial scavengers respond to just such a territorial collapse to converge on large kills.'

Scent trails are small compared to the range of sounds, which, as already implied, far outstrip the reach of perceptible light. More precisely, a beam of underwater sound of a few kHz travels with several thousand times less decrement in its strength than a beam of light.

Sound-making devices are known only in benthopelagic fishes, but are likely to be short-range means of communication (p. 427). Perhaps the only long-range sound users are sperm whales, which take much of their food at deep-sea levels.

Visual systems

Deep-sea animals with both the most complex photophores and eyes are fishes, cephalopods and decapod crustaceans, which include euphausiid shrimps. Such photophores are involved inter alia in luminescent camouflage, the means of not being seen by predators, many with the most complex kinds of eyes. But, as will already be evident, even in the twilight zone the association of elaborate eyes and ventral light organs is by no means invariable. To take but one combination, there are fishes and octopods with large eyes and no light organs.

We begin with fishes, which have been studied most, and whose visual systems will serve later for comparison and contrast with those of cephalopods, polychaete worms and decapod crustaceans.

Mesopelagic fishes

Nearly all mesopelagic fishes develop 'normal' to very large eyes compared to the size of the head. The relative orbital dimensions in species with 'normal' eyes are much as one might find in shallow-water fishes of comparable size, but they are in the minority. For instance, out of some 150 species of lantern-fishes described by Wisner (1974), which range from a few to over 16 cm in length, less than a third have eyes that are not large by comparison with the length of the head before the gill covers. Furthermore, over half of these species (mainly in the genus *Lampanyctus*) grow to a length of 10 cm or more and actually have eyes that are of much the same size or even larger than those of small species with disproportionately large eyes. Similar orbital relations obtain among the mesopelagic stomiatoid fishes (e.g. see Grey's 1964 report on the Gonostomatidae).

The larger the eye the more the quanta of light that will enter the pupil. But enlargement of the eye to scale will not result in greater illumination of the retina. If the eye simply doubles in size, the area of the round pupil and the area of the retinal image will both be enlarged—following the doubling of the diameter—by a factor of four. Generally, then, in eyes of different sizes but of identical dioptric proportions the illumination of the retina will be the same in all.

In an eye of a given size the way to admit more quanta of light to the retina is through widening the pupil, which may be seen in nocturnal birds and mammals as well as in deep-sea fishes. But in teleosts the iris is not adjustable; the lens projects through the pupil and both have much the same diameter. Moreover, the fish lens is spherical and whatever the size or increased aperture of the eye, the radius and focal length of the lens are related by Matthiessen's ratio.* Even though spherical, the fish lens is free of the optical aberrations such as occur in glass spheres of uniform refractive index. In Locket's (1977) valuable review there is a photograph of the lens of *Mora moro*, a deep-sea cod, immersed in sea water and resting on some print, which is magnified without distortion right to the edge of the lens.

The eye's sensitivity depends ultimately on the design of the retina. In deep-sea fishes much of the thickness of the retina is due to the outer segments of the visual cells, which in nearly all species are rods. The rods are not only long, but also of uniform thickness and packed closely together like the bristles of a fine brush. Indeed, in some species there are multiple-bank retinae formed of two to six layers of rods, which altogether may measure hundreds of micrometres (Fig. 132). The longer the rods and the closer their pile, the more light they will intercept over a given area of retina. But multiple-bank retinae are evidently more than devices for the efficient interception of dim light. Recent experiments and calculations by Denton and Locket (1976) on RRS *Discovery* indicate that such retinae should enable their users to discriminate the colours of light emitted by luminescent forms.

Moreover, in numerous species the retina is backed by a special layer (tapetum) that will reflect or scatter back light intercepted by the rods (see Locket, 1977). Lastly, many rods may be connected, by way of bipolar and ganglion cells, to a particular optic nerve fibre. Thus, even in very dim light, this nerve fibre may be activated through the combined responses of its visual cells. Summation of visual response, as judged by the mean number of rods linked to one optic nerve fibre, varies considerably

---

* Ideally, the distance from the centre of the lens to the retina should be 2·55 times the radius of the lens, '... and it rarely actually differs from this figure, known as Matthiessen's ratio, by more than one or two integers in the last decimal place—even in the tubular eyes of deep-sea forms, which were once called telescopic because they were thought to be radically different from ordinary fish eyes in their optical principles'—Walls, 1942.

from species to species (Munk, 1965, 1966). More recently, Locket (1977) has found that the rods are clearly and regularly associated in groups over certain parts of the retinae of scopelarchid fishes ('pearl eyes'), and of a barracudina, *Notolepis rissoi*. Over the main retina of scopelarchids, where the rods are grouped, there is an average of 23 rods in each bundle. The cellular and optical organization is such as to '... suggest that each group may function as a single macroreceptor' (Locket, 1971).

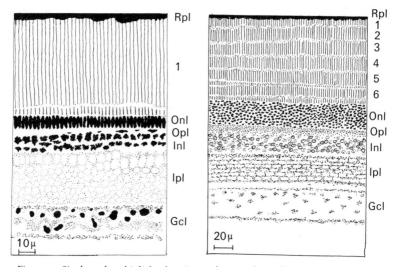

Fig. 132   Single and multiple-bank retinae of mesopelagic fishes. Left, *Vinciguerria lucetia* and right an argentinoid fish, *Nansenia groenlandica*, with six (1-6) banks of rods. Rpl, pigment layer of retina; Onl, outer nuclear layer; Opl, outer plexiform layer; Inl, inner nuclear layer; Ipl, inner plexiform layer; Gcl, ganglion cell layer. (Adapted from Munk, 1966)

'But whatever the structure and use of the eye, the efficiency with which the retina absorbs the light quanta incident on it will be of the first importance' (Denton and Warren, 1957). The photosensitive pigments, as Denton and Warren found, are golden forms of rhodopsin that absorb blue-green wavelengths (480–485 nm) most efficiently. Moreover, the eyes of mesopelagic fishes such as *Argyropelecus* spp., *Chauliodus sloani*, *Gonostoma elongatum*, *Searsia* sp., *Myctophum punctatum* and *Diretmus*, have a high density of pigment compared to the human eye. The human retinal density of rhodopsin for light of 500 nm, which is about 0·15, may be compared with the figure of 1·0 or more for a number of mesopelagic fishes. If this visual adaptation, together with the advantages the fish has in a wider pupil and more transparent eye media are considered together '...we may suppose that if a deep-sea fish and a human were both looking at the same field of blue-green light, the number of quanta absorbed/cm^2 of retina/sec would be 15–30 times greater for a deep-sea fish than for the human' (Denton and Warren, 1957). On these and other estimations are based figures indicating that deep-sea fishes can see daylight at depths below 1,000 metres in the clearest parts of the ocean (see also Clarke and Denton, 1962).

Special adaptations

Such sensitive, wide-open eyes, with the pupillary rim of the iris more or less concentric with the outline of the lens, look more or less laterally in many deep-sea fishes. In diverse forms there are correlated changes in the dioptric design of the eyes, which may involve the setting of the eyes and the retinal organization.

In species of several families the pupil is enlarged so that it is not entirely stopped by the lens, thus leaving an aphakic (lensless) space. Munk and Frederiksen (1974), using mathematical models, have considered the optical significance of aphakic apertures. In forms with apertures around the lens, as in species of *Gonostoma*, they showed that even a small gap between lens and pupil will considerably enlarge the area of retina that will receive an image at full aperture. Where the space is before the lens (rostral aphakic gap) the pupil is somewhat elliptical to oval, the major axis in the horizontal plane. This kind of eye has evolved in numerous species that include lantern-fishes (*Diaphus* spp., *Taaningichthys* and *Lampadena*), searsiids, notosudids, bathylagids, rat-trap fishes (malacosteids) and free-living males of two ceratioid angler-fishes (*Cryptosarus couesi* and *Ceratias holboelli*) (Fig. 133). Again, the aphakic gap will increase the illumination, and thus enhance sensitivity, particularly over the posterior (temporal) area of the retina. Moreover, the eyes are set forward on the head or placed so that there is a wide (binocular) overlap between their visual fields.

The binocular field illuminates a specially modified temporal part of each retina. In many species (e.g. *Nansenia*, *Bathylagus*, *Eustomias* and the two male angler-fishes) the rods of the temporal area have the longest light-sensitive parts, but Munk (1966) does not say whether there is an increase in visual-cell density. At all events, the binocularity and the lengthening of the temporal rods should give high sensitivity in the forward field of vision, coupled with good spatial perception of nearby animals. Both faculties are biologically valuable in dim surroundings.

In other species with a rostral aphakic space the specialized temporal part of the retina contains a depression or fovea, which is steep-sided (convexiclivate) in searsiids and notosudids and shallow (concaviclivate) in certain species of *Bathylagus*. Considering just the first kind of fovea, the rods opposite the depression may be thinner, longer and more tightly packed than those of surrounding regions (Locket, 1977). A convexiclivate fovea, as Pumphrey (1948) proposed, has optical properties such that a symmetrical image will quickly become asymmetrical if the image

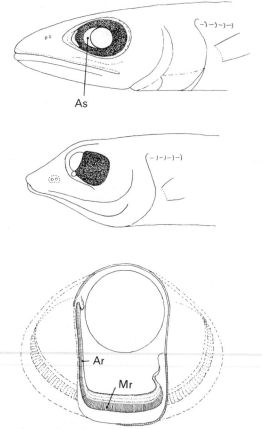

Fig. 133   Eyes of mesopelagic fishes. Above, *Notosudis*, with an egg-shaped pupil and wide forward aphakic space (As); centre, *Winteria telescopa*, with forwardly directed tubular eyes; below, a tubular eye in relation to a normal eye, Ar, accessory retina; Mr, main retina. (Redrawn from Marshall, 1971)

is not centred exactly on the fovea. Such a fovea is well fitted to keep a very sensitive visual 'fix' on a nearby organism. More recently, Harkness and Bennet-Clark (1978), who used optical models, argue that light refraction over a steep-sided convexiclivate fovea could provide sensitive and unambiguous indications of errors in the focussing of the eye. Their model applies particularly to animals with monocular vision, which could, through the sensory output of the focussing muscles, obtain cues as to the distance of an object in view. Deep-sea fishes with both foveae in the binocular field would seem to have the best of both visual worlds.

Lastly, diverse lantern-fishes, barracudinas (paralepidids) and *Omosudis*, and there are no doubt other instances, have a ventral aphakic space (Marshall, 1971). The line of keenest binocular vision is presumably downward, but when resting these fishes may often be poised at angles well away from the horizontal plane. Indeed, barracudinas are known to hover head upwards in the water (p. 356). When lantern-fishes swim normally, their ventral aphakic spaces probably become part of the system that monitors their own luminescence with respect to ambient light (p. 357).

The widest binocular fields obtain in tubular eyes, which have evolved independently in representatives of at least 11 families of deep-sea fishes (Marshall, 1971). The two main visual axes of tubular eyes are more or less parallel, and how this may well have evolved is shown by vertical sections through the eyes of two small stomiatoids, *Valenciennellus tripunctulatus* and *Vinciguerria poweriae* (Locket, 1977) (Fig. 134). Sideways, the eyes of the first species seem normal apart from a somewhat upward setting of the lens. Inwardly, however, the ventral (main) upward-looking part of the retina has much longer rods than the dorsal accessory part. The entire bipartite differentiation of the retina and the disposition of the lens indicates that the main visual axis is upward between the horizontal and vertical planes. In *Vinciguerria poweriae*, the eye seen laterally is almost round in outline but the lens has a distinct dorsal setting. Within the eye more of the main retina, compared to that of its relative, is directly below the lens. Thus, the main visual axis is much nearer the vertical than the horizontal plane. The accessory retina is closer to the lens. A further dorsal shift of the lens, combined with differentiation of the retina so that the main part is centred immediately below the lens, gives tubular eyes that look vertically upward, as in hatchet-fishes (*Argyropelecus*). Clearly, the nearer the visual axes approach the parallel condition, the wider is the binocular field. Even in fully formed tubular eyes the main retina and lens setting conform to Matthiessen's ratio (p. 381) but not so the accessory retina (see below).

'The wide binocularity of tubular eyes should enable a fish to judge the distance of nearby organisms—an obvious advantage in dim surroundings—and may also provide extra sensitivity (as do human eyes when used together). But most of the sensitivity resides in the main retina, with its very efficient light-absorbing screen of golden pigments (see Denton and Warren's 1957 figures for the density of pigment in the eyes of *Argyropelecus*). Further, the enlargement of an image on the main retina is the same as that in a normal eye of the same dioptic dimensions (Fig. 133). Such a saving of ocular space—and most mesopelagic fishes are

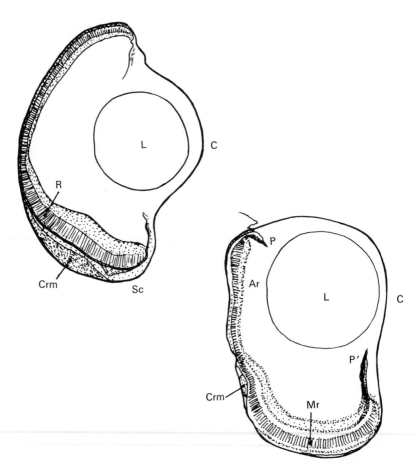

Fig. 134 Vertical sections through the eyes of two small stomiatoid fishes, *Valencien-nellus tripunctualatus* (above) and *Vinciguerria poweriae*. Note the definite dorsal displacement of the lens and the oblique setting of the pupil (between P and P′) in the latter species. C, cornea; Crm, choroid rete mirabile; L, lens; Ar, accessory retina; Mr, main retina; Sc, sclera. (Redrawn and simplified from Locket, 1977)

small—is gained at the expense of the accessory retina, which, as Munk (1965, 1966) demonstrates, is too close to the lens to be in focus. Tubular eyes are thus not bifocal. Even so, the accessory retina, as Brauer (1908) believed, could still detect movements, though only of brightly lit organisms' (Marshall, 1971).

All short, deep-bodied forms, such as the hatchet-fishes (*Argyrope-lecus*, *Opisthoproctus* and *Macropinna*, have upward-gazing tubular eyes but different diets. The hatchet-fishes, with relatively large jaws, take prey ranging from copepods to small fishes; the tiny-jawed *Opistho-proctus* seems to feed on parts of siphonophores (Marshall, 1954). More elongated fishes with upward-looking tubular eyes, notably scopelar-chids and certain evermannellids, are large-mouthed predators, able to swallow large prey.

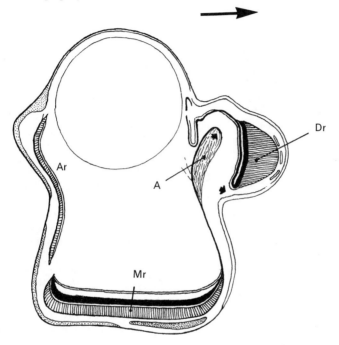

Fig. 135 Diagram of tubular eye of the argentinoid fish, *Dolichopteryx longipes* (the large arrow points laterally). The transparent part (window) of the diverticu-lum (between small solid arrows) is placed so that light lateral to the fish is reflected from the argentea (Ar) through the window and thence to the diverticulum retina (Dr). Ar, accessory retina; Mr, main retina. (Redrawn from Frederiksen, 1973)

Tubular eyes that look upward seem well placed to see prey silhouetted against the relatively strong, down-going part of the light field. But there are fishes with both small jaws (*Stylephorus* and *Winteria*) and large jaws (giganturids) that have forwardly directed tubular eyes (Fig. 133). Per-haps these fishes spend some of their time hovering head upwards in the sea. Such a habit has evidently been assumed for giganturids and *Style-phorus*. How *Stylephorus* may see and take food is considered by Locket (1977).

Apart from free-living males of the angler-fish genus *Linophyne* (p. 390) fishes with tubular eyes live from upper to lower mesopelagic levels. Such eyes are thus not adapted for vision only in the deeper and dimmer parts of the mesopelagic zone; indeed, species of the hatchet-fish *Sternoptyx*, with laterally set eyes of normal configuration, live at lower mesopelagic levels (> 600 metres), whereas *Argyropelecus* spp., all with tubular eyes, are upper mesopelagic forms.

As will be recalled, fishes with tubular eyes are alike in that none is known to undertake daily and full-scale vertical migrations to the surface waters. The conjecture was that these non-migrators, even though not having the richer supplies of food made available by vertical migration, had the right kind of sure and sensitive eyes to maximize their chances of seeing and seizing prey. Indeed, certain species (*Argyropelecus affinis* and *Scopelarchus analis*) have eyes with bright yellow lenses (Somiya, 1976; Munz, 1976). The absorption spectra of these lenses indicate that the eyes receive wavelengths longer than the main wavelength (475 nm) used, for instance, in ventral camouflage. As Munz argues, such camouflage will fail against predators with yellow lenses, whose sensitivity is displaced towards wavelengths lower than those of the background light and the strongest rays of ventral bioluminescence. In other words, the ventral lights will appear brighter than the background light.

Tubular eyes give sure and sensitive vision in but one direction. Laterally, the unfocussed accessory retina, as mentioned already, may detect the movements of well-lit forms. Much of the visual field, particularly below the fishes, seems not to be covered, but there are devices for extending the visual field. In opisthoproctids and *Gigantura*, diverticula of the retina project from the eye cup, generally in a lateral direction. The cavity of each diverticulum, which communicates with the eye cup through a slit-like opening, is lined with well-developed retinal tissues and overlaid by choroid. The setting and optical organization of the diverticula are such that they gather light outside the binocular field of vision (Munk, 1966, and Frederiksen, 1973) (Fig. 135).

There are no such retinal adjuncts in scopelarchids and evermannellids, which have other devices for extending the visual field (Locket, 1977). In scopelarchids there is a 'lens-pad' set laterally to the lens and overlapping the margin of the pupil such that light from below will pass through the pad and illuminate the dorsal part of the accessory retina (Fig. 136). Evermannellids have what Locket calls an optical fold, which overlies the lateral and ventral part of the eye. Light from below is directed through the fold to the lens of the eye. Comparing the fold with the lens-pad, Locket (1977) concludes, 'These two structures are interesting both for themselves, and as two similar solutions to a common problem, although the anatomy and optics differ in the two cases. The scopelarchid lens pad, being derived from corneal epithelium, is intraocular,

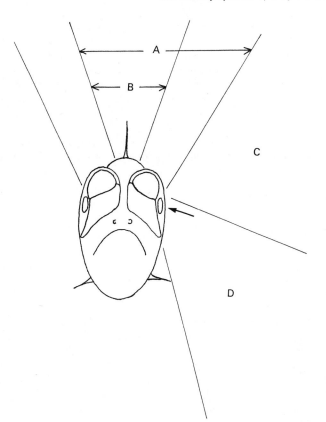

Fig. 136 Fields of view of the fish *Scopelarchus* seen from the front. The main retina
has a field of view dorsal to the fish (A), which overlaps widely with that of
the other eye to give a dorsal field of binocular vision (B). The more ventral
part of the accessory retina has a monocular lateral field (C). The ventral
field (D) is subserved by the dorsal accessory retina, which receives light
through lens pad (arrowed). (Redrawn from Locket, 1977)

but the evermannillid fold is developed in the extra-ocular "adipose eye-
lid". Both include modified cells in their system of alternate layers of
more or less refractile material but, whereas the lens pad lamellae are
about 200 nm in thickness, those of the fold are about 2·5 nm. The relation of these dimensions to the wave length of the light transmitted
suggests that the fold may direct the light according to geometrical optics,
but the lens pad may function by wave guiding.'
   The eyes of mesopelagic fishes are thus intricately and variously developed for life in a twilight world, but there are outstanding exceptions.

The virtually transparent (and abundant) species of *Cyclothone*, such as *braueri*, *signata* and *alba*, have small beady eyes set well forward on the head. The eye cup is shallow and the retina, which looks disproportionately thick, is certainly out of focus with the nearby relatively large lens. Recent studies by Locket (1977) of *C. braueri* and *C. pseudopallida*, shows that there is some regression among the rods and other cells of the retina. Thus, the eyes of *Cyclothone* spp. are both very small and not properly developed, which suggests, as Locket says, that vision may be of minor importance in their lives. Is this true also during their early life in the euphotic zone? Studies of the eyes of postlarval and juvenile *Cyclothone* would be of interest.

Bathypelagic fishes

But in the bathypelagic zone the small unfinished eyes of the black species of *Cyclothone*, which are dominant in numbers of individuals, are clearly by no means exceptional. Such eyes occur also on gulper-eels (Lyomeri), snipe-eels (Cyemidae) and whale-fishes (Fig. 137). The eyes of female angler-fishes grow little after metamorphosis and by the time they are fully adult the eyes are so small and sunken as to seem functionless (Bertelsen, 1951). In free-living male angler-fishes the eyes are well developed (except in gigantactids, which have large olfactory organs). The eyes are large (melanocetids) to very large (ceratiids). In the males of linophrynids the eyes are tubular and look forwards (Fig. 137). There must also be a wide binocular field in the ceratiids (*Ceratias* and *Cryptopsaras*), as shown inter alia by the wide rostral aphakic space and sighting grooves before the eyes (Munk, 1964). The eyes of male deep-sea angler-fishes reach their full development after metamorphosis, when in some species (e.g. melanocetids) they must help in the capture of prey. Parasitic males of ceratiid and other groups do not seem to feed after metamorphosis (p. 255). Thus, we can only presume that male angler-fishes with well-developed eyes have one 'final end in view'; the recognition of the light lure of their proper partners (see also p. 364). Apart from this special function, it is understandable that the eyes of fishes in the sunless bathypelagic zone should be small and variably regressed. But even small, not entirely deficient eyes, may be enough to detect the presence of nearby sparks of living light.

Bottom-dwelling fishes

As in the pelagic zones, the eyes of bottom-dwelling fishes tend to be small and deficient below the threshold of light (ca. 1,000 metres). Species with well-developed eyes, which are centred over the upper slope, have

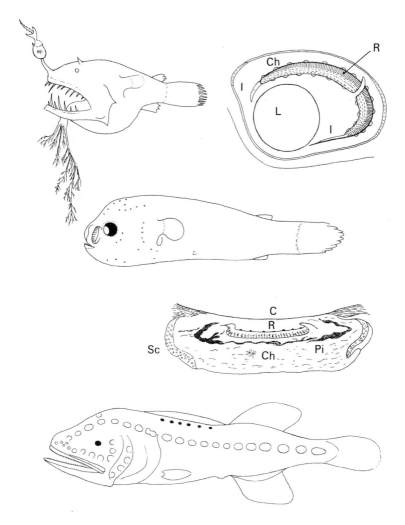

Fig. 137   The eyes of bathypelagic fishes. Above left, female of the angler-fish *Lino-phryne arborifera*, below it a dwarf male *Linophryne* with tubular eyes, shown top right in section. L, lens; Ch, choroid; R, retina; I, iris. The lower-most fish is a whale fish *Ditropichthys storeri*, above which is a section through one of its very regressed eyes. C, cornea; R, retina; Pi, retinal pigment; Sc, sclerotic cartilage. (*Linophryne* redrawn from Munk, 1966; *Ditropichthys* redrawn from Marshall, 1971)

many of the optical and retinal adaptations that obtain in mesopelagic species.

Consider, for instance, the eyes of rat-tails (Macrouridae), a dominant group of benthopelagic fishes. In slope-dwelling species, whose populations are centred mainly between depths of 250 and 1,000 metres, the eyes are very large (e.g. *Macrourus* spp., *Coelorhynchus* spp.) to fairly large (e.g. species of *Ventrifossa*, *Nezumia* and *Coryphaenoides*). Most species of *Gadomus* and *Bathygadus* have rather small eyes, but these live lower down the slope, mostly between depths of 750 and 2,000 metres. Abyssal macrourids (*Chalinura* spp., *Lionurus* spp. and *Echinomacrurus mollis*) have small eyes (Marshall, 1973). In the last species the lens is cataractic and the retina degenerate (Munk, 1966). In brief, species that live within reach of the sun possess the largest eyes. Much the same is true in brotulids, the other diverse group of benthopelagic fishes (Marshall, 1954).

But there seem always to be exceptions. When last on RRS *Discovery* I noticed that trawls from well below 1,000 metres contained large-eyed alepocephalids, such as *Leptoderma*, *Talismania* and *Conocara*. But off north-west Africa, where the *Discovery* was trawling, the waters are relatively productive and these fishes may have ranged deeply in search of their pelagic food. Where the populations of these 'exceptions' are actually centred is something for future investigation.

Benthic fishes of the upper slope also have well-developed eyes. In *Chlorophthalmus* spp. the eyes are not only large but have a prominent rostral aphakic space and an elaborate tapetum. Furthermore, in certain species, at least, the lenses are yellow (Denton, 1956). Thus, their users may well be able to 'see through' the ventral bioluminescent camouflage of euphausiids and prawns. Tripod-fishes (*Bathypterois* spp.) have small eyes and nearly all species have depth ranges that are beyond or extend beyond a level of 1,000 metres (Sulak, 1977). In *Bathypterois grallator* and *B. longipes* the eyes have some structural deficiencies (Munk, 1966). But again, even small, not entirely regressed eyes, may be used to detect nearby flashes of bioluminescence. Some may emerge from copepods, the main food of tripod-fishes.

The eyes of *Ipnops*, a near relative of the tripod-fishes, are quire unusual. Underlying the clear roofing bones of the skull are a pair of bright yellow organs, once thought to be luminous, but each is formed essentially of a flattened retina containing rods and very few other kinds of retinal elements (Munk, 1959). Munk counted about 500 fibres in each optic nerve. Moreover, in the brain of *Ipnops murrayi* the optic rectum is better developed than its homologue in a tripod-fish (*Bathypterois quadrifilis*) (Marshall and Staiger, 1975). Evidently the flattened eyes of *Ipnops* are functional, but what do they 'see'? Polynoid worms, which may be luminous, are commonly eaten by *Ipnops*.

The eyes of cephalopods

Perhaps the most striking contrast between the visual systems of deep-sea cephalopods and fishes is that small regressed eyes are rare in the former animals. In midwater space such eyes, as stated above, are common among bathypelagic fishes. But there is no bathypelagic cephalopod fauna comparable to that of fishes. The vampire (*Vampyroteuthis*) and *Bathyteuthis abyssicola*, both with well-developed eyes, are certainly well known from bathypelagic levels, but they live also above the threshold of light. If a fish loses most, or even all, of its visual function it may still, by its lateralis and olfactory systems, find partners or prey (pp. 402 and 413). Cephalopods do not have sense organs analogous to the neuromasts of fishes and seemingly little, at least in midwater species, in the way of an olfactory system. The lack of such sensors may well go far to explain why they have not been able to establish a bathypelagic fauna.

There are many striking parallels between the eyes of mesopelagic cephalopods and fishes. Both tend to have large, even very large, eyes.[*] In his report on the cephalopods of the *Valdivia* Expedition, Carl Chun (1910) remarks that the eyes are rarely spherical in outline; they are usually elliptical or oval due to the shortening of the main visual axis. Thus, in eyes of comparable diameter the cephalopod has a smaller lens than is found in the spherical eye of a fish. In both groups the lens projects well beyond the pupil, so as to intercept and focus light towards the retina (the outer corneal tissues over the lens have the same refractive index as water). Eyes with a bowl that is oval in outline may be seen, for instance, in the mesopelagic sepioid, *Spirula* (Fig. 138). In the retina are long rods (rhabdomes), which in the central area measure 0·13 to 0·14 mm in length (Chun, 1910).

In certain mesopelagic cephalopods the eyes have relatively long optical axes, as in the squid *Benthoteuthis* (Fig. 138). The eyes of *Benthoteuthis*, which are set at an angle of 45° with the long axis of the body, look upward and forward, and at the base of this line of vision in the retina is a shallow (concaviclivate) fovea. Over the foveal area, as in fishes, the rods are longer than elsewhere in the retina. In the central region Chun found there about 250,000 rods per square millimetre, measuring 0·4 to 0·5 mm in length. He remarked that these are the longest retinal rods in the animal kingdom.

In the mesopelagic octopod *Amphitretus pelagicus* further extension of the optical axis has produced a tubular eye (measuring e.g. 7 mm long and 4 mm wide in a specimen studied by Chun) (Fig. 138). The eyes are set more or less parallel to one another in the gelatinous layer of the

---

[*] The eyes of a large giant squid *Architeuthis* (18 metres or more) have a diameter of about 37 cm (Burtt, 1974).

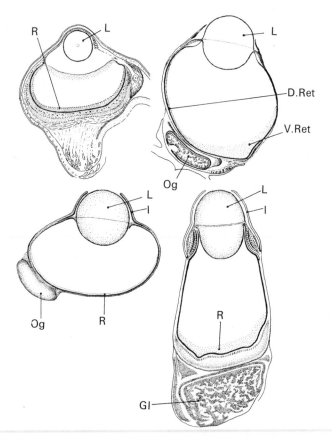

Fig. 138   The eyes of a pelagic worm and cephalopods. Top left, cross-section of the
eye of the alciopid polychaete worm *Vanadis formosa*. L, lens; R, retina.
Top right, section through eye of deep-sea squid, *Benthoteuthis megalops*.
Bottom left, section through eyes of deep-sea sepioid, *Spirula*. Bottom right,
section through tubular eye of deep-sea octopod, *Amphitretus pelagicus*.
(Redrawn from Chun, 1903)

In each of the cephalopods the extent of the retina, formed in part of long
rhabdomes, is shown dotted. L, lens; R, retina; D.Ret and V.Ret, dorsal
and ventral retina; I, iris; Og, optic ganglion; Gl, ganglionic layer.

body wall so that the dioptric parts protrude into the sea. As in fishes,
the retina is at the base of the eye (but there is no accessory retina).

There are other parallels with fishes. In the squid *Mastigoteuthis* there
appears to be a forward aphakic space in the eyes and certainly in *Histioteuthis meleagroteuthis* the lenses are yellow (Denton and Warren, 1968).
But the eyes of *Histioteuthis*, unlike those of any fish, are unequal in

size, the left eye being much the larger. Recently Munz (1976) found a yellow lens in both eyes of *H. meleagroteuthis*. Evidently this squid 'sees through' luminescent camouflage. Moreover, the smaller right eye has the retina divided into upward- and downward-looking parts while in the left eye the retina is much thickened ventrally (J. Z. Young, 1977). In both eyes, upward vision, judged by the length of the rods, must be particularly sensitive.

There is clearly much to be discovered in the eyes of mesopelagic cephalopods. As J. Z. Young (1977) remarks, the large eyes and long rhabdomes of deep-sea forms are often described. But he has found also that there are complicated subdivisions of the retina in the squid *Bathothauma*, the octopods *Japetella* and *Cirroteuthis* and probably also in *Vampyroteuthis*.

The only cephalopod known to have both small and regressed eyes is the finned octopod, *Cirrothauma murrayi*. Chun (1914) found that the eyeballs are completely closed, without lens or iris, and contain very regressed retinal tissues. He observed that it was the only known blind cephalopod.

Photosensitive vesicles of cephalopods

Light detectors, known as photosensitive vesicles, are found in many cephalopods. In the octopods there is a single vesicle near the hind end of each stellate ganglion. The vesicles of squids and sepioids are near the optic tract. For instance, in the mesopelagic squid *Pterygioteuthis* there are dorsal and ventral pairs of vesicles, both embedded in the cephalic cartilages behind the optic lobes (Fig. 139). In *Abraliopsis* spp. there are three or four sets of vesicles, but in *Sandalops* there is a single ventral pair (R. E. Young, 1977). More generally, the photosensitive vesicles of cephalopods tend to increase in size with depth. Thus, among decapods, the vesicles are small in the squid *Loligo* and the cuttlefish, *Sepia*, moderately large in *Spirula* and very large in various other mid-water genera (*Pterygioteuthis*, *Mastigoteuthis* and *Aristioteuthis*). They are largest of all in *Bathyteuthis*, which has a pair of ventral vesicles (J. Z. Young, 1977; R. E. Young, 1977).

Light-sensitive cells (which produce small electrical discharges on stimulation with light) are found over elongated vesicles in these organs (Mauro, 1977). Dorsal photosensitive vesicles, as in the midwater squid *Pterygioteuthis* and *Abraliopsis*, may enable the animal to compare its own light with that of the background and thus be essential in means to match the first kind of light to the other (R. E. Young, 1977). Photosensitive vesicles may also be involved as light detectors in controlling physiological functions such as circadian rhythms and seasonal activities, particularly reproductive behaviour and spawning (Mauro, 1977).

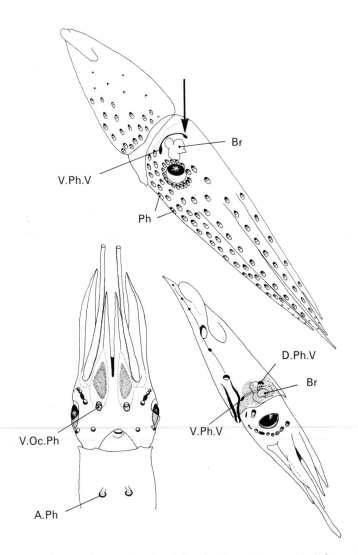

Fig. 139   Photosensitive vesicles of cephalopods. Above, *Histioteuthis dofleini*. The arrow representing downwelling light points to the dorsal photosensitive vesicles (shown black). Below right, *Pterygioteuthis microlampas* showing positions of dorsal and ventral photosensitive vesicles (D.Ph.V and V.Ph.V). Br, brain; Ph, photophores. Below left, ventral view of head of *P. microlampas*, showing positions of main photophores. V.Oc.Ph, ventral ocular photophore, A.Ph. anal photophore. (Redrawn from R. E. Young, 1977)

The eyes of pelagic polychaete worms

Large image-forming eyes, essentially like those of fishes and cephalopods, are found in one family of pelagic polychaetes, the Alciopidae (p. 104). Recently, Wald and Rayport (1977) have studied the structure of the eyes and obtained electroretinograms from *Torrea candida* (12–17 cm long) taken at the surface, and from two species of *Vanadis*, taken in a haul at about 300 metres. Besides a distinct cornea and a large spherical lens, both forms have a main retina (upright, as in cephalopods) in the focal plane of the lens and accessory retinae lying beside the lens (Fig. 138). In *Torrea* the main retina contains about 10,000 photoreceptor cells.

Electroretinograms showed that in *Torrea* the primary retina is most light-sensitive at 400 nm, but in *Vanadis* comparable measurements gave a peak sensitivity of 460 to 480 nm, the wavelengths that penetrate most deeply in clear oceanic waters. Indeed, the peak sensitivity of *Vanadis* is much like that of deep-sea fishes, a physiological convergence to add to the structural convergences mentioned above.

The eyes of deep-sea crustaceans

Little is known of the structure and function of the eyes of deep-sea copepods and ostracods. Here we shall be largely concerned with the larger midwater crustaceans, notably the mysids, euphausiids and prawns.

First, a brief comparative survey of the development of eyes in the larger deep-sea crustaceans is needed as a background. In mesopelagic amphipods, mysids, euphausiids and prawns the eyes are well developed or even large. By contrast, the eyes of bathypelagic species tend to be small or regressed. For instance, bathypelagic euphausiids either have small eyes (giant species of *Thysanopoda*) or regressed eyes (*Bentheuphausia*). Moreover, *Bentheuphausia* is without photophores while those of giant *Thysanopoda* are small (p. 108). At mesopelagic levels, on the other hand, species of the prawn genera *Sergia*, *Oplophorus* and *Systellaspis* possessing dermal photophores have relatively larger eyes than in their relatives that lack such photophores (Omori, 1974).

The relative development of the eyes in bottom-dwelling crustaceans is best known in the isopods (Menzies et al., 1973). Off Beaufort, North Carolina, 11 species taken at depths from 2,000 metres to the abyssal plain (4,500 metres) had no eyes. On the slope between 200 and 2,050 metres 3 species had eyes but 6 were not so provided. Below 500 metres in the Peru–Chile Trench all species lacked eyes. Species of *Serolis* from subantarctic and antarctic localities showed progressive deterioration of the eyes with increase in their depth ranges. In general, as Menzies and his colleagues say, species from shallow levels have multifaceted black eyes, whereas those from deep waters have pink, yellow or colourless eyes, with or without eye facets.

It is well known that the compound eyes of the larger crustaceans are much like those of insects. The eyes are generally spherical and formed of many visual units (ommatidia) set closely together so that their visual axes are more or less normal to the curvature of the eye. The facets of the eye mark the outer corneal parts of the ommatidia. Below the cornea is the main optical part, the crystalline cone, which directs light to a long receptor cluster (rhabdome and retinular cells) (see Fig. 140). In apposition eyes the ommatidia are optically isolated by screens of pigment so that each functions as a single light-receptive unit. There are no such colour-cell screens in superposition eyes, and thus receptor clusters may be excited by light from neighbouring crystalline cones as well as from their own. There is clearly a wider 'pupil' in the superposition kind of eye, which is found in nocturnal insects and presumably also in deep-sea crustaceans.

The optical aspects of image formation in the spherical eyes of deep-sea prawns were studied by M. F. Land during Cruise 77 of RRS *Discovery* (15 July–31 August 1976). Typically, these animals '... have large spherical eyes which show a strong orange glow when examined from the direction of illumination. This feature is shared by the superposition eyes of insects (especially moths) and indicate a similar optical mechanism. Superposition images were observed in lightly fixed eyes of *Oplophorus spinosus* when these were cut in half and illuminated from the side. However, the mechanism of image formation is not the same as that described in insects by Exner in his classic study. There are no high refractive index "lens cylinders" (as in moths and fireflies) but instead the light is bent to a focus by an array of mirrors—the highly reflecting surfaces of the transparent "cones" which form a square array just below the cornea. The disposition and dimensions of these mirror faces are such as to bring nearly all light from a particular direction to a single point within the rhabdome layer. This principle appeared to hold for all the decapods examined, including a benthic species (*Metapenaeus* sp.) with eyes five times as large as those of *Oplophorus*' (Land, 1976).

Beside their wide 'pupil', there are evidently other means of increasing the visual sensitivity of deep-sea prawns. In *Acanthephyra pelagica* and *Sergestes robustus*, which both live in the Mediterranean and extend northward in the Atlantic to about 60°N, Casanova (1977) made a special study of certain features of their eyes (the diameter of the eyes, the number of ommatidia along a median section of the eye, the length of the ommatidia, the length and breadth of the crystalline cone) in relation to their living space. In the first species, for instance, individuals from the Bay of Biscay had larger eyes with more and longer ommatidia containing longer crystalline cones than were found in individuals of the

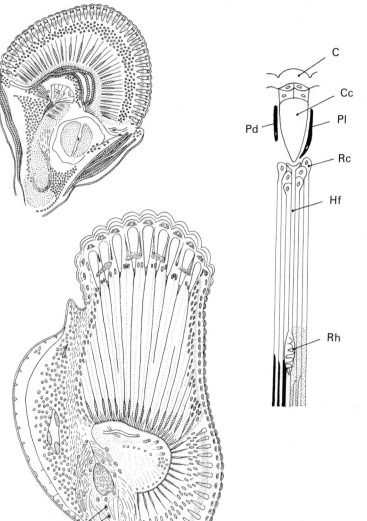

Fig. 140   The eyes of euphausiids. Above, section through simple spherical eye of
Euphausia sp. Below, section through the biolobed eye of *Stylocheiron
suhmii*. Note the parallel disposition of the larger and longer ommatidia
of the upper lobe. E.Ph, eye photophore. (Redrawn from Chun, 1896)
On the right (after Kampa, 1965) is a diagrammatic representation of an
ommatidium. C, cornea; Cc, crystalline cone; Pd and Pl, pigment of the
dark adapted and light adapted eye; Rc, retinula cell; Hf, hyaline filament;
Rh, rhabdom.

same size from the Mediterranean. As Casanova argues, in the more northerly living space of both species the intensity of midwater light is much less than that measured at the same depth (e.g. 400 metres) in the Mediterranean: hence the accentuation of the eye features, as shown under *Acanthephyra pelagica*. A larger superposition eye with an increased number of ommatidia will presumably be a better light-gatherer, while increase in the length of the ommatidia should improve the efficiency of their light-receptive parts. Lastly, the elongation of the crystalline cones may well be associated with improvements in the array of focussing mirrors (see above). All such changes should make the eyes more sensitive.

But some crustaceans have eyes without dioptric parts. Various mysid shrimps with reduced compound eyes are sometimes described as 'blind', but as Elofsson and Hallberg (1977) point our, their eye-stalks may not be lost; indeed, they may be transformed into remarkable visual organs. Thus, in the deep-sea species *Amblyops abbreviata*, taken off Korsfjord, Norway, between 400 and 500 metres, the eye-stalks are each transformed into a pair of horizontal lamellae that bear arrays of ommatidia without any trace of dioptric parts. Altogether there are about 1,000 ommatidia composed mainly of retinular cells more than 1 mm in length. Each ommatidium is formed of a cylinder of about 12 retinular cells, which on their inner sides form a carpet of microvillae. Much the same kind of eyes are found in *Pseudomma affine*. But in *Boreomysis scyphops*, taken between 2,000 and 3,000 metres, the eyes resemble two spoon bowls with the convex sides turned towards each other. The bowl bears a large number of tightly packed ommatidia, which stand on a layer of large red pigment cells. As in *Amblyops*, the ommatidia lack dioptric parts, but again the retinular cells bear microvillae, here disposed multidirectionally in clusters. Evidently, light is caught by the microvillae, which are not, as in most arthropods, disposed in regular arrays. Their multidirectional orientation suggests that they are efficient catchers ('with the photopigment dipoles arranged along the microvillar axis') of non-polarized light (Elofsson and Hallberg, 1977). How these remarkable eyes are used must be left for future investigations.

Euphausiid shrimps, as we saw (p. 238), have two main kinds of eyes. Oceanic species with spherical eyes belong mainly to the genera *Euphausia* (Fig. 140) and *Thysanopoda*: those with eyes divided into two distinct lobes form the genera *Nematoscelis*, *Nematobrachion* and *Stylocheiron*. In euphausiids with bilobed eyes the second pair of thoracic legs, which are very elongated, are probably used to grasp prey (such as copepods).

Euphausiids with bilobed eyes, as Kampa (1965) says, literally have four eyes; the two lobes are separated by a dense screen of pigment cells. The development of the upper lobe, which has longer and wider ommatidia than those of the lower lobe, varies considerably from species to species (Fig. 140). In *Stylocheiron affine* and *S. suhmii* the upper lobe consists of a few ommatidia, but in species such as *S. maximum* and *Nematoscelis difficilis* the two lobes are comparable in size. At the other extreme, in *Nematobrachion flexipes* the upper lobe is much the larger. But whatever the relative size of the lobes, the disposition of the ommatidia, as Kampa points out, follows a constant pattern. In the upper lobe, which actually looks upward and forward, the effective ommatidia are always set in parallel: in the lower lobe the ommatidia are set radially so that the visual field is mostly downward and backward.

In the euphausiid ommatidium the retinula cells are disposed around a hyaline filament that runs from the base of the crystalline cone to a rhabdome that contains a spiral thread. Most likely, the hyaline filament, according to Kampa, transmits light by internal reflections, directly and without loss to the rhabdome. The spiral of the rhabdome, as she found, is loosest in shallow-dwelling euphausiids with spherical eyes, such as *Euphausia pacifica* and *Meganyctiphanes norvegica*. In species with bilobed eyes the coil of the spiral is much tighter in the rhabdomes of the upper lobe than in those of the lower. Such a spiral, as T. I. Shaw suggested, could be one of the most efficient means of spreading the light conducted by a rod, and so be a very efficient means of intercepting the relatively few quanta of light available at depth.

In *Nematoscelis difficilis* the two lobes certainly differ in range of spectral sensitivity (Boden et al., 1961). Electroretinograms showed that the upper lobe is most sensitive to light of wavelengths from 460 to 470 nm, which accords well with the spectral sensitivity curve of the photosensitive pigment; the maximal sensitivity of the lower lobe ranges more widely over 460 to 515 nm. Evidently, the upper lobe is most sensitive to background light, while the lower lobe, which, according to Kampa (1965), should form apposition images, covers both the background light and the longer wavelengths of bioluminescence.

As we may recall, an outstanding difference between euphausiids with spherical eyes and bilobed eyes is that the former undertake daily vertical migrations: bilobed eyes are associated with non-migratory and predatory ways. Oceanic migrators belong mainly to the genera *Euphausia* and *Thysanopoda*. Their spherical eyes, as Brinton (1967) suggests, should not only provide mosaic image-formation but must respond in ways that will enable them to stay at a particular light level (isolume) during their migrations. In bilobed eyes, the upper lobe seems suited for registering the amount of ambient light, whereas the lower lobe, which may have better screening between the ommatidia, serves for mosaic vision.

The upper lobe certainly seems to have the right attributes for the sensitive perception of background light; it looks upward and is composed of long ommatidia with wide light-gathering dioptric parts that lead to sensitive receptor clusters (with tightly wound spirals in the rhabdomes). It is significant also that in mesopelagic species (*Stylocheiron maximum* and *Nematobrachion sexspinosus*) the upper lobe is either about equal in size or larger than the lower lobe: in the shallower-living epipelagic species (e.g. *S. suhmii* and *S. affine*) the upper lobe is much the smaller, composed, as we saw, of a few ommatidia. Thus it could well be that one function of the upper lobes is to enable their users to keep vertical station by reference to a particular light level—a means of correcting deviations from this level in the members of a population. We will recall that non-migrating euphausiids are neatly segregated in the water column (p. 236).

As a final conjecture, the upper lobes may be involved in concert with the long thoracic legs in the capture of prey, which is first seen and followed by the lower lobes. If the eyes are turned so that the long sensitive ommatidia of the upper lobes look forward, a binocular 'bead may be drawn' on prey that is to be grasped by the armature at the end of the thoracic legs. Here, at least, is an idea that might be tested.

The chemical senses: smell and taste

Beside their olfactory organs and taste-buds, fishes may have three other chemosensory means: single taste cells over their bodies and perhaps also in the mouth; free nerve-endings in the skin and receptors in the lateral-line system. Here we consider the familiar first two systems in fishes and then turn to their analogues in the invertebrates. Just how vertebrates smell and taste '... is less understood than sight, hearing, touch and the lateral line. However, the importance of the chemical senses in the lives of aquatic animals becomes more apparent each year. Among fishes, taste is restricted primarily to aspects of food finding and feeding, but smell, while it may serve for food detection and location, is also important in reproductive and social behaviour' (Bardach and Villars, 1974).

The olfactory system of fishes

In jaw-bearing fishes the olfactory organs are housed in a capsule on each side of the snout. Each organ consists of olfactory lamellae that are fashioned in rosette form or disposed in parallel. The perception of smell is by way of ciliated sensory cells in the epithelium of the lamellae, which may receive scents in the water that flows into the capsule through the anterior nostril (and is discharged through the posterior nostril). The sensory cells are innervated through axons that run back to glomerular synapses in an olfactory bulb, which is connected to the forebrain

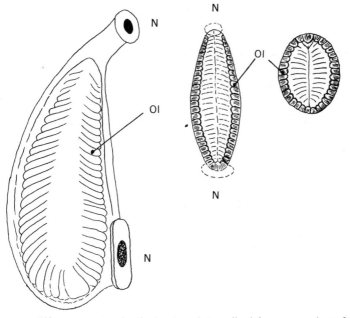

Fig. 141   Olfactory organs of a freshwater eel *Anguilla* (left), a mesopelagic fish, *Chiasmodon niger* (centre) and a benthic fish, *Bathypterois dubius* (right). N, nostril; Ol, olfactory lamellae. (From Marshall, 1954)

through an olfactory nerve. In deep-sea fishes, as in other kinds, the olfactory organs are diversely developed. At one extreme the organs are large (macrosmatic), as in the sharks and bottom-dwelling eels; at the other, they are small (microsmatic), as in female angler-fishes (ceratioids). Fishes with the best developed sense of smell have the most numerous olfactory lamellae. The freshwater eel (*Anguilla*), which spends part of its life as a deep-sea fish, is a good example (Fig. 141).

Midwater fishes

Compared to their size, mesopelagic fishes have highly to slightly developed olfactory organs. In two predatory forms, the giant-swallower, *Chiasmodon niger*, and *Evermannella*, the length of the olfactory rosettes are about half the length of those in the freshwater eel and formed of much the same number of lamellae (40–50) (Marshall, 1954) (Fig. 141). But other predatory species, capable also of swallowing large prey, such as the stomiatoids *Chauliodus*, *Echiostoma*, *Idiacanthus* and *Malacosteus*, have small or even regressed olfactory organs. In a third group (e.g.

Stomias, *Chirostomias* and *Leptostomias*) the olfactory organs, which contain about 14 to 20 lamellae, are moderately developed. The same is true of the mesopelagic eels (*Nemichthys*, *Avocettina* and *Serrivomer*), whereas their bottom-dwelling relatives (e.g. *Synaphobranchus*) have very large olfactory organs. Concerning the consumers of small prey, the larger lantern-fishes, which depend largely on crustaceans, have relatively well-developed olfactory organs. For instance, in a *Lampanyctus alatus* (length 110 mm) there are about 30 lamellae in each olfactory organ. Thus, there are both large-prey and small-prey eaters with a well-developed olfactory system. At the other extreme, the reduction or regression of this system in predatory stomiatoids is understandable. Apart from *Malacosteus*, the genera listed above, and there are others, may well bring their prey close by means of luminous lures (p. 364) and, like female ceratioid angler-fishes, largely or entirely dispense with an olfactory system. Though without a luminous barbel, *Malacosteus* has large sensitive eyes (with a forward aphakic space) with the rare capacity of seeing the red prey exposed by its cheek lights (p. 363). And its olfactory organs are so regressed that it can have little or no sense of smell.

In nearly all mesopelagic fishes known to me the development of the olfactory organs is much the same in both sexes. The exceptions are the transparent species of *Cyclothone* (e.g. *braueri*, *signata* and *alba*), which have macrosmatic males and microsmatic females. The same olfactory dimorphism is true of their bathypelagic relatives. Indeed, considering that black *Cyclothone* are dominant in numbers of individuals while most species are ceratioid angler-fishes, olfactory dimorphism is the rule in the bathypelagic fish fauna. The exceptions are found in the minor groups such as the gulper-eels and snipe-eels (Cyemidae), with small or regressed olfactory organs in both sexes.

After metamorphosis, the males of most angler-fishes rapidly become macrosmatic. When fully developed each olfactory organ contains a stack of broad sensory lamellae, which are open to the sea through two wide nostrils. The olfactory organs of female ceratioids are reduced to small papillae (Fig. 142). Matching this extreme dimorphism, the olfactory nerves, bulbs and forebrain are strongly developed in males but regressed in females (Marshall, 1967, 1971). Male angler-fishes with small olfactory organs, notably the ceratiids, have large eyes (p. 390). Indeed, the males of most species have well-developed to large eyes (p. 390), but those of gigantactids, which have small eyes, have large olfactory organs. Whether by sight and olfaction, or more rarely by one of these senses alone, the males, presumably aided by their lateral-line sense, must find their proper partners.

The males of *Cyclothone* spp. also develop a much larger olfactory complex (olfactory rosette–nerve–bulb–forebrain) than the females. In fact, female *Cyclothone* have a very reduced olfactory system (Fig. 143).

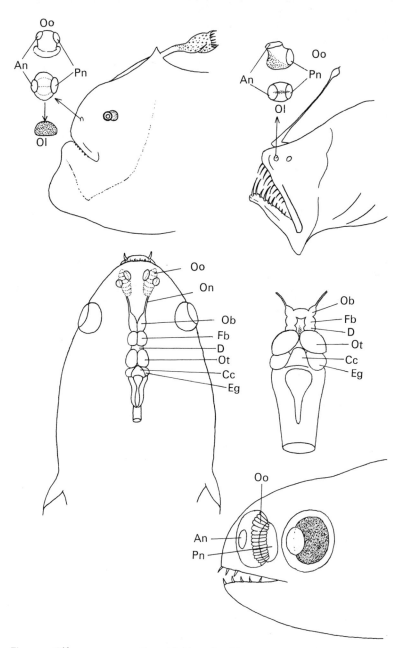

Fig. 142 Olfactory organs of ceratioid angler-fishes. Top left, *Himantolophus groenlandicus*; top right, *Melanocetus murrayi*, females; centre right, brain of female *M. murrayi*; bottom left, brain and olfactory organs of free-living male of *Oneirodes* sp; bottom right, olfactory organ and tubular eye of free-living male of *Linophryne macrorhinus*. An, anterior nostril; Cc, corpus cerebelli; D, diencephalon; Eg, eminentia granularis; Fb, forebrain; Ob, olfactory bulb; Ol, olfactory lamillae; On, olfactory nerve; Oo, olfactory organs; Ot, optic tectum; Pn, posterior nostril. (Redrawn from Marshall, 1967)

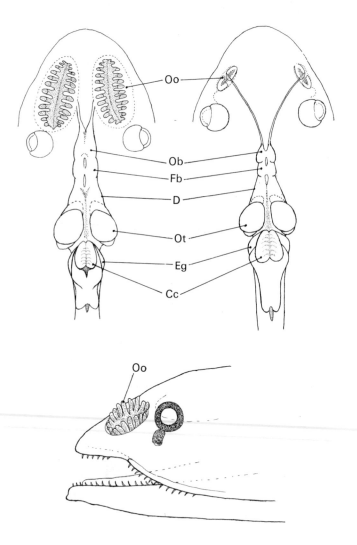

Fig. 143 Olfactory organs and brain of *Cyclothone microdon*. Top left, male; top right, female; bottom, head of male. Cc, corpus cerebelli; Eg, eminentia granularis; D, diencephalon; Fb, forebrain; Ob, olfactory bulb; Oo, olfactory organ; Ot, optic tectum. (Redrawn from Marshall, 1967)

Moreover, the males are also the smaller sex, but the difference is not so marked as in ceratioids. Mature males are about two-thirds the length of the females, which when fully grown range from about 25 to 70 mm in the various species. When the males of *Cyclothone braueri* and *C. microdon*, and presumably of other *Cyclothone*, are approaching full maturity, the olfactory organs enlarge and a forward growth of the snout produces a fleshy rostrum. Perhaps the latter forms a hydrodynamic fairing to direct water smoothly through the enlarged anterior nostrils as the fish swims (Badcock and Merrett, 1976).

Bathypelagic fishes are spread in the largest living space on earth. If, as seems most reasonable, female ceratioids and *Cyclothone* secrete pheromones* to attract their mates, why should so vast an environment have 'favoured' this kind of chemical signalling? During any time in the breeding season of a species, the chances of sexual encounter will depend on the population densities of ripe females and responsive males. Now, through far-reaching economies in their organization, bathypelagic fishes, though living in a food-poor environment, are quite numerous. Between depths of 1,000 and 2,000 metres, where they are most abundant, Bertelsen (1951) estimated that there may be no more than a mean distance of 30 metres separating one ceratioid from the next. Separation between members of a particular species will be more than this, but the figure will be least where this species is dominant. In its area of dominance any species of *Cyclothone* is likely to be at least ten times as abundant as the local ceratioid population, which suggests that individuals have a mean separation of 3 metres or less. Moreover, the distribution of both ceratioids and *Cyclothone* is likely to be clumped rather than random.

But what is the ratio between ripe males and females in a species' area or area of dominance? In a local ceratioid population, females may well be represented by at least two year groups and males by one only. Thus, females are likely to outnumber the males. But the males reach sexual maturity soon after metamorphosis, whereas the females take much longer to mature. 'Of most species, not a single mature female has been caught, and even if a contributory cause may be that these are fished less effectively than the younger specimens, we can hardly suppose that the unoccupied mature females amount to more than 2–5% of the total stock of metamorphosed females.' After reasoning thus, Bertelsen (1951) estimated that there would be at least 15–30 ripe or near ripe males to every ready female (Marshall, 1971). Moreover, as in *Cyclothone microdon* and probably in other black species, the females live longer

---

* Pheromones are 'substances which are secreted to the outside by an individual and received by an individual of the same species, in which they release a specific reaction, for example, a definite behaviour' (Butler, 1967).

than the males, some of which undergo sex reversal (Badcock and Merrett, 1976).

Even if their olfactory thresholds are as low as that of freshwater eels, how might male ceratioids and *Cyclothone* find their mates? Much depends on the behaviour of the females and on local water movements. At depths of about 2,000 metres in the North Atlantic current velocities vary from nearly zero to 15 cm/sec with a mean of about 5 cm/sec (Swallow, 1962). Dispersal of scents will thus tend to be rather slow.

In the sea, as in air, still conditions will confuse the keenest noses. But, given a breeze, male insects find a source of pheromones by flying upwind. Once a male has picked up the scent he turns into the wind, and as he flies 'his olfactory receptors continue to be very slightly stimulated, keeping him flying upwind, until rather suddenly, when he gets within a few centimetres of the female, the stimulation greatly increases and indicates her proximity' (Butler, 1967).

But how might bathypelagic fishes, which cannot see or touch ground, face or swim into a current? Blind fishes show positive rheotaxis without tactile clues (Marshall and Thinés, 1958; Poulson, 1963). If we think of the mean current velocity (about 5 cm/sec) at 2,000 metres in the North Atlantic, Poulson's observations on blind amblyopsid fishes are intriguing. He found that both species of *Amblyopsis* 'often swim in midwater and show positive rheotaxis in currents of 2–7 cm/sec without tactile reference'. The biological value of such response in ceratioids and *Cyclothone* would seem to be great. If the semicircular canals of the ear are involved, they are certainly well-developed in these fishes. Clearly, tests are needed, and female ceratioids, at least, may be caught alive.

Finally, there is the time factor. The food reserves of male ceratioids are mostly in the liver, which reaches maximum size at metamorphosis. Beside a sizeable liver, *Cyclothone* males have substantial reserves of fat. While they are seeking their mates, the males can draw on these reserves for certain periods of time.

Thus, by visual and olfactory clues, males of the above groups may find their partners. It is certainly striking that olfactory dimorphism has evolved in luminous bathypelagic fishes. There is no evidence yet of such development in non-luminous species (Marshall, 1967). In the sunless bathypelagic region, pheromones, which can be perceived at greater range than visual signals, seem apt, considering the 'favourable' factors of this environment (Marshall, 1971).

Bottom-dwelling fishes

In the benthopelagic fish fauna, which, as we recall, consists of species that live near the deep-sea floor, microsmatic forms are virtually unknown. The largest olfactory organs are found in the black squaloid sharks and eels (synaphobranchids). In the rest of the fauna the organs are well to moderately well developed.

Rat-tail fishes (Macrouridae), which together with the brotulids and the deep-sea cods (Moridae) make up most of the benthopelagic fauna, are often the most numerous fishes in local populations. Indeed, the photographic records of baited cameras between upper slope and abyssal levels (ca. 5,000 metres) show that rat-tails are the commonest fishes to be attracted. Hagfishes, which have a well-developed olfactory system, may be common at depths down to 2,000 metres, but at greater depths the main scavengers are large macrourids (Isaacs and Schwartzlose, 1975).

Macrourids tend to head into a current, but, as Heezen and Rawson (1977) saw (p. 305), lose such orientation in slack water. This response to currents (positive rheotaxis) which may bear a scent trail should lead them to the source of this trail. Moreover, rat-tails tend to live in small groups and it may well be, as Isaacs and Schwartzlose (1975) propose, that long-range homing to food is through the successive collapse of hunting territories (p. 380). Certainly, the olfactory organs of benthopelagic (but not bathypelagic) macrourids are comparatively well formed. The number of olfactory lamellae in each rosette ranges from 18 to 36 (Marshall, 1973). Deep-sea cod and brotulids, which are also attracted to bait, also have good olfactory organs.

In the benthopelagic fish fauna olfactory dimorphism seems to be confined to the halosaurs. McDowell (1973) found that mature males with large testes, unlike immature individuals, have a large and tubular anterior nostril, the formation of which is evidently accompanied by modification of the olfactory rosette. Each lamella becomes much enlarged and lobulated so that the entire rosette looks like a sprig of broccoli. In breeding males of the halosaur genus *Aldrovandia* the anterior nostril becomes a black fleshy tube and is associated with an enlarged posterior nostril (Sulak, 1977b). During the breeding season halosaur males are presumably attracted by scents from their proper partners. This is not to imply, of course, that the perception of scents plays no part in the mating of species lacking sexually dimorphic systems of olfaction.

In the benthic fauna, the rays, which have large olfactory organs, no doubt use them to good effect. The olfactory rosettes of tripod-fishes (*Bathypterois*) are either moderately developed or rather small. In the related *Ipnops* they are very small (Marshall and Staiger, 1975). Members of the angler-fish order (Lophiiformes) that may form part of the benthic fauna on the slope (*Lophius*, *Chaunax* and ogcocephalids) also have very small olfactory organs. But they have means of luring their prey.

The gustatory system of fishes

Little is known of the taste-bud system of midwater fishes. One sees what are presumably taste-buds over the lips and in the mouth. More may be inferred of the gustatory system of benthopelagic fishes, particularly of macrourids and deep-sea cod, both members of the Gadiformes. In fishes generally, each taste-bud usually consists of a cluster of 12 to 20 cells, some of which bear minute bristles (microvillae); the whole resembles a closed flower bud and is set in the epidermis (Bardach and Villars, 1974).

A mental barbel, which presumably carries taste-buds, is present in most macrourid fishes. It is absent in the midwater genera (*Cynomacrurus* and *Odontomacrurus*), which is not surprising. The tasting of food was thought to require definite contact but orientation to food by taste has now been established (Bardach et al., 1967). Even so, a small gustatory appendage is hardly adequate to detect food at midwater levels. A barbel is more suited to deliberate exploration along the bottom.

As stated elsewhere (p. 276), the rat-tails that feed on benthopelagic food organisms (Bathygadinae) have a gustatory system that differs in certain respects from those that take benthic food as well, or even predominantly (Macrourinae). In both subfamilies the ramification of certain cranial nerves, particularly of the ramus recurrens of the facial nerve (VII) over the head and forepart of the body, suggests that as in certain cod-like fishes (Gadidae) these nerves innervate taste-buds over the body and fins. In teleost fishes generally the fibres innervating taste-buds in the mouth and on the lips and, if present, on the head, barbels and body, travel in cranial nerve VII and end in an anterior and dorsal part (facial lobe) of the hind brain. Fibres from taste-buds in the pharynx and over the gill system are conveyed by cranial nerves IX and X to a posterior and dorsal part (vagal lobe) of the hind brain.

The vagal lobes appear to be best developed in the macrourine rat-tails (e.g. *Nezumia*, *Coelorhynchus* and *Coryphaenoides*), and in such forms the mouth and pharynx are rich in taste-buds: particularly over the inner and lower parts of the gill arches and on a complex of papillae before and between the pharyngeal teeth. One is reminded of similar elaborate systems in bottom-grubbing species of carp-like fishes (cypriniforms)—systems that must help them to separate food items from swallowed sediment. Macrourine rat-tails, which may also swallow sediment (p. 257), have, it seems, the same means of gustatory processing. But bathygadine rat-tails, simply possessing well-developed gill rakers to retain swallowed food, have, by contrast, few taste-buds in the mouth and pharyngeal regions.

The chemical sense of invertebrates

Like fishes, some of the more complex invertebrates have olfactory means that may be used to find food or facilitate sexual congress. Here we shall be concerned largely with some of the crustaceans.

Regarding their response to food materials, the sensitivity to amino-acids of the mesopelagic mysid *Gnathophausia ingens* was compared to that of certain other crustaceans (Fuzessery and Childress, 1975). When known quantities of amino-acids were admitted to the aquarium the feeding activities of *Gnathophausia* were evoked by smaller increments of these substances than were found necessary to elicit feeding responses from the other crustaceans, such as the pelagic galatheid crab *Pleuroncodes* and the shore-crab *Cancer*. As Fuzessery and Childress say, chemoreceptors and mechanoreceptors are likely to be important in food finding in the ocean depths.

Pelagic crustaceans with long antennae seem best fitted to receive mechanical stimuli. In particular, many mesopelagic prawns, including all sergestids and some penaeids, have long second antennae that bear similar systems of sensory setae (Ball and Cowan, 1977). In *Acetes sibogae*, the Australian representative of a shallow-water genus of sergestids, the ultrastructure of the antennal sensilla is much like that of its mesopelagic relatives. When these prawns swim, the basal part of each antenna is held directly outwards so that the long lash beyond the antennal flexure, which is about two-thirds the length of the entire antenna and nearly twice the length of the body, is streamed parallel to the long axis of the animal. Over the whole antennae Ball and Cowan identified seven types of setae. Down the long trailing part of the antenna each segment bears setae (with lateral setules) that are bent round to form an almost enclosed tube on the medial side of the antenna. These setae are not innervated, but others, which project into the tube, are innervated and are likely to be mechanoreceptors. At the tip of the antenna there is another kind of such receptor. The basal part of the antenna, including the flexure, bears four kinds of setae. Two types, borne on part or all of the antennal base, and another type at the flexure, are also innervated and have the structures (scolopales) of mechanoreceptors. But a seventh kind, also distributed along the entire basal section, are either innervated with three or eight to ten dendrites. These setae, which are 'paired unadorned cylinders with a pore at the tip' are likely to be chemoreceptors (Ball and Cowan, 1977).

Indeed, *Acetes sibogae australis* has marked chemosensory powers of tracking scent trails (Hamner and Hamner, 1977). In a series of laboratory tests the scents left by food or by blotting paper soaked in meat extract or 17 kinds of amino-acids were followed precisely. Observations (202) were made on individuals that responded to (or passed through) a visible part of a scent trail or to a trail made visible by a fluorescent dye. During the first minute of such trail presentation all the animals were able to follow them but after 6 minutes the numbers were halved. Since discrete dye trails remained in the water column for 6 minutes and prawns 3 cm in length swam at an average speed of 5.6 cm/sec, such individuals in relatively still oceanic waters could follow a scent trail left by a falling particle of food as far as 20 metres. The precise means of chemoreception have yet to be determined, but as the Hamners conclude, 'Perhaps the deep sea, with its low turbulence, is laced by a complex array of attractive or repellent chemical trails. If so, scents may prove more important for aquatic animals than they are for terrestrial animals.'

The chemoreceptors of mesopelagic prawns might also play a part in sexual congress, but sex pheromones, though produced by certain benthic decapods, are unknown in their pelagic relatives (Ball and Cowan, 1977). In the copepods, however, there is structural evidence that there may well be chemical attraction between the sexes. Thus, male calanoid copepods, which are usually somewhat smaller than the females, often have a series of sense organs (aesthetascs) on the antennules. The aesthetascs may be cylindrical, rod-like, club-shaped, vesicular, sac-like or lobate (Brodskii, 1950). These kinds of sense organs, which are found also on the antennal flagella of many decapod crustaceans, are thin-walled and supplied with the dendrites of bipolar neurones. They are strongly chemosensitive and in copepods may well enable the males to follow scent trails that are left, presumably by the females, during the reproductive season. In the relatively quiet depths of the ocean, such trails, as we have seen, will tend to persist.

If deep-sea calanoid copepods are like the freshwater species *Diaptomus pallidus*, both sexes will have chemoreceptors on their mouthparts. In *Diaptomus*, as Friedman and Strickler (1975) found, there are two types of sensillae. The first, which resemble contact chemoreceptors known as peg-in-socket sensillae, are mainly on the mandibles. The other kind, with a distal pore system, are actually formed by the setae of the mandibular palps and the maxillary limbs. Friedman and Strickler, who were unable to find mechanoreceptors, suggest that the mouthpart sensillae of *Diaptomus* are important in food selection.

Midwater cephalopods, as already implied, have no obvious olfactory organs. Nothing comparable to the olfactory organs of the coastal squid *Lolliguncula brevis* (Emery, 1977) has yet been described. But the bottom-dwelling octopods certainly have chemosensors associated with the suckers on the arms. Nixon and Dilly (1977), after noting that the suckers of *Octopus vulgaris* have a much greater repertoire of responses to chemical and tactile stimuli than do those of the cuttlefish, *Sepia officinalis*, suggest that the sensory superiority of the octopus is related to its greater involvement with the bottom. In midwaters, it looks as though there is the same contrast between the cephalopods and crustaceans as there is between cephalopods and fishes (p. 393). Cephalopods, which depend so much on their visual systems and rarely lose their eyes, are largely confined to mesopelagic levels, whereas crustaceans, with chemosensory, mechanosensory and even perhaps a 'lateral line' sense (p. 422) are at home both in the mesopelagic and bathypelagic zones. And many, as we saw (p. 397), have lost or nearly lost their eyes.

Lateral-line (distance-touch) senses in fishes and invertebrates

The ears and lateral-line organs of fishes form the acoustico-lateralis system. Both develop from similar ectodermal parts (placodes) of the embryo and both have hair-bearing sensory cells, capped by a gelatinous cupula, that are housed in fluid-filled canals. Here we consider first the lateral-line system of fishes and its likely analogues in some of the invertebrates.

Actually, the sensory units (neuromasts) of the lateral-line system of fishes are either housed in mucus-filled canals or freely exposed to the water (Fig. 144). Whatever their situation, neuromasts have the same basic structure and function: each consists of hair-bearing sensory cells that protrude into the base of a gelatinous cupula, the whole excited by displacements in the medium. Lateral-line organs, as Dijkgraaf (1962) says '... serve mainly to detect and locate moving animals (prey, enemies, social partners) at short range on the basis of current-like water disturbances ... Furthermore, there is evidence that the size of the moving object, as well as its velocity and direction of movement, are distinguished.'

In deep-sea fishes, as in other kinds, there is wide variation in the degree of development and complexity of the lateral-line system. In hagfishes it is very reduced or even absent: at the other extreme, it is highly developed in rat-tail fishes. One of the simplest systems is found in most members of the bathypelagic fauna; the ceratioid angler-fishes, gulper-eels (Lyomeri) and bob-tailed snipe-eels (Cyemidae). These fishes simply have free-ending organs on the head and body, even on the tail fin in some angler-fishes (Figs. 145 and 146). In the latter each neuromast is set at the end of epidermal outgrowths that range from low papillae to long stalks. In gulper-eels the many lateral-line outgrowths, which extend well down the tail, have the form of a cylindrical tube with an expanded disc-like head (Bertin, 1934). Their fine structure has yet to be elucidated, but much is known of the skin papillae of the snipe-eel, *Cyema atrum* (Meyer-Rochow, 1978). The papillae are flattened and have a minute

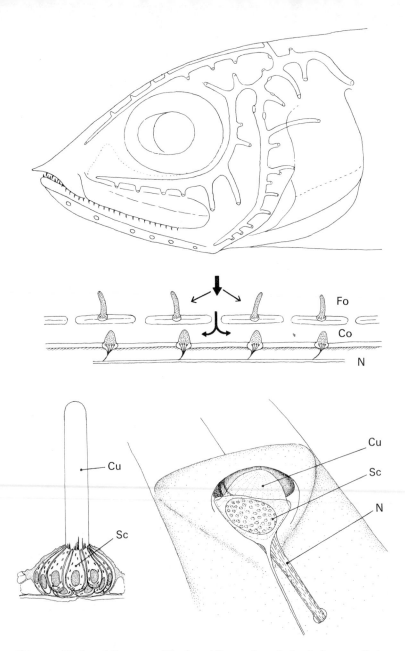

Fo

Co

N

Cu

Sc

Cu

Sc

N

Fig. 144   The lateral-line system. Top, lateral-line canals on the head of a mesopelagic
fish, *Searsia*. Centre, diagram of free-ending organs (Fo) and canal organs
(Co), cupulae stippled. Bottom left, the free-ending organ of a larval teleost
(after Iwai, 1967); Cu, cupula; Sc, sensory cells. Bottom right, organ from
the head canal of a burbot, *Lota lota*; N, nerve. (Redrawn from Flock, 1965)
(Redrawn from Marshall, 1971)

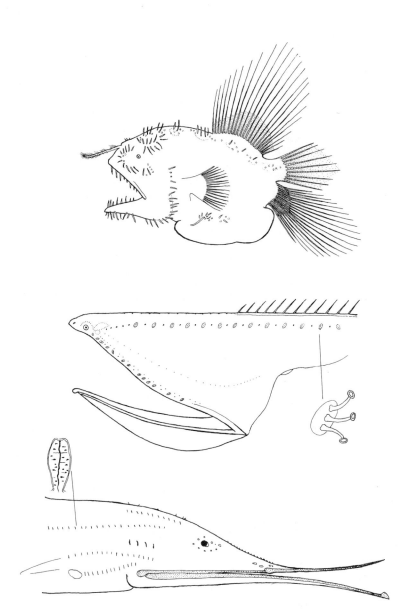

Fig. 145   Free-ending lateral-line organs. Above, a female ceratioid angler-fish *Caulophryne jordani* (after Bertelsen); centre, a gulper-eel *Eurypharynx pelecanoides*; below, a bob-tailed snipe-eel *Cyema atrum* (after Bertin).(Redrawn from Marshall, 1954)

terminal pore and an inner central rod, between which and the outer epidermis is clear gelatinous tissue. The rod appears to contain four to eight myelinated nerve fibres (or multilamellar tubes) and below its base there is a cluster of cells that may be connected to a lateral-line nerve. As he was unable to find the hair processes (stereocilia or kinocilia) that occur on neuromast organs and as the pore and gelatinous filling of the papilla is reminiscent of the jelly-filled electroreceptors (ampullary organs) of elasmobranchs, Meyer-Rochow suggests that the papillae of *Cyema* may be electroreceptors. But the complete fine structure of these curious papillae has yet to be elucidated. Lastly, *Cyclothone* spp., the commonest bathypelagic fishes, have both canal organs and free-ending organs.

In midwater fishes neuromasts set at the tips of extensive papillae seem to be confined to bathypelagic species. But such free-ending organs, as we shall see, are found also in benthic deep-sea fishes. Free-ending organs on lampreys, sharks, larval teleosts and various adult teleosts are either slightly protuberant or flush with the surface of the skin. The angler-fish *Lophius* and the cave-dwelling amblyopsid fishes, with neuromasts on low papillae, are among the exceptions in shallow-water environments (Marshall, 1971).

Neuromasts in mucus-filled canals are buffered to some extent against 'uninteresting' water displacements. Thus, they are largely shielded from displacements along the body surface, but they are well placed to be stimulated by local pressure differences, that is, by water displacements at right angles to the body surface (Dijkgraaf, 1962). The cupulae of free neuromasts will be much more subject to water 'noise', particularly to turbulent movements and to water flowing over the skin during active locomotion.

Fishes whose neuromasts are entirely or mainly free (for example, *Lepidosiren*, larval teleosts, cave-dwelling amblyopsids, gobies, *Lophius*, and so on) are slow or intermittent swimmers. Moreover, the first and third forms live in obscured or dark surroundings. For our present purpose, comparison of bathypelagic species with the amblyopsids is particularly relevant. In the true cave dwellers, *Typhlichthys* and *Amblyopsis*, the neuromasts are extended on short stalks, but in the epigean species *Chologaster cornutus* and the troglophilic *C. agassizii* there is no such provision (Fig. 146). Moreover, the cave dwellers carry more neuromasts and larger lateralis centres in the cerebellum than do *Chologaster* spp. When 40 mm *Typhlichthys* and *Amblyopsis* were stationary their cupulae moved as adult water fleas (*Daphnia magna*) swam at distances of 15 to 35 mm from these receptors. Poulson (1963) concluded that the 10–12 mm difference between the distance that a prey causes cupular movement in a stationary fish and the distance at which a gliding fish orients towards the same prey shows the advantage of decreased 'noise'.

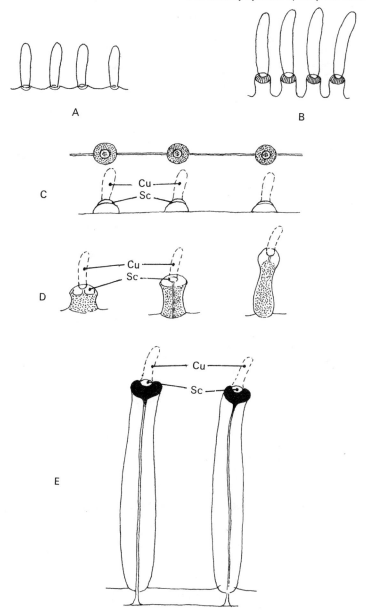

Fig. 146   Free-ending lateral-line organs of two cave fishes (A and B, redrawn from
Poulson, 1963) and three ceratioid angler-fishes. C, *Dolopichthys*, neuromasts
on papillae; D, *Cryptopsaras*, neuromasts on short stalks; E, *Neoceratias*,
neuromasts on long stalks. Cu, cupula; Sc, sensory cells. (Redrawn from
Marshall, 1971)

Bathypelagic fishes live in waters that tend to be gentle in movement. The neuromasts of an angling ceratioid or gulper-eel should thus be free to work at low levels of water noise, a factor that also seems to 'favour' olfactory signals between the sexes. Relevant experiments with an angler-fish, at least, are feasible. Meanwhile, we must note that the lateralis centres in the cerebellum and the lateral-line nerves are very well developed in ceratioids and other bathypelagic fishes (Marshall, 1967). Structurally, then, these fishes seem well equipped to detect and locate moving prey. Ceratioids feed on organisms ranging in size from copepods to prawns, squid and fishes (Bertelsen, 1951). Bertelsen also observes that *Caulophryne* and *Neoceratias*, the first with a non-luminous esca, the second without an illicium, have specially well-developed lateral-line organs. In both genera the neuromasts, which are carried at the end of

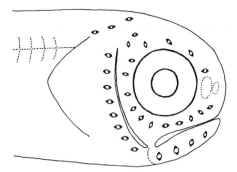

Fig. 147 Lateral-line organs (in canals) on the head of a latern-fish, *Electrona antarctica*. (From Marshall, 1954)

long filaments, are thicker on the epidermal ground than they are in species with a luminous bait (Marshall, 1971).

Many mesopelagic fishes have a mixed system of neuromasts. Thus, lantern-fishes, melamphaids, evermannellids and searsiids not only have large and complex head neuromasts (Fig. 147) which are set in wide canals, but also free neuromasts on the head. The neuromasts in the body canal are relatively small. Evidently, these fishes have a particularly sensitive head system, which no doubt enables them to detect and find prey, and keep contact with each other, in their dim twilight world.

A mixed neuromast system is found also in most benthopelagic fishes. The most striking feature is the development of very wide head canals housing large neuromasts, such as are found in rat-tails (macrourids), halosaurs and many brotulids (Fig. 148). In rat-tails the area of skin stretched over the head canals is well over half the entire surface of the head. In some species the canals open through pores: in others the system is closed. Each neuromast, which lies on an oval bolster of connective tissue, is generally circular in form and has a diameter ranging in various species from 0·5 to 3·0 mm. The many thousands of hair cells in a neuro-mast project into the base of a large gelatinous cupula, the vane of which is set across the canal. The canals are filled with a clear endolymph, the volume of which is 3·5 per cent or more of the entire body volume in *Coryphaenoides rupestris*. In chemical composition the canal endo-lymph of this species resembles blood plasma and cerebral fluid (Fänge et al., 1972). The body canals, which extend well down the tail, are much

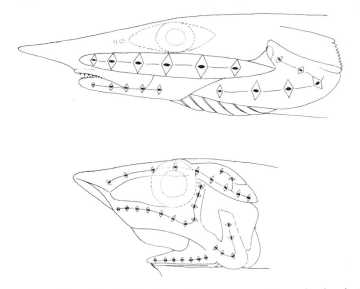

Fig. 148   Wide lateral-line head canals and large neuromasts in two benthopelagic fishes. Above, a halosaur, *Aldrovandia*; below, a macrourid, *Coelorhynchus*. (Redrawn from Marshall, 1971)

narrower than the head canals and thus have room only for relatively small neuromasts (Marshall, 1965).

Beside the canal neuromasts there are free-ending organs on the head and body. They may be found over most of the head, especially over the underparts. Over the rest of a macrourid they commonly occur along the upper and lower sides of the lateral-line canal. The head organs, which may be housed in a small circular depression or between two crescentric eminences, have a definite orientation.

Rat-tail fishes thus have a very extensive and elaborate lateral-line sys-tem. The base line of this system is much increased by the long tapering tail. As there is likely to be some direct relation between the size of a neuromast and its sensitivity, the organs of rat-tails must cover a wide range of sensitivity. The most sensitive elements are presumably the large enclosed head organs, each of which, as we saw, bears a crested cupula with the vane set across the canal. Current-like displacements of water directed more or less at right angles to the head canals, and which agitate (through pores or skin) the canal endolymph to cause a shearing motion along the vane of the cupula, will excite the neuromasts. (The canal pores are small, which lends point to the suggestion by Fänge et al., 1972, that the lack of such openings prevents clogging of the canals with sediment particles.) The combination of the sensitive head system and the long-based body system must surely often enable rat-tails to detect and find their prey, much of which consists of animals swimming near the deep-sea floor (p. 276). The same must be true of halosaurs and brotulids.

In the Chlorophthalmidae, the dominant family of benthic fishes under the warm ocean, there is also a mixed system of neuromasts (Marshall and Staiger, 1975). In *Ipnops murrayi* there are free-ending organs on the head, on the lateral-line scales of the body and over the base of the tail fin. The head organs are not only closely associated with the lateral-line canals, but extend also over the snout, the roof of the skull, the cheeks and the gill covers. The snout organs are short pigmented papillae, each bearing a whitish neuromast. The other head organs are transparent and oval in cross-section. They are variable in length (from about 0·1 to 0·5 mm) and within each a nerve runs to the neuromast at the tip. The other neuromasts on the head are housed in the canal system. But along the body canal there are no neuromasts, which here are carried on papillae set along the base of the lateral-line scales. There are similar body organs in *Bathypterois*, *Chlorophthalmus* and *Bathytyphlops*, but in the mixed system on the head the free-ending organs are not so extensive as in *Ipnops*.

Altogether there are some 3,000 free-ending organs and about 40 head organs in *Bathypterois*, *Chlorophthalmus* and *Bathytyphlops*, but in the up on its pelvic fins, one can imagine that these organs will enable it to detect and seize copepods and amphipods etc., that pass within range of its lateralis (distance-touch) system. Much the same may be true of the tripod-fishes, which depend largely on copepods (p. 305).

Once water is set in motion, the displacement persists for a considerable time. It is surely this aqueous attribute that has prompted the evolution of lateral-line systems with sensors that enable users to detect and find the sources of hydrodynamic displacements, whether caused by social partners, prey or predators. If this is granted, it would be surprising if fishes (and some amphibians) were the only aquatic animals with displacement receptors. Cephalopods, surprisingly, do not appear to have such receptors, but they do occur on arrow-worms (Chaetognatha). There is a well-developed series of free-ending organs over the body of an arrow-worm, and, as in fishes, there are hair-bearing sensory cells

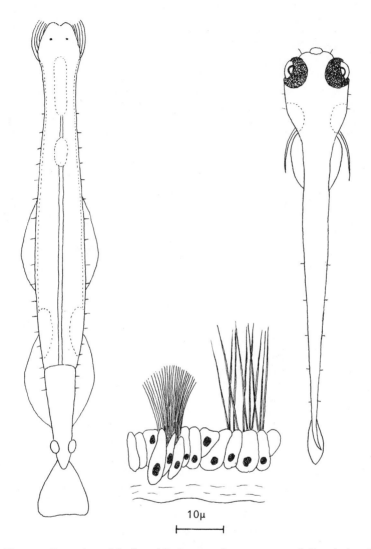

Fig. 149 Comparison of the 'lateral-line' organs of an arrow-worm (left) and a larval
fish *Oryzias* (right). (After Iwai, 1967) Between the two is shown a tuft of
setae and a fan of ciliated neurones from an arrow-worm *Spadella*.
(Redrawn from Horridge, 1966)

that are excited by nearby vibrations (Fig. 149). Through the responses of these sensors an arrow-worm is able to seize prey with great rapidity and accuracy (Horridge and Boulton, 1967). Pelagic nemerteans also have free-ending 'lateral-line' organs.

Hair cells are set on the 'fingers' of the ctenophore *Leucothoe*, while the hydromedusan *Eutonina indicans* evidently has vibration receptors, for it will bend its manubrium towards a nearby source of disturbances. Horridge (1966) also draws attention to Laverack's work on the lobster *Homarus vulgaris*, which has bristle-bearing vibration receptors in pits over the forward parts of its body, particularly over the large chelae. Moreover, he suspects that the tentacle tips of tube-dwelling animals and the opical tufts of cilia on trochophore larvae, which have a nerve supply, are sensitive to vibrations.

Even the largest midwater crustaceans may not have enough body surface for fan-bristle receptors like those of the lobster. But there are two kinds of sensillae on the integument of various prawns, which have been studied by Mauchline and his colleagues (1977). In all the oplophorid and pandalid prawns they examined there are tuft organs on the dorsal surface of the fourth abdominal segment and the telson. As each tuft organ consists of a group of open-ended tubular setae, they may well function as chemoreceptors. They found also that the integument of acanthephyrid and systellaspid prawns is completely covered by small and very delicate scales. At the base of each scale, which is shield-shaped, there is a pore in the integument. The scales point forward over the fore half of the body and backward over the rear half. The authors conclude, 'They probably have a sensory function as distance receptors, monitoring water currents and disturbances in the environment surrounding the shrimp.'

Lastly, could the long, trailing whiplash parts of the antennae of sergestid prawns act as a lateral-line system? As we saw, along these parts there are paired setae on each segment that bend round to form a virtually closed tube that is medial to the animal as it swims. Each segment also bears small setae (with lateral setules) that project into the tube and are found along the entire length of the trailing flagellum. These 'tube' setae are mechanoreceptors, and as Ball and Cowan (1977) say, they could provide an accurate means of prey localization. The antennal setal tubes and their sensors certainly remind one of the long lateral-line canals and segmental neuromasts of fishes. Such intriguing structural convergences ought to have some common functional ground.

Ears and sounds in fishes

Though some fishes virtually have no eyes or olfactory organs, all have some kind of acoustico-lateralis system. More precisely, while the loss or marked regression of the lateral-line organs is very rare, and known only in hagfishes (Myxinidae), all fishes have inner ears. Hagfishes are again exceptional in that the two vertical semicircular canals found in other jawless fishes are represented by one canal with two ampullae, each containing a band of sensory hair-cells (crista). Though they may not perceive sounds, the ears of hagfishes, like those of other fishes, have certain life-sustaining functions: they are (presumably) involved in the maintenance and regulation of muscular tone, they monitor angular accelerations of the body and they bear gravity receptors.

Angular accelerations are detected by hair-bearing cells in the ampullae of the semicircular canals, which, as in the lateralis system, are capped by a gelatinous cupula. Fishes monitor their orientation with respect to the direction of gravity through the utricular chambers of the ears, which act as static organs. Any movement away from an even position produces a shear between the sensory organ (macula) of the utriculus and the ear-stone (otolith) that is suspended over the hair cells. Where there is enough illumination, fishes use downwelling light to keep a visual watch on their position of equilibrium.

Both visual and static ways of keeping equilibrium are, of course, available to mesopelagic fishes. Concerning the former, the downward and strongest part of the light field is no doubt registered by their very sensitive eyes. Fishes of the sunless bathypelagic zone have only their static receptors for maintaining a position of equilibrium. Benthic species that live below the threshold of light have fins as well as ears 'to tell them that they are the right way up'.

Bathypelagic fishes certainly have well-developed utricular receptors and semicircular canals. This is very noticeable in small fishes, such as dwarf male angler-fishes and *Cyclothone* spp., in which well over half the volume of the neurocranium is devoted to the otic capsules. Space within the latter is largely taken up by the canals and utriculus. The lower saccular part of the ear is small, which seems to be the rule in bathypelagic and mesopelagic fishes (Bierbaum, 1914, and personal observations) (Fig. 150). But in most benthopelagic fishes, that is macrourids, deep-sea cods and brotulids, this part of the ear is very large and so is its otolith (Fig. 150). Since the sacculus is the hearing part of the ear, it is in keeping that fishes of these three families have means of sound production.

Sound-producing mechanisms

Teleost fishes make sounds in two main ways: the action of relevant muscles either vibrates the swimbladder wall or causes stridulation between suitably shaped parts of the skeleton. Elasmobranchs, which have no swimbladder and an unsuitable skeleton, make no more than incidental sounds, as when they swim or feed (Marshall, 1962). But the

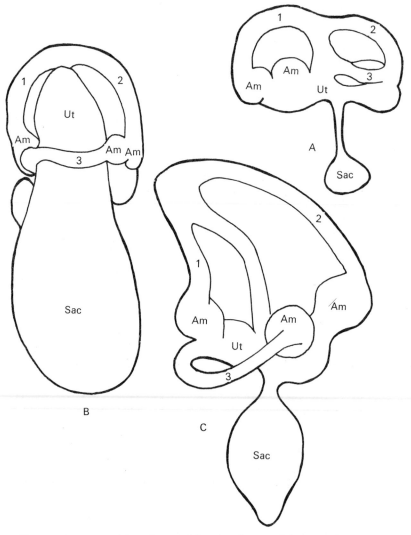

Fig. 150   Inner ears of three deep-sea fishes. A and C are of the mesopelagic species *Chauliodus sloani* and the hatchet-fish *Sternoptyx diaphana*, B is from the rat-tail *Hymenocephalus italicus*. 1, 2, 3, semicircular canals; Ut, utriculus; Sac, sacculus; Am, ampulla. (After Bierbaum from Marshall, 1954)

gas-filled swimbladder found in many mesopelagic fishes is simply a hydrostatic organ; it is not provided with drumming muscles. Nor does it seem likely that these fishes have skeletal parts that are suitable for stridulation. This is true also of bathypelagic and benthic species, which in any event have no swimbladder (Marshall, 1960).

Except for the sharks, chimaeroids and alepocephalids, benthopelagic deep-sea fishes have a well-developed swimbladder (Marshall, 1960, 1965a), but in the halosaurs and notacanths the swimbladder has no drumming muscles. This is also true of the deep-sea cods (Moridae), though they may well grate together their pharyngeal teeth, which are moved by large muscles that originate close before the forward wall of the swimbladder. Such stridulatory sounds, as certainly occur in certain haemulids and carangids, will be picked up and 'given body' by the swim-bladder. Moreover, the swimbladder forks anteriorly so that a circular pad on the forward part of each fork fits against a membrane-covered foramen in the rear wall of the otic capsule. The sacculus, as we have seen, contains a large otolith. Teleosts with some kind of coupling between the inner ears and the swimbladder have, at the very least, extrasensitive hearing. But what do morids hear? (Marshall, 1971).

Most kinds of morids and macrourids are centred over the continental slope at depths between 200 and 2,000 metres. The slope-dwelling macrourids, which also have a very large otolith in each sacculus, have paired drumming muscles on the forward part of the swimbladder (Fig. 151). These muscles, which are homologous with those of drum-fishes, gurnards and other sound-making fishes, are confined to the males. Dissections of about 50 species representing all genera showed that both attachments of each muscle are either on the forward wall of the swim-bladder or one attachment is to the adjacent body wall (Marshall, 1973).

Brotulid fishes, also most diverse over the upper continental slope, have more elaborate sonic devices. In the oviparous forms, which comprise about a hundred of some 175 species in all (Mead, Bertelsen and Cohen, 1964), the male alone has large drumming muscles. These are attached to the otic capsule and the thick forward wall of the swimbladder. The compliance of the system probably resides in the modified ribs of the first three vertebrae, which converge to make forward suspensory bars for the swimbladder and to serve as points of attachment for all or part of the sonic muscles (Fig. 151). The viviparous forms are less well known, but in some species, at least, both sexes have drumming muscles, and one or more pairs of the first three sets of ribs form attachments for the swimbladder and its muscles.

Since macrourids and brotulids are easily the most diverse groups of benthopelagic fishes in the deep sea, sound signalling is thus prevalent in the fauna of the upper slope. But in the bathygadine group of macrourids, most of which tend to live at levels of 1,000 metres or more,

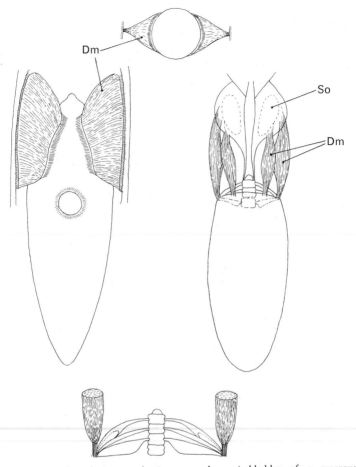

Fig. 151 Sound-producing mechanisms on the swimbladder of a macrourid
Malacocephalus laevis (left) and a brotulid *Monomitopus metriostoma*
(right). The drumming muscles (Dm) of the macrourid are inserted on the
body wall and the forepart of the swimbladder; those of the brotulid attach
to the swimbladder and modified ribs (below) and the otic capsules. Note
the large saccular otolith (So) of the inner ear. (From Marshall, 1971)

and in the abyssal genera of the very diverse subfamily Macrourinae (for
example, *Lionurus* and *Nematonurus*), neither sex has drumming
muscles. Moreover, compared to their sonic relatives from the upper
slope, these deeper-living macrourids have very small saccular otoliths.
Sound-making and large (hearing) otoliths would thus seem to be associated.*

* This is also true in sciaenids (Schneider, 1962).

More precisely, even without the deep-sea cods, almost half of the ben-thopelagic fauna of the continental slopes (that is, some 250 macrourids and 150 brotulids) must be sound producers. In the macrourids and ovi-parous brotulids the male is the vocal sex, but note also that *both* sexes have large sacculi. In fact, sound making and correlated hearing is con-fined to species with large eyes, fitted for vision in a twilight world. More-over, some of the morids and macrourids have a light gland on the belly. Most likely, sound signals and luminescence, if present, are involved in courting and mating activities. Male rivalry, as in the cod, a relative of the macrourids, may also be expressed in sounds. But why should abyssal macrourids, which have small, probably regressed eyes and no light glands, be without sound-making devices? One might think that in a sun-less environment sound signals would be a good means of assembling the sexes. It seems, though, that directional help from a sound field is virtually confined to a (near-field) displacement region described by a radius of $1/r^2$ from the source. Abyssal macrourids, as shown by a detailed photographic survey at depths around 2,500 metres, live in small groups (2 to 5 fishes) separated by much more than the radius of the near field.* Individuals of slope-dwelling species, which can be very abundant, are probably very often within near-field earshot of one another (Marshall, 1971). Their relative the cod (*Gadus morhua*) is able to find a sound source emitting a pure tone of 75 Hz over distances up to 5.3 metres in the horizontal plane (Schuijk, 1974).

But how is sexual encounter managed in abyssal rat-tails? They not only live in small groups, but recent observations of the population densi-ties of two abyssal species, *Lionurus carapinus* and *Nematonurus armatus*, obtained during dives of the DSRV *Alvin* off the Hudson Canyon between 1,700 and 2,800 metres, show that these two species are by no means scarce (Cohen and Pawson, 1977). The first species, which was among the five most common species and observed on all but the deepest dive, had an abundance, expressed as individuals per thousand square metres, ranging from 0 to 2.37. The second species was the third most abundant fish and observed on five out of seven dives. Its density per 1,000 m² ranged from 0 to 3.93 with the highest value at the greatest depth (2,797 metres). At depths below 2,400 metres it was the dominant fish species. Thus, the meeting of the sexes in these two species is not so hazardous as might be supposed. Indeed as observations from deep submersibles and continuous series of stereo-mosaic photo-graphs both show, population densities gained by these methods are con-siderably more than would be calculated from catches made by trawls over the same distances.

---

* Investigations show that the fundamental frequency of swimbladder sounds correspond to the frequency of muscular contractions, which are not likely to be more than 150/sec.

Early life histories 12

The life history of organisms spans all the phases of their existence. Here we concentrate on the time between germ-cell production and the appearance of young that are beginning to take adult shape.

The early stages are crucial in the life history of many-celled marine animals. In diverse invertebrates and fishes there are heavy losses of eggs and young, which seem to be matched by great fecundity. Thinking particularly of teleost fishes, 'Despite the high mortality, the broadcasting of many buoyant eggs and larvae by currents enables a species to cover and exploit its living space, and even, when possible, to extend it' (Marshall, 1971). This kind of early life is common in midwater animals of the deep sea. The exceptions follow the rule of the animals on the deep-sea floor, where the invertebrates, at least, produce relatively large and few eggs that hatch into advanced young. The larval stages may even be passed within the egg. The high fecundity–small egg–larval phase pattern, or the low fecundity–large egg–suppression of larval phase pattern is found in so many deep-sea animals that we must consider whether there is room for intermediate types. There are certainly two main patterns, but whatever the way, natural selection will tend to preserve the fitter or fittest modes of early life history in competing species.

The mean geological life history of marine species is about 6 million years among the invertebrates, according to Durham (1967): Valentine (1977) suggests a mere one million years. In the geological scale species come and go so quickly that one wonders how often their extinction (in stable times) is due to persistent failures in their early life histories. Species must maintain their 'fitness for change' if they are to survive in future, perhaps different, environments (Bronowski, 1977). Such adaptability, which no doubt resides in genetic means of maintaining variability, would seem to be more feasible in early life-history patterns than in adult structure. There seem to be more variables, as in fecundity, the manner of spawning (whether the eggs are all shed at once—semelparity—or in batches at intervals—iteroparity), the time and extent of the spawning season, the size, number and kinds of eggs and young, the length of each phase in the life history, the location of nursery grounds, the length of the entire life history, and so forth. But the fitness of the entire system may be marred by one weak link in the chain of events.

These reproductive arrangements, which we must consider, form the strategy (generalship) of any life history. As Slobodkin (1964) says, evolution is like a game, and the players must somehow stay in the game. In games theory, a 'strategy' is the complete programme of what a player will do in every conceivable situation (Rapoport, 1960). Strategy thus seems a reasonable concept in the consideration of life histories. One may also think in terms of specific arrangements or tactics towards reproductive success. Indeed, two valuable reviews (Stearns, 1976, and Pianka, 1976) concentrate on life history tactics.

Studies of life histories cannot cover every conceivable situation. In variable environments the problem is particularly complex. Deep-sea environments are relatively constant, and in such stable (so-called 'predictable') conditions, MacArthur (1962) has argued that reproductive effort can be modest. In unstable or turbulent environments, where stochastic (chance) events prevail, high fecundity is needed. But we must remember that the early life of many midwater and some bottom-dwelling animals of the deep sea is passed in turbulent nursery grounds near the surface. Though these surroundings are part of the euphotic, productive layer, the latter also nourishes many predators. Thus, fecundity must be more than enough to cover the inroads of these 'many predators'.

The early life of midwater animals

There is much we do not know of the early life of mesopelagic and bathypelagic animals. Even so, a comprehensive review would fill a sizeable book. Our concern is with selected members of the dominant groups of invertebrates and fishes, beginning with the copepods.

Much of what is known of the life histories of oceanic copepods is summarized by Sekiguchi (1974). Their earliest stages are not well known, but there are three main types: the fertilized eggs are shed free into the sea (as in *Calanus*, *Rhincalanus* and *Pleuromamma*, all with deep-sea species); before hatching the eggs are carried in sacs attached close to the genital opening (as in *Euchaeta*, *Pareuchaeta* and *Valdiviella*, also with deep-sea representatives), and thirdly the eggs, which are adhesive, are laid on some suitable surface (as in *Aetideus* and *Xanthocalanus*, the latter containing bottom-dwelling species, some deep-sea). Copepods hatch as a small (typically less than 1 mm) nauplius larva, with an oval unsegmented body bearing three pairs of appendages (antennules, antennae and mandibles). The nauplius passes through a series of moults into the metanauplius phase, which is followed by further moults through copepodite stages (I–V) before the adult form is attained. In the relatively few species where the structure of the naupliar mandible is known, Sekiguchi (1974) realized that the basal joint of this appendage either carried a masticatory blade (used in chewing food) or was without one. In the first group the chewing process becomes ready for use in nauplius stage IV, as in cold-water species of both hemispheres such as *Calanus cristatus* (northern), *Calanus acutus* and *Rhincalanus gigas* (southern). These copepods both overwinter and spawn in the depths, and there are well-marked ontogenetic migrations. For instance, in the northern North Pacific, free eggs and nauplii of *Calanus cristatus* were not fished above a depth of 500 metres, but later stages (copepodites I–IV) were commonest at all seasons in the uppermost 200 metres. The pre-adult stage (copepodite V) had a wide vertical distribution (centres of maximum concentration in spring being 0–50 metres and 400–500 metres, in summer 200–300 metres and 1,550–1,700 metres, and in winter 0–100 metres and 800–1,000 metres). Adults were concentrated below 500 metres both summer and winter (Sekiguchi, 1975). In the Southern Ocean such ontogenetic migrations between deep winter and shallow summer levels tend to maintain populations within their proper bounds (see p. 252).

Copepods without a masticatory process on the mandibles of their nauplii belong to genera such as *Aetideus, Pareuchaeta, Paraugaptilus, Valdiviella* and *Xanthocalanus*, all with deep-sea representatives. In these the chewing process begins to form at a later stage (nauplius IV or more usually, copepodite I). Many of these copepods live from warm temperate to tropical regions. Four species of *Pareuchaeta* taken in the northern North Pacific seem to stay at mesopelagic levels for their entire life, and Sekiguchi 1974) presumes the same is true of other copepods without naupliar masticatory processes. He contrasts these forms with the ontogenetic migrators of cool waters, which are herbivorous '... and have to begin their naupliar lives seeking food in the superficial productive layer'. The non-migrators, if they are like the shallow-water *Acartia*, have enough egg yolk to carry them through the naupliar stages, and do not begin feeding until copepodite stage I, when the mandible is fully formed. Moreover, deep-sea members of this group (e.g. *Euchirella, Valdiviella* and *Paraugaptilus*), develop egg cases enclosing a few large yolky eggs. (The eggs of *Valdiviella insignis* measure 1·92 × 1·40 mm— Herring, 1974a).

Though much is unknown, one can begin to see certain main features in the life-history strategies of deep-sea copepods. There has to be a balance between fecundity and size of the eggs. Large eggs, with substantial reserves of nutrients, enable the start of trophic life to be post-poned until a relatively late larval stage, when the young are large enough not to be restricted, as are nauplii, to minute kinds of food organisms. Though fecundity is necessarily reduced, it is apparently enough to enable many mesopelagic and bathypelagic copepods to spend their entire life span in the depths. Species with an actively feeding nauplius stage lay many small eggs, but the life-history pattern must be such that the nauplii are able to feed on small species of phytoplankton etc. Thus, in the northeast Atlantic Ocean, *Calanus finmarchicus* overwinters around a level of 1,000 metres as copepodite stages (mainly V), which moult into adults in spring. The adults ascend and breed so that many of the resulting nauplii are at hand to feed on the spring bloom of phyto-plankton. But most species of deep-sea copepods live under the euphotic zone in subtropical and tropical regions, where plankton productivity is more or less continuous throughout the year. We shall need to consider the breeding seasons, if any, of such deep-sea animals.

The life-history strategy must maintain the populations of a species in the face of a moving environment. Clearly, the pattern must enable enough of the young to grow up in the right ecological surroundings. Current speeds are likely to be greater at mesopelagic than at bathype-lagic levels, but many species from both zones spend their early life in the still livelier surface waters. How do species keep in their living space?

These and other aspects of life-history strategies will emerge at later stages of this chapter. Indeed, the deep-sea euphausiids, which are a dominant group in the micronekton and relatively well investigated, will widen our approach. Like the copepods, they hatch as a tiny nauplius larva (< 0·6 mm) with the same three pairs of appendages (Fig. 152), but the antennae and mandibles, unlike those of copepods, have no setae for filtering minute kinds of food from the water. The nauplius has no mouth and subsists on yolk reserves from the egg, as must, it seems, the next metanauplius stages. The metanauplius changes markedly to form the first calyptopis larva (0·7–2·52 mm), which now has filtering appen-dages. The body has a distinct carapace region and tail, which is becoming segmented. Mandibles, maxillae and maxillipeds are well formed but there are no signs as yet of the thoracic legs and abdominal pleopods. Then come further growth stages and moults to form a series of furcilia larvae, which become more and more like the adults (Fig. 152). The com-pound eyes take shape, complexity and mobility while the thoracic and abdominal limbs successively acquire their proper form. By the time the last furcilia has moulted into the postlarval stage, all the appendages are formed and the telson is assuming its adult form and armature (Mauch-line and Fisher, 1969).

There are some 85 species of euphausiids and most (57) shed their eggs into the sea. The remaining species, comprising, inter alia, the genera

Developments in Deep-Sea Biology

Nematobrachion, Nematoscelis and *Stylocheiron*, protect the embryos (some species in a pouch) until the larvae hatch. The eggs are attached to posterior pairs of thoracic legs (Mauchline and Fisher, 1969). These authors found that the volume of the gravid ovary is about 10 per cent of the total body volume, and they estimate that the fecundity varies at least between 3 and 800 eggs. In the three genera cited above the fecundity

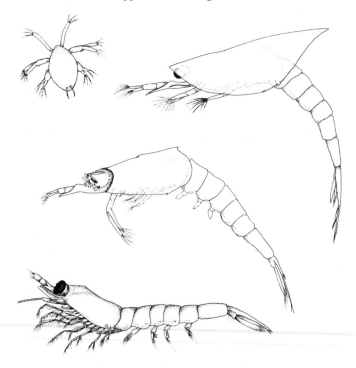

Fig. 152 Stages in the early life history of a deep-water euphausiid, *Thysanopoda acutifrons* and, above left, the nauplius of *Thysanoessa inermis*. From top to bottom, calyptopis stage (length 3 mm), early furcilia stage and late furcilia stage. (After Einarsson, from Marshall, 1954)

is 3–20 in *Stylocheiron* spp. and probably 200–400 in *Nematobrachion* and *Nematoscelis*. In a particular species the fecundity increases with the size of the mature female.

Epipelagic species of euphausiids that reach a maximum length of 25 mm and live from warm temperature to tropical regions, mature at a maximum age of one year, when they breed and most likely die. The life histories of mesopelagic and bathypelagic euphausiids are not well known, but certain contrasts may be made. Species of *Stylocheiron*, as

432

we saw, have the lowest fecundity of euphausiids, and in *S. suhmii*, a small lower epipelagic species (length 7 mm), three is one estimate. The fecundity of the mesopelagic species *S. elongatum* (length 18·5 mm) is unknown, but it presumably produces much larger eggs than its relative, for the first calyptopis measures 2·52 mm, whereas the same stage of *S. suhmii* is 1·28 mm. The size of the second calyptopis of *Nematoscelis tenella*, a mesopelagic species, is much the same (1·60 mm) as those of its epipelagic relatives (see Mauchline and Fisher, 1969). The fecundity of *Nematoscelis* spp., as already stated, is probably 200–400 eggs. How do *Stylocheiron* spp. manage on such a low fecundity? Do they spawn oftener than *Nematoscelis* spp. or are they more numerous? The latter does not seem likely from investigations in the equatorial Pacific Ocean (Roger, 1971). This work shows also that the young stages (furcilia) of mesopelagic species (*Thysanopoda orientalis*, *T. monacantha*, *T. pectinata*, *Stylocheiron maximum* and *Nematobrachion boopsis*) live at shallower levels than the adults. It looks as though the life history begins in the surface layers and there is a gradual descent to the adult level during the furcilia stages. During their early life they will drift to the west in the South Equatorial Current but after descending to levels between 100 and 200 metres, they will return east in the Cromwell Counter Current (see also p. 38). Below this, between 200 and 300 metres there is a deep equatorial current, again flowing in an easterly direction. As the two opposed currents above 200 metres are presumably stronger than the third, one can see how equatorial populations of euphausiids do not tend to be transported to the west. Actually, all the mesopelagic species listed above have a more or less wide distribution in central and equatorial waters between latitudes 40°N and 40°S. Indeed, the distribution of *Stylocheiron maximum* and *Nematobrachion boopsis* extends beyond these limits, the first between about 60°N and 60°S, the second between 40°N and 54°S. Populations in the central water masses must circulate in the gyres.

The 'giant' bathypelagic species of *Thysanopoda* undoubtedly spawn in the depths, where they spend their whole life. There are 200 to 300 eggs in the ovaries of *Thysanopoda cornuta* and *T. egregia*, which is comparable to the fecundity of epipelagic species of this genus. But the eggs of the giants are much larger (> 1 mm) than those of their epipelagic relatives (0·23–0·50 mm) (Mauchline, 1972). The larvae, as one might expect, are also larger (see also p. 261). Nemoto (1977) suggests that the young stages of bathypelagic *Thysanopoda* soon adopt the (mainly copepod) diet of the adults, which in food-poor surroundings seems advantageous. Large young should not only have the speed to seize large prey but also the capacity to deal with a relatively wide size range of food organisms. Mauchline (1972) estimates that the large bathypelagic euphausiids live 3 to 7 times as long as the small annual epipelagic species.

The increased longevity may mean that they spend much of their life span in potential or actual sexual maturity, which gives the sexes time to meet. As Mauchline concludes, a long-lived bathypelagic form with strictly limited fecundity will have a relatively slow replacement rate. Indeed, all seems to be in keeping with the need to run life histories on impoverished resources.

We should also return to the timing of life-history stages. To recapitulate, most deep-sea animals, whether pelagic or benthic, live under surface waters that are virtually seasonless in productivity. Thus, one might predict that there would be no particular breeding seasons. Recent evidence supports this conjecture, but there are exceptions (p. 462). In the seasonally productive temperate to polar regions it must be advantageous for young euphausiids to begin feeding when and where there is most food of the right kind (such as we saw in the northeastern Atlantic populations of *Calanus finmarchicus*, which are of prime importance as whale food). Of euphausiids, there is no better instance than the whale food (krill) of the Southern Ocean, *Euphausia superba*. Adult krill live at epipelagic levels south of the Antarctic Convergence, where they are circumpolar but most widespread and abundant in the Atlantic Sector. They spawn near the surface in shelf and oceanic waters from January to March. The large eggs (about 0·7 mm in diameter) sink, and in the open ocean probably reach the lower levels of the warm deep water at about 1,800 metres. After hatching the nauplii move upwards, and as they climb develop through the metanauplius stages—in time to begin feeding during the summer blooms of phytoplankton. Presumably the yolk reserves are enough to fuel a few weeks' climb through as much as 1,800 metres. Then follow two further calyptopis and six furcilia stages before they reach adolescence. During the next two years the krill grow to a maximum length of about 6 cm, when they live mainly in the uppermost 200 metres of the water column. (See Everson, 1977, for a valuable review of krill biology.)

Both copepods and euphausiids produce eggs that are denser than sea water. In species that spawn freely in the sea, the sinking of the eggs, their hatching and the possible return of young stages to upper nursery grounds, are crucial times in the early life history. But if the females carry the eggs until they hatch, as in diverse species of both groups, the young may be released in suitably chosen surroundings.

This brings us to the deep-sea prawns, which live successfully both as free-spawners and egg-protectors: they are, as we have seen (p. 110), dominant groups of the micronekton. The free-spawners include the sergestids, nearly all of which are mesopelagic. The eggs hatch into typical nauplius larvae, which in *Sergia lucens*, and probably in all other sergestids, do not feed but subsist on yolk reserves. Then follow three spiny elaphocaris stages, which bear mouthparts that filter phytoplankton and suspended matter. The spines project symmetrically from the carapace and tail (Fig. 153). Those on the carapace acquire secondary spines during the elaphocaris stages, when the young roughly double in size (to about 2 mm) and stalked eyes develop. Besides retarding the rate of sinking, the spines may well deter small predators. During the succeeding acanthosoma stages the form is more shrimp-like and some or all of the thoracic legs function in swimming, which is usually tail-first and upside-down. In the postlarval (mastigopus) stages the abdominal segments also acquire swimming legs (pleopods) and the motion is now head-up and forward. After a final moult the postlarva becomes an adolescent (Marshall, 1954; Omori, 1974).

The life history of sergestids is not well known, but one species, *Sergia lucens*, which is very common in Suruga Bay, Japan, has been carefully investigated (Omori, 1974). By day it lives between depths of 150 and 300 metres and towards nightfall migrates up to the surface waters. Indeed, it spawns near the surface at night, and the season lasts from late May to mid-November, with a maximal period from June to August. Omori found the spawning to begin when the temperature between 20 and 50 metres exceeded 18° C. The spherical eggs are a light greenish-grey and measure about 0·25 mm in diameter. There are 1,700 to 2,300 in one brood. The eggs of a deeper-living (500–700 metre) species, *Sergestes similis*, are of much the same size. In Monterey Bay, California, this sergestid spawns from June to July and from December to January. The age and size at maturity of both prawns is 10–12 months and more than 36 mm. The life span of *Sergia lucens* is about 15 months.

The eggs of caridean prawns are attached to special setae on the abdominal appendages of the female. The larvae that emerge are either protozoeae or zoeae. We shall consider the Oplophoridae, and one species, *Acanthephyra quadrispinosa*, has been well studied by Aizawa (1974) in the waters off Japan and in the laboratory. It is known from all three oceans and reaches a body length of about 60–70 mm. Off Japan by day it lives between levels of 400 and 1,000 metres: by night, most of the population has migrated to depths between 200 and 300 metres.

The eggs of *Acanthephyra quadrispinosa*, which are as scarlet as the adults, are ovoid with a major axis of 0·8–1·1 mm and a minor axis of 0·6–0·8 mm (Fig. 37). The egg clutch carried by a female depends on her size and in Aizawas' counts ranged from 359 to 1,543. For instance, females of 60, 65 and 75 mm in body length carried 500–1,000, 570–1,250 and 1,000–1,5000 eggs respectively. Aizawa collected egg-bearing females in every season but those with eggs containing eyed larvae near to hatching were found only in winter and spring. The spawning season was estimated to extend from April to November or December with a peak from summer to autumn. In Bermudan waters eggs near hatching were found

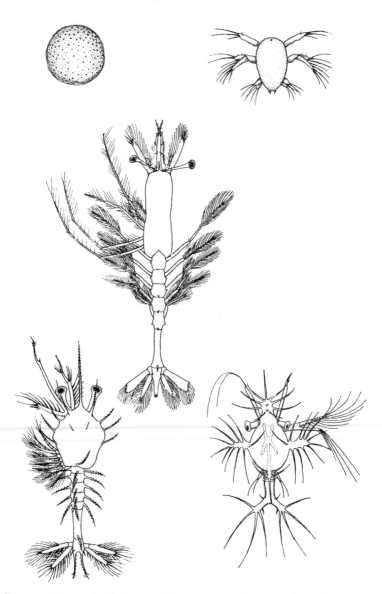

Fig. 153   Stages in the life history of deep-sea prawns. Top, egg and nauplius; centre, third postlarva of *Gennadas elegans*; bottom left, first postlarva of *G. elegans*; bottom right, second protozoea of *Sergia lucens*. (Redrawn from Omori, 1974)

on females of *Acanthephyra purpurea* during investigations lasting from April to September.

Aizawa thinks that the hatching of *A. quadrispinosa* in Japanese waters is influenced by the temperature cycles. In Sagami Bay the temperature at 200 metres, the night-time level of this prawn, reaches a maximum of 13–14°C in winter, which is a few degrees more than that at other seasons. Hatching is thought to occur at night, when the migrating females mass at levels around 200 metres. Larvae hatching from eggs kept in the laboratory (at 12–14°C) soon began to feed on the nauplii of brine-shrimps (*Artemia*) (Fig. 37). Moulting of the larvae into the second, third and fourth stages occurred 3, 7 and 17 days after hatching. The second larval stage has a body length of 4·0 mm. Stages beyond the fifth were taken at sea. When they have reached about 12 mm the young enter the postlarval phase and at lengths near 30 mm have an adult appearance. If hatching is prevalent at the night-time level of the adults there must be a descent to daytime levels during the early life history.

Catches taken in Japanese waters from January 1964 to October 1966 were used to study growth changes. Measurements of body length showed that *Acanthephyra quadrispinosa* grows at a remarkably uniform rate until, two years after hatching, it becomes adult at a length of 60–70 mm, when growth ceases. The life span is three or more years (Aizawa, 1974). The growth of the body, expressed as a daily rate, is 0·07 mm, which in *Sergia lucens* is nearly twice as much (Omori, 1974). But this sergestid, as we saw, reaches maturity at a length of about 40 mm in 10 to 12 months. To reach maturity (from 4 to 60 mm) in one year, *A. quadrispinosa* would need to grow at about twice its natural rate. In Hawaiian waters a small surface-dwelling sergestid, *Lucifer chacei*, grows at a rate of 0·32 mm per day to reach a mature size (> 8 mm) in 7 to 8 months. Such growth is evidently sustainable in warm and productive surroundings, but is presumably impossible at cool and less productive deep-sea levels. Antarctic krill (*Euphausia superba*), which is somewhat smaller than *A. quadrispinosa*, also reaches maturity at 50–60 mm in two years. But nearly all of its growth takes place during the summer months from October to March, when its daily growth increment may be nearly twice that of the prawn. The daily growth increment in weight is about 0·8 mg, whereas in the North Atlantic the comparable figure for *Thysanopoda acutifrons*, a mesopelagic euphausiid, also with a two-year cycle, is 0·356. The growth rate of krill is so much faster than that of this and other euphausiids (see Mauchline and Fisher, 1974) that Everson (1977) wonders whether its life span has been adequately covered by past investigations. But it lives in very productive waters and stays with its food: energy is not diverted to vertical migrations.

The fecundity of certain oplophorid prawns is much lower than that of *Acanthephyra quadrispinosa*. The species *Hymenodora glacialis*, *H.*

frontalis, Systellaspis debilis, Oplophorus gracilirostris and *O. spinosus* are of much the same size but the females carry much fewer (15–30) and larger eggs (2·3–4·6 × 2·0–2·6 mm) than their relative (Omori, 1974). And, whereas the newly hatched larvae of *A. quadrispinosa* are 4 mm in length, the corresponding stage of *Systellaspis debilis* is about three times this size, with '... a saw-like rostrum, well-formed eyes and antennae and the complex of thoracic and abdominal limbs. Not a great deal of change is needed to turn this young stage into an adolescent' (Marshall, 1954). The early larval stages are 'telescoped' within the eggs and protected by the females. When the young are eventually set free in the sea, they may still carry yolk reserves, and when these have been used, they have the size, mobility and capacity to prey on a relatively wide range of food organisms. Such young must surely have a greater ability to escape predators than have the early stages of species that produce many small eggs (e.g. *Acanthephyra quadrispinosa* and *A. purpurea*).

Deep-sea oplophorid prawns produce either many small eggs or few large eggs. There seem to be no intermediates. Besides *Acanthephyra quadrispinosa* and *A. purpurea*, the small-egg producers include *A. curtirostris, A. acutifrons, A. architelsonis, Notostomus auriculatus* and *N. elegans* (range of egg sizes, 0·76–1·12 × 0·56–0·82 mm): besides the five species listed in the previous paragraph, the large-egg producers include, *Systellaspis cristata, S. braueri, Ephyrina bifida* and *Pasiphaea hoplocerca* (range of egg sizes, 2·24–4·64 × 1·60–3·52 mm) (Herring, 1974). Both strategies, the one based on safety in numbers of eggs and larvae and the other on the safety of abbreviated development and the production of a few advanced young, seem equally acceptable (to processes of selection) as ways of maintaining local populations. At all depth horizons there are species producing either small or large eggs (Foxton, 1970). Evidently there is either a high mortality or a low mortality strategy, but why is the middle excluded? Why are there no species with middling numbers of eggs that hatch into moderately advanced largae?

The sergestid prawns 'rely' exclusively on safety in numbers, but the euphausiids seem to be either low or high fecundity types, and perhaps the same is true of the copepods (see pp. 429–31). The deep-sea mysids, if they are like *Eucopia* (Herring, 1974; Casanova, 1977), have a low fecundity. Indeed, the brooding of a few advanced young in the marsupium is typical of peracarid crustaceans (pp. 460–1).

All the members of two widely successful and major groups, the arrow-worms and cephalopods, emerge from the egg as advanced young. The hermaphrodite, but not self-fertilizing, arrow-worms produce relatively large eggs that hatch into young with much the look of their parents including prospective shape, fin pattern and jaw structure. For instance, the small (ca. 8 mm) epipelagic species, *Pterosagitta draco*, lays a jelly-

like mass containing 200 to 300 eggs (diameter 0·36–0·40 mm) that produce young measuring about 2 mm eleven hours after hatching. Species
of the deep-sea genus *Eukrohnia*, which range from about 30 to 45 mm
in length, produce much larger eggs. Near the time of laying the eggs
are described as 'large and few' and in the species *E. hamata* the female
retains the clutch in a kind of brood pouch formed by flexion of the lateral
fins (Ghirardelli, 1968). In the Southern Ocean the early life history of
Eukrohnia hamata and *Sagitta maxima* seems to be something like that
of *S. gazellae* (p. 103). Their eggs are presumably shed at depths where
the young (2–5 mm) were taken (David, 1955). In the subtropical and
tropical belts little is known of the sequence of events in the early life
history of deep-sea arrow-worms, but it may well be that the nursery
grounds of the young are well above the concentrations of their parents.

The early life history of deep-sea cephalopods is even more obscure.
There is '... almost complete ignorance of the eggs and early larval
stages...' of oceanic squids (Clarke, 1966). One common ommastrephid
squid, *Todarodes pacificus*, lays a gelatinous egg mass on the bottom
containing 300 to 4,000 very yolky eggs (0·7–0·8 mm diameter). The
newly hatched larvae are unable to swim clear of the bottom. Reviewing
this and other evidence, particularly his finding of gravid female squid
in sperm whales, which are believed to forage near the upper continental
slope, Clarke (1966) suggests that the laying of egg masses at slope levels
may well be a common habit among oceanic squids. Moreover, the
sepioid *Spirula* of slope waters, whose ripe ova measure 1·7 mm and the
newly hatched larvae 2·8 mm, seems also to be a bottom spawner (Bruun,
1943).

If the young of deep-sea cephalopods are born in the depths, they must
subsequently ascend to the upper waters, for there is evidence of ontogenetic migrations. For instance, Clarke and Lu (1975) found that the
largest individuals of squid such as *Abraliopsis affinis*, *Histioteuthis
meleagroteuthis* and *Cranchia scabra* were taken at deeper levels than
their smaller relatives. *Liocranchia reinhardti*, another cranchiid squid,
was common in *Discovery* nets at 11°N, 20°W, and after analysing 490
specimens, Lu and Clarke (1975) found that four out of five of those taken
deeper than 200 metres were smaller than 6 mm. They suggest that after
hatching in deep water the young ascend to the surface layer, where after
a period of growth, there follows a descent of individuals larger than
45 mm in mantle length. Off California, the cranchiid *Galiteuthis phyl-
lura* also shows ontogenetic descent. Larval life is passed at increasingly
deeper levels between the upper 100 metres and a level of 600 metres,
until, at a mantle length of 60 mm, larval development is over and the
animals descend as subadults to 900 metres or more (Roper and Young,
1975). In the octopod *Japetella diaphana* (Fig. 154) most of the smaller
individuals (< 10 mm) were sampled at the above *Discovery* station

between 50 and 200 metres: most of the larger ones were deeper than 600 metres. Lu and Clarke's (1975) tentative reconstruction is that egg laying and hatching occurs in the upper 200 metres, from where, after a brief period of growth, the young descend rapidly to 500–600 metres and then more gradually to 1,000–1,250 metres as they grow to maturity. The large animals ascend to spawn. Roper and Young (1975) suggest that in species with unusually large larvae, such as cranchiids and chiroteuthids, there is a gradual ontogenetic descent. But in other species of

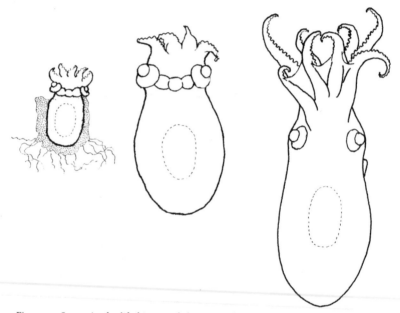

Fig. 154   Stages in the life history of the mesopelagic octopod, *Japetella diaphana*. From left to right, larva, with gelatinous 'hairy' coat (about 5 mm); later larva (about 15 mm); young female (about 50 mm). (After Thore, from Marshall, 1954)

pelagic cephalopods the sequence is that '... larvae occur in near surface waters and at a particular size abruptly descend to the adult habitat. An exception to this pattern occurs in *Vampyroteuthis infernalis* where the young occupy greater depths than the adults.'

Large, long-lasting larval stages, which go with wide distributions of the adult phase (p. 465), have evolved also in benthic invertebrates (pp. 465–7), and in fishes. The classic type in fishes is the leptocephalus larva of eels and related groups, but we turn first to the more usual kind of early life history.

Except for certain small squaloid sharks (e.g. *Isistius*), midwater fishes of the deep ocean are teleosts. Most fishes are teleosts, and their success may well be due in part to their evolution of means to produce small buoyant eggs. 'During the latter stages of maturation, follicle cells secrete dilute body fluids, containing half or less of the salt content of sea water, into the developing eggs. Part of this store of buoyant fluid is retained beneath the skin of the larvae, and they thus float with ease. (These buoyancy chambers are soon replaced by the swimbladder.) Denton (1963) even suggests that the primary biological significance of dilute body fluids in marine teleosts resides in their use for producing buoyant eggs' (Marshall, 1971). Indeed, some lay eggs that are slightly over-buoyant, which may well be true of deep-sea species. The early larval stages of both mesopelagic and bathypelagic fishes are well known from the surface waters, but not their eggs. An exception seems to prove the rule: in the Strait of Messina between Italy and Sicily turbulent hydrographic conditions sometimes lead to the upward transport of the eggs of both midwater and bottom-dwelling deep-sea fishes, which have been described by Sanzo (1933). Normally, it would seem that midwater fishes, whether a lantern-fish at 500 metres or a ceratioid angler at 1,500 metres, spawn where they live. The eggs develop and hatch as they float upward and so do the still over-buoyant larvae. Newly hatched larvae usually subsist for a time on yolk reserves, so the optimum arrangement would be for the postlarvae (the larval stages after the yolk has been resorbed) to reach the surface nursery ground when they are ready to feed.

This reconstruction of a common strategy in the early life history of midwater fishes thus depends essentially on negative evidence—the lack of records of eggs in surface waters—but there seems to be some positive support in Ahlstom and Counts' (1958) thorough study of *Vinciguerria* spp. in the eastern Pacific (Fig. 155). Species of this small stomiatoid genus live high up at mesopelagic levels, and at one station samples taken in the upper 140 metres of the eggs and larvae of *Vinciguerria lucetia* are interesting. Most of the eggs (30–90 metres) are below most of the larvae (20–40 metres), which fits the idea of larvae hatching from rising buoyant eggs. More evidence is needed. The sequence of events after the postlarvae have begun to feed in the surface layer is much better known. If they find enough food of the right kind they grow and develop, and if, near the metamorphosis stage, they are among the few to have escaped predation etc., there is soon a rapid change to the juvenile stage, when they begin to look like their parents. Near and after metamorphosis the young move down towards the adult habitat. Ahlstrom and Counts (1958) describe metamorphosis '... as the period during which marked changes occur in body proportions and body structures without any marked increase in standard length. The length may not increase at all, or it may even diminish.' Where there is shrinkage during metamorphosis, which

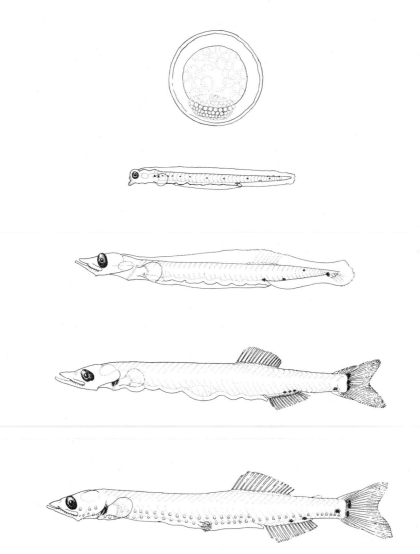

Fig. 155   Stages in the life history of a mesopelagic fish, *Vinciguerria lucetia* in the
eastern Pacific. From top to bottom, egg; postlarva (2·2 mm); postlarva (6·0
mm); postlarva (13·1 mm); metamorphosis stage (15·0 mm). (Redrawn from
Ahlstrom and Counts, 1958)

is greatest in fishes with leptocephalus larvae, the vertebral column must ossify during the juvenile stage.

The eggs of diverse midwater fishes range from 0·5 to 1·65 mm in diameter (Marshall, 1953; Hopkins and Baird, 1977). They are thus of much the same size or much smaller than the eggs of oplophorid prawns (p. 438). Herring (1974) found small prawn eggs have less lipid reserves (ca. 15 per cent of the wet weight) and are denser (ca. 1·05–1·075) than are large prawn eggs (lipid ca. 40 per cent, density ca. 1·03). They are all denser than sea water, but as we saw (p. 435), they are carried by the females until they hatch. The eggs of copepods and euphausiids are also denser than sea water, but again are carried until hatching by the females of some species (p. 432). In species that freely shed their eggs after fertilization, the sequence of events in the early life history has had to be shaped, as it were, around non-buoyant spawn. The eggs may be shed near the surface (e.g. *Calanus finmarchicus* in the northeastern Atlantic, p. 431, *Sergia lucens*, p. 435), and if there is much sinking, the larvae need a good supply of yolk for their climb to nursery grounds (e.g. *Euphausia superba*, which produces fish-sized eggs). Cephalopods enclose their heavy, yolk-dense eggs in a gelatinous mass, which in many species may be laid on the deep-sea floor. But the pelagic octopod, *Japetella*, may be a near-surface spawner (p. 440). There are thus diverse life-history strategies in pelagic invertebrates, which may be seen as alternate ways of reproducing in oceanic space with non-buoyant eggs. Teleost fishes with their buoyant eggs have the one basic strategy outlined above. The few exceptions will be considered later.

Lanter-fishes (Myctophidae) are very diverse (ca. 300 species), widely distributed and among the best known of midwater fishes. During extended surveys half the total catch of young fishes will often consist of myctophids (Moser and Ahlstrom), but only recently have free eggs been netted and recognized. The eggs of *Lampanyctodes hectoris*, a common lantern-fish over shelf and slope at subtropical latitudes of the Southern Hemisphere, were identified by Robertson (1977) in a surface plankton sample in New Zealand waters. The eggs are slightly ovoid (0·73–0·83 × 0·65–0·72 mm) with a narrow perivitelline space and a segmented yolk bearing a single oil globule. When close to maturity in the ovaries the eggs of other myctophids (*Lobianchia dofleini*, *Electrona rissoi*, *Symbolophorus californiense*, *Triphoturus mexicanus* and *Lampanyctus steinbecki*) range from 0·55 to 0·84 mm in diameter (Robertson, 1977).

Lobianchia dofleini is best known from the Mediterranean and North Atlantic Ocean (50°–25°N): it occurs also in the South Atlantic, southeastern Pacific and southern Indian Ocean. In the Mediterranean, Tåning (1918), after analysing many samples of young stages taken from 1908 to 1909 and in 1910, found the main spawning period was from winter to summer. It is a rather small lantern-fish (maximum length ca. 50 mm), and in individuals from 31 to 40 mm in length Tåning counted 330 to 484 large eggs in the ovaries. In other lantern-fishes of about this size (*Benthosema glaciale, Hygophum benoiti, Myctophum punctatum*), the fecundity was generally less than 1,000 eggs: in larger species *Symbolophorus veranyi* (120 mm) and *Ceratoscopelus maderensis* (70 mm) there were over 2,500 eggs. It thus looks as though all species produce eggs of much the same size. Moreover, the larger the female of a given species, the greater should be her fecundity. Thus in Tåning's figures for *Lobianchia dofleini*, there were counts of 330–376 large eggs in females measuring from 31 to 33 mm in standard length: in one of 40 mm there were 484 eggs.*

The sequence of stages from postlarval to adolescent in *Lobianchia* may be seen in Figure 156. Larval stages, being small (2–3 mm) and fragile, are rarely recoverable from plankton nets. Some of the marked changes during metamorphosis are evident when the late postlarval stage is compared to the metamorphosis stage, which, though of the same size, has acquired a more elegant shape, larger eyes and adult pectoral fins. Metamorphosis stages of lantern-fishes, like those of other midwater fishes, retain some of the postlarval pigment pattern, but the prospective adult pigmentation appears as a dark peppering of small melanophores over the head, back and forward part of the body. Development towards the adult constellation of light organs is much more advanced, but sexual differences in the luminescent system are completed during the adult phase.

Karnella and Gibbs (1977) have used juvenile to adult stages of *Lobianchia dofleini* to trace the life history in Bermudan waters. Their analysis of the seasonal abundance and percentage composition in the catches of juveniles (10–24 mm), subadults and adults showed that this species has a one-year cycle and breeds from January (possibly December) to June, mostly in the winter months. Contrary to earlier investigations, which suggested that the populations of *L. dofleini* in the western North Atlantic were expatriates and derived from the transatlantic drift of young hatched in eastern North Atlantic spawning grounds, they found clear evidence that this lantern-fish breeds not only off Bermuda but further north in slope water off New England. Breeding populations thus live in the western North Atlantic (spawning season, January to June), the eastern North Atlantic (spawning season March to October) and in the Mediterranean (spawning season mainly February to June). As Karnella and Gibbs conclude, 'The need is great for intensive

---

* If the mature ova fill the entire volume of an ovary, one would expect the fecundity to increase as the cube of the ovarian dimensions. Actually, the increase may be nearer a squared factor.

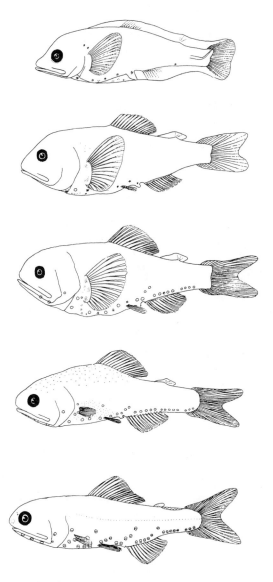

Fig. 156   Stages in the life history of a lantern-fish, *Lobianchia dofleini*. Lengths from top to bottom, 5·5 mm, 8·0 mm, 11·5 mm, 11·0 mm and adult stage. Just above the adult is the metamorphosis stage. (Redrawn from Tåning, 1918)

studies of different populations in different geographic areas of the world ocean, the kinds of studies that would provide information on life histories and data useful for relating behaviour to environmental factors, while at the same time contributing material for determination of the genetic–systematic status of geographic populations.'

A one-year life history is probably common among small and medium-sized mesopelagic fishes that live from warm temperate to tropical regions. Larger species, such as those of the stomiatoid genera *Chauliodus*, *Astronesthes* and *Gonostoma*, may require more than one year to reach sexual maturity (Hopkins and Baird, 1977). For instance, off Japan the females of *Gonostoma gracile* (length 125 mm) live for at least two years: the males mature in one year and after breeding, most of them turn into females (Kawaguchi and Marumo, 1967). In temperate to cold regions, mesopelagic fishes, like euphausiids and prawns, need more than one year, also presumably related to the restricted growing season, to become mature.

Extensive surveys in warm temperate to tropical regions show that the breeding period of lantern-fishes may cover two or more seasons, with a peak spawning time in one. Under nursery grounds that are more or less productive throughout the year, there would seem to be no pressure to concentrate spawning in one short period. Even so, extended spawning seasons vary from place to place, as we saw in *Lobianchia dofleini*. Off Bermuda coccoliths dominate the phytoplankton for most of the year, and one prevalent species, *Coccolithus huxleyi*, may be most abundant in late autumn or early winter (Deevey and Brooks, 1977), just before the peak spawning period of *L. dofleini*. But these investigators also found that numbers of copepods—prime food for young fishes—were least in the water column in November 1969 and May 1970. Evidently, maximal conditions change from year to year and in any event, one should not consider the spawning period of one species in isolation. In Bermudan waters there are at least 15 breeding species of lantern-fishes (Gibbs et al., 1971) and it may well be that their spawning seasons tend to be spaced so as to reduce competition in nursery grounds etc. This idea was developed by Goodyear et al. (1972) when considering marked differences in the spawning season of various Mediterranean midwater fishes with surface-dwelling larvae. They suggest that non-synchronous breeding periods should help to reduce competition among the larvae and juveniles of these species.

Perhaps Jespersen and Tåning (1926) were thinking on these lines when they wrote of such exclusive aspects in the distribution and life history of *Vinciguerria* spp. in the Mediterranean and neighbouring Atlantic waters. Their observation of the three species from Atlantic stations is particularly interesting. *Vinciguerria attenuata*, which has incipient tubular eyes, lives deepest. *V. nimbaria* (=*sanzoi*) shallowest, and *V.*

poweriae at intermediate levels. What is most remarkable is that the post-larval stages may also be segregated in depth. In *poweriae* and *nimbaria* at least, the postlarvae were concentrated at about 50 metres and 20 metres, respectively. Apart from partitioning resources, the disposition of the young at different levels raises questions. If the eggs are spawned at adult levels, is their rising and hatching such that the relative levels of the postlarvae reflect that of their parents, or do the postlarvae of each species seek their own optimal light intensity?

Stages in the development of *Vinciguerria lucetia* in the eastern Pacific, redrawn from Ahlstrom and Counts (1958), may be seen in Fig. 155. Their findings on the depth distribution of the eggs and larval stages of this species have already been cited (p. 441). The larvae are 'thin and threadlike at hatching', which indicates that they carry little yolk and soon enter the postlarval phase. The photophores appear as white bodies at the beginning of metamorphosis, at the end as pigmented, seemingly functional, organs. Though small fishes (ca. 40 mm), *Vinciguerria* species produce relatively large eggs (measurements by Ahlstrom and Counts gave the following ranges of egg diameters in mm: *V. lucetia* 0·58–0·74, *V. nimbaria* 0·64–0·72, *V. poweriae* 0·75–0·85). Their fecundity is thus bound to be modest, as Jespersen and Tåning (1926) found. In a 30 mm female of *V. poweriae* they counted 300 (large) eggs; in a 42 mm *V. attenuata* there were 250 eggs.

In fishes, as in other animals, mortality is greatest during the early life history. The death rate is highest during the egg and larval stages, per-haps 5–10 per cent per day in very fecund fishes (Cushing, 1975). But as fish grow beyond postlarvae to mature adults, their natural mortality, due largely to predation, declines. In midwater fishes the most critical phase must be passed during the ascent of eggs and larvae towards the surface waters. Thus, the deeper the spawning level the more will ascending young be exposed to predation. If, as seems likely, midwater fishes spawn where they live, one might expect the deeper dwellers to be more fecund than those that live nearer the surface. Though data on fecundity are limited, it is relevant to note that the deeper-dwelling (lower mesopelagic to bathypelagic) lantern-fishes (*Lampadena* spp., *Taaningichthys* spp. and certain *Lampanyctus*) are large and thus among the most fecund, members of their family, which is not to say that large species are absent from shallower levels. There is closer evidence when the mesopelagic *Cyclothone braueri* (Fig. 157) is compared to *C. microdon* from lower mesopelagic to bathypelagic levels. In the eastern North Atlantic (30°N, 23°W), Badcock and Merrett (1976) found that ripe *C. braueri* (24–29 mm) contained 200–300 eggs, whereas in *C. microdon* (45–49 mm) the number was ca. 2,200–3,300. As both species may be judged to produce eggs of much the same size (ca. 0·5 mm), the greater fecundity of the larger species is understandable.

Conjecturing freely from Badcock and Merrett's data that spawning in *C. braueri* is centred at 600 m and in *C. microdon* at 1,200 metres, the twofold difference in depth is associated with a tenfold difference in fecundity. Moreover, the population density of *C. microdon* is about half that of *C. braueri*, judging from the number caught. In well-investigated fishes, mortality, as one might expect, is found to be greatest and very heavy during the egg stage, and no doubt the same is true of *Cyclothone*. At any one time during the spawning season it seems likely that the number of spawners will represent no more than a few per cent of the entire local stock of mature individuals. Eggs may thus be thin on the spawning ground, but mortality is still likely to be heavy. In ascending the water column between the level of spawning and the surface, mortality at a given depth will be related above all to the density of predators,

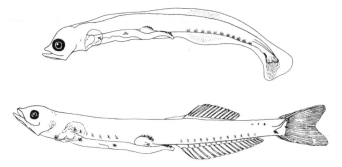

Fig. 157   Postlarval stages of *Cyclothone braueri*. Above, 4·8 mm; below, 13·2 mm. (Redrawn from Jespersen and Tåning, 1926)

which will depend on the biomass of midwater animals and its proportion of predators.

The biomass increases logarithmically towards the surface, whereas the proportion of predators decreases, perhaps in similar fashion. There are, of course, other factors. For instance, soon after the eggs turn into mobile larvae mortality should decrease: on the other hand, lateral transport of the young will delay their ascent to the nursery grounds and prolong their exposure to predators. Much is uncertain, but whatever the integrated effect of all mortality factors over the ontogenetic water column, it looks as though *microdon*, with less than half the population density of *braueri*, needs ten times the fecundity of the latter to maintain its deeper centred adult stocks.

Turning to surer matters, Badcock and Merrett (1976) found that the life span of *Cyclothone braueri* seems to contain a single spawning cycle, whereas *C. microdon* may well spawn more than once during a longer life cycle (more than one year, presumably). Moreover, *microdon*, like *Gonostoma gracile* pp. 261–2), undergoes sex reversal; males, the smaller sex, turn into females. There was no evidence of such change in *braueri*. It would thus seem that the yearly population fecundity of *microdon*, being considerably more than that represented by ripe one-year-old females, is much enhanced. A fecundity of two or three thousand eggs in *Cyclothone microdon*, which is presumably much the same in its other bathypelagic congeners, seems relatively modest, but below 1,000 metres these are much the commonest fishes. Less numerous and larger fishes, such as gulper-eels and some of the ceratioid angler-fishes, would thus seem to have every need of their bulk to be adequately fecund. Such bathy-pelagic fishes, as we have seen (pp. 262–4), have a relatively simple organization, which internally, coupled with the absence of a swim-bladder, should leave extra abdominal space for the ripe eggs. In the gulper-eel, *Eurypharynx pelecanoides*, a female of 600 mm contained about 33,000 well-developed ova, which were 0·9 mm in diameter, yellow and with four or five yellow oil globules (Raju, 1974). Concerning angler-fishes, most of the female specimens with attached males have mature or maturing ovaries. In a *Ceratias holboelli* of 650 mm standard length there were nearly 5 million immature eggs (0·2–0·4 mm): in a smaller species, *Edriolychnus schmidti*, a female of 63 mm contained 9,100 ripening eggs. Mature females of ceratioid species with free-living males are rarely caught, but Bertelsen (1951, and in Mead et al., 1964) has figures, besides those just cited, for *Dolopichthys*. In a 114 mm female of *D. longicornis* there were eggs of two size groups: about 10,800 ripening eggs of 0·5–0·75 mm and much the same number of immature eggs (0·1–0·3 mm). There is thus evidence of increased fecundity in both small and large fishes of the bathypelagic zone, which must find enough nourishment in an impoverished environment to provide food reserves for thousands of developing eggs. Even so, their somatic organization is so economically integrated that a higher proportion of their limited energy budget can be diverted to gonadal development than in fishes from more productive environments.

The eggs of gulper-eels hatch into leptocephalus larvae, which in the deep sea are found also in true (anguilliform) eels, halosaurs and nota-canths. Leptocephalus larvae have a transparent, laterally compressed body, ranging from narrow to broad leaf shapes. If the eggs of gulper-eels are like those of the mesopelagic (true) eels, *Nessorhamphus*, the larvae can live for a time on considerable reserves of yolk (Fig. 158). In eels, after the yolk has gone, the larvae grow into leptocephali and become planktonic—in the European freshwater eel (*Anguilla anguilla*) for about 2½ to 3 years—before metamorphosis. Their growth and sur-vival for long periods is puzzling, for food is never found in the gut. After close study of leptocephali of the bandtooth conger eel, *Ariosoma*

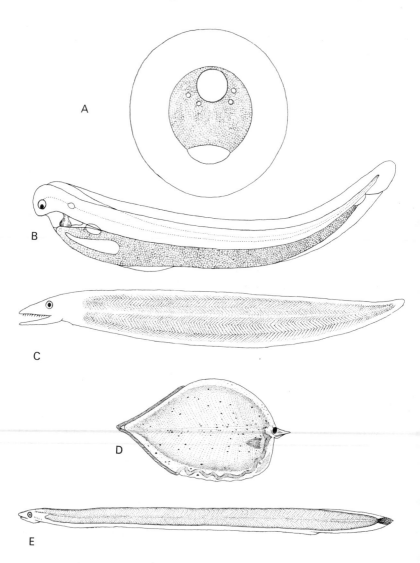

Fig. 158   A–C, stages in the life history of a deep-sea eel, *Nessorhamphus ingolfianus*;
A, egg (2·5 mm diameter); B, newly hatched larva (6 mm); C, leptocephalus
larva. D, leptocephalus of bob-tailed snipe-eel *Cyema atrum* (about 70 mm);
E, leptocephalus of deep-sea eel, *Synaphobranchus* (length 250 mm). (A–C,
after Schmidt; D, after Bertin; E, after Bruun)

balearicum, Hulet (1978) suggested that before metamorphosis lepto-cephali derive their nutrition from dissolved and minute particulate material in the sea, and there is preliminary evidence (from tracer work) of their uptake of solutes (Hulet, in litt.). The buccal lining of the leptocephalus of *Ariosoma*, like that over the tentacles of beard-worms, bears myriads of microvillae. Certainly, the leptocephalus seems an ideal kind of larva. It is transparent ('In the full light of a shipboard laboratory only the eyes of a live and swimming individual are visible' Hulet, 1978). If detected by its movements, a writhing leptocephalus must be difficult to seize and swallow by any predator of about its own size, which in deep-sea fishes ranges from a length of a few centimetres to nearly 2 metres.

The leptocephali of gulper-eels are best known in *Eurypharynx*, whose larval life is believed to last for more than 2 years. Compared with deep-sea eels, the size at metamorphosis (30–40 mm) is small, but already the gape is enormous, able no doubt to engulf relatively large prey. Gulper-eel leptocephali seem to be commonest at levels between 100 and 200 metres and after metamorphosis the young presumably begin their descent to adult bathypelagic habitats. The leptocephali of *Cyema atrum*, a bob-tail snipe-eel, also bathypelagic when adult, have a broad leaf shape (Fig. 158). Larval life may take two years (see the review by Mead et al., 1964; Raju, 1974).

The eggs of deep-sea anglers hatch into the commoner kind of fish larva. The newly hatched larvae are globular and short tailed. Bertelsen (1951) found that male larvae of *Himantolophus groenlandica*, 'presumably just hatched', were from 2·2 to 2·5 mm in length. Even at this stage the sexes are distinguishable, but only by a single feature, the presence on the snout of females of a small papilla that will become the illicium and light lure of the adult. As they feed and grow in the upper waters, the larval stages of both sexes (apart from the developing illicium) still look much the same, but during metamorphosis the other sexually dimorphic characters become prominent. Females, now with a stalked illicium and escal bulb, are clearly going to retain their globular form as adults (not in *Gigantactis*): males have a slimmer body form and are well advanced in features such as the size and structure of eyes and olfactory organs that so distinguish the sexes as adults. Bertelsen (1951) found also that the metamorphosing young sank to depths between 1,000 and 3,000 metres. His analysis of material from the well-investigated North Atlantic showed that most of the commonest species, at least, are summer spawners: the exceptions are linophrynid anglers, which are spring spawners, and whose larval stages tend to live at deeper levels (100–200 metres) than those of the majority.

Bertelsen's analysis of North Atlantic collections led him also to conclude that larval life in ceratioids probably lasts less than 2 months: then comes a comparatively rapid metamorphosis. Thereafter, the free-living existence of males may last for about 6 months, whereas the females of the best known species (e.g. *Melanocetus johnsoni* and *Cryptosaras couesi*) take at least 6 months to attain a (subadult) length of 20 to 25 mm. Comparable reconstruction by Hopkins and Baird (1977) for the common mesopelagic stomiatoid, *Valenciennellus tripunctulatus*, which they assume to be an annual species is: larval to metamorphosis stages (2–4 mm to 7–10 mm), possibly 15 to 30 days; juvenile to subadult stages (7–10 mm to 20–23 mm), 6–8 months. The difference would seem to be

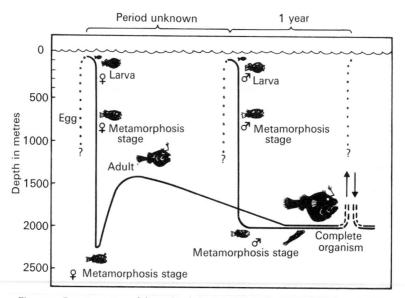

Fig. 159   Reconstruction of the cycle of changes during the life history of a ceratioid angler-fish. The larvae live near the surface and there is a gradual descent on metamorphosis. (From Bertelsen, 1951)

in much shorter larval life of the stomiatoid. Further discussion would be premature without comparative data for other species: meantime it is reasonable to suggest that the shorter the larval life, which involves high mortality, the higher the number of metamorphosis stages should be descending to maintain adult populations.

The slow rate of growth of female ceratioids after metamorphosis and the development of the ovaries indicates that they '... take more than one, presumably several, years to reach maturity' (Bertelsen, 1951) (Fig. 159). Males that are not parasitic on the females are progenetic (progenesis is '... produced by precocious sexual maturation of an organism still in a morphologically juvenile stage'—Gould, 1977). Such males, which are found in *Melanocetus, Himantolophus, Gigantactis* and some oneirodids, have large testes, even in late larval and metamorphosis stages. Not long after metamorphosis, when they are a few centimetres in length and ready to mate, they have also developed pincer-like jaws, likely means, it is thought, to nip on to the skin of the female during breeding. Progenesis in males that become parasitic on the female (cera-tiids and linophrynids) is not elicited until they become attached (Fig. 160), when the testes develop, and so do the ovaries of their partners (Bertelsen, 1951). Besides these reproductive strategies, one by non-para-sitic, the other by obligatory parasitic means, Pietsch (1976) suggests that there is a third, facultative kind. In *Caulophryne* and the oneirodid *Lepta-canthichthys*, both non-parasitized and parasitized females may have de-veloped ovaries. Free-living males are at present unknown. Pietsch thinks it probable that attachment occurs whenever grown individuals of the two sexes meet, regardless of their sexual readiness.

As well as economizing in the energy budget of a species and reducing the trophic competition between the sexes (p. 261), the progenesis of small males that become temporarily or permanently attached to their partners is surely an apt way of reproducing in the largest and most deserted deep-sea environment. A mated female becomes a cross-fertiliz-ing 'hermaphrodite'. But how do the sexes meet? Sensory means are con-sidered elsewhere (p. 408). Bertelsen also considers two other factors, the relative numbers of males and females and the population density of metamorphosed ceratioids. Analysis of the catches of metamorphosed males and females suggests that the former are somewhat more numerous (54–60%), and concerning mature forms, he conjectures that unmated females make up no more than 2–5 per cent of the total stock of metamor-phosed females. If so, the catches indicate that the males, most of which will be ready to mate, are at least 15 to 30 times as numerous as their waiting partners. Regarding the population density, his calculations (based on the number of metamorphosed ceratioids caught per hour in the S.200 net) suggest that the mean distance between one individual and the next is about 30 metres at depths between 1,000 and 2,000 metres, where the stock is densest. But this figure covers all species taken. As Bertelsen concludes, in any one species, despite the great excess of males, the mean distance between a male and the nearest female is likely to be much more than 30 metres. The males must search actively for their mates, and to do so they seem to be well equipped.

To conclude, the reproductive strategies of deep-sea angler-fishes appear to be closely fitted to trophic conditions where the sexes complete most of their growth. All or much of the growth in males is in productive surface waters. After metamorphosis, parasitic males stop feeding and are ready to mate: non-parasitic males have well-advanced testes but may continue to feed and grow to some extent (p. 455). Females grow up in

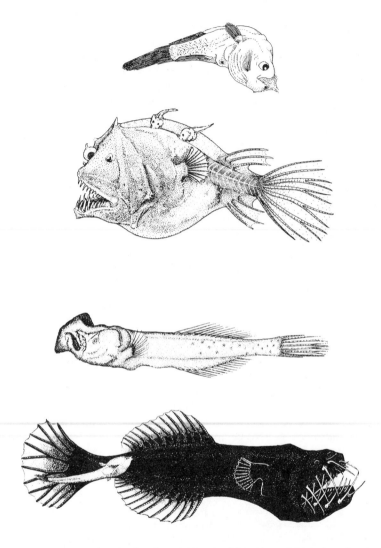

Fig. 160   Parasitic dwarf males of ceratioid angler-fishes. Above, *Edriolychnus* showing
a male enlarged and a female with two attached males. (From a specimen
in the Institute of Oceanic Sciences)
Below, *Neoceratias spinifer* showing a male enlarged and a female with
attached male. (Redrawn from Bertelsen, 1951)

food-poor bathypelagic waters. As we saw, they grow slowly after meta-morphosis, and they need, as it were, to grow to a relatively large size to become adequately fecund. It is thus understandable that the females may take several years to reach maturity. The males grow up in richer surroundings that contain diverse small mesopelagic fishes with an annual life cycle. Ceratioid males are still smaller subannual fishes. More-over, the fertile distribution of ceratioids is limited to subtropical and tropical regions between ca. 40°N and ca. 35°S (Bertelsen, 1951), where plankton production and growth continues throughout the year. Here their reproductive strategies have evolved, and they are evidently not adaptable to the restricted growing season in colder waters. The large females that have been taken north and south of the fertile area are vegetative expatriates, even if bearing a parasitic male. It is significant also that small annual fishes of mesopelagic levels (e.g. the stomiatoids *Valenciennellus* and *Vinciguerria* and the lantern-fishes *Diogenichthys*, *Notolychnus* and *Gonichthys*) are unknown from subantarctic and antarctic waters. The few mesopelagic fishes that pass their entire life in Antarctic waters are either large (> 100 mm, e.g. the argentinoid *Bathylagus antarcticus*, the lantern-fishes *Electrona antarctica* and *Gymnoscopelus braueri* and the barracudina *Notolepis coatsi*) or medium-sized (50–100 mm, e.g. the lantern-fish *Protomyctophum anderssoni*), with a life cycle of two years or more. The same is true of the thirty or more species that are endemic or common in subantarctic waters (see the catch and size records of Parin et al., 1974).

The early life of benthopelagic animals

Virtually nothing is known of the early life of the invertebrates that swim near the deep-sea floor. The copepods, such as *Xanthocalanus*, may well lay their eggs at the bottom (p. 429): certainly their young stages are eaten by small-mouthed rat-tail fishes (*Nezumia* spp.), which also contain benthic polychaete worms. Half-grown individuals of the finned octo-pods (*Cirroteuthis*) are taken in trawls, but the eggs and young have yet to be discovered. More is known of the early life of the fishes. The eels, halosaurs and notacanths, as already stated, have leptocephalus larvae, which may be taken in the upper waters and presumably arise from eggs shed near the deep-sea floor. In the North Atlantic *Synaphobranchus kaupi*, a common cosmopolitan eel, spawns in the Sargasso Sea. After hatching, the leptocephalus larvae drift (at depths between 100 and 300 metres) for some 18 to 22 months and grow to about 12 cm before meta-morphosis and descent to the slope (Bruun, 1937).

Considering their diversity, wide distribution and abundance, very little is known of the early life history of rat-tail fishes (Macrouridae).

Free eggs of *Hymenocephalus italicus, Coelorhynchus coelorhynchus* and *C. fasciatus* measure from 1·0 to 1·21 mm in diameter, while those of *Nezumia sclerorhynchus* are somewhat larger (1·60 mm). The large northern North Atlantic species, *Macrourus berglax*, which must lay eggs of about 4 mm in diameter, has a total fecundity of about 25,000, but the number of ripe eggs in a batch, as in another large species, *Coryphaenoides rupestris*, is about 15 to 16 thousand (Marshall, 1973). Where are the eggs laid? In three of the four species cited above, the free eggs were taken in the Strait of Messina, where turbulent waters sometimes transport midwater animals to the upper levels of the oceans. The eggs of *Coelorhynchus fasciatus* were caught in a tow net attached near a bottom trawl. After reviewing such evidence and examining the relatively few young macrourids taken by Dana Expeditions, which have done so much to trace the life histories of midwater animals, I was inclined to think that rat-tails lay their eggs near the bottom and that the early larval life is passed below the 'seasonal' thermocline at levels of about 200 metres (Marshall, 1965). If this is so, one might expect the deepest-living species, like those at midwater levels (p. 447), to be the most fecund, but evidence seems to be equivocal. On the one hand, in the North Atlantic, *Sphagemacrurus grenadae*, which is smaller than *Nezumia sclerorhynchus* and is centred at deeper levels (860–1,215 metres, cf. 450–730 metres) produces eggs of much the same size (ca. 1·5 mm).

Macrourus berglax and *Coryphaenoides rupestris* from shallower (200–600) or about the same levels (400–1,200 metres) respectively, have a ready fecundity of 15 to 16 thousand eggs, at least five times the expected figure for the two *Nezumia* species. The same lack of correlation between depth range and fecundity seems to be true of benthic fishes. A 139 mm individual of the tripod-fish *Bathypterois quadrifilis* from 1,200 metres contained about 4,500 eggs, whereas in a 106 mm *Ipnops murrayi* from about 2,000 metres there were only 900 eggs (Mead et al., 1964). Nielsen (1966) found that *Ipnops* and related species produce less than 2,000 eggs. On the other hand, and returning to rat-tails, two deep-water species have a rather high fecundity. In *Cetonurus globiceps*, from about 1,000–2,000 metres, one estimate of fecundity is about 10,000 eggs (Mead et al., 1964), while the figure for *Lionurus carapinus*, with populations centred below 1,500 metres, is 12,000 to 16,500 (Haedrich and Polloni, 1976).

Perhaps the benthic fishes spend their early life near the bottom (see p. 458), but what of rat-tails? Larval and juvenile stages can be found at midwater levels (Marshall, 1965). Indeed, at stations in the north-eastern North Atlantic (60°N, 20°W and 53°N, 20°W) Merrett (1978) has found the larval and juvenile stages of at least three species from near surface to midwater levels, and there is a correlation between size of young and depth, suggestive of ontogenetic migrations. Even so, why are the larval stages of rat-tails not so common in the surface layer as, say, those of ceratioid angler-fishes? Perhaps the former appear in the upper waters only where there are ascending currents. Judging from the diet of small-mouthed adult rat-tails, such as *Nezumia* spp., there is enough near-bottom sustenance in the form of young and adult copepods

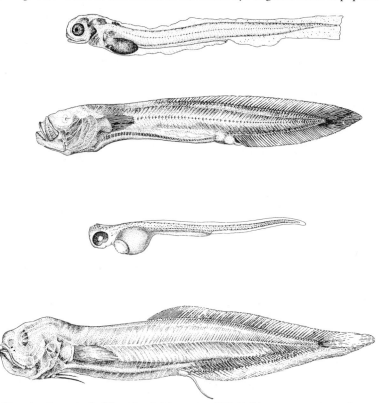

Fig. 161   Stages in the life cycle of viviparous ophidioid fishes. Top, ovarian embryo (7·5 mm) of *Nybelinia erikssoni* and, just below, adult (70 mm). Bottom, ovarian embryo (5 mm) of *Barathronus bicolor* and, below, adult (91 mm). (Redrawn from Nielsen, 1969)

to support the larvae and juveniles of rat-tails and of other bottom-dwelling fishes. In shallow waters, at least, fish eggs, presumed to be neutrally buoyant, have been taken close to the bottom (Oug, 1977). Such sampling is needed in the deep sea. And it must also be clear how little we know of the fecundity, spawning seasons and spawning grounds of bottom-dwelling species.

The early life of oviparous brotulid fishes (Ophidioidea) is even more obscure than that of rat-tails, but again may be passed near the deep-sea floor. There are also viviparous ophidioid fishes. Those of the family Aphyonidae are progenetic, reaching maturity at a small size (< 100 mm) while retaining larval features, such as a transparent, scaleless skin, pedunculate pectoral fins and a lightly ossified skeleton. The males produce spermatophores, which are stored in the ovaries of the females (Nielsen, 1969). Species of the benthopelagic genera *Barathronus* and *Aphyonus* produce larger egg clutches (1,200–1,600) than those of the (probably) midwater genera *Meteoria, Nybelinia, Sciadonus* and *Leucochlamys*, in which the number of eggs varies from 20 to 70. Nielsen found a number of free embryos (length 4–9 mm) in the ovaries of *Barathronus bicolor* and *Nybelinia erikssoni* (see Fig. 161). Clearly the young are born at an advanced stage.

Viviparity with sperm storage seems such an apt way of reproducing in the deep sea that one expects it might have become a common strategy. Apart from a few zoarcids (e.g. *Parabrotula*), it is confined to the ophidioid fishes. Moreover, the oviparous deep-sea brotulids are much more successful (in number of species, ubiquity and biomass) than their viviparous relatives (see also Nielsen et al., 1968). Actually, the most prominent live-bearing fishes in the benthopelagic fauna are the various kinds of squaloid sharks (*Centroscymnus, Centrophorus, Etmopterus, Oxynotus*, etc.) that are common over the continental slope. All appear to be ovoviviparous: the young are formed from large yolky eggs retained in the oviduct of the female.

Early life histories in the benthos
Reproductive strategies in benthic and other deep-sea fishes

Under the subtropical and tropical belts the most diverse and commonest of the benthic deep-sea fishes are chlorophthalmids (Sulak, 1977a). Their larval life is virtually unknown but may well be passed close to the deep-sea floor. In the tripod-fish *Bathypterois phenax* and the related *Ipnops murrayi* the maturing eggs are close to 1 mm in diameter, and we saw earlier that fecundity in these two genera is low. Sulak found that mature eggs stripped from the second species sank in sea water samples with salinities adjusted to the range between 33‰ and 37‰. Since trawling surveys indicate that adults may be widely separated from each other, he argues that unless there are spawning aggregations or self-fertilization occurs '... the eggs must remain on the bottom for some length of time to ensure fertilization'. All chlorophthalmids are synchronous hermaphrodites (p. 214) but it is unlikely, as Sulak continues, that self-fertilization is their normal mode of reproduction. If it were, the populations of widely distributed species, such as *Bathypterois atricolor*, ought to be fragmented into clone-like local stocks, each with a distinct character complex. But there is no such evidence: even in tripod-fishes with very wide distributions there is comparable variety in their characters from one region to the next.

Where tripod-fishes are very common, for example in the Gulf of Panama and the Tongue of the Ocean off the Bahamas (p. 305), one might conjecture that self-fertilization is unnecessary. Where mature individuals are so thin on the ground that they may fail to meet, self-fertilization could be a last-resort means of maintaining local populations. That the chlorophthalmid fishes go to the extra expense of energy in developing testes in synchrony with ovaries almost forces one to this conclusion. Indeed, bathysaurids, which are predacious benthic fishes, and whose populations are thus likely to be still more thinly dispersed than, say, those of copepod-consuming tripod-fishes, are also synchronous hermaphrodites.

Reproductive arrangements are otherwise in other benthic deep-sea fishes. In the polar and temperate regions of both hemispheres the main forms are sea-snails (Liparidae) and eel-pouts (Zoarcidae), which have separate sexes and produce large eggs, entailing low fecundity. The type specimen of the sea-snail *Careproctus ovigerum*, caught at a depth of 2,900 metres, is a mature male bearing in its mouth a mass of eggs, each about 4·5 mm in diameter and containing well-developed embryos. The sea-snail-like appearance of the embryos suggests that this species is a mouth-brooder, and so may be *Careproctus kermadecensis*. In the female type specimen of this species, Nielsen (1964) found about 850 eggs, most of which are 1 mm in diameter. There are also a few large eggs (8 mm diameter), which may well indicate, as Nielsen says, that this species is also a mouth-brooder. Such large eggs would need to be brooded a few at a time. The deep-sea (lycodine) species of eel-pouts probably lay their eggs on the bottom (Mead et al., 1964).

Away from the deep-sea floor, synchronous hermaphroditism is unknown in benthopelagic fishes. They are more mobile than benthic species, and at comparable depths, judging from trawling records, they are generally more abundant. Moreover, in the two most diverse and often most commonly represented groups, the rat-tails and brotulids, the sexes are separate and sonically dimorphic: the males alone have drumming muscles on the swimbladder (p. 425). Except for a perianal light gland in *Chlorophthalmus* spp. (Somiya, 1977), the chlorophthalmids seem to be without means of signalling.

Moving up to the lower midwaters, the ceratioid anglers and black *Cyclothone*, the dominant groups of bathypelagic fishes are the most sexually dimorphic of all deep-sea fishes. The attraction of males by luminescent (p. 390) and pheromonic signals (p. 405) from the females, is integral to their highly adapted life-history patterns. In the two dominant mesopelagic fish groups, the stomiatoids and lantern-fishes, the sexes are also separate, and in both, particularly the latter, may be dimorphic in their luminescent systems (p. 367). But in the alepisauroid fishes (Mead et al., 1964) and the notosudids (Bertelsen et al., 1977), synchronous hermaphroditism is the rule. Apart from a few species in the first group, these are non-luminous fishes and without any other evident means of signalling. They are also medium-sized to large kinds of predatory fishes, and mature individuals, which are non-migrators and able to take large prey (p. 129), are not often caught. No doubt they are able to evade midwater nets, but it is true also that large predators are much less numerous than their prey. Mature individuals of a species may thus be dispersed in well-separated hunting grounds. As in the chlorophthalmids, it is thus conceivable that self-fertilization is a last-resort means of maintaining thinly distributed populations.

Relevant to the present discussion, Ghiselin (1969) has argued that hermaphroditism should be advantageous where it is hard to find a mate. Heath (1977) has considered simultaneous hermaphroditism from a cost and benefit standpoint. Though the energy cost in gonad development is higher in a pair of synchronous hermaphrodites than in a male and female of a gonochoristic species, Heath concludes that where reproductive contacts are few or production of ova is limited by physical constraints, the former reproductive means may be advantageous. Both factors may well apply in the deep sea. Even so, there is the obstinate fact that species of *Chlorophthalmus* are quite abundant (Mead et al., 1964). Indeed, these species are centred on the more productive upper regions of the slope, and as already stated, they have luminous means of communication. Further investigation of these exceptions should be illuminating.

Invertebrates

The most active benthic invertebrates are decapod crustaceans. After swimming in search of food or mates, avoiding enemies and so forth, they resume their benthic habit (p. 205). Many are large caridean prawns (e.g. of the genera *Acanthephyra* and *Parapandalus*), which no doubt have much the same kind of early life history as their midwater relatives, except that the young presumably start their free swimming existence near the bottom. Deep-sea species of the shrimp genus *Sclerocrangon* brood the embryos in large eggs attached under the abdomen of the female.

Some of the peracarid crustaceans, especially amphipods, are also active forms. In peracarids relatively few advanced young are produced from a small clutch of eggs brooded in the female's marsupium (Fig. 162). For instance, in 19 species of deep-sea isopods, mostly from slope levels,

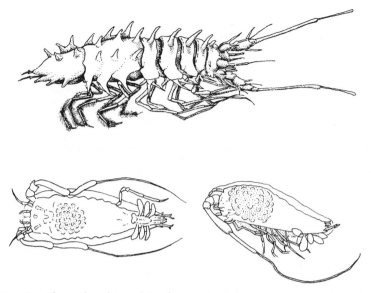

Fig. 162   Above, abyssal isopod *Storthyngura benti*; below, two embryos from the brood-pouch of a related species *S. novae-zelandiae*. (Redrawn from Wolff, 1962)

the number of eggs ranged from 2 to 80 (mostly 5 to 25) in females from 1·3 to 25·0 mm in length. There were more eggs in the larger individuals (Wolff, 1962). After an immature stage preceding gonad development, there are preparatory and brooding stages in the life cycle of isopods and amphipods. In the preparatory stage the oostegites (medial plates attached to thoracic limbs that form the brood chamber) are rudimentary but there is heavy yolk development in the ovaries: in the brooding stage the oostegites are fully developed and the eggs are released into the marsupium.

Reproductive cycles

Using the above three stages, Rokop (1977a) has investigated the reproductive condition of two isopods (*Eurycope californiensis* and *Ilyarachna profunda*) and one amphipod (*Harpiniopsis excavata*) from a series of samples taken (with otter trawl and epibenthic sled) at 13-week intervals (from October 1970 to October 1971), from the same station (32° 26·5′N, 117° 28·5′W) at a depth of 1,240 metres in the San Diego Trough off California. The two isopods are small (3–4 mm) sexually dimorphic species with long ambulatory and natatory appendages that are known only from depths between 450 to 1,300 metres off southern California. The amphipod has been recorded at mid-latitudes of the Atlantic and eastern Pacific Ocean at depths from 400 to 5,100 metres. In all three species the proportions of immature, preparatory and brooding stages in the samples shows clearly that breeding is all the year round. There are no signs of seasonal peaks of reproduction.

Rokop (1974) found also that most of the other common invertebrates in his samples reproduced throughout the year. Some, such as the two isopods and the amphipod considered above, reproduce asynchronously. The cycles of gamete formation in the individuals of a population are so out of phase with each other that at any one time a relatively constant proportion of the stock is in a breeding state. For instance, in the isopod *Eurycope californiensis* 33 to 55 per cent of the females were brooding over the period investigated. A similar mode of reproduction obtains in the brittle-stars *Ophiomusium lymani* and *Ophiacantha normani*. In the latter species at any one time some females contained small oocytes, others had medium-sized oocytes and in the rest they were large.

There are also continuous reproducers. All the adult individuals in a local stock are reproductively active throughout the year, but only a few eggs and a small quantity of sperm are released at any one spawning. Rokop found such reproduction in the bivalve molluscs *Nuculana pontonia* and *Nucula darella* and in the polychaete worm *Fauveliopsis glabra*. Analysis of the data on oocyte-size distribution or the stages of spermatogenesis in the individuals of a stock, indicates that in each individual gamete formation is continuous, with but partial spawning at any one time. In *Nuculana pontonia* this conjecture is supported by the constant addition of young to the stock. In all seasons small individuals with a shell length of less than 4 mm formed 15 to 24 per cent of the population.

Year-round reproduction is not entirely the rule in the San Diego Trough. The brachiopod *Frieleia halli* and the scaphopod mollusc *Cadulus californicus* are seasonal breeders (Rokop, 1977b). Egg formation in the brachiopod begins early in the atumn and leads each year to spawning between winter and spring: in the scaphopod oogenesis begins at much the same time (as in 1971) or even earlier (in 1970) and there was a spawning period (in 1971) between late summer and early autumn. The brachiopod produced less than 1,000 eggs, much less than the 8,000 to 60,000 eggs produced by shallow-water relatives of equivalent size. Since the eggs are also smaller than those of other species of articulate brachiopods Rokop suggests that they probably hatch into free-swimming larvae that feed near the bottom before settlement. The large eggs (up to 240 μm) of the scaphopod, which are considerably larger than those (ca. 160 μm) of relatives (*Dentalium* spp.) that produce planktonic larvae, are likely either to develop directly into juvenile stages or to hatch into larvae that exist on substantial reserves of yolk until they settle (lecithotrophic larvae).

In the deep sea there are no physical markers of the seasons. Near the bottom at the 1,200 metre station in the San Diego Trough the yearly fluctuations in temperature and salinity are only 0·3C° and 0·02‰. Periodicity in the food supply seems the only possible external factor that might be linked to seasonal spawning, but Rokop (1977) says that the annual cycle in surface productivity off San Diego is both modest in seasonal change and by no means regular in time from year to year. In the brachiopod *Frieleia halli*, on the other hand, the reproductive cycle was regular and predictable over 3 years. A more likely (and ultimate) cause of seasonal reproduction, as Rokop argues, may simply be to ensure maximum fertilization. The two seasonally spawning species are much less abundant than the year-round reproducers and are either sessile or of limited mobility. The population densities and mobility of the year-round species are presumably adequate for their non-epidemic mode of reproduction.

Though based on investigations in one place, Rokop's conclusion that most deep-sea invertebrates of the benthos reproduce throughout the year seems reasonable. Hartman and Fauchald's (1971) studies of the diverse small polychaete worms that form much of the benthic fauna in North Atlantic regions (between 100 and 5,000 metres) led them to suggest also that most species are year-round reproducers. Compared to their shallow-water relatives, invertebrates of the deep-sea benthos also have a strictly limited fecundity (p. 464), which must be enough to keep local population densities in a steady state (i.e. on the average, each individual in a population must be replaced once during its life history). Low fecundity and year-round spawning imply trickle recruitment of young to the stocks, which seems in keeping with the low carrying capacity of deep-sea environments. Moreover, at any one time and place, there are bound to be recruits from more than one species seeking living space on the bottom.

Early life and ecology

In continental shelf (and other land-fringing) waters, the larval life of most benthic invertebrates is bound to the life of the phytoplankton. Diverse species of small flagellates and diatoms ($< 50\,\mu$m) are the main food of the pelagic larvae produced by polychaete worms, molluscs, bryozoans, crustaceans, echinoderms and so forth (Thorson, 1946)). Thorson (1950) estimated that such planktotrophic kinds of larvae occur in about 70 per cent of the species of benthic invertebrates that live in shallow seas. The percentage is most (90–95) in subtropical and tropical regions, but in shelf waters of polar regions species with planktotrophic larvae are in a small minority: in the deep sea they are still rarer.

Thorson's (1950) review of the literature convinced him that pelagic larvae are virtually absent in the deep-sea benthos, but more recent discoveries have proved otherwise. Ockelmann's (1965) studies of bivalve molluscs in the northeast Atlantic are particularly important. He found that the type of larval life could be predicted if either the size of the ripe eggs or the larval shell (prodissoconch) was known. In species with planktotrophic larvae the diameter of the ripe eggs ranges from 48 to 85 μm, that of the larval shell from 50 to 150 μm. In species that produce lecithotrophic larvae, which live on their yolk during a short pelagic existence (p. 462), these two dimensions are 90 to 140 μm and 135 to 230 μm, respectively. Species with the largest ripe eggs (150 to 200 μm) have an embryonic shell of 230 to 500 μm and develop directly into juvenile stages. Ockelmann concluded that lecithotrophic development is dominant in deep-sea bivalves and that direct development is more a feature of shelf-dwelling species.

Using Ockelmann's criteria of larval types, Knudsen's (1970) studies support his conclusion. Out of 48 species of bathyal and abyssal bivalves (many are nuculoids), he judged 32 to have lecithotrophic larvae, 4 to have planktotrophic larvae and 12 to be direct developers. In 10 species of elasipod sea-cucumbers Hansen (1975) thinks that the eggs are large enough to indicate a lecithotrophic development. Indeed, in one group (the psychropotids) very large eggs (1·7–4·4 mm) seem to be correlated with a long pelagic life as juveniles.

Except for certain types (pp. 465–6), there is little direct knowledge of planktotrophic larvae in the deep-sea benthos. There are certainly species with such larvae that live from lower-shelf to upper-slope levels. The deep-sea red crab (*Geryon quinquidens*) from upper-slope depths, produces typical zoea and megalopa larvae (Perkins, 1973). Slope regions have also been colonized by the seaward-drifting larvae of shelf-dwelling invertebrates. Such settlements form sterile 'pseudopopulations' (Mileikovsky, 1971).

Larval stages of the wood-boring bivalve *Xylophaga* and the loricate larvae of *Priapulus atlanticus* were dominant among the invertebrates that had recolonized a box of azoic sediment left at 1,760 metres for 2 months (Grassle, 1977). If food is available the bivalves grow and reproduce quickly, and veliger larvae are produced that may live for considerable periods. The loricate larvae are presumably never pelagic, which will also be true of young emerging from directly developing eggs, as in porcellanasterid sea-stars (Madsen, 1961). Pogonophorans also produce large eggs and the resulting embryos are brooded in the tube until they become worm-shaped, though still bearing ciliated bands (Southward, 1975, and Fig. 163). The emerging young are light enough to be swept along by bottom currents (Southward, 1971).

The predominance in the deep-sea benthos of lecithotrophic larvae, all

presumably with yolk reserves, to last for some period of pelagic existence near the bottom, invites consideration. Many of the larvae probably
come from the denser clusters of adults and without dispersal there could
soon be overcrowding. As Scheltema (1972) says, even if the pelagic phase
lasts only one or two days the larvae may be transported for several kilometres. Measurements of current velocities over the slope and under the
Gulf Stream off the northeast United States, show that neutrally buoyant
larvae could be carried for distances of 2·6 to 38·0 kilometres in a day.
Thus, dispersal of the lecithotrophic larvae of deep-sea bivalves may well

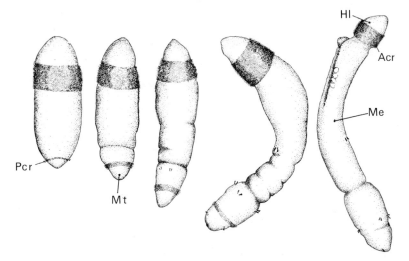

Fig. 163   Early stages in the life history of the pogonophore *Siboglinum fjordicum*. In
each may be seen a head lobe (Hl), an anterior ciliated ring (Acr), a mesosome
(Me), a posterior ciliated ring (Pcr) and a metasome (Mt). (Redrawn from
Webb)

have led to the wide distribution of many species (probably over 35 per
cent) on both sides of the Atlantic. Scheltema contrasts such broad distribution with the much more restricted distributions of the asellote isopods, which develop directly into juveniles and thus have no dispersal
phase. He thinks it likely that less than 5 per cent of asellote species have
distributions covering both sides of the Atlantic. In wider comparative
treatment, based on their epibenthic sled samples between New England and Bermuda, Sanders and Grassle (1971) include the brittle-stars
(ophiuroids), which have still broader distributions than the bivalves and
perhaps produce longer-lived pelagic larvae.

Long-lived pelagic larvae are certainly produced by some invertebrates
of the deep-sea benthos. In that they are large and planktonic they seem

to be the invertebrate counterparts of fish leptocephalus larvae. But, unlike leptocephali, (p. 451), they have means of feeding on microscopic plants and other small organisms. One of the most remarkable is *Planktosphaera pelagica*, which attains a diameter of at least 1 cm, has a complete gut, and is propelled by cilia disposed over elaborate looping tracts. *Planktosphaera* is a gigantic tornaria larva and may belong to one of the abyssal species of acorn-worms (Enteropneusta), such as *Glandiceps abyssicola* (Hadfield, 1975). Some of the sipunculid worms produce a *Pelagosphaera* larva, also reaching about a centimetre in diameter and with a complete gut (Fig. 60). There is a 'head', which can be extended or withdrawn, and which bears a girdle of large cilia (metatroch) used both in collecting food and in flotation (Åkesson, 1961). The dispersal of *Pelagosphaera* larvae, known from world-wide localities in warm and temperate waters, is presumably correlated to the very wide distribution of adult sipunculid worms (see Cutler, 1977). Studies of at least five species of *Pelagosphaera* larvae in the North Atlantic, which drifted in the major east–west current systems (for 250 days or more), suggest that transatlantic crossings are feasible (Rice, 1975).

Little is known of the early life of reptant decapod crustaceans in the deep sea. Of the true crabs (Brachyura), we have seen that *Geryon* has the usual larval stages, but direct development is possible in the deeper-living species. The life of the crayfish-like eryonids is less obscure simply because they produce large, long-lived larvae. Bernard's (1953) study of the Dana Collections led him to assign certain larvae to the two shallower-living genera, *Polycheles* and *Stereomastis*. A zoea stage is followed by an *Eryoneichus* larva, with a spiny, inflated carapace (Fig. 89). *Eryoneichus* stages, which reach 3 cm or more in length, were taken in midwater nets fishing between depths of 600 and 5,000 metres. In the gut of several larvae taken between 500 and 2,500 metres there were various microscopic plants (coccoliths, diatoms, dinoflagellates and silicoflagellates), radiolarians, foraminiferans and masses of siphonophore nematocysts. Evidently the larvae have all the tropic means for a long pelagic existence. Bernard found that as the larvae grew they moved into deeper waters, some reaching depths well below the levels of the adult environments.

Some of the inarticulate brachiopods have a relatively long, free-swimming larval phase. Just before the young settle, the valves are developing and they are feeding and moving with the aid of a well-formed lophophore (Rudwick, 1970). In the cosmopolitan deep-sea inarticulate, *Pelagodiscus atlanticus*, known from depths between 400 and 5,400 metres, the larval stage (Fig. 164) is still better adapted for a long pelagic existence. Like the adults, *Pelagodiscus* larvae, which bear a fringe of protecting setae, are known from localities in all three oceans.

It seems clear, then, that benthic deep-sea invertebrates with larvae

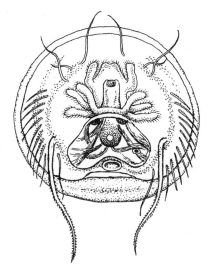

Fig. 164   Larval form of the brachiopod *Pelagodiscus atlanticus*.

specially adapted for a long pelagic existence have very wide distributions. But species producing these larvae represent a small proportion
of the entire fauna. Perhaps the same is true of the species that produce
small planktotrophic larvae, which are associated with high fecundity.
Larval mortality is very heavy in such forms. In the deep sea Thorson
(1950) felt that the ascent of planktotrophic larvae to the productive surface waters, followed later by the descent of advanced stages, was simply
not feasible. Certainly the great majority of species in the deep-sea benthos have become independent of phytoplanktonic food through the
evolution of lecithotrophic or direct development. It is clear also that the
phytoplankton can be exploited by large, specially adapted larvae.
Clearer still are the gaps in our knowledge.

SECTION IV
ASPECTS OF BIOGEOGRAPHY

Ecological biogeography 13

The aspects of biogeography that will concern us are largely ecological, and there are good reasons for the decision. The first is lack of space. Proper biogeographical surveys demand close knowledge of the biological units (taxonomy) and good coverage (sampling) of the relevant geographic areas. More than most biological disciplines, taxonomy is time-consuming: it requires years of experience in the critical handling of both biological collections and literature. Moreover, there are not many taxonomists, and still fewer able to deal with deep-sea collections. Deep-sea biological surveys are also time-consuming and expensive. There have been great strides in deep-sea exploration in the last 30 years, but even so, taxonomists pursuing the biogeographical implications of their work soon became aware of the gaps in oceanic (and taxonomic) coverage. But there are so many reports, dealing partly or entirely with deep-sea biogeography, that even a select review of those covering well-known organisms and well-explored regions would take much more space than is here available.

Conditions of life in the deep sea are also involved in the present ecological emphasis. Each species has a living space defined by a unique complex of physical and biological factors. Each of these factors may be seen as one of the dimensions of an n-dimensional hypervolume modelling the unique environment (ecological niche) of a species (Hutchinson, 1957). For instance, the ecological niche of a midwater species migrating up each day from an oxygen-minimum layer can be seen as described largely by five dimensions representing the ranges of oxygen tension, temperature, pressure, light intensity and food availability. Except for the light intensity (migrators tend to stay close to an isolume) the physical dimensions are extensive. Food availability, if represented by biomass, will also be a major dimension, for oxygen minimum layers occur in very productive regions. Indeed, the food dimension in the deep sea, whether eutrophic or oligotrophic, has a great influence on zoogeography. Hence earlier interest (pp. 219–311) and the present ecological emphasis.

Most deep-sea animals also 'live in a hypervolume' of stable dimensions. The headquarters of deep-sea animals, as stressed before, are below the warm ocean, or more exactly, their fertile distribution is between the subtropical convergences of the two hemispheres. Thus, food supplies, even if small, are more or less constant throughout the year. Except for the full-scale migrators of the mesopelagic zone—and they move from one stable environment to another—the adults also live in a virtually unchanging physical environment. The early life of most deep-sea animals is also passed in such surroundings.

The physical factors are so constant, particularly at the deeper levels, that deep-sea zoogeographers almost expect that many of their animals will have very wide distributions. One part of the ocean should be as good as another (so long as there are means of dispersal). Diverse species, whether midwater or bottom-dwelling, certainly have populations in two or more oceans, but the representation of any one species varies considerably from place to place. In their endeavour to elucidate the quantitative as well as the qualitative aspects of distribution, biogeographers are bound to be especially concerned with ecological factors.

Again, considering the constant conditions of life, there were also expectations that there would be relatively little genetic variability in deep-sea animals. But investigations so far have proved otherwise (pp. 499–501). On a larger scale, there is also very considerable taxonomically related diversity. More precisely, for a given number of individuals sampled in a particular way there are more species in the deep-sea benthos than in comparable communities of shallow waters. Supporters of the stability-time hypothesis argue, inter alia, that conditions of life in deep-sea headquarters have been stable ('predictable') for so long that high diversity was bound to evolve (see also p. 145).

Study of latitudinal gradients in species diversity is one of the classical aspects of zoogeography. As expressed by Fischer (1960), 'The diversity of biotas, on land and in the sea, is greatest in climates of high and constant temperatures, such as are found over much of the tropics, and decreases progressively in the fluctuating and cold climates normally associated with high latitudes.' But, as I say elsewhere (Marshall, 1963), '... in expressing this rule Fischer was simply concerned with the neritic province of the ocean. To adapt this rule to include the deep ocean, the first part would read, "The diversity of biotas on land and in the ocean, is greatest in or *under* climates of relatively high and constant temperatures etc. ...." (Here I follow Fischer and others in thinking it reasonable to speak of hydrospheric, as well as atmospheric, climate.)'

There are also studies concerned to trace how patterns of distribution are related to the disposition of water masses, thermal fronts, and so forth. Some mesopelagic animals are confined within a water mass, but others cross boundaries and still others live only in near-continental waters (p. 484). Trophic factors are most clearly evident in the central water masses (p. 478). At bathypelagic levels, at least for the more complex animals, the trophic dimension is paramount (p. 487).

Since the time of the *Challenger* Expedition (1872–76) zoogeographers have been concerned to trace the zonation of the benthic deep-sea fauna. Until about 30 years ago, these studies involved only the larger benthic invertebrates (megafauna), and there is considerable agreement that on the largest scale there are three faunas: bathyal (to 2,000 metres), abyssal (2,000–6,000 metres) and hadal (> 6,000 metres). Now there is also the minifauna (US 'macrofauna') and the meiofauna to be considered. Certainly these two parts of the faunal size-spectrum each contain many more species than the largest and in both these there is also a more gradual decline in species diversity with depth than in the megafauna. How far, then, do the terms bathyal, abyssal and hadal have meaning in the depth ranges of invertebrates smaller than a few millimetres? There may well be zonation but this will be of a different order. Concerning the megafauna, there are indications, as stated already that ecological factors, such as trophic and competitive interaction, are involved in gradients of zonation. But we shall be concerned mostly with the relations between large-scale patterns of distribution and the trophic dimensions of benthic 'hypervolumes'.

Pelagic distributions

Oceanic diatoms

Consideration of outstanding types of diatom distribution in the Atlantic Ocean leads well, both by comparison and contrast, to similar treatment of mesopelagic fishes. Though species in a diatom community may come and go rather quickly, the dominant members form certain species assemblages, each associated with individual oceanic features. These key assemblages are revealed after analysis of diatom remains in samples of deep-sea sediment. Considerable quantities of such remnants reach the bottom in the faeces of herbivorous copepods etc., and largely because this fall is relatively rapid, distributions of key species on the deep-sea floor are a sedimentary reflection, as it were, of distributions directly above in the euphotic zone (see also p. 288). Both the abundance and distributions of such species also reflect the patterns of primary productivity and nutrient content of the surface waters. Using Q-mode factor analysis to process the abundance of 42 species in 37 core-sample tops, Maynard (1976) found six types of diatom species assemblages in the Atlantic Ocean.

A subantarctic assemblage, with northern bounds near the Subantarctic Convergence and dominated largely by *Fragilariopsis kerguelensis*, is the diatomaceous counterpart of a main type of mesopelagic faunal distribution, represented later by fishes (Fig. 165). Certain mesopelagic fishes, like the gyre-margin assemblage of diatoms (*Pseudoeunotia doliolus* etc.), flourish in waters peripheral to the great gyres (p. 483). A warm-water diatom assemblage, dominated by *Coscinodiscus nodulifer*, and described by Maynard as '... abundant in the Equatorial and Benguela Divergence Zones and off the southeast coast of Africa', corresponds largely with the Amazonian and Guinean provinces of mesopelagic fishes (Backus and Craddock, 1977). An assemblage of subtropical species in the central North Atlantic (between 53° and 44°N) more or less corresponds with Backus and Craddock's Northern Gyre Province. These authors' Atlantic Subarctic Province and the northerly half of their Azores–Britain Province covers what Maynard calls a subpolar assemblage (dominated by *Thalassiosira gravida*).

Diatoms are indicators of the richer oceanic waters, which overly the population centres of numerous mesopelagic fishes (and invertebrates). Even so, the degree of correspondence between the distribution of diatom assemblages and that of mesopelagic fishes is striking. There is also a negative correlation. In Maynard's charts of factors it is clear that stations under the northern and southern Atlantic gyres sampled few or no diatom remains. But such poorly producing central water masses contain mesopelagic fish species that tend to be smaller, or even dwarfed, by comparison with their relatives in richer waters (p. 480). In turning to mesopelagic fishes one must thus remember that the trophic dimension has a considerable influence on their patterns of distribution.

Mesopelagic fishes

During the last 15 years R. H. Backus and his colleagues at the Woods Hole Oceanographic Institution have used a 10-foot (3-metre) Isaacs–Kidd midwater trawl over much of the Atlantic Ocean in order to study the distribution patterns of mesopelagic fishes (see Backus et al., 1970; Backus and Craddock, 1977). During daylight the net was generally lowered to the maximal depth of sound scattering (in a 'deep scattering layer'), as determined by a 12 kHz echo-sounder. By night, when most of the fishes have risen to upper regions where there are marked changes of temperature with depth, both the echo-sounder and a bathythermograph were used to determine the depth of trawling. 'Experience has shown that sound-scattering maxima, temperature inversions and the bottom of thermoclines and surface isothermal layers mark planes of concentration of midwater fish' (Backus et al., 1970). As a result of such fishing methods and wide coverage of the Atlantic, they have not only

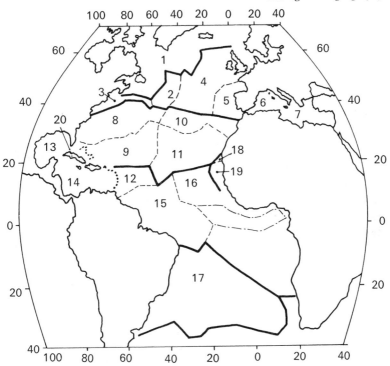

Fig. 165   Chart of Atlantic Ocean pelagic faunal provinces. 1, Atlantic Subarctic
Province; 2, Northern Gyre; 3, Slope Water; 4, Azores–Britain Province;
5, Mediterranean outflow; 6, Western Mediterranean Sea; 7, Eastern
Mediterranean Sea; 8, Northern Sargasso Sea; 9, Southern Sargasso Sea;
10, Northern North African Subtropical Sea; 11, Southern North African
Subtropical Sea; 12, Lesser Antillian Province; 13, Gulf of Mexico;
14, Caribbean Sea; 15, Amazonian Province; 16, Guinean Province;
17, South Atlantic Subtropical Sea; 18, Northern Mauritanian Upwelling;
19, Southern Mauritanean Upwelling; 20, Straits of Florida.
Provinces 2–7 form the North Atlantic Temperate Region, 8–11 the North
Atlantic Subtropical Region, 18–19, the Mauritanian Upwelling Region and
12, 14, 15, 16 and 20 the Atlantic Tropical Region. Provinces 1, 13 and 17
are also classed as regions. The heavier boundaries are regional ones, the
thinner ones provinces. (After Backus and Craddock, 1977)

delimited pelagic faunal provinces but also determined their characteristic sound-scattering levels. More precisely, passage from province to
province is a transition between individual assemblages of mesopelagic
fishes and reverberation levels (see Fig. 165).

   In their coverage of the Atlantic, the Woods Hole investigators distinguish between subarctic, temperate, subtropical and tropical regions.

(The Mauritanian upwelling off northwest Africa and the Gulf of Mexico are also regarded as regions.) Apart from the last, each region is divided into two or more provinces. Consideration at the provincial level is beyond our present scope, which, in keeping with earlier treatment of diatom assemblages, is concerned with regional distributions. Backus and Craddock (1977) use the lantern-fishes as regional indicators. Fifty-four species out of 75 sampled altogether live in the subtropical and tropical belts and are subdivided thus: 18 in the Tropical Region, 13 in the Subtropical Region, 18 in both the foregoing regions and 5, more narrowly, in the Tropical Region and the equatorward halves of the north and south Subtropical Regions (tropical–semisubtropical species).

The Atlantic Tropical Region (consisting of Lesser Antillean, Amazonian and Guinean Provinces and the Caribbean Sea and the Straits of Florida, see Fig. 165) is spanned by the thermal equator and is characterized by very warm surface waters (mostly 22°–29°C) overlying a relatively shallow thermocline. Eighteen species of lantern-fishes, as stated above, live in this region, but certain, such as *Diaphus dumerili*, are swept northward in the Gulf Stream to the northern Sargasso Sea and beyond (see Fig. 165). The North and South Atlantic Subtropical Regions are each largely circumscribed by a great mid-ocean gyre. The surface waters are a few degrees cooler than those of the tropical region but relatively warm water penetrates deeper, so that at depths from about 200 to 1,000 metres the ocean cools both equatorward and poleward from the subtropics. Thirteen of the 75 Atlantic lantern-fishes sampled live in the North Atlantic Subtropical Region: that is, they occur principally in the Sargasso and North African Subtropical Seas (see Fig. 165). Nearly all live also in the South Atlantic Subtropical Region but not in the intervening Tropical Region. These species, which Backus and Craddock call 'bipolar' or 'biantitropical' (e.g. *Lepidophanes gaussi*), are a cogent reminder that the distributions of many mesopelagic animals are far from continuous. But 18 species of lantern-fishes, such as *Ceratoscopelus warmingi*, live equally well in tropical and subtropical regions of the Atlantic. Five, as already stated, are more narrowly tropical–semisubtropical: they live, as Backus and Craddock say, in the area of the northeast and southeast trade winds and intervening doldrums. Much smaller still is the Mauritanian Region, characterized by periods of upwelling (see p. 65). Here live one, perhaps two, endemic species of lantern-fishes.

'The North Atlantic Temperate Region (consisting of Slope Water, Northern Gyre, Azores–Britain Province, Mediterranean Outflow and Western and Eastern Mediterranean Seas) stretches from the northern edge of the North Atlantic's subtropical sea to the polar front. The region is bounded approximately by 9° and 15° isotherms at 200 metres. Although there is considerable variation from province to province, the temperate region is generally more productive than the tropics and subtropics (Ryther, 1963). Related to this superior productivity is the region's greener colour and greater turbidity. Seasonal effects are stronger in the temperate North Atlantic than in the tropics and subtropics. In some respects, the Eastern Mediterranean sea verges on the subtropical; nevertheless, its fauna is clearly temperate' (Backus and Craddock, 1977; see also Fig. 165). A few kinds of lantern-fishes are restricted to the North Temperate Region, but most of the eighteen species live also in northern subtropical waters (e.g. *Ceratoscopelus maderensis*) or in the Subarctic Region (e.g. *Benthosema glaciale*). Again, some species (8) are 'bipolar', that is, they live also at comparable latitudes in the South Atlantic Ocean.

Lastly, the Atlantic Subarctic Region lies across the Polar Front from the North Atlantic Temperate Region. 'The temperature gradient across the front is strong except in the east where the front becomes ill-defined, then disappears. The Subarctic Region is subject to great seasonal change, the most significant part of which, no doubt, is that in solar radiation' (Backus and Craddock, 1977). This region is poor in species; indeed, there are no endemic lantern-fishes.

The most southerly reach of the Woods Hole investigations is at about the latitude of the La Plata River, South America. Distributions in more southerly regions of the Atlantic have been recently investigated by the Russians (Parin et al., 1974). Actually the stations were in the southwestern quarter of the Atlantic from about the Equator to latitude 60°S. Considering only the lantern-fishes, there are a number of species (called peripheral–central) that live just north of the Subtropical Convergence in a southerly belt of the Central Water Mass. For instance, *Gonichthys barnesi* has a circumglobal distribution between latitudes 30° and 40°S. Further south there is a subantarctic assemblage between the Subtropical and Antarctic Convergences. Parin and his colleagues list about 20 species, of which most, such as *Protomyctophum tenisoni* and *Symbolophorus boops*, are endemic to the region. South of the Antarctic Convergence there are 5 or more endemic species of lantern-fishes (e.g. *Electrona antarctica* and *Gymnoscopelus braueri*), which contrasts with the absence of such species in waters north of the northern Polar Front. Distinctions between antarctic, subantarctic and peripheral–central assemblages become blurred off the South American coast where the Falklands Current may carry members of each assemblage northward of their headquarters.

This zoogeographic sketch of the Atlantic has involved over 120 species of lantern-fishes, many of which live elsewhere. To consider only the most widely distributed species, one has only to consult a recent report on the lantern-fishes of the eastern Pacific Ocean (Wisner, 1974) to realize that about half the Atlantic warm-water species and more than three-quarters of those from cooler southern Atlantic waters are more or less circumglobal in their respective zones of the ocean. Evidently, the circumglobal spread of species is easier in southerly waters, where there are no continental interruptions. And what is true of lantern-fishes applies also to other common groups of mesopelagic animals.

Returning to the Atlantic, Backus and Craddock (1977) emphasize that their zoogeographic scheme applies to mesopelagic fishes in general. The Russian investigations (and earlier ones) support them. The two authors conclude, '... the literature of Atlantic biogeography suggests that the scheme applies to mesopelagic animals as a whole and probably to epipelagic plants and animals as well'. The references to plants is surely appropriate when we recall the opening section on distribution patterns in assemblages of diatoms (p. 473).

The Water masses and distributions of mesopelagic animals

It should be clear that the distributions of some lantern-fishes cross the boundaries of water masses: at the other extreme, there are species with headquarters within a particular water mass. For instance, widely distributed species live in both subtropical and tropical waters. 'Specialist' species with narrower distributions will be largely our present concern.

In the present context little more needs to be said of subantarctic and antarctic species. Their distributions are evidently limited by the sharp thermal front at the Antarctic Convergence and the broader temperature discontinuity at the Subtropical Convergence. They live below the West Wind Drift, which presumably accounts for their marked tendency, as already noted, to be circumpolar in distribution. Just how their life-history strategies are related to their conveyor-belt environment remains to be investigated.

Our concern centres first on species that live in part or most of the Central Water Masses of subtropical regions (Fig. 166). Regarding those with restricted distributions, there are species in the South Atlantic, as we saw, that live north of the Subtropical Convergence in a southerly belt of central water. The southerly limit is the convergence, but what determines the northern bounds? How do these species manage to stay within the southerly sweep of the great South Atlantic Gyre? More is known of conditions in the Sargasso Sea, which forms the westerly half of the North Atlantic Central Water Mass.

The Sargasso Sea, with a perimeter defined by certain temperature–salinity characteristics, is divided into northern and southern provinces, which are separated by a thermal front (see Fig. 165). In the upper few hundred metres the northern part is cooler, less stable and more productive than the southern part. Each province holds an individual assemblage of mesopelagic fishes (Backus et al., 1969).

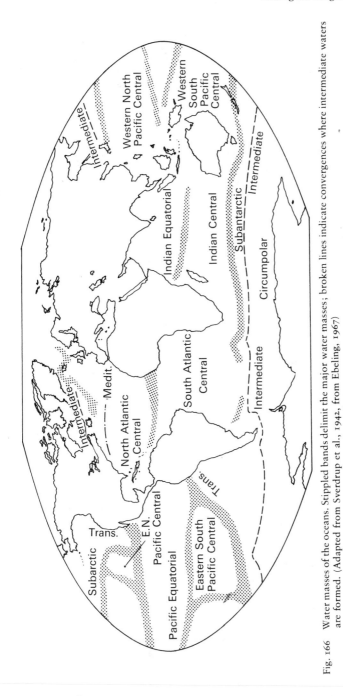

Fig. 166   Water masses of the oceans. Stippled bands delimit the major water masses; broken lines indicate convergences where intermediate waters are formed. (Adapted from Sverdrup et al., 1942, from Ebeling, 1967)

Off Bermuda, which is in the northern province of the Sargasso Sea, there are likely to be more than 200 species of mesopelagic fishes. Many (59) are lantern-fishes, nine of which are numerically dominant. Seven of these dominant species, such as *Lobianchia dofleini* and *Notolychnus valdiviae*, are small or diminutive (Gibbs et al., 1971). Here and else-

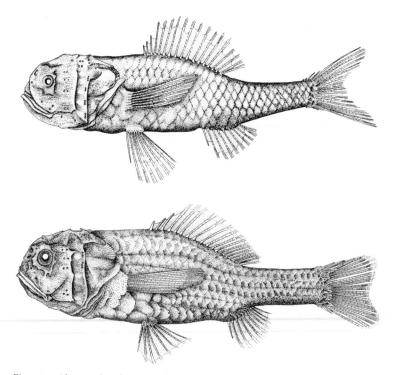

Fig. 167   Above a dwarf species of *Melamphaes*, *M. pumilus* (length about 25 mm); below, a large species *M. suborbitalis* (length about 95 mm). (Redrawn from Ebeling and Weed, 1962)

where in central waters, there is a distinct tendency for resident mesopelagic fishes to be smaller than their relatives in waters of greater productivity.

This tendency was first realized by A. W. Ebeling, who has integrated his findings in a paper on the zoogeography of tropical deep-sea animals (Ebeling, 1967). Ebeling's special studies concern the systematics and zoogeography of the Melamphaidae, a family of xenoberycoid fishes found largely at mesopelagic levels over most of the ocean. The largest genus *Melamphaes* consists of about 20 species and 4, such as *M. simus* and

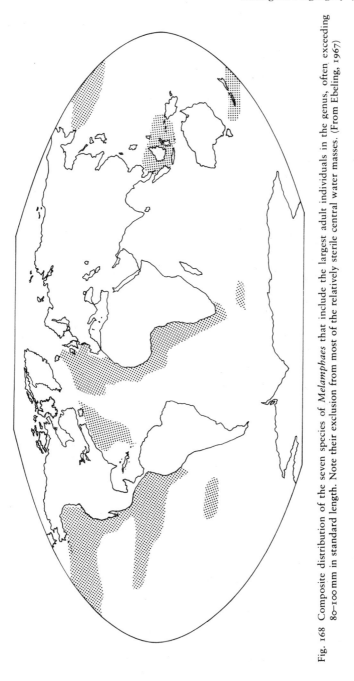

Fig. 168 Composite distribution of the seven species of *Melamphaes* that include the largest adult individuals in the genus, often exceeding 80–100 mm in standard length. Note their exclusion from most of the relatively sterile central water masses. (From Ebeling, 1967)

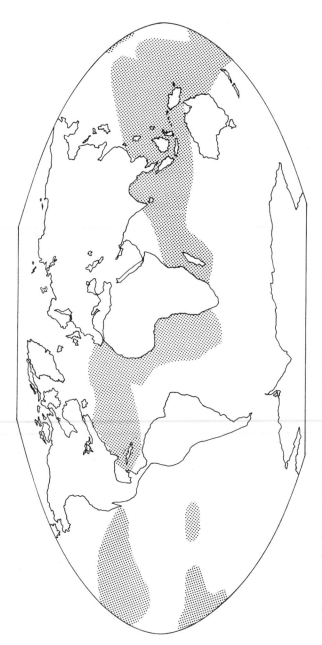

Fig. 169 Composite distribution of the four species of *Melamphaes* comprising a group of 'dwarf' individuals less than 28 mm in standard length. (From Ebeling, 1967)

M. pumilis, whose adults seldom exceed 25 mm in standard length, are dwarfs (Fig. 167). Adults of other species are mature at lengths beyond 70 mm (Fig. 167). *Melamphaes pumilis* is confined to the western North Atlantic, predominantly to the northern province of the Sargasso Sea. In the North Atlantic *M. simus* is restricted largely to peripheral regions of eastern central waters, but it lives also, again almost entirely, in the Central Water Masses of the North Pacific and Indo-Pacific regions. The large species of *Melamphaes* are centred in more productive waters, both peripheral to the gyres, especially in equatorial regions, and in temperate regions (Fig. 168). Evidently, the dwarf species are suited to conditions of life below the clear, deep-blue 'deserts' of the subtropics (Fig. 169).

Like the lantern-fishes, melamphaids depend largely on zooplanktonic food, but the latter fishes, not like the former, being inveterate migrators in search of food in upper and richer waters, may well display more sharply the effects of contrasting trophic conditions. Even so, among the species of *Chauliodus* (viper-fishes), which are migrators, there are dwarfed forms in central waters. In the Atlantic there are two small, central water mass species, *C. danae* and *C. minimus*, the first in the North, the second in the South Atlantic. Both reach a length of 150 mm, whereas *C. sloani*, a species from richer peripheral waters, grows to 350 mm. Actually, *C. sloani* lives in all three oceans on the periphery of the central gyres: only in the Indian Ocean and Southwest Pacific does it occur in some parts of the Central Water Masses. But, as Parin and Novikova (1974) also say, *C. sloani* may well represent a species complex. In the present context it must be added that species of *Chauliodus*, being predatory fishes, tend to feed at higher, less productive levels of the food pyramid that melamphaids and lantern-fishes. Even though migratory, the dwarfed species seem to have 'felt the pinch' of the Central Water Masses.

Through study of 'dwarf' residents in the Central Water Masses, one is thus led to consider species that are centred in more productive waters. With such aspects of Ebeling's work in mind, Cohen (1973) has analysed the Indian Ocean midwater fish fauna to see how many species are restricted to each of the three main Equatorial, Central and Subantarctic Water Masses. Of 50 species of lantern-fishes, most of which occur elsewhere, 18 appear to be centred in the Equatorial Water Mass, 6 in the Central Water Mass and 5 in the Subantarctic Water Mass. Species with wider distributions are subdivided thus: 2 range from the Arabian Sea to the subantarctic; 8 live in the Equatorial and Central Water Masses and 11 in central and subantarctic regions. Cohen estimates that about 42 per cent of lantern-fish species probably transcend water mass boundaries.

Near-continental distributions

During this survey of distributions in certain groups of mesopelagic fishes we have considered species that are centred in waters peripheral to the great subtropical gyres. There are also species with narrower peripheral distributions in near-continental waters over the upper continental slope. For instance, Baird (1971) remarks that species of the hatchet-fish genus *Polyipnus* (Fig. 43), which occur from warm temperate to tropical regions, live close to land nearly everywhere. Evidently, these species and others, such as certain notosudid fishes, have spawning grounds near the continents and oceanic islands (Bertelsen et al., 1976). Near-continental waters may also contain the right kinds and quantities of small zooplankters needed by small-mouthed mesopelagic fishes such as *Polyipnus* spp. and certain lantern-fishes (Baird et al., 1975). Closer investigations of these most 'specialized' members of the mesopelagic fauna should be well worth while.

Oxygen minimum layers and mesopelagic distributions

As we have seen (p. 33), oxygen concentrations are minimal in the upper midwaters of the ocean. In the northern Indian Ocean and the eastern tropical Pacific, both regions of high productivity, midwater values of oxygen content are low enough to have profound effects on the distribution of midwater fishes. Indeed, as minimal values of oxgen concentration are 0·5 millilitre per litre or even less, it is remarkable that active pelagic animals, at least those of the larger and more complex kinds, are able to exist at all in water so deprived of dissolved oxygen. But Kanwisher and Ebeling (1957) caught numerous midwater fishes, particularly hatchet-fishes, in an oxygen minimum layer in the eastern tropical Pacific where tensions were less than 0·25 ml/l. Moreover, many fishes had a swimbladder containing almost pure oxygen. At midwater levels in the Arabian Sea, where oxygen values were below 0·08 ml/l, there are substantial numbers of gonostomatid and myctophid fishes (see also pp. 485–6).

Concerning the larger and more complex crustaceans, experiments have shown that euphausiids and the large mysid, *Gnathophausia ingens*, are very efficient in extracting and using the little oxygen available in solution. Off southern California the mysid lives in midwaters with oxygen tensions as low as 0·20 ml/l. Live individuals from this region, as Belman and Childress (1976) found, are very effective in three respects: they are so active that their gills, which have a relatively large surface area to absorb oxygen, are bathed by about eight body volumes of water per minute; they consume oxygen at a relatively low rate (0·034 ml/g/hr), and they are efficient (50 to 80%) in the utilization of oxygen. At the large copepod level, Childress (1971) found that *Gaussia princeps*, which may also be caught in the oxygen minimum layer off Southern California, can tolerate anaerobic conditions for about 12 hours and also consumes oxygen at much lower rates than do shallow-water forms. It is an active migrator, and the oxygen debt accumulated during its day-time stay in the minimum layer can be burned off, as Childress says, when it migrates up to waters of higher oxygen content.

Regarding the fishes that live in the same oxygen minimum layer, meta-bolism in the large berycoid *Anoplogaster cornuta* (fang-tooth) has been investigated by Gordon and his colleagues (1976) (Fig. 170). It consumed oxygen slowly down to the smallest tensions they could measure with

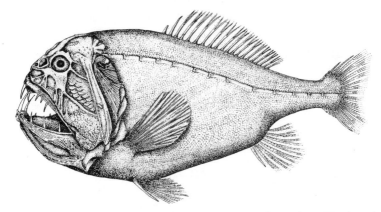

Fig. 170   The fang-tooth fish *Anoplogaster cornuta*. (Redrawn from Woods and Sonoda, 1973)

their equipment. When the fish is in the minimum layer, the authors suggest that it may either stop moving or may even be capable of anaerobic metabolism (two of their fishes lived for short periods in the respirometer after the oxygen levels had become immeasurably low and survived after transfer to air-saturated sea water).

Thus, various mesopelagic animals, ranging from copepods to fishes, are able to live in oxygen minimum layers. But many others, judging from their distributions, do not have this facility. In the Indian Ocean there is a hydrochemical front south of the Equator (at about 10°S) that separ-ates the northerly Monsoon Gyre, bearing low quantities of oxygen and high nutrient concentrations, from the well-oxygenated, low-nutrient-containing waters of the Subtropical Gyre. The front slopes northwards and downwards from about 100 metres such that at midwater levels in the Arabian Sea and Bay of Bengal the oxygen content at the minimal depth is 0·5 ml/l or less (Wyrtki, 1973). In the Arabian Sea Wyrtki attributes the oxygen minimum layer to the isolation and stagnation of North Indian Intermediate Water coupled with the high productivity in the northern Indian Ocean. Here north of 3°N, an oxygen minimum layer (< 1·0 ml/l) extends from a depth above 200 metres to 1,200 metres or more.

The marked influence of the oxygen minimum layer on the mesopelagic fish fauna of the Arabian Sea is well summarized by Goodyear and Gibbs (1969), 'Judging from data on stomiatoids, very few mesopelagic fishes have been able to adapt to conditions in the Arabian Sea. Perhaps six species in the families Chauliodontidae, Stomiatidae, Melanostomiatidae and Malacosteidae occur there, and only one of these, *Stomias affinis*, has not speciated to some degree. By contrast, approximately 50 species in these same families are found in the adjacent Equatorial or Central Waters of the Indian Ocean.' In their special study on stomiatoid fishes of the *Astronesthes cyaneus* species group, Goodyear and Gibbs found that in *A. lamellosus*, which lives in oxygen-poor waters of the Arabian Sea and Bay of Bengal, the gill filaments are about three times as long as in three other species from well-oxygenated waters. Evidently, the long gill filaments are correlated with life in an oxygen minimum layer. The same adaptation is found in the viper-fish *Chauliodus pammelas*.

Turning now to the warm eastern Pacific Ocean, it seems that lantern-fish distributions are also disturbed by oxygen-poor water. Wisner (1974) draws attention to the absence of certain species in the region between Baja California and central Chile. Species such as *Diogenichthys atlanticus* (but not *D. laternatus*), and four species of *Centrobranchus* were not taken in this area, where there is an oxygen minimum layer (1·0 ml O_2/l or less), but they do occur elsewhere in the eastern Pacific in well-oxygenated waters. Wisner is inclined to correlate these and similar gaps in lantern-fish distribution to the presence of oxygen-poor water, which seems reasonable, particularly in view of parallel conditions in the Indian Ocean.

Distribution patterns of the larger bathypelagic animals

In the bathypelagic zone the trophic dimension seems to be the dominant mediator of distribution in organisms ranging in size from large euphausiids to fishes. Earlier, I argued that the food-poor waters of this zone have elicited marked simplifications in the organization of the resident fishes. In the present context we can best appreciate the trophic influence on distribution by tracing the large-scale changes of zoogeographic disposition in a descent from epipelagic to bathypelagic levels in the Pacific Ocean.

Patterns of distribution in epipelagic waters have been derived by

McGowan (1972) from studies of organisms that include euphausiids, chaetognaths, pteropods, foraminifera, some genera of copepods, ommastrephid squid, tuna and salmon. There are rather few basic patterns, which are: subarctic and subantarctic, north and south transition zones (peripheral–central), north and south central, equatorial, eastern tropical Pacific and warm-water cosmopolites. There is considerable overlap between these faunal regions, but by means of refining analysis McGowan has shown that in each region there is a central 'core' of typical species. As in the Atlantic, there are species with discontinuous 'antitropical' distributions.

In the epipelagic Pacific the above patterns first became clear in Brinton's (1962) report on the distribution of the euphausiid shrimps. At mesopelagic levels he subdivides 13 species into those with subarctic (1), central (4), central–equatorial (7) and cosmopolitan (1) distributions. Regarding the seven species that live in both central and equatorial waters, he remarks that they are lacking in the belt between latitudes 10°N and 20°N where there is a pronounced oxygen minimum layer. If we recall the transition-zone (peripheral–central) species of mesopelagic fishes (p. 477), it looks as though major patterns of distribution at mesopelagic levels are much as those in the epipelagic zone. Indeed, McGowan (1972) states that the distributions of Pacific lantern-fishes and sergestid prawns (and diatoms) tend to fit epipelagic patterns. Furthermore, the major faunal regions in the mesopelagic Atlantic have been subdivided into provinces (Fig. 165).

Though the bathypelagic zone is less well explored than the mesopelagic, it has been covered well enough to reveal major features of distribution (if any). Concerning the giant bathypelagic euphausiids in the Pacific (*Thysanopoda cornuta*, *T. egregia*, *T. spinicaudata* and *Bentheuphausia amblyops*) Brinton (1962) states that they were most often caught where there was most sampling. But he continues thus, 'Conditions characterized by cooled water and enrichment prevail in most of the regions where the giant euphausiids have been found: (1) the northeastern Pacific near 40°N and the outer margins of the California and Peru Currents where warm water masses impinge upon colder currents, and (2) near the equator or at the margins of the Equatorial Countercurrent— places where upwelling and thermal anticlines are found in the open ocean (Cromwell, 1953).' Thus, bathypelagic euphausiids tend to be concentrated under the more productive (eutrophic) regions of the ocean.

But there are zoogeographic patterns at bathypelagic levels, as may be seen in Mukhacheva's (1966 and 1974) reports on the distribution of *Cyclothone* species. Considering only the bathypelagic forms, four (*acclinidens*, *pallida*, *microdon* and *obscura*) live in parts of all three oceans, while two (*livida* and *pygmaea*) have a restricted distribution. Catch records of C. *acclinidens* fall between latitudes 40°N and 40°S in the subtropical and tropical belts. But it is most abundant in eutrophic regions, as in the Arabian Sea, Bay of Bengal, equatorial Indian Ocean, and off California and Panama in the eastern Pacific. *Cyclothone pallida* has much the same broad distribution: *C. microdon* has a wider distribution in the Atlantic Ocean (subarctic to antarctic) and a narrow one in the Indian and Pacific Oceans (20°S to antarctic waters). Both species are also most abundant in eutrophic regions. *Cyclothone obscura* is largely an equatorial species, distributed in the Atlantic Ocean between latitudes 20°N and 20°S, in the Indian Ocean between the Equator and 20°S and in the Arabian Sea and Bay of Bengal, in the Pacific between 10°N and 10°S. It is thus largely confined to eutrophic regions, which is interesting when one considers that it is the deepest-living species of *Cyclothone* (centred below 1,500 metres). At this level, where non-biological conditions are virtually constant everywhere, one part of the ocean seems just as physically suitable as another. But, as *C. obscura* lives furthest below the productive euphotic zone and is relatively large (maximum length 66 mm), it looks as though this species has come to stay under equatorial eutrophic regions in order to obtain enough food. The other three-ocean species, whose populations are centred at higher levels, are not so narrowly confined: they live under eutrophic and oligotrophic waters but are most abundant under the former.

Of the two restricted species, *Cyclothone livida* is largely confined to eutrophic waters off the Africa coast between northwest regions and the Gulf of Guinea. *Cyclothone pygmaea*, a dwarf species, lives in the Mediterranean, largely an oligotrophic sea. Its small size (25 mm) seems in keeping with its impoverished environment (see also p. 262). Even so, like its larger, widely distributed relatives, it flourishes best under richer waters, as in northern parts of the Western Mediterranean and in the Alboran Sea (see Jespersen and Tåning, 1926).

The dominant bathypelagic fishes in number of species (ca. 100) are the ceratioid anglers. About half belong to the family Oneirodidae, which Pietsch (1974) has revised. More recently, Bertelsen and Pietsch (1977) have studied Atlantic material collected by the F.R.V. *Walther Herwig*. While the further exploration reported in the latter paper has extended the range of certain species, it seems that most are restricted to warm regions of two or one of the three oceans. One species only, *Oneirodes eschrichti* (Fig. 103), is almost cosmopolitan. Concerning the better-known species of *Oneirodes*, Pietsch concludes that they live in eutrophic regions '... only rarely occurring in waters of low productivity'. This is well shown on the distribution chart of *O. eschrichti*, where nearly all records come from eutrophic waters.

The commonest species of deep-sea angler-fishes, *Cryptopsaras couesi* and *Melanocetus johnsoni*, appear to be centred at higher levels (800–1500 metres) than their oneirodid relatives. In the North Atlantic, the distribution of both, as judged by catches of larval stages, covers both western and eastern halves. To analyse their distribution, Bertelsen (1951) considers the eastern North Atlantic on its own but divides the western half into northwestern and southeastern areas, which correspond largely and respectively with the north and south provinces of the Sargasso Sea as defined by Backus and Craddock (1977) (Fig. 165). The larvae of *Cryptosarus couesi* are much commoner in the western North Atlantic, the percentage catches in the northwestern, southwestern and eastern areas, being 55, 35 and 10 respectively. The corresponding percentage catches of *Melanocetus johnsoni* are 12, 15 and 68, showing it to be concentrated in the eastern North Atlantic (small numbers of both species occur in the Caribbean Sea). The first species is thus centred in the western North Atlantic, where it is evidently commoner in the more productive northwestern part of this sector (see also p. 478). Catches of the easterly *Melanocetus johnsoni* also come largely from the more productive regions (see Bertelsen, 1951). Here then, are two species with opposed concentrations in the warm waters of the North Atlantic, both no doubt related in part to physical conditions, particularly to isothermal configurations, according to Bertelsen. But in the headquarters of both, contrasting trophic dimensions determine relative abundance.

To draw this section together, we turn first to the marked differences in patterns of distribution between the faunas of the mesopelagic and bathypelagic zones. One striking contrast concerns the absence of a subtropical assemblage at bathypelagic levels. In the mesopelagic fish fauna small or dwarf species in subtropical central waters contrast with larger relatives in richer peripheral or off-lying waters. Evidently, production in the (oligotrophic) central water masses is not enough to support an underlying association of bathypelagic fishes, or even of giant euphausiids. The widely distributed bathypelagic animals live in both equatorial and subtropical waters, but they tend to be concentrated in eutrophic areas.

Perhaps *Cyclothone obscura* is one of the few instances of a bathypelagic animal with a circumglobal equatorial distribution, a type commoner in the mesopelagic fauna. Again, the distribution is eutrophic, even sharper than that of its more widely distributed relatives.

Though physical conditions are so constant and stable at bathypelagic levels, the prediction that the larger animals should tend strongly to consist of circumglobal species is far from safe.* A surprising number of species are restricted to one ocean, more precisely they are centred in an eutrophic part of one ocean. As in *Cyclothone livida* (p. 488), this

---

* Bathypelagic copepods, which are below the size range considered here, do tend strongly to have circumglobal distributions (see p. 77). The prediction would rather be that most species are most abundant under the more productive waters.

is true of certain oneirodid angler-fishes in the transitional waters off California and in the eastern tropical Pacific (Pietsch, 1974).

Lastly, it is curious that in the eutrophic waters of the subantarctic region there appear to be no endemic species of bathypelagic fishes below the well-marked mesopelagic fauna (see p. 477). But this appears not to be true of the subarctic Pacific, also eutrophic, where live two endemic species of angler-fishes, *Oneirodes thompsoni* and *O. bulbosus*. Pietsch (1974) indicates that both species are centred at levels between 600 and 850 metres. If so, they are lower mesopelagic fishes, which may account in part for their close association with the subarctic water mass. The reflection of epipelagic patterns of distribution at mesopelagic levels may well be related to the widespread habit of daily vertical migrations in mesopelagic animals—migrations in search of food, which also support non-migrating predators (p. 248). The deeper food-poor habitat of bathypelagic animals, which are largely or entirely non-migratory, tends to keep them concentrated below the more productive regions of the euphotic zone.

Benthopelagic distributions

Animals that live near the deep-sea floor range from copepods to large cephalopods and fishes (pp. 272–82). Patterns of distribution are best known in fishes, and here we shall be concerned largely with the rat-tails (Macrouridae). Except for the Arctic Sea, they have been trawled in all major regions of the ocean, but most of the 250 or more species live from warm temperate to tropical waters. The populations of nearly all kinds are centred at slope levels down to 2,000 metres, which is generally true of benthopelagic (and benthic) fishes. Both the population density and species diversity of bottom-dwelling fishes decline markedly, and exponentially, down the slope. Off the northeast coast of America near the Hudson Canyon dives in the DSRV *Alvin* between 1,768 and 2,797 metres have shown that more species may be seen than are caught in trawls. Based on species seen, both visually and photographically, during dives, Cohen and Pawson (1977) equate the number of fish species recorded per hour with $58.85e^{-1.32x}$, where x is the depth. The curve ranges from 5 to 6 species per hour at 1,700–1,800 metres to just over one per hour at 2,800 metres (total numbers range from 20 to 7 species). These figures are particularly interesting in that they cover the transition from the diverse bathyal fish fauna (200–2,000 metres) to the few abyssal species that are centred at levels below 2,000 metres. Among macrourids, *Coelorhynchus*, represented by more than 50 species, is a generic indicator of the bathyal fauna, while *Chalinura*, with about 7 species, is an indicator of the abyssal fauna.

Though there are many more species of rat-tails under the warm ocean than elsewhere, most of these species are concentrated in particular regions. The greatest concentration is in the Sulu Sea (Philippine–Indonesian region), where about 75 species are known (Gilbert and Hubbs, 1920). Over 50 species have been trawled off Japan and over 20 in both the Hawaiian and Panamanian regions. In the western North Atlantic, three out of four species (over 40 in all) are represented in the Gulf of Mexico and the Caribbean Sea. The Indian Ocean, where about 25 species are known, seems to hold a poor macrourid fauna. In all centres of concentration, especially the first, there are endemic species. Elsewhere, Marshall (1965) concluded, 'Certain semi-enclosed parts of the ocean, such as the Sulu Sea, the Gulf of Mexico and the Caribbean, and isolated islands, such as the Hawaiian group, have provided the means for considerable speciation.' In the Gulf and Caribbean there seem to be three kinds of species: endemic (e.g. *Coryphaenoides mexicanus*) species that have originated there and spread to neighbouring regions (e.g. *Coelorhynchus caribbaeus*), and wide-ranging adaptable species (e.g. *Chalinura mediterranea* and *Cetonurus globiceps*) (Marshall, 1973).

Of some 65 species of benthopelagic macrourids known from the Atlantic Ocean, 9 only have been taken elsewhere. There are 3 three-ocean species, *Nematonurus armatus*, *Malacocephalus laevis* and *Trachonurus sulcatus*. The first, an abyssal species, ranges over an open circumglobal environment of virtually constant physical properties: it is also an opportunistic species with a marked propensity to forage well above the bottom (p. 274). *Malacocephalus laevis* and *Trachonurus sulcatus* are slope-dwellers with a wide distribution attributable to the production of long-lived pelagic young that may grow to a considerable size before metamorphosis (see Marshall, 1964). Two other species, *Squalogadus modificatus* and *Echinomacrurus mollis*, the second abyssal, may well be widespread (in two oceans) because they are partly bathypelagic in habit.

Apart from the widely distributed forms, 18 species of macrourids live on both sides of the Atlantic. Some, such as the bathygadines (6 species), produce large pelagic young, which may be long lived. But, except for *Hymenocephalus italicus*, there is no evidence that the other species may have such young. If most macrourids pass their entire life history near the deep-sea floor (p. 456), this may be why many slope-dwellers are restricted to one side of an ocean. One may speculate that the amphi-Atlantic occurrence of some species (e.g. of *Nezumia*) may be due to continental drift, to one part of a species population drifting away from the other. It is certainly easier to understand the amphi-Atlantic distribution of abyssal species, such as those of the genus *Chalinura*. There seems to be no reason why they should not spread from one side of the Mid-Atlantic Ridge to the other. Lastly, the trans-Atlantic spread of the two species

(*Macrourus berglax* and *Coryphaenoides rupestris*) that live from subarctic to temperate regions also seems understandable. They '... have
had the advantage, as it were, of the suitable depths (less than 1,000
metres) that extend—via the Faroe–Iceland–Greenland ridge, and then
round Greenland to Labrador—from one side of the Atlantic to the other'
(Marshall, 1965).

Benthic distributions
Zonation of the benthic fauna

The fauna of the deep-sea floor, as we saw earlier (Chapter 5) ranges
widely in size from meiofaunal species measuring fractions of a millimetre to megafaunal species spanning a metre or more. Between meiofauna and megafauna are animals (the minifauna) about 1 to 10 mm in
major dimensions. The meiofauna is concentrated in the uppermost few
centimetres of sediment, which is also the habitat of a great many
members of the minifauna, such as polychaete worms, bivalves and
tanaid crustaceans. There are also infaunal species in the megafauna, but
most are epifaunal, attached to, or moving over, the deep-sea floor (see
pp. 162–215). Little is known of the large-scale distribution of the meiofauna, while investigations of the minifauna have been concentrated in
the western North Atlantic. Here we shall be necessarily and largely concerned with the ecological zoogeography of the megafauna, as sampled
by dredge and trawl over the entire ocean from *Challenger* times (1872–
76) onwards. First, though, some general treatment of zonation is possible and relevant.

In his *Zoogeography of the Sea*, Ekman (1953) placed the upper limit
of the deep-sea benthic fauna (the megafauna) between depths of 200
and 400 metres. Between this limit and a depth of about 1,000 metres
an archibenthal fauna could be distinguished from the deeper abyssal
fauna. But, as Ekman said, 'Investigations into the position of the greatest
faunal change seem not to have been undertaken and will have to wait
until our knowledge of the deep-sea fauna is more complete.' More recently, specialists of megafaunal groups in the deep-sea benthos have set
the transition between an upper-slope (bathyal) fauna and a deeper
abyssal fauna at a level of about 2,000 metres. At about 6,000 metres
the abyssal fauna gives way to the hadal fauna of the deep-sea trenches
(see for instance, Madsen, 1961, and Hansen, 1975). Concerning the
number of stations with published records of holothurians, Hansen has
divided the deep-sea floor into three depth intervals: 1,000–2,000 metres,
2,000–6,000 metres and 6,000–11,000 metres and under each he gives the
percentage area of sea floor represented, the number of stations per interval and the number of stations per unit of area, which in descending order

are: 3·4, 254 and 74·7; 78·4, 267 and 3·4; 1·3, 58 (40 Russian) and 44·6. Thus, the abyssal region (2,000–6,000 metres) which covers about three-quarters of the deep-sea floor, has been least well sampled, which is hardly surprising but needs to be kept in mind.

Closer to our present concern, Haedrich and his colleagues (1975) have concentrated on the zonation and faunal composition of the bottom-dwellers on the continental slope south of New England, USA. Their trawls, which took benthopelagic and benthic fishes as well as epifaunal invertebrates (of the megafauna), came from depths between 140 and 2,000 metres. By means of transect and cluster analysis, they concluded that the fauna is distributed in three zones; shallow (141–285 metres), middle (393–1,095 metres) and deep (1,270–1,928 metres). The faunal boundary between shallow and middle zones, which is well marked, corresponds with the lower part of the transition from shelf to slope levels, which, as the authors suggest, may also be a transition from a region with seasonal temperature fluctuation to one of little annual variation. The boundary between middle and deep zones corresponds with a change in gradient (between upper and lower slope) and seemingly also with changing conditions of temperature: the middle zone lies within the main thermocline, whereas the lower is in isothermal waters. Fishes and echinoderms were the most abundant taxa, the former predominating in shallow and middle zones, the latter at deeper levels.

More recently, Gardiner and Haedrich (1978) have analysed the zonation of the above benthic megafauna by using a test based on Maxwell–Boltzman statistics. The test indicates whether the distributions of the boundaries of species disposed along a gradient are random or clustered (zoned). Gardiner and Haedrich find that zonation is marked in the megafauna living between 200 and 3,000 metres, but below 3,000 metres little zonation is detectable. If the same conclusion is more or less true elsewhere, then most of the benthic megafauna of the deep sea is zoned, for most species live between depths of 200 and 3,000 metres.

Off the Carolinas, USA, an intensive photographic survey down to depths beyond 5,000 metres showed that large epifaunal invertebrates are zoned in narrow bands parallel to the depth contours (Rowe and Menzies, 1969; Menzies et al., 1973). Over the upper continental slope (200–800 metres), these bands, including animals such as the solitary coral *Flabellum goodei*, the quill worm, *Hyalinoecia artifex*, and the seastar, *Astropecten americanus*, were correlated with marked changes in sediment texture (size of particles) and variations of temperature, the latter related to changes in the movement and proximity of the Gulf Stream. At depths greater than 1,000 metres, where temperatures were stable and differences in sediment texture slight, banded distribution seemed to be related to zonation in supplies of detrital food. But Rowe and Menzies suggest that banding may arise from the interaction between species such that narrow bands are a function of which animals can or cannot survive successfully together with the brittle-star *Ophiomusium lymani*, a dominant species down to depths of 3,000 metres or more.

Rex (1977) also considers that biological interactions play a part in zonation, or rather in rates of zonation (see also p. 303). He has used cluster and factor analysis to follow depth-correlated changes in the species composition of deep-sea gastropod assemblages in the Woods Hole transect between New England and Bermuda. Taxa dominated by croppers and predators (e.g. echinoderms and gastropods) are zoned more rapidly than infaunal deposit-feeders (e.g. bivalves and polychaete worms). Populations of the latter may be so thinned by predators that there is more overlap in the depth ranges of constituent species, thus diminishing the overall rate of faunal change with depth. To support his hypothesis, Rex shows that the differing rates of zonation in epifaunal predators and croppers, gastropods and infaunal polychaete-bivalve deposit-feeders correspond to their levels in the food pyramid. In the gastropods the sharpest faunal change came at the transition between shelf and slope. The rate of faunal change from upper slope to abyssal levels seemed proportional to the rate of change of depth, thus being most on the slope, moderate on the continental rise and least at abyssal levels. The entire pattern of faunal change may reflect the increasing uniformity of the environment with increasing depth. Perhaps the uniformity factor coupled with poor trophic conditions has much to do with the fading of zonation at abyssal levels.

Members of the minifauna (bivalves and polychaetes) are thus less rapidly zoned than predators and croppers of the megafauna. One might expect that zonation in the meiofauna, most of which are lower in the food pyramid than the minifauna, to be still more gradual. Comparable analysis of the meiofauna is not available, but is understandable when one realizes that this fauna is very diverse, little known taxonomically, and distributed in small-scale patterns. For instance, half the species of meiobenthic copepods that Coull (1972) collected are undescribed and few were duplicated in samples from 18 stations at depths beyond 400 metres.

Even so, as Coull noticed in copepods and Tietjen (1976) found in free-living nematodes off North Carolina, the deep-sea meiofauna is definitely zoned, particularly according to the type of sediment. Tietjen recognizes a sand zone (50–500 metres), separated by a transition zone with few endemic species, from a clayey-silt zone (800–2,500 metres). More precisely, 106 out of 209 species taken were restricted to one of four sedimentary habitats. There were 35 species in quartz-algal sands (50–100 metres), 17 in foraminiferan sands (250–500 metres), 5 in sandy-silt (500–800 metres) and 49 in clayey-silts (800–2,500 metres). Species diversity, measured by the Shannon–Weaver function and (the correlated) species richness by $(S - 1)/\ln N$, where S is the number of species and N the number of individuals, is higher in the sandy sediments of the outer shelf and upper slope than in the clayey-silt sediments of middle and lower slope. Tietjen attributes this to the greater number of habitats available in sandy sediments, which apparently became less and less available with depth in clayey-silts. The relationship between life form and type of sediment was covered earlier (pp. 290–1).

The transition zone between sandy and silty sediments (500–800 metres), which contains so few endemic species (5) and across which faunal change is greatest, is where the Gulf Stream impinges directly on the bottom from time to time. It will be recalled that the zonation of large benthic invertebrates also changed across this region. Transitions in the fauna of the New England slope also coincide with changes in thermal conditions (p. 493). In temperate regions, where the change from seasonal conditions on the shelf to a seasonless regime on the slope is very marked, there also seems to be a particularly sharp transition between shallow and deep-sea faunas.

Concerning sedimentary influences on zonation, it is hardly surprising that some members of the meiofauna are associated with certain kinds of deposit. There are also less specialized forms. The effects of biological interactions on zonation are at present clearer in the megafauna. Lastly, we must bear in mind that there have been few special investigations on zonation in the benthic deep-sea fauna. What we are beginning to know encourages further exploration.

The trophic dimension and the megafauna

Except near hydrothermal vents, the benthic fauna of the deep sea is energized by organic matter that falls from above.* There is, of course, fallout in the region of hydrothermal vents, but here productivity seems to depend largely on the production of hydrogen sulphide, which is an energy source for rich crops of chemosynthetic bacteria. In one way or another, the bacteria support abundant populations of large invertebrates, some giants compared to their relatives on less productive parts of the deep-sea floor (pp. 306–7). But the zoogeography of hydrothermal vents is at present no more than an intriguing prospect: here we are concerned with the 'under-energized' remainder of the ocean bottom.

Apart from the sporadic fall of a large corpse, any region of the bottom receives most of its organic material from the relatively swift and direct fall of faecal waste from copepods, euphausiids and other zooplankters, no doubt mainly from those that live in the epipelagic zone, whether

---

* Animals living on the abyssal plains may also receive organic matter that is swept down the slope from time to time in turbidity currents (p. 284).

permanently or during the night. There is a much slower fall of finer organic material, which is evidently a minor source of benthic nourishment. Distribution patterns of plant productivity, zooplankton biomass and mid-water particle concentrations are alike (p. 285). Thus the amount of organic waste reaching any particular part of the bottom is related to productivity directly above in the epipelagic zone. Naturally, the longer the fall of any given quantity of organic matter, the less reaches the bottom. It is hardly surprising that the most impoverished and widespread expanses of sediment are in central subtropical regions: they lie under the least productive (central) water masses and over the deeper reaches of the ocean basins.

Sokolova (1972), as we saw, divides deep-sea sediments into eutrophic and oligotrophic, which may be separated by those of a transitional nature. Of the greater quantities of organic waste that fall on eutrophic sediments, part may be buried while still in an active, usable state: some of the rest is decomposed at the surface. Nearly all of the lesser fall of organic matter on oligotrophic sediments is decomposed at the surface: the buried part is in a drastically transformed and inactive condition. Thus, eutrophic sediments are the richer in kinds of organic matter usable by micro-organisms and detritus-feeders. If some of this material is resuspended near the bottom, it will reinforce the falling particles intercepted by suspension-feeders. In quantities of organic carbon, which range from 0·25 to 1·5 per cent of the dry weight of organic matter, eutrophic sediments are some 15 times richer than those of oligotrophic regions. The teeth of tertiary sharks and accumulations of manganese nodules are good indicators of oligotrophic sediments.

Beard-worms (Pogonophora), the only megafaunal forms known to exist directly on dissolved organic substances, are confined largely to eutrophic regions. As we saw earlier (p. 177), they are concentrated near hydrothermal vents, on the continental slope and in the richest trenches (Kurile–Kamchatka and South Sandwich). Earlier investigations on specific patterns of distribution along eutrophic pathways of the slope were centred in the western North Atlantic (Southward and Brattegard, 1968; Southward, 1971b). In a paper drawing on these investigations, Cutler (1975) has analysed the distribution patterns of pogonophores and sipunculid worms on the contintinental slope off the Carolinas, USA. For 14 out of 27 species found in this region, there seems to be a faunal barrier southeast of Cape Lookout, North Carolina (around 34°N and between 150 and 2,500 metres). Cutler suggests that the barrier may result from the effects of bottom dispersal on larval dispersal. North of the barrier off the coast of Virginia, bottom currents flow south at depths between 200 and 3,000 metres: south of the barrier, off South Carolina, the flow inside the Blake Plateau (from 200 to 600 metres) is northward, but outside it is southward (from 800 to 3,000 metres). Presumably some species are unable to get past the region of reversed flow. Isopod crustaceans also show a distinct distribution off the Carolinas (Menzies et al., 1973).

Concerning the trophic distribution of deposit-feeders, suspension-feeders and carnivores in the invertebrate megafauna, we are most indebted to M. N. Sokolova. In her 1972 paper she has used analysis of 228 samples collected by Soviet research vessels in the Pacific and Indian Oceans at depths ranging from 3,000 to 6,000 metres. Relevant catches by other deep-sea expeditions were also considered.

Sponges (hexactinellids and tetraxonids) and serpulid polychaetes were the main kinds of suspension-feeders under analysis. Both are more or less widely distributed in both eutrophic and oligotrophic regions of the two oceans. For benthic invertebrates living on oligotrophic sediments, Sokolova suggests that the most effective mode of feeding is on suspended particles: for deposit-feeders, as there is little or no subsurface nutriment, only surface-croppers can prevail. There is also the fact that sponges and serpulid worms are widely adaptable to deep-sea conditions. In particular, both became small where food is scarce. Certain deep-water sponges assume small, apparently neotenous forms (Koltun, 1972). They must be especially small in the deep basins under the central water masses, where serpulid worms are also dwarfed. At the other extreme, these worms grow to giant size near hydrothermal vents. Some suspension-feeders, such as the sea-lilies, are not so adaptable to the poorest regions of the deep-sea floor.

Deposit-feeders of various taxa tend to be confined to eutrophic pathways around the subtropical gyres and in more poleward waters: they are rare on the great expanses of oligotrophic sediment under the central water masses. Thus, in the Indian Ocean, large deposit-feeders, such as sea-stars of the family Porcellanasteridae, various sea-urchins (e.g. Pourtalesiidae and Urechinidae) and molpadiid sea-cucumbers, are largely confined to the Arabian Sea, the Bay of Bengal, the equatorial region and waters south of the Subtropical Convergence. In the Pacific Ocean these animals are centred in the equatorial region, in peripheral (off continental) waters and in subarctic waters. Sea-cucumbers, such as elpidiids and deimatids, and bivalve molluscs of the families Malletiidae and Nuculanidae, are most prevalent on eutrophic sediments, but have broader distributions. Here we may recall that many elasipod sea-cucumbers are light enough to swim over the bottom, and are thus not confined to browsing in one small area. On somewhat impoverished sediments, they may still make a living. The bivalves, like sponges and serpulid worms, may also become dwarfed in impoverished conditions. Bivalves are one of the main groups of the minifauna in central subtropical waters (Hessler, 1972).

Looking more closely at porcellanasterid sea-stars, the maps of distribution in Madsen's (1961) report show that nearly all findings are in eutrophic regions. For instance, the Indo-Malay region, which is a main centre of distribution, is one of the most productive parts of the ocean. Species of the genera *Porcellanaster* and *Ericmaster*, the only ones known from bathyal to abyssal levels, are almost entirely confined to eutrophic pathways off the continents: there are hardly any records elsewhere, not even in equatorial regions.

There are also deposit-feeders among the larger kinds of polychaete worms. In the abyssal Pacific Levenstein (1971/72) has considered the ecology and zoogeography of species belonging to the genera *Fauveliopsis* (Flabelligeridae), *Travisia* and *Kesun* (Opheliidae). He describes them as 'non-selective detritus-feeders swallowing sediment' and concludes that their distribution patterns show good correlation with those of eutrophic sediments.

Carnivorous members of the invertebrate megafauna have much the same kind of eutrophic distribution as the detritus-feeders. Sokolova (1972) lists predatory sea-stars (Brisingidae, Astropectinidae, Benthopectinidae, Pterasteridae and Goniasteridae) as instances of animals largely confined to eutrophic sediments. Certain species of the brittle-star genus *Amphiophiura*, which are predators and scavengers, have a somewhat wider distribution, while that of sea-anemones is still wider. In a more recent paper, Sokolova (1976) remarks that it is small actinians that penetrate into oligotrophic regions.

Turning to other anthozoan cnidarians, Pasternak (1977) in a paper on the black corals (Antipatharia) states that the ranges of most deep-sea corals are much narrower than the eutrophic areas surrounding the continents. One species of black coral, *Bathypathes lyra*, which is an abyssal form known from depths between 3,500 and 6,000 metres, is unusual in that a number of specimens have come from oligotrophic areas of the Central Pacific. Indeed, Pasternak concludes that this species has a '... preference for areas where food conditions are far from optimal for the majority of other corals'.

For final instances, we turn again to Levenstein's (1971/72) paper on certain large polychaete worms in the abyssal Pacific. Species of the aphroditid genus *Mallicephala* are active carnivores on other invertebrates and it is thus not surprising that they are confined to eutrophic regions.

In conclusion, it would seem that the most widespread species in the invertebrate megafauna, found in both oligotrophic and eutrophic parts of the ocean, are to be sought among the sponges, serpulid worms and other widely adaptable forms. Species largely confined to, or centred in eutrophic regions, are evidently in the majority. Such species, if known widely in all three oceans, are described as cosmopolitan, but this is not strictly true. They are cosmopolitan in the eutrophic pathways of the ocean.

Variation, diversity and zoogeography

Cosmopolitan species live in most parts of the ocean, but it seems that few are virtually similar everywhere in taxonomic characters. On close scrutiny species thought to be cosmopolitan prove to consist of a few closely related species, each with an individual geographic range. Some species are polytypic, composed, like man, of races or subspecies that are distinguished, inter alia, by their distribution. Most relevant to present discussion, the range of variation in species may be related to their geographic range. This aspect impressed Ebeling and Weed (1963) in their study of melamphaid fishes (pp. 480–3). Cosmopolitan species occurring in 7 to 9 water masses showed significant variation in about 50 per cent of the (13–19) characters they analysed. In species recorded from 4 to 5 water masses this percentage fell to 20, and lower still to 10 in those from 2 to 3 water masses. The widely adaptable species are the most variable.

The relationship between taxonomic variation and genetic variation is not necessarily close, but there is presumably considerable correlation. Study of genetic variation in marine organisms has followed the pioneer work of Lewontin and Hubby (1966), who put the dogma 'DNA makes RNA makes protein' into electrophoretic practice. There are corresponding pairs of genes in the chromosome pairs, one half of each pair being of paternal, the other of maternal, origin. At a particular gene locus, the gene pair may exist in a variety of different forms (alleles) in different individuals. Both the paternal and maternal alleles bear a code for each protein in an individual. Thus, variability in the form of proteins at gene loci should give an indication of genetic variability. As different forms of a protein have different mobilities in an electric field, tests of this protein by electrophoresis will show whether it exists in one or more states. If the paternal and maternal alleles are different, the protein will display at least two different mobilities in an electric field. By extracting the same protein from enough individuals of a population (and applying the same electrophoretic test), one can estimate how much allelic variations the population holds. Parallel tests are run on other proteins, thus providing estimates of the variability at the gene loci that code for these proteins (At each locus a gene is represented by the two parental alleles, but the same gene can occur in many allelic forms among the members of a population.) In most species studied the degree of allelic variability is considerable. For instance, more than 50 per cent of the genes tested in members of a population may exist in different allelic forms. Allelic variability in individuals may occur in 25 per cent of the genes tested, but the usual level is 5 to 15 per cent.

In the deep sea, such studies have centred on members of the megafauna. Concerning the pelagic fauna, Gould (1975 and 1976), quoting unpublished work by J. W. Valentine, has given advance information on euphausiids. In species from high, middle and low latitudes, genetic variability increases steadily towards the tropics. There is a parallel increase in species diversity. Are the two trends related?

At epipelagic and mesopelagic levels, euphausiids and many of the smaller fishes have adopted an annual life cycle in the subtropical and tropical belts; in temperate and cold regions, where the growing season is short, the cycle extends for two years or more (p. 446). If the rate of evolution in annual species is quicker than it is in those with longer generation times, then it is not surprising that in and under the warm ocean there are more species of euphausiids and fishes than in colder waters. Moreover, evolution towards high species diversity may well engender further diversity. The more the species that coexist in a given area, the more the biological interactions, especially in food webs. Through such interaction, one part of a species may become isolated from another and diverge to form a new species. It is, of course, true that evolution towards high diversity is limited by the diversity of ecological niches. But there is high niche diversity in subtropical and tropical waters and I have argued elsewhere (pp. 249–50) that it is largely mediated through daily vertical migrations. Species-packing in the water column seems to be 'arranged' by interactions between migrators and non-migrators.

To return to Valentine's euphausiids, it may well be that genetic variability and species diversity are related. The genome of a species not only determines its structure down to fine levels, but also its physiological, ecological and ethological characters. One may expect genetic variability to be high where biological interactions are compex. If the euphausiids are epipelagic species, then the complexity of their biological surroundings increases greatly during the night. If one thinks of genetic variability simply in terms of adaptation to long-term environmental variability, then species living in temperate regions should be more variable than those in subtropical and tropical waters. But so far, the evidence is otherwise.

There are as yet no studies of latitudinal genetic variability in closely related members of the benthic deep-sea fauna. In unrelated forms from shallow waters, individual variability (heterozygosity) is most (22%) in a tropical clam (*Tridacna*), least in a brachiopod and a sea-star from cold waters (1–4%) and at an intermediate level (7%) in a horse-shoe crab (*Limulus*) from middle latitudes. Deep-sea members of the invertebrate megafauna taken from the San Diego Trough off southern California at a depth of 1,244 metres were about as variable as the clam. In the cosmopolitan brittle-star, *Ophiomusium lymani*, about 50 per cent of the 15 gene loci of 257 individuals were polymorphic, while individual variability existed at 17 per cent of these loci (Ayala and Valentine, 1974). In four species of sea-stars, allelic variation at 24 individual loci (bearing codes for enzymes and other soluble proteins) was about the same as that in the brittle-star. In one species, *Nearchaster aciculosus*, it was as high as 71 per cent (Valentine et al., 1975). Ayala and Valentine conclude that species from environments with more stable food supplies, such as the tropics and the deep-sea, have high genetic variability compared to those from shallow waters in temperate latitudes, where productivity is seasonal. Valentine and his colleagues stress that food supplies are diverse in stable environments. Thus, adaptations to trophic resource variability may involve much genetic variability.

We saw earlier that taxonomic variability in mesopelagic fishes (melamphaids) was greater in the more widely distributed species. The same does not seem to be true of genetic variability in benthic invertebrates. *Ophiomusium lymani*, with a much greater geographic and bathymetric range than the sea-stars cited above, has about the same genetic variability. A more relevant comparison would be with a closely related species of brittle-star. And is the genetic variability the same everywhere in the range of a cosmopolitan species ? In their study of genetic variability of deep-sea echinoderms, Murphy and her colleagues (1976) suggest that populations of *Ophiomusium lymani* may be genetically isolated (the proportion of heterozygotes tends to be lower than that expected from Hardy–Weinberg equilibrium). Thus, variability may change from place to place. More generally, these authors contend that selection pressures are often weaker within a species range than at the periphery. In the centre, where ecological conditions are optimal and interspecific competition is well established, relaxation of selection pressure allows of greater genetic variability. At the periphery, where gene flow is restricted and individuals are close to the limits of tolerable conditions, selection pressures are more intense. Rates of reproduction may tend to be negative, thus tending to eliminate genetic variability and produce a homozygous population.

Phenotypic uniformity in the individuals of peripherally isolated populations is central to Mayr's (1963) elaboration of his Founder Principle (see also p. 307). In the deep sea, where distributions of benthic species in the megafauna (and pelagic fauna) are on a large scale, the principle should be widespread, but we must await further investigations along the lines pursued by Murphy and her colleagues (1976). What levels of genetic variability exist in the benthic meiofauna and minifauna, among which small-scale, mosaic patterns of distribution are prevalent ? Here should lie increased understanding of evolutionary processes.

It is fitting to end with one of the newest and most intriguing 'developments in deep-sea biology'. Ideas are emerging to be refined or refuted in the light of further developments. In the meantime, stimulating ideas, even if too 'oceanic', are much better than no ideas at all. But is 'no ideas at all' conceivable, especially in deep-sea biology ?

Bibliography

1 Ahlstrom, E. H. and Counts, R. C. (1958) 'Development and distribution of *Vinciguerria lucetia* and related species in the eastern Pacific' *Fishery Bulletin* 139, Vol. 58, pp. 363–416.
2 Aizawa, Y. (1974) 'Ecological studies of micronektonic shrimps (Crustacea, Decapoda) in the western North Pacific' *Bull. Ocean Research Institute, Univ. Tokyo* 6, pp. 84.
3 Åkesson, B. (1961) 'Some observations on *Pelagosphaera* larvae (Sipunculoidea)' *Galathea Report* 5, pp. 7–17.
4 Alexander, R. McN. (1966) 'Physical aspects of swimbladder function' *Biol. Rev.* 41, pp. 141–176.
5 Aleyev, Yu. G. (1977) *Nekton* Dr W. Junk b.v., The Hague, English translation, 435 pp.
6 Alldredge, A. (1976) 'Appendicularians' *Scientific American* 235, pp. 94–105.
7 Allen, J. A. and Turner, J. F. (1974) 'On the functional morphology of the family Verticordiidae (Bivalvia) with descriptions of new species from the abyssal Atlantic' *Phil. Trans. Roy. Soc. Lond. B.* 268, pp. 401–536.
8 Andriashev, A. P. (1965) 'A general review of the Antarctic fish fauna' pp. 491–550. In *Biogeography and Ecology in Antarctica* Van Mieghem, J. and Van Oye, P. (ed.) Dr W. Junk b.v., The Hague, pp. 762.
9 Angel, M. V. (1971/72) 'Planktonic oceanic ostracods—Historical, present and future' *Proc. Roy. Soc. Edn.* 73, 22, pp. 213–228.
10 Angel, M. V. and Fasham, M. J. R. (1974) 'Sond Cruise 1965: Further factor analysis of the plankton data' *J. Mar. Biol. Ass. UK.* 54, pp. 879–894.
11 Angel, M. and Angel, H. (1974) *Ocean Life* Octopus Books, London, 72 pp.
12 Ayala, F. J. (1977) In *Evolution* Dobzhansky, T., Ayala, F. J.,

Stebbins, G. L. and Valentine, J. W., W. H. Freeman and Company, San Francisco, 572 pp.

13 Ayala, F. J. and Valentine, J. W. (1974) 'Genetic variability in a cosmopolitan deep water ophiuran, *Ophiomusium lymani*' *Mar. Biol.* 27, pp. 51–57.

14 Azam, F. and Hodson, R. E. (1977) 'Size distribution and activity of marine microheterotrophs' *Limnol. Oceanogr.* 22, 3, pp. 492–501.

15 Backus, R. H., Craddock, J. E., Haedrich, R. L., Shores, D. L., Teal, J. M., Wing, A. S., Mead, G. W. and Clarke W. D. (1968) '*Ceratoscopelus maderensis:* Peculiar sound-scattering layer identified with this myctophid fish' *Science* 160, 3831, pp. 991–993.

16 Backus, R. H., Craddock, J. E., Haedrich, R. L. and Shores, D. L. (1969) 'Mesopelagic fishes and thermal fronts in the western Sargasso Sea' *Mar. Biol.* 3, pp. 87–106.

17 Backus, R. H., Craddock, J. E., Haedrich, R. L. and Shores, D. L. (1970) 'The distribution of mesopelagic fishes in the equatorial and western North Atlantic Ocean' pp. 20–40. In *Proceedings of an International Symposium on Biological Sound-Scattering in the Ocean* Farquahar, G. B. (ed.) Maury Center for Ocean Science, Washington, DC.

18 Backus, R. H. and Craddock, J. E. (1977) 'Pelagic faunal provinces and sound-scattering levels in the Atlantic Ocean' pp. 529–548. In *Oceanic Sound-Scattering Prediction* Andersen, N. R. and Zahuranec, B. J., (ed.) Plenum Press, New York and London.

19 Badcock, J. (1970), 'The vertical distribution of mesopelagic fishes collected on the Sond Cruise' *J. Mar. Biol. Ass. UK.* 50, 4, pp. 1001–1044.

20 Badcock, J. and Merrett, N. R. (1976) 'Midwater fishes in the eastern North Atlantic—I. Vertical distribution and associated biology in 30°N, 23°W, with developmental notes on certain myctophids' *Prog. Oceanogr.* 7, pp. 3–58.

21 Bainbridge, R. (1952) 'Underwater observations on the swimming of marine plankton' *J. Mar. Biol. Ass. UK.* 31, 1, pp. 107–112.

22 Baird, R. C. (1971) 'Systematics, distribution and zoogeography of the marine hatchetfishes (family Sternoptychidae)' *Bull. Mus. Comp. Zool. Harvard* 142, pp. 1–28.

23 Baird, R. C., Hopkins, T. L., and Wilson, D. F. (1975) 'Diet and feeding chronology of *Diaphus taaningi* (Myctophidae) in the Cariaco Trench' *Copeia*, pp. 356–365.

24 Baker, A. de C. (1970) 'The vertical distribution of euphausiids near Fuerteventura, Canary Islands ("Discovery" SOND Cruise, 1965)' *J. Mar. Biol. Ass. UK.* 50, 2, pp. 301–342.

25 Ball, E. E. and Cowan, A. N. (1977) 'Ultrastructure of the antennal

sensilla of *Acetes* (Crustacea, Decapoda, Natantia, Sergestidae)' *Phil. Trans. Roy. Soc. Lond. B.* 277, 957, pp. 429–457.

26 Bardach, J. E., Todd, J. H. and Crickmer, R. (1967) 'Orientation by taste in fish of the genus *Ictalurus' Science* 155, pp. 1276–1268.

27 Bardach, J. E. and Villars, T. (1974) 'The chemical senses of fishes' pp. 49–104. In *Chemoreception in Marine Organisms* Grant, P. T. and Mackie, A. M. (ed.), Academic Press, London, 295 pp.

28 Barker Jørgensen, C. (1976) 'August Pütter, August Krogh, and modern ideas on the use of dissolved organic matter in aquatic environments' *Biol. Rev.* 51, pp. 291–328.

29 Barnard, J. L. (1961) 'Gammaridean Amphipoda from depths of 400 to 6000 metres' *Galathea Report* 5, pp. 23–128.

30 Barnard, J. L. (1962) 'South Atlantic abyssal amphipods collected by RV *Vema*' In *Abyssal Crustacea. Vema research series* 1, pp. 1–78.

31 Barnes, A. T. and Case, J. F. (1972) 'Bioluminescence in the mesopelagic copepod *Gaussia princeps* (T. Scott)' *J. Exp. Mar. Biol. Ecol.* 8, pp. 53–71.

32 Barnes, A. T. and Case, J. F. (1974) 'The luminescence of lanternfish (Myctophidae): spontaneous activity and responses to mechanical, electrical, and chemical stimulation' *J. Exp. Mar. Biol. Ecol.* 15, pp. 203–221.

33 Barraclough, W. E., LeBrasseur, R. J. and Kennedy, O. D. (1969) 'Shallow scattering layer in the subarctic Pacific Ocean: detection by high frequency echo sounder' *Science* 166, pp. 611–613.

34 Bé, A. W. H. (1966) *Distribution of Planktonic Foraminifera in the World Oceans* Second International Oceanographic Congress, Moscow, pp. 26.

35 Beebe, W. and Vander Pyl, M. (1944) 'Eastern Pacific expeditions of the New York Zoological Society. XXXIII, Pacific Myctophidae (fishes)' *Zoologica NY* 29, 2, pp. 59–95.

36 Beers, J. R. and Stewart, G. L. (1969) 'The vertical distribution of micro-zooplankton and some ecological observations' *J. Cons. Int. Explor. Mer* 33, 1, pp. 30–44.

37 Belman, B. W. and Childress, J. J. (1976) 'Circulatory adaptations to the oxygen minimum layer in the bathypelagic mysid *Gnathophausia ingens*' *Biol. Bull. Woods Hole* 150, 1, pp. 15–37.

38 Belyaev, G. M. (1970) 'Ultra abyssal holothurians of the genus *Myriotrochus* (order Apoda, fam. Myriotrochidae)' *Tr. Inst. Okeanol. USSR.* 86, pp. 458–488.

39 Belyaev, G. M. (1972) 'Hadal bottom fauna of the world ocean' *Inst. Oceanol. Moscow* 1966. English transl. Israel Program for Scientific Translations, Jerusalem, pp. 281.

40 Benović, A. (1973) 'Diurnal vertical migration of *Solmissus albescens*

(Hydromedusae) in the Southern Adriatic' *Mar. Biol.* **18**, pp. 298–301.

41 Bernal, J. D. (1967) *The Origin of Life* Weidenfeld and Nicolson, London, 345 pp.

42 Bernard, F. (1953) 'Decapoda Eryonidae (*Eryoneicus* et *Willemoesia*)' *Dana Report* **37**, pp. 93.

43 Bernikov, R. G., Vinogradova, L. A., Gruzov, L. N., Paliy, N. F., Sedykh, K. A., Sukhoruk, V. I., Chmyr, V. D., Fedoseev, A. F. and Yakovlev, V. N. (1966) *Productive Zones of the Equatorial Atlantic Ocean* (Abstract) Second International Oceanographic Congress, Moscow, pp. 40–41.

44 Berrill, N. J. (1950) *The Tunicata*, Roy. Soc., London, 354 pp.

45 Bertelsen, E. (1951) 'The ceratioid fishes' *Dana Report* **39**, pp. 1–276.

46 Bertelsen, E., Krefft, G. and Marshall, N. B. (1976) 'The fishes of the family Notosudidae' *Dana Report* **86**, pp. 1–114.

47 Bertelsen, E. and Pietsch, T. W. (1977) 'Results of the research cruises of FRV *Walther Herwig* to South America. XLVII. Ceratioid anglerfishes of the Family Oneirodidae collected by FRV *Walther Herwig*' *Arch. Fisch. Wiss.* **27**, 3, pp. 171–189.

48 Bertin, L. (1934) 'Les poissons apodes appartenant au sous-ordre des Lyomeres' *Dana Report* **3**, pp. 1–56.

49 Bierbaum, G. (1914) 'Untersuchungen über den Bau der Gehörorgane von Tiefseefischen' *Z. Wiss. Zool.* **3**, 3, pp. 281–380.

50 Birstein, J. A. and Vinogradov, M. E. (1966) 'Change of Feeding Habits as the Cause of the Taxonomic Isolation of the Deep-Sea Fauna' Second International Congress, Moscow, pp. 45–56.

51 Birstein, J. A. and Zarenkov, N. A. (1970) 'On the bottom decapods (Crustacea, Decapoda) of the Kurile–Kamchatka Region' *Tr. Inst. Okeanol. USSR.* **86**, pp. 420–426.

52 Biscaye, P. E. and Eittreim, S. L. (1977) 'Suspended particulate loads and transports in the nepheloid layer of the abyssal Atlantic Ocean' *Marine Geology* **23**, pp. 155–172.

53 Bishop, J. K., Edmond, J. M., Ketten, D. R., Bacon, M. P. and Silker, W. B. (1977) 'The chemistry, biology and vertical flux of particulate matter from the upper 400 m of the equatorial Atlantic Ocean' *Deep-Sea Res.* **24**, 6, pp. 511–548.

54 Blaxter, J. H. S. (1976) 'Light as an ecological factor: II' pp. 189–210. In *The 16th Symposium of the British Ecological Society 1974* Blackwell Scientific Publications, Oxford.

55. Blaxter, J. H. S., Wardle, C. S. and Roberts, B. L. (1971) 'Aspects of the circulatory physiology and muscle systems of deep-sea fish' *J. Mar. Biol. Ass. UK.* **51**, pp. 991–1006.

56 Boden, B. P. and Kampa, E. M. (1964) 'Planktonic bioluminiscence' *Oceanogr. Mar. Biol. Ann. Rev.* 2, pp. 341–371.

57 Boden, B. P., Kampa, E. M. and Abbott, B. C. (1961) 'Photoreception of a planktonic crustacean in relation to light penetration in the sea' pp. 189–196. In *Progress in Photobiology. Proc. 3rd Int. Congr. Photobiology.*

58 Bogdanov, Yu. A. (1965) 'Suspended organic matter in the Pacific' *Oceanology* 5, pp. 77–85.

59 Bone, Q. (1973) 'A note on the buoyancy of some lantern-fishes (Myctophoidei)' *J. Mar. Biol. Ass. UK.* 53, pp. 752–761.

60 Bone, Q. and Roberts, B. L. (1969) 'The density of elasmobranchs' *J. Mar. Biol. Ass. UK.* 49, 4, pp. 913–937.

61 Brauer, A. (1908) 'Die Tiefsee-Fische II Anatomische Teil' *Wiss. Ergebn. Valdivia* 15, 2, pp. 1–266.

62 Brinton, E. (1962) 'The distribution of Pacific euphausiids' *Bull. Scripps Institution of Oceanography* 8, 2, pp. 51–270.

63 Brinton, E. (1967) 'Vertical migration and avoidance capability of euphausiids in the California Current' *Limnol. Oceanogr.* 12, pp. 451–483.

64 Broch, H. (1957) 'Pennatularians (*Umbellula*)' *Rep. Swedish Deep-Sea Expedition 1947–1948* 2 (Zool.), pp. 347–364.

65 Brodskii, K. A. (1950) 'Calanoida of the far eastern seas and polar basins of the USSR.' *Opred. Faune SSSR* 35, pp. 3–441. English transl. Israel Program for Scientific Translations, Jerusalem.

66 Bronowski, J. (1977) *A Sense of the Future.* The MIT Press, Cambridge, Mass., and London, England, 286 pp.

67 Brooks, A. L. (1977) 'A study of the swimbladder of selected mesopelagic fish species' pp. 565–590. In *Ocean Sound Scattering Prediction, Marine Science* Vol. 5 Andersen, N. R. and Zahuranec, B. J. (ed.) Plenum Press, New York and London, 859 pp.

68 Brundage, W. L., Buchanan, C. L. and Patterson, R. B. (1967) 'Search and serendipity' pp. 75–88. In *Deep-Sea Photography.* Hersey, J. B. (ed.) The Johns Hopkins Press, Baltimore, 310 pp.

69 Bruun, A. F. (1937) 'Contributions to the life histories of deep-sea eels: Synaphobranchidae' *Dana Report* 9, pp. 31.

70 Bruun, A. F. (1943) 'The biology of *Spirula spirula*' *Dana Report,* 24, pp. 44.

71 Bruun, A. F. (1956) 'Animal life of the deep-sea bottom' pp. 149–195. In *The Galathea Deep Sea Expedition 1950–1952* Bruun, A. F. et al. (ed.) George Allen and Unwin, London, 296 pp.

72 Buchsbaum, R. and Milne, L. J. (1960) *Living Invertebrates of the World* Hamish Hamilton, London, 303 pp.

73 Buckminster Fuller, R. (1961) 'Conceptuality of fundamental

structures' pp. 66–88. In *Structure in Art and in Science* Kepes, G. (ed.) Studio Vista, London, 189 pp.

74 Bulatov, R. P. and Stepanov, V. N. (1971) 'The discovery and investigation of east-tropical gyres in the Indian Ocean' *Proc. Joint Oceanogr. Assembly (Tokyo, 1970)*, pp. 389–392.

75 Bullard, E. (1969) 'The origin of the oceans' *Scientific American* **221**, 3, pp. 66–75.

76 Bunge, M. (1973) *Method, Model and Matter* D. Reidel Publishing Company, Dordrecht, Holland, 196 pp.

77 Burnett, B. R. (1977) 'Quantitative sampling of microbiota of the deep-sea benthos—I. Sampling techniques and some data from the abyssal central North Pacific' *Deep-Sea Res.* **24**, 8, pp. 781–790.

78 Burtt, E. T. (1974) *The Senses of Animals* Wykeham Publications, London, 157 pp.

79 Butler, C. G. (1967) 'Insect pheromones' *Biol. Rev.* **42**, 1, pp. 42–87.

80 Buzas, M. A. and Gibson, T. G. (1969) 'Species diversity: benthonic foraminifera in Western North Atlantic' *Science* **163**, pp. 72–75.

81 Cairns-Smith, A. G. (1971) *The Life Puzzle* Oliver and Boyd, Edinburgh, 165 pp.

82 Canetti, E. (1962) *Crowds and Power* Victor Gollancz, London, 495 pp.

83 Carey, A. G. (1972) 'Food sources of sublittoral, bathyal and abyssal asteroids in the northeast Pacific Ocean' *Ophelia* **10**, pp. 35–47.

84 Carlgren, O. (1956) 'Actiniaria from depths exceeding 6000 metres' *Galathea Report* **2**, pp. 9–16.

85 Carson, Rachel L. (1951) *The Sea Around Us* Staples Press Ltd, London, 230 pp.

86 Casanova, J. P. (1977) 'La Faune Pélagique Profonde (Zooplancton et Micronecton) de la Province Atlanto-Mediterranéenne' Thèse présentée à l'Université de Provence (Aix-Marseille 1).

87 Cassie, R. M. (1963) 'Microdistribution of plankton' *Oceanogr. Mar. Biol. Ann. Rev.* **1**, pp. 223–252.

88 Childress, J. J. (1971) 'Respiratory rate and depth of occurrence of midwater animals' *Limnol. Oceanogr.* **16**, pp. 104–106.

89 Childress, J. J. (1977) 'Effects of pressure, temperature and oxygen on oxygen-consumption rate of the midwater copepod *Gaussia princeps*' *Mar. Biol.* **39**, 1, pp. 19–24.

90 Childress, J. J. and Nygaard, M. H. (1974) 'The chemical composition of midwater crustaceans as a function of depth of occurrence off Southern California' *Mar. Biol.* **27**, pp. 225–238.

91 Chindonova, Yu. G. (1959) 'The nutrition of certain groups of abyssal macro-plankton in the northwestern area of the Pacific Ocean' *Tr. Inst. Okeanol. USSR.* **30**, pp. 166–189.

92 Chun, C. (1910) 'Die cephalopoden' *Wiss. Ergebn. Valdivia* 18, pp. 1–552.

93 Chun, C. (1914) 'Cephalopoda from the *Michael Sars* North Atlantic Deep-Sea Expedition' *Rep. Michael Sars N. Atl. Deep Sea Exped.* 3, pp. 1–28.

94 Church, R. (1971) 'Deepstar explores the ocean floor' *National Geographic Magazine* January 1971, pp. 110–129.

95 Clark, R. B. (1964) *The Dynamics of Metazoan Evolution* Clarendon Press, Oxford, 313 pp.

96 Clarke, G. L. (1970) 'Light conditions in the sea in relation to the diurnal vertical migrations of animals' *Proceedings of an International Symposium on Biological Sound Scattering in the Ocean* Maury Center for Ocean Science, Washington DC, pp. 41–50.

97 Clarke, G. L. and Backus, R. H. (1956) 'Measurements of light penetration in relation to vertical migration and records of luminescence of deep-sea animals' *Deep-Sea Res.* 4, pp. 1–14.

98 Clarke, G. L. and Denton, E. J. (1962) 'Light and animal life' pp. 456–568. In *The Sea* Hill, M. N. (ed.) Vol. 1, Ch. 10, Interscience Publishers Ltd, London.

99 Clarke, M. R. (1966) 'A review of the systematics and ecology of oceanic squids' *Adv. Mar. Biol.* 4, pp. 91–300.

100 Clarke, M. R. (1971) 'Biological samplers' pp. 60–70. In *Deep Oceans* Herring, P. J. and Clarke, M. R. (ed.) Arthur Barker, London, 320 pp.

101 Clarke, M. R. (1977) (a) 'A brief review of sampling techniques and tools of marine biology' pp. 439–469. In *A Voyage of Discovery* Angel, M. (ed.) Supplement to *Deep-Sea Research* Pergamon Press, 696 pp.

102 Clarke, M. R. (1977) (b) 'Beaks, nets and numbers' pp. 89–126. In *The Biology of Cephalopods* Nixon, M. and Messenger, J. B. (ed.) *Symp. Zool. Soc. Lond.*, 615 pp.

103 Clarke, M. R. (1978) 'Buoyancy control as a function of the spermaceti organ in the sperm whale' *J. Mar. Biol. Ass. UK.* 58, 1, pp. 27–72.

104 Clarke, M. R. and Herring, P. J. (1971) 'Animal miscellany: A survey of oceanic families' pp. 164–204. In *Deep Oceans* Herring, P. J. and Clarke, M. R. (ed.) Arthur Barker, London. 320 pp.

105 Clarke, M. R. and Lu, C. C. (1975) 'Vertical distribution of cephalopods at 18°N, 25°W in the North Atlantic' *J. Mar. Biol. Ass. UK.* 55, 1, pp. 165–182.

106 Clarke, M. R. and Lu, C. C. (1974) 'Vertical distribution of cephalopods at 30°N, 23°W in the North Atlantic' *J. Mar. Biol. Ass. UK.* 54, 4, pp. 969–984.

107 Clarke, T. A. (1973) 'Some aspects of the ecology of lanternfishes

(Myctophidae) in the Pacific Ocean near Hawaii' *Fishery Bulletin* 71, 2, pp. 401–434.

108 Clarke, W. D. (1961) 'A giant specimen of *Gnathophausia ingens* (Dohrn, 1870) (Mysidacea) and remarks on the asymmetry of the paragnaths in the suborder Lophogastrida' *Crustaceana* 2, 4, pp. 313–324.

109 Cohen, D. M. (1964) 'Bioluminescence in the Gulf of Mexico Anacanthine fish *Steindachneria argentea*' *Copeia*, pp. 406–409.

110 Cohen, D. M. (1973) 'Zoogeography of the fishes of the Indian Ocean' pp. 451–463. In *Ecological Studies. Analysis and Synthesis* Vol. 3, Zeitzschel, B. (ed.) Springer-Verlag, Berlin.

111 Cohen, D. M. (1977) 'Swimming performance of the gadoid fish *Antimora rostrata* at 2,400 metres' *Deep-Sea Res.* 24, pp. 275–277.

112 Cohen, D. M. and Pawson, D. L. (1977) 'Observations from the DSRV Alvin on populations of benthic fishes and selected larger invertebrates in and near DWD-106. Baseline Report of Environmental Conditions in Deepwater Dumpsite 106.' *NOAA Dumpsite Evaluation Report* 77–1, Vol. 2 Biological Characteristics US Dept. Commerce NOAA, Nat. Ocean Survey.

113 Collard, S. B. (1970) 'Forage of some eastern Pacific midwater fishes' *Copeia*, pp. 348–354.

114 Corliss, J. B. and Ballard, R. D. (1977) 'Oases of life in the cold abyss' *National Geographic Magazine*, 152, 2, pp. 441–453.

115 Corner, E. D. S., Denton, E. J. and Forster, G. R. (1969) 'On the buoyancy of some deep-sea sharks' *Proc. Roy. Soc. Lond. B.* 171, pp. 415–427.

116 Coull, B. C. (1972) 'Species diversity and faunal affinities of meiobenthic Copepoda in the deep sea' *Mar. Biol.* 14, pp. 48–51.

117 Coull, B. C., Ellison, R. L., Fleeger, J. W., Higgins, R. P., Hope, W. D., Hummon, W. D., Rieger, R. M., Sterrer, W. E., Thiel, H. and Tietjen, J. H. (1977) 'Quantitative estimates of the meiofauna from the deep-sea off North Carolina, USA.' *Mar. Biol.* 39, 3, pp. 233–240.

118 Crane, J. M. (1965) 'Bioluminescent courtship display in the teleost *Porichthys notatus*' *Copeia*, pp. 239–241.

119 Culliney, J. L. and Turner, R. D. (1976) 'Larval development of the deep-water wood boring bivalve *Xylophaga atlantica* (Richards) (Mollusca, Bivalvia, Pholadidae)' *Ophelia* 15, 2, pp. 149–162.

120 Cushing, D. H. (1951) 'The vertical migration of planktonic Crustacea' *Biol. Rev.* 26, pp. 158–192.

121 Cushing, D. H. (1973) (a) *The Detection of Fish* Pergamon Press, Oxford, 200 pp.

122 Cushing, D. H. (1973) (b) 'Production in the Indian Ocean and the transfer from the primary to the secondary level' pp. 475–544. In

Ecological Studies. Analysis and Synthesis Vol. 3, Zeitzschel, B. (ed.) Springer-Verlag, Berlin.

123 Cushing, D. H. (1975) Marine Ecology and Fisheries Cambridge University Press, 278 pp.

124 Cutler, E. B. (1975) 'Zoogeographical barrier on the continental slope off Cape Lookout, North Carolina' Deep-Sea Res. 22, pp. 893–901.

125 Cutler, E. B. (1977) 'The bathyal and abyssal Sipuncula' Galathea Report 14, pp. 135–156.

126 D'Aoust, B. G. (1970) 'Physical constraints on vertical migration by mesopelagic fishes' pp. 86–99. In Proceedings of an International Symposium on Biological Sound Scattering in the Ocean Farquhar, G. B. (ed.) Maury Center for Ocean Science, Washington, DC.

127 David, C. N. and Conover, R. J. (1961) 'Preliminary observations on the physiology and ecology of luminescence in the copepod Metridia lucens' Biol. Bull. Mar. Biol. Lab. Woods Hole 121, pp. 92–107.

128 David, P. M. (1955) 'The distribution of Sagitta gazellae' Discovery Report 27, pp. 235–278.

129 David, P. M. (1965) (a) 'The Chaetognatha of the Southern Ocean' pp. 296–323. In Biogeography and Ecology in Antarctica Van Mieghem, J. and Van Oye, P. Dr W. Junk, b.v., The Hague, 762 pp.

130 David, P. M. (1965) (b) 'The surface fauna of the ocean' Endeavour 24, pp. 95–100.

131 Dayton, P. K. and Hessler, R. R. (1972) 'Role of biological disturbance in maintaining diversity in the deep sea' Deep-Sea Res. 19, pp. 199–208.

132 Deevey, G. B. and Brooks, A. L. (1977) 'Copepods of the Sargasso Sea off Bermuda: species composition, and vertical and seasonal distribution between the surface and 2,000 m' Bull. Mar. Sci. 27, 2, pp. 256–291.

133 Denton, E. J. (1956) 'Recherches sur l'absorption de la lumière par le cristallin des poissons' Bull. Inst. Océanogr. Monaco 1071, pp. 1–10.

134 Denton, E. J. (1963) 'Buoyancy mechanisms of sea creatures' Endeavour 22, pp. 3–8.

135 Denton, E. J. (1970) 'On the organization of reflecting surfaces in some marine animals' Phil. Trans. Roy. Soc. Lond. B. 258, pp. 285–313.

136 Denton, E. J. (1971) 'Examples of the active transport of salts and water to give buoyancy in the sea' pp. 277–288. In 'A Discussion on Active Transport of Salts and Water in Living Tissues' Keynes, R. D. (org.) Phil. Trans. Roy. Soc. Lond. B. 262, pp. 83–342.

137 Denton, E. J. (1974) 'Trevor Ian Shaw (obituary)' pp. 359–380. In

Biographical Memoirs of Fellows of the Royal Society London, The Royal Society.

138 Denton, E. J. and Gilpin-Brown, J. B. (1973) 'Floatation mechanisms in modern and fossil cephalopods' *Adv. Mar. Biol.* 11, pp. 197–268.

139 Denton, E. J., Gilpin-Brown J. B. and Wright, P. G. (1972) 'The angular distribution of light produced by some mesopelagic fish in relation to their camouflage' *Proc. Roy. Soc. Lond. B.* 182, pp. 145–158.

140 Denton, E. J. and Herring, P. J. (1976) 'Filter pigments in photophores' *Cruise Report No 46, RRS Discovery cruise 77* Institute of Oceanographic Sciences.

141 Denton, E. J. and Marshall, N. B. (1958) 'The buoyancy of bathypelagic fishes without a gas-filled swimbladder' *J. Mar. Biol. Ass. UK.* 37, pp. 753–767.

142 Denton, E. J. and Shaw, T. I. (1961) 'The buoyancy of gelatinous marine animals' *J. Physiol. Lond.* 161, pp. 14–15.

143 Denton, E. J. and Warren, F. J. (1957) 'The photosensitive pigments in the retinas of deep-sea fish' *J. Mar. Biol. Ass. UK.* 36, pp. 651–662.

144 De Witt, F. A. (1972) 'Bathymetric distribution of two common deep-sea fishes, *Cyclothone acclinidens* and *C. signata* off Southern California' *Copeia* 88.

145 De Witt, F. A. and Cailliet, G. M. (1972) 'Feeding habits of two bristlemouth fishes, *Cyclothone acclinidens* and *C. signata* (Gonostomatidae)' *Copeia* pp. 868–871.

146 Dijkgraaf, S. (1962) 'The functioning and significance of the lateral line organs' *Biol. Rev.* 38, pp. 51–105.

147 Dineley, D. (1973) *Earth's Voyage Through Time* Hart-Davis, Mac-Gibbon, London, 320 pp.

148 Dinet, A. (1973) 'Distribution quantitative du méiobenthos profond dans la région de la dorsale de Walvis (Sud-Ouest Africain)' *Mar. Biol.* 20, pp. 20–26.

149 Donaldson, H. A. (1975) 'Vertical distribution and feeding of sergestid shrimps (Decapoda: Natantia) collected near Bermuda' *Mar. Biol.* 31, 1, pp. 37–50.

150 Donaldson, H. A. (1976) 'Chemical composition of sergestid shrimps (Decapoda: Natantia) collected near Bermuda' *Mar. Biol.* 38, 1, pp. 51–58.

151 Dugdale, R. C. (1976) 'Nutrient cycles' pp. 141–172. In *The Ecology of the Seas* Cushing, D. H. and Walsh, J. J. (ed.) Blackwell Scientific Publications, Oxford, 467 pp.

152 Dunlap, C. R. (1970) 'A reconnaissance of the deep scattering layers in the eastern tropical Pacific and the Gulf of California' pp. 395–408. In *Proceedings of an International Symposium on Biological*

Sound Scattering in the Ocean Farquhar, G. B. (ed.) Maury Center for Ocean Science, Washington, DC.

153 Durham, J. W. (1967) 'The incompleteness of our knowledge of the fossil record' *J. Paleont*, **41**, pp. 559–565.

154 Ebeling, A. W. (1967) 'Zoogeography of tropical deep-sea animals' pp. 593–613. In *Proceedings of the International Conference on Tropical Oceanography* 1965 University of Miami, Institute of Marine Sciences.

155 Ebeling, A. W. and Cailliet, G. M. (1974) 'Mouth size and predator strategy of midwater fishes' *Deep-Sea Res*. **21**, 11, pp. 959–968.

156 Ebeling, A. W., Cailliet, G. M., Ibara, R. M. and De Witt, F. A. Jr (1970) 'Pelagic communities and sound scattering off Santa Barbara, California' pp. 1–19. In *Proceedings of an International Symposium on Biological Sound Scattering in the Ocean* Farquhar, G. B. (ed.) Maury Center for Ocean Science, Washington, DC.

157 Ebeling, A. W. and Weed, W. H. (1963) 'Melamphaidae III, Systematics and distribution of the species in the bathypelagic fish genus *Scopelogadus* Vaillant' *Dana Report* **60**, pp. 1–58.

158 Edmunds, M. (1974) *Defence in Animals* Longman, London, 357 pp.

159 Edwards, A. S. and Herring, P. J. (1977) 'Observations on the comparative morphology and operation of the photogenic tissues of myctophid fishes' *Mar. Biol*. **41**, pp. 59–70.

160 Ekman, S. (1953) *Zoogeography of the Sea* Sidgwick and Jackson, London, 417 pp.

161 Elofsson, R. and Hallberg, E. (1977) 'Compound eyes of some deep-sea and fjord mysid crustaceans' *Acta Zoologica* **58**, 3, pp. 169–177.

162 Elton, C. S. (1966) *The Pattern of Animal Communities* Methuen and Co., London, 432 pp.

163 Emery, D. G. (1975) 'The Histology and Fine structure of the olfactory organ of the squid, *Lolliguncula brevis* Blainville' *Tiss. Cell*. **7**, pp. 357–367.

164 Emlen, J. M. (1973) *Ecology: an Evolutionary Approach* Addison-Wesley Publishing Company, Reading, Massachusetts, 493 pp.

165 Epply, R. W., Holm-Hansen, O. and Strickland, J. D. H. (1968) 'Some observations on the vertical migration of dinoflagellates' *J. Phycol*. **4**, pp. 333–340.

166 Ericson, D. B. and Wollin, G. (1967) *The Ever-Changing Sea* Alfred A. Knopf, New York, 354 pp.

167 Everson, I. (1977) *The Living Resources of the Southern Ocean* Southern Ocean Fisheries Survey Programme, FAO Rome, 156 pp.

168 Fage, L. (1956) 'Les pycnogonides (excl. le genre *Nymphon*)' *Galathea Report* **2**, pp. 167–182.

169 Fänge, R., Larsson, A. and Lidman, U. (1972) 'Fluids and jellies of

the acousticolateralis system in relation to body fluids in *Coryphaenoides rupestris* and other fishes' *Mar. Biol.* 17, 2, pp. 180–185.

170 Fauvel, P. (1959) 'Classe des annélides polychètes' *Traite de Zoologie* 5, Masson, Paris, pp. 13–223.

171 Fell, H. B. (1966) 'The ecology of ophiuroids' pp. 129–143. In *Physiology of Echinodermata* Boolootian, R. A. (ed.) Interscience, New York.

172 Fenchel, T. (1974) 'Intrinsic rate of natural increase: the relationship with body size' *Oecologia* 14, pp. 317–326.

173 Filatova, Z. A. (1959) 'Bivalve molluscs of the abyssal zone of the north-western Pacific' *Proceedings of the XVth International Congress of Zoology*, London 1958, pp. 221–222.

174 Filatova, Z. A. (1971) 'On some mass species of the bivalve molluscs from the ultra-abyssal of the Kurile–Kamchatka Trench' *Tr. Inst. Okeanol. USSR.* 92, pp. 46–60.

175 Filatova, Z. A. and Zenkevitch, L. A. (1966) 'The quantitative distribution of the deep-sea bottom fauna in the Pacific Ocean' *Second International Oceanographic Congress*, Moscow, 1966, pp. 114–115.

176 Fischer, A. G. (1960) 'Latitudinal variations in organic diversity' *Evolution* 14, pp. 64–81.

177 Forster, G. R. (1971) 'Line fishing on the continental slope. III Midwater fishing with vertical lines' *J. Mar. Biol. Ass. UK.* 51, 1, pp. 73–77.

178 Foxton, P. (1964) 'Seasonal variations in the plankton of Antarctic waters' *Biologie Antarctique First, SCAR Symposium* Hermann, Paris, pp. 311–318.

179 Foxton, P. (1970) (a) 'The vertical distribution of pelagic, decapods (Crustacea; Natantia) collected on the SOND cruise 1965. I The Caridea' *J. Mar. Biol. Ass. UK.* 50, 4, pp. 939–960.

180 Foxton, P. (1970) (b) 'The vertical distribution of pelagic decapods (Crustacea; Natantia) collected on the SOND cruise 1965. II The Penaeidea and general discussion' *J. Mar. Biol. Ass UK.* 50, 4, pp. 961–1000.

181 Foxton, P. and Roe, H. S. J. (1974) 'Observations on the nocturnal feeding of some mesopelagic decapod Crustacea' *Mar. Biol.* 28, pp, 37–49.

182 Frankenberg, D. and Menzies, R. J., (1968) 'Some quantitative analyses of deep-sea benthos off Peru' *Deep-Sea Res.* 15, pp. 623–626.

183 Frederiksen, R. D. (1973) 'On the retinal diverticula in the tubular-eyed opisthoproctid deep-sea fishes *Macropinna microstoma* and *Dolichopteryx longipes*' *Vidensk. Meddr. Dansk Naturh. Foren,* 136, pp. 233–244.

184 Friedman, M. M. and Strickler, J. R. (1975) 'Chemoreceptors and feeding in calanoid copepods (Arthropoda: Crustacea)' *Proc. Natn. Acad. Sci. USA* **72**, pp. 4185–4188.

185 Friedrich, H. (1969) *Marine Biology* Sidgwick and Jackson, London, 474 pp.

186 Fryer, G. and Iles, T. D. (1972) *The Cichlid Fishes of the Great Lakes of Africa* Oliver and Boyd, Edinburgh, 641 pp.

187 Fuzessery, M. M. and Childress, J. J. (1975) 'Comparative chemosensitivity to amino-acids and their role in the feeding activity of bathypelagic and littoral crustaceans' *Biol. Bull. Woods Hole* **149**, pp. 522–538.

188 Gardiner, A. C. (1933) 'Vertical distribution in *Calanus finmarchicus*' *J. Mar. Biol. Ass. UK.* **18**, pp. 575–610.

189 Gardiner, F. P. and Haedrich, R. L. (1978) 'Zonation in the deep benthic megafauna' *Oecologia* **31**, pp. 311–317.

190 Gardiner, L. F. (1975) 'The systematics, postmarsupial development and ecology of the deep-sea family Neotanaidae (Crustacea: Tanaidacea)' *Smithsonian Contributions to Zoology* No 170, pp. 1–265.

191 Gardiner, M. S. (1972) *Biology of the Invertebrates* McGraw-Hill, New York, 954 pp.

192 Geistdoerfer, P. (1975) *Ecologie Alimentaire de Macrouridae (Téléostéens, Gadiformes)* Thèse de Doctorat D'Etat Es-Sciences Naturalles, l'Université de Paris VI, 315 pp.

193 George, J. D. and Southward, E. C. (1973) 'A comparative study of the setae of Pogonophora and Polychaetous Annelida' *J. Mar. Biol. Ass. UK.* **53**, pp. 403–424.

194 Gerlach, A. S. (1971) 'On the importance of marine meiofauna for benthos communities' *Oecologia* **6**, pp. 176–190.

195 Ghirardelli, E. (1968) 'Some aspects of the biology of the chaetognaths' *Adv. Mar. Biol.* **6**, pp. 271–375.

196 Ghiselin, M. T. (1969) 'The evolution of hermaphroditism among animals' *Quart. Rev. Biol.* **44**, pp. 189–208.

197 Gibbs, R. H., Goodyear, R. H., Keene, M. J. and Brown, D. W. (1971) *Biological Studies of the Bermuda Ocean Acre. II Vertical Distribution and Ecology of the Lanternfishes (Family Myctophidae)* Report to US Navy Underwater Systems Center, Smithsonian Institution, Washington, DC.

198 Gilbert, C. H. and Hubbs, C. L. (1920) 'The macrouroid fishes of the Philippine Islands and the East Indies' *Bull. US Nat. Mus.* (**100**) 1, 7, pp. 369–588.

199 Gislén, T. (1951) 'Crinoidea with a survey of the bathymetric distribution of the deep-sea crinoids' *Rep. Swedish Deep-Sea Expedition 1947–1948* **2** (Zool.), 3, pp. 51–59.

200 Glover, R. S. (1967) 'The continuous plankton recorder survey of the North Atlantic' *Symp. Zool. Soc. Lond.* **19**, pp. 189–210.

201 Golovan, G. G. (1974) 'Preliminary data on the composition and distribution of the bathyal ichthyofauna (in the Cap Blanc area)' *Oceanology* **14**, 2, pp. 288–290.

202 Goodyear, R. H. and Gibbs, R. H. (1969) 'Systematics and zoogeography of stomiatoid fishes of the *Astronesthes cyaneus* species group (family Astronesthidae) with descriptions of three new species' *Arch. Fischereiwiss* **20**, 2/3, pp. 107–131.

203 Goodyear, R. H., Zahuranec, B. J., Pugh, W. L. and Gibbs, R. H. (1972) 'Ecology and vertical distribution of Mediterranean midwater fishes' pp. 91–229. In *Mediterranean Biological Studies Final Report* Report to US Navy Office of Naval Research, Contract No N00014–67–A–399–000–7.

204 Gordon, D. C. (1970 (a) 'A microscopic study of organic particles in the North Atlantic Ocean' *Deep-Sea Res.* **17**, pp. 175–185.

205 Gordon, D. C. (1970) (b) 'Some studies on the distribution and composition of particulate organic carbon in the North Atlantic Ocean' *Deep-Sea Res.* **17**, pp. 233–253.

206 Gordon, I. (1955) 'Crustacea Decapoda' *Rep. Swedish Deep-Sea Expedition 1947–1948* **2** (Zool), 19, pp. 239–245.

207 Gordon, M. S., Belman, B. W. and Chow, P. H. (1976) 'Comparative studies on the metabolism of shallow-water and deep-sea marine fishes. IV Patterns of aerobic metabolism in the mesopelagic deep-sea fangtooth fish *Anoplogaster cornuta*' *Mar. Biol.* **35**, pp. 287–293.

208 Gould, S. J. (1975) 'A threat to Darwinism' *Natural History* **84**, 10, pp. 4–9.

209 Gould, S. J. (1976) 'Paleontology plus ecology as paleobiology' pp. 218–236. In *Theoretical Ecology* May, R. M. (ed.) Blackwell Scientific Publications, Oxford, 317 pp.

210 Gould, S. J. (1977) *Ontogeny and Phylogeny* Harvard University Press, Cambridge, Mass., London, England, 501 pp.

211 Grassle, J. F. (1977) 'Slow recolonisation of deep-sea sediment' *Nature* **265**, pp. 618–619.

212 Grassle, J. F., Sanders, H. L., Hessler, R. R., Rowe, G. T. and McLellan, T. (1975) 'Pattern and zonation: a study of the bathyal megafauna using the research submersible *Alvin*' *Deep-Sea Res.* **22**, 7, pp. 457–482.

213 Greenslate, J. (1974) 'Microorganisms participate in the construction of manganese nodules' *Nature* **249**, pp. 181–183.

214 Grey, M. (1964) 'Family Gonostomatidae' *Memoir Sears Foundation for Marine Research* **1**, 4, pp. 78–240.

215 Greze, V. N. (1970) 'The biomass and production of different trophic levels in the pelagic communities of South Seas' pp. 458–467. In

Marine Food Chains Steele, J. H. (ed.) Oliver and Boyd, Edinburgh, 552 pp.

216 Grice, G. D. (1972) 'The existence of a bottom-living calanoid copepod fauna in deep water with descriptions of five new species' *Crustaceana* **23**, 3, pp. 219–242.

217 Grice, G. D. and Hart, A. D. (1962) 'The abundance, seasonal occurrence and distribution of epizooplankton between New York and Bermuda' *Ecol. Monog.* **32**, 4, pp. 287–307.

218 Grice, D. G. and Hulseman, K. (1965) 'Abundance, vertical distribution and taxonomy of calanoid copepods at selected stations in the north-east Atlantic' *J. Zool.* **146**, pp. 213–262.

219 Grice, G. D. and Hulseman, K. (1967) 'Bathypelagic calanoid copepods of the western Indian Ocean' *Proc. US Nat. Mus.* **122**, 3583, pp. 1–67.

220 Gross, F. and Zeuthen, E. (1948) 'The buoyancy of plankton diatoms: a problem of cell physiology' *Proc. Roy. Soc. B.* **135**, 880, pp. 382–389.

221 Hadfield, M. G. (1975) 'Hemichordata' pp. 185–240. In *Reproduction of Marine Invertebrates, Vol. II. Entoprocts and Lesser Coelomates* Giese, A. C. and Pearse, J. S. (ed.) Academic Press, London.

222 Haedrich, R. L. (1974) 'Pelagic capture of the epibenthic rattail *Coryphaenoides rupestris*' *Deep-Sea Res.* **21**, pp. 977–979.

223 Haedrich, R. L. and Henderson, N. R. (1974) 'Pelagic food of *Coryphaenoides armatus*, a deep benthic rattail' *Deep-Sea Res.* **21**, pp. 739–744.

224 Haedrich, R. L. and Nielsen, J. G. (1966) 'Fishes eaten by *Alepisaurus* (Pisces, Iniomi) in the south eastern Pacific Ocean' *Deep-Sea Res. Oceanogr. Abstr.* **13**, pp. 909–919.

225 Haedrich, R. L. and Polloni, P. T. (1976) 'A contribution to the life history of a small rattail fish, *Coryphaenoides carapinus*' *Bull. Southern California Academy of Science* **75**, pp. 203–211.

226 Haedrich, R. L. and Rowe, G. T. (1977) 'Megafaunal biomass in the deep sea' *Nature* **269**, 5624, pp. 141–142.

227 Haedrich, R. L., Rowe, G. T. and Polloni, P. T. (1975) 'Zonation and faunal composition of epibenthic populations on the continental slope south of New England' *J. Mar. Res.* **33**, 2, pp. 191–212.

228 Haigh, K. K. R. (1970) 'Geographic, seasonal, and annual patterns of midwater scatterers between latitudes 10° and 68° north in the Atlantic' pp. 268–280. In *Proceedings of an International Symposium on Biological Sound Scattering in the Ocean* Farquhar, G. B. (ed.) Maury Center for Ocean Science, Washington, DC.

229 Hamner, W. M. (1974) 'Blue-water plankton' *National Geographic Magazine* October 1974, pp. 530–545.

230 Hamner, P. and Hamner, W. M. (1977) 'Chemosensory tracking

of scent trails by the planktonic shrimp *Acetes sibogae australis'* *Science* **195**, 4281, pp. 886–888.

231 Haneda, Y. (1938) 'Über den leuchtfisch *Malacocephalus laevis* (Lowe)' *Jap. J. Med. Sci. III Biophysics* **3**, pp. 355–366.

232 Haneda, Y. (1941) 'The luminescence of some deep-sea fishes of the families Gadidae and Macrouridae' *Pacific Science* **5**, 4, pp. 372–378.

233 Haneda, Y. (1957) 'Observations on luminescence in the deep-sea fish *Paratrachichthys prosthemius'* *Sci. Rep. Yokosuka City Mus.* **2**, pp. 15–22.

234 Hansen, B. (1956) 'Holothuroidea from depths exceeding 6000 metres' *Galathea Report* **2**, pp. 33–54.

235 Hansen, B. (1975) 'Systematics and biology of the deep-sea holothurians. Part 1. Elasipoda' *Galathea Report* **13**, pp. 1–262.

236 Hansen, B. and Madsen, F. J. (1956) 'On two bathypelagic Holothurians from the South China Sea, *Galatheathuria* n.g. *aspera* (Théel) and *Enypiastes globosa* n. sp.' *Galathea Report*, **2**, pp. 55–60.

237 Hansen, K. and Herring, P. J. (1977) 'Dual bioluminescent systems in the angler fish genus *Linophryne'* *J. Zool. Lond.* **181**, 1, pp. 103–124.

238 Hansen, V. Kr. and Wingstrand, K. G. (1960) 'Further studies on the non-nucleated erythrocytes of *Maurolicus mülleri*, and comparisons with the blood cells of related species' *Dana Report* **54**, pp. 1–15.

239 Harbison, G. R., Briggs, D. C. and Madin, L. P. (1977) 'The associations of Amphipoda Hyperiida with gelatinous zooplankton II Associations with Cnidaria, Ctenophora and Radiolaria' *Deep-Sea Res.* **24**, 5, pp. 465–488.

240 Harding, G. C. H. (1974) 'The food of deep-sea copepods' *J. Mar. Biol. Ass. UK.* **54**, 1, pp. 141–156.

241 Hardy, A. C. (1956) *The Open Sea* Collins, London, 335 pp.

242 Hardy, M. G. (1962) 'Photophore and eye movement in the euphausiid *Meganyctiphanes norvegica* (G. O. Sars)' *Nature* **196**, pp. 790–791.

243 Harkness, L. and Bennet-Clark, H. C. (1978) 'The deep fovea as a focus indicator' *Nature* **272**, 5656, pp. 814–816.

244 Hartman, O. and Emery, K. O. (1956) 'Bathypelagic coelenterates' *Limnol. Oceanogr.* **1**, pp. 304–312.

245 Hartman, O. and Fauchald, K. (1971) 'Deep water benthic polychaetous annelids off New England to Bermuda and other North Atlantic areas. Part II' *Allan Hancock Monographs in Marine Biology* **6**, 327 pp.

246 Harvey, E. N. (1952) *Bioluminescence* Academic Press, New York, 649 pp.

247 Hasler, Grethe R. (1976) 'The biogeography of some marine planktonic diatoms' *Deep-Sea Res.* **23**, pp. 319–338.

248 Heath, D. J. (1977) 'Simultaneous hermaphroditism; Cost and benefit' *J. Theor. Biol.* **64**, pp. 363–373.

249 Heezen, B. C. and Hollister, C. D. (1971) *The Face of the Deep* Oxford University Press, 659 pp.

250 Heezen, B. C. and Rawson, M. (1977) 'Influence of abyssal circulation on sedimentary accumulations in space and time' *Marine Geology* **23**, pp. 173–196.

251 Heron, A. C. (1972) 'Population ecology of a colonising species: the pelagic tunicate *Thalia democratica* II Population growth rate' *Oecologia* **10**, pp. 294–312.

252 Herring, P. J. (1967) 'The pigments of plankton at the sea surface' *Symp. Zool. Soc. Lond.* No 19, pp. 215–236.

253 Herring, P. J. (1973) 'Depth distribution of the carotenoid pigments and lipids of some oceanic animals. 2 Decapod Crustaceans' *J. mar. Biol. Ass. UK.* **53**, pp. 539–562.

254 Herring, P. J. (1974) (a) 'Size, density and lipid content of some decapod eggs' *Deep-Sea Res.* **21**, pp. 91–94.

255 Herring, P. J. (1974) (b) 'New observations on the bioluminescence of echinoderms' *J. Zool. Lond.* **172**, pp. 401–418.

256 Herring, P. J. (1976) (a) 'Bioluminescence in decapod Crustacea' *J. Mar. Biol. Ass. UK.* **56**, 4, pp. 1029–1048.

257 Herring, P. J. (1976) (b) 'Bioluminescence in the Evermannellidae' *J. Zool.*

258 Herring, P. J. (1977) (a) 'Bioluminescence in an evermannellid fish' *J. Zool. Lond.* **181**, pp. 297–307.

259 Herring, P. J. (1977) (b) 'Bioluminescence of marine organisms' *Nature* **267**, pp. 788–793.

260 Herring, P. J. (1977) (c) 'Luminescence in cephalopods and fish' pp. 127–159. In *The Biology of Cephalopods* Nixon, M. and Messenger, J. B. (ed.) *Symp. Zool. Soc. Lond.* No 38, 615 pp.

261 Herring, P. J. (1977) (d) 'Oral light organs in *Sternoptyx* and the bioluminescence of hatchet-fishes' pp. 553–567. In *A Voyage of Discovery*, Angel, M. (ed.) suppl. to *Deep-Sea Res.* Pergamon Press, Oxford, 696 pp.

262 Herring, P. J. and Locket, N. A. (1978) 'The luminescence and photophores of euphausiid crustaceans' *J. Zool. Lond.* **186**, 4, pp. 431–462.

263 Hersey, J. B. and Backus, R. H. (1962) 'Sound scattering by marine organisms' pp. 498–539. In *The Sea Vol.* 1 *Physical Oceanography* Interscience Publishers, New York and London.

264 Hessler, R. R. (1972) (a) 'Deep water organisms for high pressure aquarium studies' pp. 151–163. In *Barobiology and the Experi-*

mental Biology of the Deep Sea, Brauer, R. W. (ed.) North Carolina Sea Grant Program.

265 Hessler, R. R. (1972) (b) 'The structure of deep benthic communities from central oceanic waters' pp. 79–93. In *The Biology of the Oceanic Pacific* Mitler, C. B. (ed.) Oregon State University Press.

266 Hessler, R. R. and Jumars, P. A. (1974) 'Abyssal community analysis from replicate box cores in the central North Pacific' *Deep-Sea Res.* **21**, pp. 185–209.

267 Hessler, R. R. and Sanders, H. L. (1967) 'Faunal diversity in the deep sea' *Deep-Sea Res.* **14**, pp. 65–78.

268 Hickling, C. F. (1925) 'A new type of luminescence in fishes I' *J. Mar. Biol. Ass. UK.* **13**, pp. 914–929.

269 Hickling, C. F. (1926) 'A new type of luminescence in fishes II' *J. Mar. Biol. Ass. UK.* **14**, pp. 495–507.

270 Holm-Hansen, O. (1970) 'Microbial distribution in ocean water relative to nutrients and food sources' pp. 147–155. In *Proceedings of an International Symposium on Biological Sound Scattering in the Ocean* Farquhar, G. B. (ed.) Maury Center for Ocean Science, Washington, DC.

271 Holm-Hansen, O. and Booth, C. R. (1966) 'The measurement of adenosine triphosphate in the ocean and its ecological significance' *Limnol. Oceanogr.* **11**, pp. 510–519.

272 Hopkins, T. L. and Baird, R. C. (1977) 'Aspects of the feeding ecology of oceanic midwater fishes' pp. 325–360. In *Oceanic Sound Scattering Prediction* Andersen, N. R. and Zahuranec, B. J. (ed.) Plenum Press, New York and London, 859 pp.

273 Horridge, G. A. (1966) 'Some newly discovered underwater vibration receptors in invertebrates' pp. 395–406. In *Some Contemporary Studies in Marine Science* Barnes, H. (ed.) George Allen and Unwin, London, 716 pp.

274 Horridge, G. A. and Boulton, P. S. (1967) 'Prey detection by Chaetognatha via a vibration sense' *Proc. Roy. Soc. Lond. B.* **168**, 1013, pp. 413–419.

275 Hulet, W. H. (1978) 'Structure and functional development of the eel leptocephalus *Ariosoma balearicum* (De La Roche, 1809)' *Phil. Trans. Roy. Soc. Lond. B.* **282**, 987, pp. 107–138.

276 Hutchinson, G. E. (1957) 'Concluding remarks' *Cold Spring Harbor Symp. Quant. Biol.* **22**, Population Studies: Animal ecology and demography, pp. 415–427.

277 Hutchinson, G. E. (1959) 'Homage to Santa Rosalia or why are there so many kinds of animals?' *Amer. Naturalist* **93**, pp. 137–145.

278 Hutchinson, G. E. (1975) 'Variations on a theme by Robert MacArthur' pp. 492–521. In *Ecology and Evolution of Communities*

Cody, M. L. and Diamond, J. M. (ed.) Harvard University Press, Cambridge, Mass. 545 pp.

279 Hyman, L. H. (1955) *The Invertebrates: Echinodermata Vol. IV* McGraw-Hill Book Company, New York, 763 pp.

280 Hyman, L. H. (1959) *The Invertebrata: Smaller Coelomate Groups Vol. V* McGraw-Hill Book Company New York, 782 pp.

281 Isaacs, J. D. (1969) 'The nature of oceanic life' *Scientific American* **221**, 3, pp. 146–165.

282 Isaacs, J. D. and Schwartzlose, R. (1975) 'Active animals of the deep-sea floor' *Scientific American* **233**, 4, pp. 84–91

283 Isaacs, J. D., Tout, S. A. and Wick, G. L. (1974) 'Deep scattering layers: vertical migration as a tactic for finding food' *Deep-Sea Res.* **21**, pp. 651–656.

284 Ivanov, A. V. (1963) *Pogonophora* Academic Press, London, 479 pp.

285 Iwamoto, T. (1975) 'The abyssal fish *Antimora rostrata* (Günther)' *Comp. Biochem. Physiol.* **52B**, pp. 7–11.

286 Jacob, F. (1974) *The Logic of Living Systems* Allen Lane, London, 348 pp.

287 Jacobs, W. (1935) 'Das Schweben der wasser Organismen' *Ergebnisse der Biologie II Berlin*, pp. 131–218.

288 Jacobs, W. (1937) 'Beobachtungen über das Schweben der Siphonophoren' *Z. Vergl. Physiol.* **24**, 4, pp. 583–601.

289 James, P. T. (1973) 'Distribution of dimorphic males of three species of *Nematoscelis* (Euphausiacea)' *Mar. Biol.* **19**, pp. 341–347.

290 Jannasch, H. W. and Wirsen, C. O. (1977) 'Microbial life in the deep sea' *Scientific American* **236**, 6, pp. 42–65.

291 Jespersen, P. and Tåning, A. V. (1926) 'Mediterranean Sternoptychidae' *Rep. Danish Oceanogr. Exped. Medit.* **2** (Biol.) A **12**, pp. 1–59.

292 Jones, F. R. H. and Marshall, N. B. (1953) 'The structure and functions of the teleostean swimbladder' *Biol. Rev.* **28**, pp. 16–83.

293 Jumars, P. A. and Hessler, R. R. (1976) 'Hadal community structure: implications from the Aleutian Trench' *J. Mar. Res.* **34**, 4, pp. 547–560.

294 Kampa, E. M. (1965) 'The euphausiid eye—a re-evaluation' *Vision Research* **5**, pp. 475–481.

295 Kampa, E. M. (1970) 'Photoenvironment and sonic scattering' pp. 51–59. In *Proceedings of an International Symposium on Biological Sound Scattering in the Ocean* Farquhar, G. B. (ed.) Maury Center for Ocean Science, Washington DC. 629 pp.

296 Kanwischer, J. and Ebeling, A. W. (1957) 'Composition of the swimbladder gas in bathypelagic fishes' *Deep-Sea Res.* **4**, pp. 211–217.

297 Karnella, C. and Gibbs, R. H. (1977) 'The lanternfish *Lobianchia dofleini*: An example of the importance of life-history information

in prediction of oceanic sound scattering' pp. 361–380. In *Oceanic Sound Scattering Prediction (Marine Science Vol. 5)* Andersen, N. R. and Zahuranec, B. J. (ed.) Plenum Press, New York and London, 859 pp.

298 Kawaguchi, K. and Marumo, R. (1967) 'Biology of *Gonostoma gracile* (Gonostomatidae) I. Morphology, life history and sex reversal' *Information Bulletin on Planktonology in Japan*, pp. 53–67. Commemoration number of Dr Y. Matsue.

299 King, J. E. and Hida, T. S. (1954) 'Variations in zooplankton abundance in Hawaiian waters' *Spec. Scient. Rep. US Fish Wild. Serv. Fisheries* 118, pp. 1–66.

300 King, P. E. (1973) *Pycnogonids* Hutchinson, London, 144 pp.

301 Kinzer, J. (1970) 'On the contribution of euphausiids and other plankton organisms to deep scattering layers in the eastern North Atlantic' pp. 476–489. In *Proceedings of an International Symposium on Biological Sound Scattering in the Ocean* Farquhar, G. B. (ed.) Maury Center for Ocean Science, Washington, DC. 629 pp.

302 Kirkegaard, J. B. (1956) 'Pogonophora, *Galathealinum bruuni* n. gen. n. sp., a new representative of the class' *Galathea Report* 2, pp. 79–84.

303 Klyashtorin, L. B. and Yarzhombek, A. A. (1973) 'Energy consumption in active movements of planktonic organisms' *Oceanology* 13, 4, pp. 697.

304 Knudsen, J. (1961) 'The bathyal and abyssal Xylophaga (Pholadidae, Bivalvia)' *Galathea Report* 5, pp. 163–209.

305 Knudsen, J. (1970) 'The systematics and biology of abyssal and hadal Bivalvia' *Galathea Report* 11, 82 pp.

306 Koltun, V. M. (1970) 'Sponges of the Arctic and Antarctic; a faunistic review' *Symp. Zool. Soc. Lond.* No 25, pp. 285–297.

307 Koltun, V. M. (1972) 'Sponge fauna of the northwestern Pacific from the shallows to hadal depths' pp. 177–233. In *Fauna of the Kurile–Kamchatka Trench and its Environment* Bogorov, V. G. (ed.) *Tr. Inst. Okeanol.* 86, English transl.

308 Kramp, P. L. (1956) (a) 'Hydroids from depths exceeding 6,000 metres' *Galathea Report* 2, pp. 17–20.

309 Kramp, P. L. (1956) (b) 'Pelagic fauna' pp. 65–86. In *The Galathea Deep Sea Expedition* Bruun, A. F., Greve, Sv., Mielche, H. and Spärck, R. (ed.) George Allen and Unwin, London, 296 pp.

310 Kramp, P. L. (1959) '*Stephanoscyphus* (Scyphozoa)' *Galathea Report* 1, pp. 173–188.

311 Kuhn, W., Ramel, A., Kuhn, H. J. and Marti, E. (1963) 'The filling mechanism of the swimbladder' *Experientia* 19, 10, pp. 457–511.

312 Land, M. F. (1976) 'Optics of the eyes of mid-water crustaceans' pp. 17–18. *Cruise Report No 46. Distribution and Physiology of*

Plankton and Benthos in the Region of the Canary Islands Institute of Oceanographic Sciences.

313 Lasker, R. (1970) 'Utilization of zooplankton energy by a Pacific sardine population in the Californian current' pp. 265–284. In *Marine Food Chains* Steele, J. H. (ed.) Oliver and Boyd, Edinburgh.

314 Lawrence, J. M. (1975) 'On the relationship between marine plants and sea urchins' *Oceanogr. Mar. Biol. Ann. Rev.* 13, pp. 213–286.

315 Lawry, J. V. (1974) 'Lantern fish compare downwelling light and bioluminescence' *Nature* 247, pp. 155–157.

316 Le Danois, E. (1948) '*Les profondeurs de la mer. Trente ans de recherches sur la fauna sousmarine au large des côtes de France*' Payot, Paris, 303 pp.

317 Lemche, H. and Wingstrand, K. G. (1959) 'The anatomy of *Neopilina galatheae* Lemche, 1957 (Mollusca, Tryblidiacea)' *Galathea Report* 3, pp. 9–72.

318 Lemche, H., Hansen, B., Madsen, F. J., Tendal, O. S. and Wolff, T. (1976) 'Hadal life as analyzed from photographs' *Vidensk. Meddr. Dansk. Naturh. Foren.* 139, pp. 263–336.

319 Levenstein, R. Ya. (1971/72) 'Ecology and zoogeography of some Polychaeta representatives of the abyssal Pacific' *Proc. Roy. Soc. Edin.* 73, pp. 171–181.

320 Lévi, C. (1964) 'Spongiaires des zones bathyale abyssale et hadal' *Galathea Report* 7, pp. 63–112.

321 Lewontin, R. C. and Hubby, J. L. (1966) 'A molecular approach to the study of genic heterozygosity in natural populations. II Amount of variation and degree of heterozygosity in natural populations of *Drosophila pseudoobscura*' *Genetics* 54, pp. 595–609.

322 Lighthill, M. J. (1969) 'Hydromechanics of aquatic animal propulsion' *Ann. Rev. Fluid Mechanics* 1, pp. 413–446.

323 Litvinova, N. M. and Sokolova, M. N. (1971) 'Feeding of deep-sea ophiuroids of the genus *Amphiophiura*' *Okeanologija* 11, 2, pp. 293–301.

324 Lock, A. R. and McLaren, I. A. (1970) 'The effect of varying and constant temperatures on size of a marine copepod' *Limnol. Oceanogr.* 15, pp. 638–640.

325 Locket, N. A. (1971) 'Retinal anatomy in some scopelarchid deep-sea fishes' *Proc. Roy. Soc. Lond. B.* 178, pp. 161–184.

326 Locket, N. A. (1977) 'Adaptations to the deep-sea environment' pp. 67–192. In *The Visual System in Vertebrates* Crescitelli, F. (ed.) *Handbook of Sensory Physiology* Vol. VII/5 Springer-Verlag, Berlin, 813 pp.

327 Longhurst, A. R. (1976) 'Vertical migration' pp. 116–140. In *The Ecology of the Seas* Cushing, D. H. and Walsh J. J. (ed.) Blackwell Scientific Publications, Oxford, 467 pp.

328 Lonsdale, P. (1977) 'Clustering of suspension feeding macrobenthos near abyssal hydrothermal vents at oceanic spreading centres' *Deep-Sea Res.* **24**, 9, pp. 857–864.

329 Lowenstein, O. (1957) 'The acoustico-lateralis system' pp. 155–186. In *The Physiology of Fishes Vol. II Behaviour*, Brown, M. E. (ed.) Academic Press, New York, 526 pp.

330 Lu, C. C. and Clarke, M. R. (1975) 'Vertical distribution of cephalopods at 40°N, 53°N and 60°N at 20°W in the North Atlantic' *J. Mar. Biol. Ass. UK.* **55**, 1, pp. 143–164.

331 MacArthur, R. H. (1962) 'Some generalized theories of natural selection' *Proc. Nat. Acad. Sci.* **48**, pp. 1893–1897.

332 MacDonald, A. G. (1975) *Physiological Aspects of Deep-Sea Biology* Cambridge University Press, 450 pp.

333 MacFadyen, A. (1963) *Animal Ecology, Aims and Methods (2nd Ed.)* Pitman and Sons, London, pp. 344.

334 MacIntyre, F. (1974) 'The top millimetre of the ocean' *Scientific American* **230**, 5, pp. 62–77.

335 MacKintosh, N. A. (1937) 'The seasonal circulation of the Antarctic macroplankton' *Discovery Reports* **16**, pp. 365–412.

336 Madin, L. P. and Harbison, G. R. (1977) 'The association of Amphipoda Hyperiida with gelatinous zooplankton I Associations with Salpidae' *Deep-Sea Res.* **24**, 5, pp. 449–464.

337 Madsen, F. J. (1957) (a) '*Primnoella krampi* n. sp. A new deep-sea Octocoral' *Galathea Report* **2**, pp. 21–22.

338 Madsen, F. J. (1956) (b) 'Echinoidea, Asteroidea and Ophiuroidea from depths exceeding 6,000 metres' *Galathea Report* **2**, pp. 23–32.

339 Madsen, F. J. (1961) (a) 'The Porcellanasteridae. A monographic revision of an abyssal group of sea-stars' *Galathea Report* **4**, pp. 33–176.

340 Madsen, F. J. (1961) (b) 'On the zoogeography and origin of the abyssal fauna' *Galathea Report* **4**, pp. 177–218.

341 Magnus, D. B. E. (1967) 'Ecological and ethological studies and experiments on the echinoderms of the Red Sea' pp. 635–664. In *Proceedings of the International Conference on Tropical Oceanography 1965* University of Miami, Institute of Marine Sciences.

342 Marshall, N. B. (1951) 'Bathypelagic fishes as sound scatterers in the ocean' *J. Mar. Res.* **10**, 1, pp. 1–17.

343 Marshall, N. B. (1953) 'Egg size in arctic, antarctic and deep-sea fishes' *Evolution* **7**, 4, pp. 328–341.

344 Marshall, N. B. (1954) *Aspects of Deep-Sea Biology* Hutchinson, London, 380 pp.

345 Marshall, N. B. (1955) 'Alepisauroid fishes' *Discovery Report* **27**, pp. 303–336.

346 Marshall, N. B. (1960) 'Swimbladder structure of deep-sea fishes in

relation to their systematics and biology' *Discovery Report* 31, pp. 1–122.

347 Marshall, N. B. (1961) 'A young *Macristium* and the ctenothrissid fishes' *Bull. Brit. Mus. Nat. Hist. Zoology* 7, 8, pp. 353–370.

348 Marshall, N. B. (1962) (a) 'Observations on the Heteromi, an order of teleost fishes' *Bull. Brit. Mus. Nat. Hist. Zool.* 9, 6, pp. 249–270.

349 Marshall, N. B. (1962) (b) 'The biology of sound producing fishes' *Symp. Zool. Soc. Lond.* 7, pp. 45–60.

350 Marshall, N. B. (1963) 'Diversity, distribution and speciation of deep-sea fishes' *Systematics Association, Publication Number 5. Speciation in the Sea* pp. 181–195.

351 Marshall, N. B. (1964) 'Bathypelagic macrourid fishes' *Copeia* 1, pp. 86–93.

352 Marshall, N. B. (1965) 'Systematic and biological studies of the macrourid fishes (Anacanthini-Teleostii)' *Deep-Sea Res.* 12, pp. 229–322.

353 Marshall, N. B. (1967) 'The olfactory organs of bathypelagic fishes' pp. 57–70. In *Aspects of Marine Zoology* Marshall, N. B. (ed.) *Symp. Zool. Soc. Lond.* 19, 270 pp.

354 Marshall, N. B. (1970) 'Swimbladder development and the life of deep-sea fishes' pp. 69–73. In *Proceedings of an International Symposium on Biological Sound Scattering in the Ocean* Farquhar, G. B. (ed.) Maury Center for Ocean Science, Washington DC.

355 Marshall, N. B. (1971) *Explorations in the Life of Fishes* Harvard University Press, Cambridge, Mass., 204 pp.

356 Marshall, N. B. (1972) 'Swimbladder organization and depth ranges of deep-sea teleosts' pp. 261–272. In *The Effects of Pressure on Organisms. Symp. Soc. Exper. Biol.* No 26, Cambridge University Press, 516 pp.

357 Marshall, N. B. (1973) 'Family Macrouridae' pp. 496–665. In *Fishes of the Western North Atlantic. Memoir Sears Foundation for Marine Research* 1, 6.

358 Marshall, N. B. (1974) 'Midwater fishes of the dark ocean' *Spectrum, British Science News* No 118/1, pp. 2–5.

359 Marshall, N. B. and Bourne, D. W. (1964) 'A photographic survey of benthic fishes in the Red Sea and the Gulf of Aden, with observations on their population density, diversity and habits' *Bull. Mus. Comp. Zool. Harvard Univ.* 132, 2, pp. 223–244.

360 Marshall, N. B. and Cohen, D. M. (1973) 'Order Anacanthini (Gadiformes)' pp. 479–495. In *Fishes of the Western North Atlantic. Memoir Sears Foundation for Marine Research* 1, 6.

361 Marshall, N. B. and Merrett, N. R. (1977) 'The existence of a benthopelagic fauna in the deep sea' pp. 483–498. In *A Voyage of Discovery* Angel, M. (ed.) Pergamon Press, Oxford, 696 pp.

362 Marshall, N. B. and Staiger, J. C. (1975) 'Aspects of the structure, relationships and biology of the deep-sea fish *Ipnops murrayi* (family Bathypteroidae)' *Bull. Mar. Sci.* **25**, 1, pp. 101–111.

363 Marshall, N. B. and Thinés, G. L. (1958) 'Studies of the brain, sense organs and light sensitivity of a blind cave fish (*Typhlogarra widdowsoni*) from Iraq' *Proc. Zool. Soc. Lond.* **131**, 3, pp. 441–456.

364 Matthews, F. D., Dumkaer, D. M., Knapp, L. W. and Collette, B. B. (1977) 'Food of western North Atlantic tunas (*Thunnus*) and lancetfish (*Alepisaurus*)' *NOAA Technical Report NMFS SSRF—706*, US Department of Commerce.

365 Mauchline, J. (1972) 'The biology of bathypelagic organisms, especially Crustacea' *Deep-Sea Res.* **19**, pp. 753–780.

366 Mauchline, J., Aizawa, Y., Ishimaru, T., Nishida, S. and Marumo, R. (1977) 'Integumental sensilla of pelagic decapod crustaceans' *Mar. Biol.* **43**, 2, pp. 149–156.

367 Mauchline, J. and Fisher, L. R. (1969) 'The biology of the euphausiids' *Adv. Mar. Biol.* **7**, pp. 1–454.

368 Mauro, A. (1977) 'Extra-ocular photoreceptors in cephalopods' pp. 287–308. In *The Biology of Cephalopods* Nixon, M. and Messenger, J. B. (ed.) *Symp. Zool. Soc. Lond.* **38**, 615 pp.

369 Maximova, M. P. (1971) 'Nutrients in the troposphere and subtroposphere of the Indian Ocean and their relation to productivity' *Proc. Joint Oceanogr. Assembly* (Tokyo, 1970).

370 Maynard, N. G. (1976) 'Relationship between diatoms in surface sediments of the Atlantic Ocean and the biological and physical oceanography of overlying waters' *Paleobiology* **2**, pp. 99–121.

371 Mayr, E. (1963) *Animal Species and Evolution* Harvard University Press, Cambridge, Mass., 797 pp.

372 McDowell, S. B. (1973) 'Order Heteromi (Notacanthiformes)' pp. 1–228. In *Fishes of the Western North Atlantic. Memoir Sears Foundation for Marine Research* **1**, 6.

373 McElroy, W. D. and Seliger, H. H. (1962) 'Origin and evolution of bioluminescence' In *Horizons in Biochemistry* Kasha, M. and Pullman, B. (ed.) Academic Press, New York, pp. 91–101.

374 McGowan, J. A. (1972) 'The nature of oceanic ecosystems' pp. 9–28. In *The Biology of the Oceanic Pacific* Miller, C. B. (ed.) Oregon State University Press.

375 McLaren, I. A. (1963) 'Effects of temperature on growth of zooplankton and the adaptive value of vertical migration' *J. Fish. Res. Bd Canada* **20**, pp. 685–727.

376 Mead, G. W., Bertelsen, E. and Cohen, D. M. (1964) 'Reproduction among deep-sea fishes' *Deep-Sea Res.* **11**, pp. 569–596.

377 Meglitsch, P. A. (1967) *Invertebrate Zoology* Oxford University Press, 961 pp.

378 Menzel, D. M. (1974) 'Primary productivity, dissolved and particulate organic matter, and the sites of oxidation of organic matter' pp. 659–678. In *The Sea, Vol. 5 Marine Chemistry* Goldberg, E. D. (ed.) Wiley-Interscience, 895 pp.

379 Menzies, R. J. (1962) 'The isopods of the abyssal depths in the Atlantic Ocean' In *Abyssal Crustacea. Vema Research Series* No 1, pp. 79–206.

380 Menzies, R. J., George, R. Y. and Rowe, G. T. (1973) *Abyssal Environment and Ecology of the World Oceans* John Wiley, New York, London, 488 pp.

381 Merrett, N. R. (1978) 'On the identity and pelagic occurrence of larval and juvenile stages of rattail fishes (Family Macrouridae) from 60°N, 20°W and 53°N, 20°W' *Deep-Sea Res.* 25, pp. 147–160.

382 Merrett, N. R. and Roe, H. S. J. (1974) 'Patterns of selectivity in the feeding of certain mesopelagic fishes' *Mar. Biol.* 28, pp. 115–126.

383 Meyer, D. L. (1973) 'Feeding behaviour and ecology of shallow-water unstalked crinoids (Echinodermata) in the Caribbean Sea' *Mar. Biol.* 22, 2, pp. 105–130.

384 Meyer, D. L. and Macurda, J. (1977) 'Adaptive radiation of the comatulid crinoids' *Paleobiology* 3, 1, pp. 74–82.

385 Meyer-Rochow, V. B. (1978) 'Skin papillae as possible electroreceptors in the deep-sea eel *Cyema atrum* (Cyemidae: Anguilloidei)' *Mar. Biol.* 46, pp. 277–282.

386 Mileikovsky, S. A. (1971) 'Types of larval development in marine bottom invertebrates, their distribution and ecological significance: a re-valuation' *Mar. Biol.* 10, pp. 193–213.

387 Millar, R. H. (1959) 'Ascidiacea' *Galathea Report* 1, pp. 189–209.

388 Mironov, A. N. (1971) 'Soft sea-urchins of the family Echinothuriidae collected by the RV *Vityaz* and *Academik Kurchatov* in the Pacific and Indian Oceans' pp. 317–325. In *Fauna of the Kurile–Kamchatka Trench. Tr. Inst. Okeanol.* 92.

389 Monniot, C. and Monniot, F. (1975) 'Feeding behaviour of abyssal tunicates' pp. 357–362. In *Ninth European Marine Biology Symposium* Barnes, H. (ed.) 760 pp.

390 Mortensen, Th. (1927) *Handbook of the Echinoderms of the British Isles* Oxford University Press, 471 pp.

391 Morton, J. E. (1958) *Molluscs* Hutchinson University Library, London, 232 pp.

392 Motoda, S., Taniguchie, A. and Ikeda, T. (1974) 'Plankton ecology in the western North Pacific ocean: Primary and secondary productivities' *Indo-Pac. Fish. Council Proc.* Sec. III 15th Session, Wellington, 1972, pp. 86–110.

393 Mukhacheva, V. A. (1966) 'The composition of species of the genus *Cyclothone* (Pisces, Gonostomatidae) in the Pacific Ocean' pp.

98–146. In *Fishes of the Pacific and Indian Oceans: Biology and Distribution* Rass, T. S. (ed.) Israel Program for Scientific Publications.

394 Mukhacheva, V. A. (1974) 'Cyclothones (Genus *Cyclothone*, fam. Gonostomatidae Pisces) of the world ocean and their distribution' *Tr. Inst. Okeanol. USSR.* **96**, pp. 189–254.

395 Munk, O. (1959) 'The eyes of *Ipnops murrayi* Günther 1887' *Galathea Report* **3**, pp. 79–87.

396 Munk, O. (1964) 'The eyes of some ceratioid fishes' *Dana Report* **61**, pp. 1–16.

397 Munk, O. (1965) '*Omosudis lowei* Günther, 1887, a bathypelagic fish with an almost pure-cone retina' *Vidensk. Meddr. Dansk. Naturh. Foren.* **128**, pp. 341–355.

398 Munk, O. (1966) 'Ocular anatomy of some deep-sea teleosts' *Dana Report* **70**, pp. 1–62.

399 Munk, O. and Frederiksen, R. D. (1974) 'On the function of aphakic apertures in teleosts' *Vidensk. Meddr. Dansk. Naturh. Foren.* **137**, pp. 65–94.

400 Muntz, W. R. A. (1976) 'On yellow lenses in mesopelagic animals' *J. Mar. Biol. Ass. UK.* **56**, 4, pp. 963–976.

401 Murina, V. V. (1974) 'Contributions to the knowledge of the fauna of sipunculid worms from the South Atlantic based on the data of the *Academik Kurchatov* Expedition in 1971' pp. 228–239. In *Biological Investigations in the Atlantic Sector of the Antarctic Ocean. Tr. Inst. Okeanol.* **98**.

402 Murphy, L. S., Rowe, G. T. and Haedrich, R. L. (1976) 'Genetic variability in deep-sea echinoderms' *Deep-Sea Res.* **23**, 4, pp. 339–348.

403 Murray, J. and Hjort, J. (1912) *The Depths of the Ocean* Macmillan, London, 821 pp.

404 Nafpaktitis, B. G. (1968) 'Taxonomy and distribution of the lantern-fishes, genera *Lobianchia* and *Diaphus* in the North Atlantic' *Dana Report* **73**, pp. 1–131.

405 Nash, L. K. (1963) *The Nature of the Natural Sciences* Little, Brown and Company, Boston, 406 pp.

406 Newman, W. A. (1974) 'Two new deep-sea Cirripedia (Ascothoracica and Acrothoracica) from the Atlantic' *J. Mar. Biol. Ass. UK* **54**, pp. 437–456.

407 Nichols, J. and Rowe, G. T. (1977) 'Infaunal macrobenthos off Cap Blanc, Spanish Sahara' *J. Mar. Res.* **35**, 3, pp. 525–536.

408 Nicol, J. A. C. (1958) 'Observations on luminescence of pelagic animals' *J. Mar. Biol. Ass. UK.* **37**, pp. 705–722.

409 Nicol, J. A. C. (1960) (a) 'Spectral composition of the light of the lanternfish (*Myctophum punctatum*)' *J. Mar. Biol. Ass. UK.* **39**, pp. 27–32.

410 Nicol, J. A. C. (1960) (b) 'Studies on luminescence on the subocular

light-organs of stomiatoid fishes' *J. Mar. Biol. Ass. UK.* 39, pp. 529–548.

411 Nicol, J. A. C. (1962) 'Animal luminescence' In *Advances in Comparative Physiology and Biochemistry* pp. 217–273. Lowenstein, O. E. (ed.) 1, Academic Press, London.

412 Nicol, J. A. C. (1967) 'The luminescence of fishes' *Symp. Zool. Soc. Lond.* 19, pp. 27–55.

413 Nicol, J. A. C. (1971) 'Physiological investigations of oceanic animals' pp. 225–246. In *Deep Oceans* Herring, P. J. and Clarke, M. R. (ed.) Arthur Barker, London, 320 pp.

414 Nielsen, J. G. (1964) 'Fishes from depths exceeding 6,000 m' *Galathea Report* 7, pp. 113–124.

415 Nielsen, J. G. (1969) 'Systematics and biology of the Aphyonidae (Pisces, Ophidioidea)' *Galathea Report* 10, pp. 7–88.

416 Nielsen, J. G., Jespersen, A. and Munk, O. (1968) 'Spermatophores in Ophidioidea (Pisces, Percomorphi)' *Galathea Report* 9, pp. 239–254.

417 Nielsen, J. G. and Munk, O. (1964) 'A hadal fish (*Bassogigas profundissimus*) with a functional swimbladder' *Nature* 204, 4958, pp. 594–595.

418 Nilsson-Cantell, C. A. (1955) 'Cirripedia' *Rep. Swedish Deep-Sea Expedition 1947–1948*, 2 (Zoology), No 17, pp. 215–220.

419 Nixon, M. and Dilly, P. N. (1977) 'Sucker surfaces and prey capture' pp. 447–512. In *The Biology of Cephalopods* Nixon, M. and Messenger, J. B. (ed.) *Symp. Zool. Soc. Lond.* 38, 615 pp.

420 Ockelmann, K. W. (1965) 'Developmental types in marine bivalves and their distribution along the Atlantic coast of Europe' *Proc. First Eur. Malacol. Congress* London 1962, pp. 25–35.

421 O'Day, W. T. (1973) 'Luminescent silhouetting in stomiatoid fishes' *Contributions in Science, Nat. Hist. Mus., Los Angeles County* No 246, pp. 1–8.

422 O'Day, W. T. and Fernandez, H. R. (1974) '*Aristostomias scintillans* (Malacosteidae): A deep-sea fish with visual pigments apparently adapted to its own bioluminescence' *Vision Res.* 14, pp. 545–550.

423 Ohno, S. (1970) *Evolution by Gene Duplication* Springer-Verlag, Berlin, 160 pp.

424 Omori, M. (1974) 'The biology of pelagic shrimps in the ocean' *Adv. Mar. Biol.* 12, pp. 233–324.

425 Oug, E. (1977) 'Faunal distribution close to the sediment of a shallow marine environment' *Sarsia* 63, 2, pp. 115–122.

426 Packard, A. (1972) 'Cephalopods and fish: the limits of convergence' *Biol. Rev.* 47, pp. 241–307.

427 Pantin, C. F. A. (1968) *The Relations Between the Sciences* Cambridge University Press, 206 pp.

428 Parin, N. V., Andriashev, A. P., Borodinula, O. D. and Tchuvasov, V. M. (1974) 'Midwater fishes of the southwestern Atlantic Ocean' *Tr. Inst. Okeanol. Moscow* **98**, pp. 76–140.

429 Parin, N. V., Becker, V. E., Borodinula, O. D., Karmovskaya, E. S., Fedoryako, B. I., Shcherbachev, J. N., Pokhilskaya, G. N. and Tchuvasov, V. M. (1977) 'Midwater fishes in the western tropical Pacific Ocean and the seas of the Indo-Australian Archipelago' *Tr. Inst. Okeanol. Moscow* **107**, pp. 68–188.

430 Parin, N. V. and Golovan, G. A. (1976) 'Pelagic deep-sea fishes of the families characteristic of the open ocean collected over the continental slope off West Africa' *Tr. Inst. Okeanol. Moscow* **104**, pp. 250–276.

431 Parin, N. V. and Novikova, N. S. (1974) 'Taxonomy of viper fishes (Chauliodontidae, Osteichthyes) and their distribution in the world ocean' *Tr. Inst. Okeanol. USSR.* **96**.

432 Parsons, T. R. (1976) 'The structure of life in the sea' In *The Ecology of the Seas* Cushing, D. H. and Walsh, J. J. (ed.) Blackwell Scientific Publications, Oxford, 467 pp.

433 Parsons, T. and Takahashi (1973) *Biological Oceanographic Processes* Pergamon Press, Oxford, 186 pp.

434 Pasternak, F. A. (1977) 'Antipatharia' *Galathea Report* **14**, pp. 157–164.

435 Pawson, D. L. (1975) 'Some aspects of the biology of deep-sea echinoderms (Abstract)' *Second Echinoderm Conference, Rovinoj 1975.*

436 Paxton, J. R. (1972) 'Osteology and relationships of the lanternfishes (Family Myctophidae)' *Bull. Nat. Hist. Mus. Los Angeles County, Science* **13**, pp. 1–81.

437 Pechenik, L. N. and Troyanovskii (1971) 'Trawling resources on the North Atlantic Continental Slope: Israel Program for Scientific Publications' *Murmanskoe Knizhnoe Izdatel'stvo*, 1970, 71 pp.

438 Perkins, H. C. (1973) 'The larval stages of the deep-sea red crab *Geryon quinquedens* Smith reared under laboratory conditions (Decapoda: Branchyrhyncha)' *Fishery Bulletin* **71**, 1, pp. 69–82.

439 Pianka, E. R. (1976) (a) 'Competition and niche theory' pp. 114–141. In *Theoretical Ecology* May, R. M. (ed.) Blackwell Scientific Publications, Oxford, 317 pp.

440 Pianka, E. R. (1976) (b) 'Natural selection of optimal reproductive tactics' *Amer. Zool.* **16**, pp. 775–784.

441 Pickard, G. L. (1963) *Descriptive Physical Oceanography* Pergamon Press, Oxford, 200 pp.

442 Picken, L. E. R. (1957) 'Stinging capsules and designing nature' pp. 56–72. In *New Biology* Johnson, M. L., Abercrombie, M. and Fogg, G. E. (ed.) Penguin Books, London.

443 Pietsch, T. W. (1974) 'Osteology and relationships of ceratioid angler fishes of the family Oneirodidae, with a review of the genus *Oneirodes* Lütken' *Bull. Nat. Hist. Mus. Los Angeles County, Science* 18, pp. 1–112.

444 Pietsch, T. W. (1976) 'Dimorphism, parasitism and sex: Reproductive strategies among deep-sea ceratoid angler fishes' *Copeia* 4, pp. 781–793.

445 Poulson, T. L. (1963) 'Cave adaptation in Amblyopsid fishes' *Amer. Midl. Nat.* 70, pp. 257–290.

446 Pugh, P. R. (1974) 'The vertical distribution of the siphonophores collected during the SOND cruise, 1965' *J. Mar. Biol. Ass. UK.* 54, 1, pp. 25–90.

447 Raju, N. S. (1974) 'Three new species of the genus *Monognathus* and the leptocephali of the Order Saccopharyngiformes' *Fishery Bulletin* 72, 2, pp. 547–562.

448 Rapoport, A. (1960) *Fights, Games, and Debates* University of Michigan Press, 400 pp.

449 Rex, M. A. (1976) 'Biological accommodation in the deep-sea benthos: comparative evidence on the importance of predation and productivity' *Deep-Sea Res.* 23, 10, pp. 975–988.

450 Rex, M. A. (1977) 'Zonation in deep-sea gastropods: The importance of biological interactions to rates of zonation' pp. 521–530. In *Biology of Benthic Organisms* Keegan, B. F. and Céidigh, P. O. (ed.) Pergamon Press, 630 pp.

451 Rice, M. E. (1975) 'Sipuncula' pp. 67–127. In *Reproduction of Marine Invertebrates Vol. II Entoprocts and Lesser Coelomates* Academic Press, London.

452 Richards, F. A. (1975) 'The Cariaco Basin (Trench)' *Oceanogr. Mar. Biol. Ann. Rev.* 13, pp. 11–67.

453 Ricker, W. E. (1969) 'Food from the sea' In *Resources and Man* Cloud (ed.) Freeman and Co, Chicago.

454 Riley, G. A. (1970) 'Particulate organic matter in sea water' *Adv. Mar. Biol.* 8, pp. 1–118.

455 Roberston, D. A. (1977) 'Planktonic eggs of the lantern-fish *Lampanyctodes hectoris* (family Myctophidae)' *Deep-Sea Res.* 24, 9, pp. 849–852.

456 Robilliard, G. A. and Dayton, P. K. (1969) 'Notes on the biology of the chaenichthyid fish *Pagetopsis macropterus* from McMurdo Sound, Antarctica' *Antarctic Journal of the United States* 4, 6, pp. 304–306.

457 Roe, H. S. J. (1972) (a) 'The vertical distributions and diurnal migrations of calanoid copepods collected on the SOND cruise, 1965. I The total population and general discussion' *J. Mar. Biol. Ass. UK.* 52, 2, pp. 277–314.

458 Roe, H. S. J. (1972) (b) 'The vertical distributions and diurnal migrations of calanoid copepods collected on the SOND cruise, 1965. II Systematic account: families Calanidae up to and including the Aetideidae' *J. Mar. Biol. Ass. UK.* **52**, 2, pp. 315–344.

459 Roe, H. S. J. (1972) (c) 'The vertical distributions and diurnal migrations of calanoid copepods collected on the SOND cruise, 1965. III Systematic account: families Euchaetidae up to and including Metridiidae' *J. Mar. Biol. Ass. UK.* **52**, 3, pp. 525–552.

460 Roe, H. S. J. (1972) (d) 'The vertical distributions and diurnal migrations of calanoid copepods collected on the SOND cruise, 1965. IV Systematic account: families Lucicutiidae to Candaciidae. The relative abundance of the numerically most important genera' *J. Mar. Biol. Ass. UK.* **52**, 4, pp. 1021–1044.

461 Rofen, R. R. (1966) 'Fishes of the Western North Atlantic Family Paralepididae' *Memoir Sears Foundation for Marine Research* **1**, 5, pp. 205–461.

462 Roger, C. (1971) 'Distribution verticale des euphausiacés (crustacés) dans les courants équatoriaux de l'Océan Pacifique' *Mar. Biol.* **10**, pp. 133–144.

463 Roger, C. (1973) 'Recherches sur la situation trophique d'un group d'organismes pélagiques (Euphausiacea). II Comportements nutritionnels' *Mar. Biol.* **18**, pp. 317–320.

464 Roger, C. and Grandperrin, R. (1976) 'Pelagic food webs in the tropical Pacific' *Limnol. Oceanogr.* **21**, 5, pp. 731–735.

465 Rokop, F. J. (1974) 'Reproductive patterns in the deep-sea benthos' *Science* **186**, pp. 743–745.

466 Rokop, F. J. (1977) (a) 'Patterns of reproduction in the deep-sea benthic crustaceans: a revaluation' *Deep-Sea Res.* **24**, 7, pp. 683–692.

467 Rokop, F. J. (1977) (b) 'Seasonal reproduction of the brachiopod *Frieleia halli* and the scaphopod *Cadulus californicus* at bathyal depths in the deep sea' *Mar. Biol.* **43**, 3, pp. 237–246.

468 Roper, C. F. E. and Brundage, W. L. (1972) 'Cirrate octopods with associated deep-sea organisms: New biological data based on deep benthic photographs (Cephalopoda)' *Smithsonian Contributions to Zoology* **121**, pp. 1–46.

469 Roper, C. F. E. and Young, R. E. (1975) 'Vertical distribution of pelagic cephalopods' *Smithsonian Contributions to Zoology* **209**, 51 pp.

470 Rothschild, B. J. (1972) 'Fishery potential from the oceanic regions' pp. 95–106. In *The Biology of the Oceanic Pacific* Miller, C. B. (ed.) Corvallis, Oregon State University Press.

471 Rowe, G. T. and Menzies, R. J. (1969) 'Zonation of large benthic invertebrates in the deep sea off the Carolinas' *Deep-Sea Res.* **15**, 6, pp. 531–537.

472 Rudwick, M. J. S. (1970) *Living and Fossil Brachiopods* Hutchinson University Library, London, 199 pp.

473 Ryland, J. S. (1970) *Bryozoans* Hutchinson University Library, London, 175 pp.

474 Ryther, J. H. (1969) 'Photosynthesis and fish production in the sea. The production of organic matter and its conversion to higher forms of life vary throughout the world ocean' *Science* 166, pp. 72–76.

475 Ryther, J. H., Hall, J. R., Pease, A. K., Bakun, A. and Jones, M. M. (1966) 'Primary organic production in relation to the chemistry and hydrography of the western Indian Ocean' *Second International Oceanographic Congress, Moscow—1966* p. 311.

476 Saidova, H. M. (1966) 'Benthic Foraminifera of the Pacific Ocean and the regularities of their distribution' *Second International Oceanographic Congress, Moscow—1966* pp. 311–312.

477 Sanders, H. L. and Grassle, J. F. (1971) 'The interactions of diversity, distribution and mode of reproduction among major groupings of the deep-sea benthos' *Proc. Joint Oceanogr. Assembly (Tokyo, 1970)* pp. 260–262.

478 Sanders, H. L. and Hessler, R. R. (1969) 'Ecology of the deep-sea benthos' *Science* 163, pp. 1419–1424.

479 Sanzo, L. (1933) 'Uova, larve e stadi giovanili di Teleostii' *Fauna e Flora del Golfo di Napoli, Macruridae* 38, pp. 255–265.

480 Scheltema, R. S. (1972) 'Reproduction and dispersal of bottom-dwelling deep-sea invertebrates: A speculative summary' pp. 58–68. In *Barobiology and the Experimental Biology of the Deep Sea* Brauer, R. W. (ed.) North Carolina Sea Grant Program, University of North Carolina.

481 Schmidt-Nielsen, K. (1975) *Animal Physiology* Cambridge University Press, 699 pp.

482 Schneider, H. (1962) 'The labyrinth of two species of drumfish (Sciaenidae)' *Copeia* pp. 336–338.

483 Scholander, P. F. (1954) 'Secretion of gases against high pressures in the swimbladder of deep-sea fishes II The rete mirabile' *Biol. Bull. Woods Hole* 107, pp. 260–277.

484 Schroeder, W. C. (1955) 'Report on the results of exploratory otter-trawling along the continental shelf and slope between Nova Scotia and Virginia during the summers of 1952 and 1953' *Deep-Sea Res. Suppl.* 3, pp. 358–372.

485 Schuijf, A. (1974) 'Directional hearing of cod (*Gadus morhua*) under approximate free field conditions' In *Field studies of Directional Hearing in Marine Teleosts* Schuijf, A. Dissertation, University of Utrecht.

486 Sekiguchi, H. (1974) 'Relation between the ontogenetic vertical

migration and the mandibular gnathobase in pelagic copepods' *Bull. Fac. Fish. Mie. Univ.* 1, pp. 1–10.

487 Sekiguchi, H. (1975) 'Seasonal and ontogenetic vertical migrations in some common copepods in the northern region of the North Pacific' *Bull. Fac. Fish. Mie. Univ.* 2, pp. 29–38.

488 Shcherbinin, A. D. (1971) 'On the interrelation of water circulation and structure in the Indian Ocean' *Proc. Joint Oceanogr. Assembly (Tokyo 1970)* pp. 392–394.

489 Shulenberger, E. and Hessler, R. R. (1974) 'Scavenging abyssal benthic amphipods trapped under oligotrophic central North Pacific gyre waters' *Mar. Biol.* 28, pp. 185–187.

490 Sieburth, J. M. (1976) 'Bacterial substrates and productivity in marine ecosystems' *Ann. Rev. Ecol. Syst.* 7, pp. 259–285.

491 Silen, L. (1951) 'Bryozoa' *Rep. Swedish Deep-Sea Expedition 1947–1948*, 2 (Zool.), No 3, pp. 63–69.

492 Slobodkin, L. B. (1964) 'The strategy of evolution' *American Scientist* 52, pp. 342–357.

493 Smayda, T. J. (1970) 'The suspension and sinking of phytoplankton in the sea' *Oceanogr. Mar. Biol. Ann. Rev.* 8, pp. 353–414.

494 Smayda, T. J. (1971) 'The ocean flora' pp. 130–149. In *Deep Oceans* Herring, P. J. and Clarke, M. R. (ed.) Arthur Barker, London, 320 pp.

495 Smith, K. L. and Hessler, R. R. (1974) 'Respiration of benthopelagic fishes: In situ measurements at 1230 metres' *Science* 184, 4132, pp. 72–73.

496 Smith, K. L. and Teal, J. M. (1973) 'Deep-sea benthic community respiration: an in situ study at 1850 metres' *Science* 179, 4070, pp. 282–283.

497 Sokolova, M. N. (1970) 'Weight characteristics of meiobenthos from different parts of the deep-sea trophic regions of the Pacific Ocean' *Okeanology* 10, 2, (Engl. transl.) pp. 348–356.

498 Sokolova, M. N. (1972) 'Trophic structure of deep-sea macrobenthos' *Mar. Biol.* 16, pp. 1–12.

499 Sokolova, M. N. (1976) 'Trophic zonality of deep-water macrobenthos as an element of the biological structure of the ocean' *Oceanology* 16, 2 (English edition), pp. 188–190.

500 Somiya, H. (1976) 'Functional significance of the yellow lens in the eyes of *Argyropelecus affinis*' *Mar. Biol.* 34, pp. 93–99.

501 Somiya, H. (1977) 'Bacterial bioluminescence in chlorophthalmid deep-sea fish: A possible interrelationship between the light organ and the eyes' *Experientia* 33, pp. 906–908.

502 Southward, A. J., Robinson, S. G., Nicholson, D. and Perry, T. J. (1976) 'An improved stereocamera and control system for close-up photography of the fauna of the continental slope and outer shelf' *J. Mar. Biol. Ass. UK.* 56, pp. 247–257.

503 Southward, E. C. (1971) (a) 'Recent researches on the Pogonophora' *Oceanogr. Mar. Biol. A.* **9**, pp. 193–220.

504 Southward, E. C. (1971) (b) 'Pogonophora of the northwest Atlantic: Nova Scotia to Florida' *Smithsonian Contributions to Zoology* **88**, pp. 1–29.

505 Southward, E. C. (1975) 'Pogonophora' pp. 129–156. In *Reproduction of Invertebrates Vol. II Entoprocts and Lesser Coelomates* Giese, A. C. and Pearse, J. S. (ed.) Academic Press, London.

506 Southward, E. C. and Brattegard, T. (1968) 'Pogonophora of the northwest Atlantic: North Carolina region' *Bull. Mar. Sci.* **18**, 4, pp. 836–875.

507 Southward, E. C. and Southward, A. J. (1967) 'The distribution of Pogonophora in the Atlantic Ocean' *Symp. Zool. Soc. Lond.* **19**, pp. 145–158.

508 Squires, D. F. (1967) 'The evolution of the deep-sea coral family Micrabaciidae' pp. 502–510. In *Proceedings of the International Conference on Tropical Oceanography 1965* University of Miami, Institute of Marine Sciences.

509 Stearns, S. C. (1976) 'Life-history tactics: a review of the ideas' *Quart. Rev. Biol.* **51**, 1.

510 Stebbins, G. L. (1977) In *Evolution* Dobzhansky, T., Ayala, F. J., Stebbins, G. L. and Valentine, J. W., W. H. Freeman and Company, San Francisco, 572 pp.

511 Steele, J. H. (1976) 'Patchiness' pp. 98–115. In *The Ecology of the Seas* Cushing, D. H. and Walsh, J. J. (ed.) Blackwell Scientific Publications, Oxford, 467 pp.

512 Steele, J. H. and Yentsch, C. S. (1960) 'The vertical distribution of chlorophyll' *J. Mar. Biol. Ass. UK.* **39**, pp. 217–226.

513 Strickland, J. D. H. (1968) 'A comparison of profiles of nutrient and chlorophyll concentrations taken from discrete depths and by continuous recording' *Limnol. Oceanogr.* **13**, pp. 388–391.

514 Strickland, J. D. H. (1972) 'Research on the marine planktonic food web at the Institute of Marine Resources: A review of the past seven years of work' *Oceanogr. Mar. Biol. Ann. Rev.* **10**, pp. 349–414.

515 Sulak, K. J. (1977) (a) 'The systematics and biology of *Bathypterois* (Pisces, Chlorophthalmidae) with a revised classification of benthic myctophiform fishes' *Galathea Report* **14**, pp. 49–108.

516 Sulak, K. J. (1977) (b) '*Aldrovandia oleosa*, a new species of the Halosauridae, with observations on several other species of the family' *Copeia* **1**, pp. 11–19.

517 Swallow, M. (1962) 'Deep currents in the oceans' *Discovery* **23**, 6, pp. 17–22.

518 Swift, M. C. (1976) 'Energetics of vertical migration in *Chaoborus trivittatus* larvae' *Ecology* **57**, pp. 900–914.

519 Tåning, A. V. (1918) 'Mediterranean Scopelidae (*Saurus, Aulopus, Chlorophthalmus* and *Myctophum*)' *Rep. Danish Oceanogr. Exped. Medit.* **2**, (Biol.) A.7.

520 Tendal, O.S. (1972) 'A monograph of the Xenophyophoria' *Galathea Report*, **12**, pp. 1–99.

521 Tendal, O. S. and Hessler, R. R. (1977) 'An introduction to the biology and systematics of Komokiacea (Textulariina, Foraminiferida)' *Galathea Report* **14**, pp. 165–194.

522 Tett, P. B. and Kelly, M. G. (1973) 'Marine bioluminescence' *Oceanogr. Mar. Biol. Ann. Rev.* **11**, pp. 89–175.

523 Thiel, H. (1975) 'The size structure of the deep-sea benthos' *Int. Rev. Ges. Hydrobiol.* **60**, 5, pp. 575–606.

524 Thompson, D'Arcy W. (1942) *On Growth and Form* Cambridge University Press.

525 Thorson, G. (1946) 'Reproduction and larval development of Danish marine bottom invertebrates' *Medd. Komm. Danmarks Fisk—Havunders. Serie Plankton* **4**, 1, pp. 1–523.

526 Thorson, G. (1950) 'Reproduction and larval ecology of marine bottom invertebrates' *Biol. Rev.* **25**, 1, pp. 1–45.

527 Thorson, G. (1971) *Life in the Sea* Weidenfeld and Nicolson, London, 256 pp.

528 Thurston, M. H. (1976) (a) 'The vertical distribution and diurnal migration of the Crustacea Amphipoda collected during the SOND cruise, 1965. I The Gammaridea' *J. Mar. Biol. Ass. UK.* **56**, 2, pp. 359–382.

529 Thurston, M. H. (1976) (b) 'The vertical distribution and diurnal migration of the Crustacea Amphipoda collected during the SOND cruise, 1965. II The Hyperiidea and general discussion' *J. Mar. Biol. Ass. UK.* **56**, 2, pp. 383–470.

530 Tietjen, J. H. (1971) 'Ecology and distribution of deep-sea meiobenthos off North Carolina' *Deep-Sea Res.* **18**, pp. 941–957.

531 Tietjen, J. H. (1976) 'Distribution and species diversity of deep-sea nematodes off North Carolina' *Deep-Sea Res.* **23**, 8, pp. 755–768.

532 Timonin, A. G. (1971) 'The structure of plankton communities of the Indian Ocean' *Mar. Biol.* **9**, pp. 281–289.

533 Trueman, E. R. (1975) *The Locomotion of Soft-Bodied Animals* Edward Arnold, London, 200 pp.

534 Turekian, K. K. (1968) *Oceans* Prentice-Hall Inc., New Jersey, 120 pp.

535 Turekian, K. K., Cochran, J. K., Kharker, D. P., Cerrato, R. M., Vaisnys, J. R., Sanders, H. L., Grassle, J. F. and Allen, J. A. (1975) 'The slow growth rate of a deep-sea clam determined by ^{228}R chronology' *Proc. Nat. Acad. Sci. USA.* **72**, pp. 2829–2832.

536 Valentine, J. W. (1973) *Evolutionary paleoecology of the Marine Biosphere* Prentice-Hall Inc., New Jersey, 511 pp.

537 Valentine, J. W. (1977) 'The geological record' pp. 314–348. In *Evolution* Dobzhansky, T., Ayala, F. J., Stebbins, G. L. and Valentine, J. W., W. H. Freeman, San Francisco, pp. 572.

538 Valentine, J. W., Hedgecock, D. and Barr, L. G. (1975) 'Deep-sea asteroids: high genetic variability in a stable environment' *Evolution* 29, pp. 203–212.

539 Vinogradov, M. E. (1968) *Vertical Distribution of the Oceanic Zooplankton* Nauka, Moscow, pp. 320.

540 Vinogradova, N. G., Kudinova-Pasternak, R. K., Moskalev, L. I., Muromtseva, T. L. and Fedikov, N. F. (1974) 'Some regularities of the quantitative distribution of the bottom fauna of the Scotia Sea and the deep-sea trenches of the Atlantic sector of the Antarctic' *Tr. Inst. Okeanol. USSR.* 98, pp. 157–182.

541 Vogel, S. (1974) 'Current induced flow through the sponge, *Halichondria*' *Biol. Bull. Woods Hole* 147, pp. 443–450.

542 Vogel, S. and Bretz, W. L. (1972) 'Interfacial organisms: passive ventilation in velocity gradients near surfaces' *Science* 175, pp. 210–211.

543 Voss, G. L. (1967) 'The biology and bathymetric distribution of deep-sea cephalopods' pp. 511–538. In *Proceedings of the International Conference on Tropical Oceanography 1965* University of Miami, Institute of Marine Sciences.

544 Waddington, C. H. (1975) *The Evolution of an Evolutionist* Edinburgh University Press, 328 pp.

545 Wainwright, S. A., Biggs, W. D., Currey, J. D. and Gosline, J. M. (1976) *Mechanical Design in Organisms* Edward Arnold, London, 423 pp.

546 Wald, G. and Rayport, S. (1977) 'Vision in annelid worms' *Science* 196, pp. 1434–1439.

547 Walls, G. L. (1942) 'The vertebrate eye and its adaptive radiation' *Cranbrook Institute of Science, Bulletin* 19, 785 pp.

548 Warner, G. F. (1977) *The Biology of Crabs* Elek Science, London, 202 pp.

549 Wassersug, R. J. and Johnson, R. K. (1976) 'A remarkable pyloric caecum in the evermannellid genus *Coccorella* with notes on gut structure and function in alepisauroid fishes (Pisces, Myctophiformes)' *J. Zool. Lond.* 179, pp. 273–289.

550 Waterman, T. H. (1948) 'Studies on the deep-sea angler-fishes (Ceratioidea). III The comparative anatomy of *Gigantactis longicirra* Waterman' *J. Morph.* 82, pp. 81–149.

551 Wiebe, P. H., Boyd, S. H. and Winget, C. (1976) 'Particulate matter sinking to the deep-sea floor at 2,000 m in the Tongue of the Ocean,

Bahamas, with a description of a new sedimentation trap' *J. Mar. Res.* **34**, 3, pp. 341–354.

552 Wiebe, P. H., Burt, K. H., Boyd, S. H. and Morton, A. W. (1976) 'A multiple opening/closing net and environmental sensing system for sampling zooplankton' *J. Mar. Res.* **34**, 3, pp. 313–325.

553 Wigley, R. L. and Emery, K. O. (1967) 'Benthic animals, particularly *Hyalinoecia* (Annelida) and *Ophiomusium* (Echinodermata), in sea-bottom photographs from the continental slope' pp. 235–250. In *Deep-Sea Photography* Hersey, J. B. (ed.) The Johns Hopkins Press, Baltimore.

554 Williams, G. C. (1966) *Adaptation and Natural Selection* Princeton University Press, 307 pp.

555 Wilson, D. S. (1975) 'The adequacy of body size as a niche difference' *Amer. Naturalist* **109**, 970, pp. 769–784.

556 Wisner, R. L. (1974) 'The taxonomy and distribution of lantern-fishes (Family Myctophidae) of the eastern Pacific Ocean' *NORDA Report* 3, Navy Ocean Research and Development Activity, Bay St Louis, Mississippi, 229 pp.

557 Wolff, T. (1956) 'Crustacea Tanaidacea from depths exceeding 6,000 metres' *Galathea Report* **2**, pp. 187–242.

558 Wolff, T. (1961) (a) 'Description of a remarkable deep-sea hermit crab, with notes on the evolution of the Paguridea' *Galathea Report* **4**, pp. 11–32.

559 Wolff, T. (1961) (b) 'Animal life from a single abyssal trawling' *Galathea Report* **5**, pp. 129–162.

560 Wolff, T. (1962) 'The systematics and biology of bathyal and abyssal Isopoda' *Galathea Report* **6**.

561 Wolff, T. (1971) 'Archimède Dive 7 to 4160 metres at Madeira: observations and collecting results' *Vidensk. Meddr. Dansk. Naturh. Foren.* **134**, pp. 127–147.

562 Wolff, T. (1976) 'Utilization of seagrass in the deep sea' *Aquat. Bot.* **2**, pp. 161–174.

563 Wolff, T. (1977) 'Diversity and faunal composition of the deep-sea benthos' *Nature* **267**, pp. 780–785.

564 Woods, L. P. and Sonoda, P. M. (1973) 'Order Berycomorphi (Beryciformes)'. In *Fishes of the Western North Atlantic. Memoir Sears Foundation for Marine Research* 1, 6, pp. 263–396.

565 Wyrtki, K. (1973) 'Physical oceanography of the Indian Ocean' In *The Biology of the Indian Ocean* Zeitzschel, B. (ed.) Chapman and Hall, London, Springer-Verlag, Berlin.

566 Yaldwyn, J. C. (1965) 'Antarctic and Subantarctic decapod Crustacea' pp. 324–332. In *Biogeography and Ecology in Antarctica* van Mieghem, J. and van Oye, P. (ed.) Dr W. Junk, The Hague, 762 pp.

567 Yentsch, C. S. (1971) 'The harvest—primary production' pp. 150–163. In *Deep Oceans* Herring, P. J. and Clarke, M. R. (ed.) Arthur Barker, London, 320 pp.

568 Young, J. Z. (1977) 'Brain, behaviour and evolution of cephalopods' pp. 377–434. In *The Biology of Cephalopods* Nixon, M. and Messenger, J. B. (ed.) *Symp. Zool. Soc. Lond.* **38**, 615 pp.

569 Young, R. E. (1977) 'Ventral bioluminescent countershading in midwater cephalopods' pp. 161–190. In *The Biology of Cephalopods* Nixon, M. and Messenger, J. B. (ed.) *Symp. Zool. Soc. Lond.* **38**, 615 pp.

570 Young, R. E. and Roper, C. F. E. (1977) 'Intensity regulation of bioluminescence during countershading in living midwater animals' *Fishery Bulletin* **75**, 2, pp. 239–252.

571 Youngbluth, M. J. (1975) 'The vertical distribution and diel migration of euphausiids of the eastern South Pacific' *Deep-Sea Res.* **22**, 8, pp. 519–536.

572 Zalkina, A. V. (1970) 'Vertical distribution and diurnal migration of some Cyclopoida (Copepoda) in the tropical region of the Pacific Ocean' *Mar. Biol.* **5**, pp. 275–282.

573 Zeitzschel, B. (1971) 'Primary production, phytoplankton standing stock and phytoplankton composition in the eastern tropical Pacific' *Proc. Joint Oceanogr. Assembly (Tokyo 1970)*, pp. 467–468.

574 Zelickman, E. A. (1972) 'Distribution and ecology of the pelagic hydromedusae, siphonophores and ctenophores of the Barents Sea, based on perennial plankton collections' *Mar. Biol.* **17**, pp. 256–264.

575 Zevina, G. B. (1972) 'Cirripedia of the genus *Scalpellum* in the northwest part of the Pacific Ocean' pp. 268–292. In *Fauna of the Kurile–Kamchatka Trench and its Environment* Bogorov, V. G. (ed.) *Tr. Inst. Okeanol.* **86**, English transl.

Index of genera and species

The page numbers in *italic* are pages on which the item is illustrated.

Abralia trigonura (deep-sea squid, Mollusca), luminescence and background light, 360; light needed for camouflage, 361

Abraliopsis (deep-sea squid, Mollusca), *118*, *120*; buoyancy, 324; light organs, 119, 360; luminescence and background light, 360; photosensitive vesicles, 395
affinis, 119; ontogenetic migrations, 439; vertical migrations, 121

Abyssobrotula galatheae (brotulid fish), *140*

Abyssocladia (sponge, Demospongia), 167

Acanthephyra (caridean prawn, Oplophoridae Decapoda), 111; benthic forms, 134, 205; size, 113
curtirostris, acutifrons, architelsonis, small eggs, 438
multispina, reflexion of light, 362
pelagica, structure of eyes and depth range, 398, 400
purpurea, small eggs, 438
quadrispinosa, adult female and life history stages, *112*; egg size, 438; growth, 437; life history, 435–7
stylorostralis, a bathypelagic species, 114

Acanthometra (Acantharia, Protozoa), flotation, 317, *318*

Acanthoplegma krohni (Radiolaria, Protozoa), *73*

Acartia (Copepoda, Crustacea), egg yolk, 430

Acetes sibogae (sergestid prawn, Crustacea), sensory structure of antennae and tracking of scent trails, 411–12

Aegina (Narcomedusae), vertical distribution, 87

Aeginura (Narcomedusae), vertical distribution, 87

Aequorea aequorea (hydroid, medusae), 86

Aetideus (Copepoda, Crustacea), adhesive eggs, 429; non-migratory, 430

Agalma (physonect siphonophore), with many swimming bells, 88
elegans, 320

Aglantha digitale (Trachymedusae), colour and distribution, 86–7; vertical distribution, 87

Aglaophenia tenuissima (Hydroida, Cnidaria), distribution, 178

Alcyonium digitatum (dead-men's fingers, Alcyonaria, Cnidaria), as suspension feeder, 178

Aldrovandia (halosaur fish, Halosauridae), lateral line organs on head, 419; sexual dimorphism of anterior nostril, 409
rostrata, depth distribution, 138, *139*

Alepisaurus (alepisauroid fish), food, 25
ferox (lancet-fish), *129*; size, 129, 248

Alepocephalus (fish, Alepocephalidae), body form, 128; food, 276

Alomosoma (echiurid worm), depth distribution, 156

Amblyops abbreviata (mysid shrimp, Crustacea), structure of eyes, 400

Amblyopsis (cave fish, Amblyopsidae), positive rheotaxis, 408; lateral line system, 416

Ammodiscus profundissimus (Foraminifera, Protozoa), *147*

Ammolynthus haliphysema (Xenophyophoria, Protozoa), *160*

Amphihelia (ivory coral, Madreporaria, Cnidaria), 184

Amphiophiura (brittle-stars, Ophiuroidea, Echinodermata), distribution, 498; food of abyssal species, 190

Amphitretus pelagicus (deep-sea octopod, Mollusca), eyes, 121; structure of tubular eyes, 393, *394*

Amphiura chiajei (brittle-star, Ophiuroidea, Echinodermata), not luminescent, 375

Amphiura filiformis, luminescence and escape-response, 375

Anguilla (freshwater eel, Anguillidea), olfactory organ, 403, 403
anguilla, life history, 449

Anoplogaster cornuta (fang-tooth, berycoid fish), 485; oxygen consumption, 485

Anotopterus (fish, Alepisauroidea), dentition, 129; diet, 248; size, 129

Antimora (deep-sea cod, Moridae), may have light organ, 373
rostrata, size, 138; swimming speed, 279–80

Antipathes (black coral, Antipatharia, Cnidaria), attached animals, 186

Aphanopus (scabbard fish, Trichiuroidea), 133; halosaur fish in gut, 135

Apseudes galatheae (Tanaidacea, Crustacea), 153

Architeuthis (giant squid, Mollusca), ammoniacal uplift, 324; size, 106; size of eyes, 393

Argentina (argentinoid fish), body form, 127

Argonauta (paper nautilus, pelagic octopod), attached to *Crambionella*, 91

Argyropelecus (hatchet-fish; Sternoptychidae, Stomiatoidea), density of retinal pigment; depth range; length of swimbladder rete mirabile, 332; tubular eyes, 127, 248, 385–6
aculeatus, feed on ostracods, *Conchoecia* spp., 80
affinis, 124; luminescent camouflage, 352, 354, 362; yellow lens in eye, 248, 388
hemigymnus and *aculeatus*, differences in diet, 245

Ariosoma balearicum (bandtooth conger eel), structure of leptocephali, 449, 451

Aristioteuthis (deep-sea squid, Mollusca), photosensitive vesicles, 395

Aristostomias (rat-trap fish, Malacosteidae), large cheek lights in males, 367
scintillans, light from cheek photophores, 363

Asbestopluma (sponge, Demospongia), form, 165
occidentalis, depth record, 165

Asteronyx excavata (brittle-star, Ophiuroidea, Echinodermata), 191
lovenii, depth distribution and attachment, 190

Astronesthes (stomiatoid fish, Astronesthidae), length of life history, 446
cyaneus, long chin barbel in males, 367–8
lamellosus, gill structure, 486
niger and *Neoceratias* (female ceratoid angler-fish), contrasts in organization, 265, 267

Astropecten americanus (sea-star, Asteroidea, Echinodermata), depth range, 493

Astrophyton muricatum (basket star, Ophiuroidea, Echinodermata), feeding posture, 190

Astrorhiza (Foraminifera, Protozoa), structure and bottom-dwelling habit, 145

Atolla (jellyfish, Coronatae, Scyphozoa), 87; colour and structure, 91
wyvillei, luminescence, 345

Augaptilus (Copepoda, Crustacea), feathery setae, 322

Aulacoctena (comb-jelly, Ctenophora), deep-sea genus, 94

Aulostomatomorpha (fish, Alepocephalidae), long tubular snout, 128

Aurelia (jellyfish, Scyphozoa), oxygen consumption, 82

Avocettina (snipe-eel, Avocettinidae), olfactory organs, 404

Barathronus (ophidioid fish), egg clutches, 458
bicolor, embryo and adult, 457, 458

Barbourisia rufa (whale-fish, Cetomimiformes), size and colour, 131

Bassogigas (brotulid fish), 140
profundissimus, size and structure of swimbladder, 338

Bathophilus (stomiatoid fish, Melanostomiatidae), short, squat species, 126
longipinnis, 125

Bathothauma (deep-sea squid, Mollusca), subdivision of retina, 395

Bathycalanus princeps (Copepoda, Crustacea), relative buoyancy, 322

Bathychordaeus (Larvacea, pelagic tunicate), size and food, 96

Bathycrinus (sea-lily, Crinoidea, Echinodermata), off east coast of America on red clay, 295
carpenteri, 187

Bathyctena (comb-jelly, Ctenophora), deep-sea genus, 94

Bathygadus spp. (rat-tailed fishes, Macrouridae), rather small eyes, 392

Bathylagus (argentinoid fish), aphakic space and retinal organization, 383; fovea in retina, 383; high water content, 268
antarcticus, 455

Bathylychnops (argentinoid fish), body form, 127

Bathymicrops (chlorophthalmid fish), structure and depth distribution, 213

Bathynemertes hardyi (pelagic nemertean worm), colour, 105

Bathynomus (Isopoda, Crustacea), size, 153

Bathypathes lyra (black coral, Antipatharia, Cnidaria), distribution, 498

Bathypathes patula, 185
Bathypterois (tripod-fishes, Chlorophthalmidae), depth distribution, 305; lateral line system, 420; olfactory rosettes, 409; small or regressed eyes, 392
  atricolor, absence of cloning, 459
  bigelowi, distribution, 213, 214
  dubius, olfactory organ, 403
  grallator, 214
  longipes, depth distribution, 213
  phenax, size of mature eggs, 458
  quadrifilis, fecundity, 456
Bathyraja pallida (skate, Rajidae), depth distribution and oily liver, 212
  richardsoni, depth distribution and oily liver, 212
Bathysaurus (benthic deep-sea fish, Bathysauridae), oily liver, 339
Bathysiphon lanosum (Foraminifera, Protozoa), 147
Bathyspadella edentata (arrow-worm, Chaetognatha), may be bathypelagic, 103
Bathyteuthis (deep-sea squid, Mollusca), buoyant liver, 325
  abyssicola, vertical migrations, 121; well-developed eyes, 393
Bathytyphlops (deep-sea fish, Chlorophthalmidae), lateral line system, 420
Bentheuphausia (Euphausiacea, Crustacea), absence of light organs, 362; regressed eyes, 397
  amblyops, distribution, 487; regressed eyes and no light organs, 109
Benthopecten armatus (sea-star, Asteroidea, Echinodermata), 193
Benthosema glaciale (lantern-fish, Myctophidae), distribution, 477; fecundity, 444
Benthoteuthis (deep-sea squid, Mollusca), structure of eyes, 393, 394
Beroë cucumis (comb-jelly, Ctenophora), distribution, 94; ionic uplift, 319
Biddulphia aurita (diatom), 55
Bolinichthys (lantern-fish, Myctophidae), sexual dimorphism in head lights, 368
Bonapartia (stomiatoid fish, Gonostomatidae), 43; silvery, 123
Boreomysis scyphops (mysid shrimp; Crustacea), structure of eyes, 400
Branchiocerianthus (giant hydroid, Cnidaria), in southwest Pacific trenches, 309; size, 178
Brisinga endacacnemos (sea-star, Asteroidea, Echinodermata), depth distribution and size, 192, 193
Brosmiculus (deep-sea cod, Moridae), light organ, 138, 373
Bugula (moss animal, Bryozoa), depth record, 167–8

Cadulus californicus (scaphopod mollusc), reproductive cycle, 462
Calanus (Copepoda, Crustacea), egg laying, 429
  acutus, seasonal migrations in Antarctic, 252, 430
  cristatus, ontogenic migrations, 430; and sound scattering layer, 223
  finmarchicus, ontogenic migrations, 431, 434, 443; sinking rate, 321–2; whale-food, 79
  helgolandicus and *carinatus*, size and vertical distribution off Canary Islands, 242
Calciosolenia sinuosa (coccolithophore), 51
Callitheuthis (deep-sea squid, Mollusca), ammoniacal uplift, 324
Calocalanus (Copepoda, Crustacea), feathery setae, 322
Careproctus amblystomopsis (sea-snail, Liparidae), depth record, 213
  kermadecensis, fecundity and egg size, 459
  ovigerum, mouth-brooding, 459
Carinaria (Heteropoda, Mollusca), predation, 107
Caulophacus (hexactinellid sponge), form, 165
Caulophryne (ceratioid angler-fish), 133, facultative male parasitism, lateral line system, 318; no light lure, 364
  jordani, lateral line system in female, 415
Cavolinia tridentata (Pteropoda, Mollusca), 99
Centobranchus spp. (lantern-fishes, Myctophidae), distribution, 486
Centrophorus (squaloid shark), live-bearing, 458
  squamosus, capture at midwater level, 274
Centrophryne (ceratioid angler-fish), small-eyed males, 134
Centroscyllium fabricii (deep-sea shark, Squaloidea), luminescence and distribution, 140
Centroscymnus (squaloid shark), live bearing, 458
  coelolepis, distribution, 140–1
Ceratias (ceratioid angler-fish), luminous secretion, 350
  holboelli, aphakic space in eye of male, 383; binocular field of eyes, 390; buoyancy, 338; fecundity, 449; food requirements, 269
Ceratium fusum (dinoflagellate), 53
  ranipes, 53
  tripos, 53
Ceratoscopelus maderensis (lantern-fish,

Ceratoscopelus maderensis—contd.
Myctophidae), distribution, 477; fecundity, 444; lipid content, 335; and sound scattering layers, 356
townsendi, 123
warmingi, distribution, 476
Cestus (comb-jelly, Ctenophora), ionic uplift, 319
veneris (Venus' girdle), swimming, 94
Cetonurus globiceps (rat-tailed fish, Macrouridae), distribution, 491; fecundity, 456
Chaetoceros decipiens (diatom), 55
Chaetoderma (Aplacophora, Amphineura, Mollusca), deposit feeder, 149
Chalinura spp. (rat-tailed fishes, Macrouridae), abyssal forms, 490; distribution, 137, 491; no light organ, 371; small eyes, 392
Challengeria (Radiolaria, Protozoa), 72
Challengeron armatum (Radiolaria, Protozoa), 74
Chauliodus (viper-fish, Chauliodontidae, Stomiatoidea), angling dorsal ray, 367; length of life history, 446; light organs, 126; migratory forms, 327; olfactory organs, 403; reflexion of light, 362; water content, 268
danae, diet and migrations, 245; distribution, 483
macouni, luminescence, 370
minimus, distribution, 483
sloani, distribution, 483; inner ear, 424; luminescent camouflage, 352, 354, 362; retinal pigment, 382
Chaunax (lophiiform fish, Chaunacidae), small olfactory organs, 409
Chelophyes appendiculata (calycophoran siphonophore), wide distribution, 90
Chiasmodon niger (giant swallower, Chiasmodontidae, Pisces), 133; olfactory organ, 403
Chirostomias (stomiatoid fish, Melanostomiatidae), moderately developed olfactory organs, 404
pliopterus, 125
Chiroteuthis (deep-sea squid), 120
Chlidonophora chuni (lamp-shell, Brachiopoda), 170
Chlorophthalmus (green-eyed fishes, Chlorophthalmidae), eyes, 392; distribution, 215; lateral line system, 420
albatrossis and *nigromarginatus*, perianal light organ, 374
Chologaster cornutus and *agassizii* (amblyopsid fishes), lateral line system, 416
Chondrocladia (sponge, Demospongia), form, 165, 167
concrescens, depth record, 165, 166

Chondrophyllia coronata (sea-anemone, Actiniaria, Cnidaria), distribution, 183
Chrysogorgia (horny coral, Gorgonacea, Cnidaria), 181
Chunella (sea-pen, Pennatulacea, Cnidaria), form, 182
Cirroteuthis (finned octopod, Mollusca), appearance, 121; buoyancy, 339; retinal subdivisions, 395
Cirrothauma murrayi (finned octopod, Mollusca), appearance, 121; regressed eyes, 395; reduction of tissues, 339
Cladorhiza (sponge, Demospongia), form, 165, 167
longipinna, 166
septemdentalis, 166
Clione limacina (gymnosomatous pteropod, Mollusca), food of baleen whales, 100
Coccolithus huxleyi (coccolithophore), abundant off Bermuda, 446
pelagicus, 51
Coccorella atrata (alepisauroid fish, Evermannellidae), 129; light organs, 357
Coelorhynchus (rat-tailed fish, Macrouridae), bathyal forms, 490; eyes, 392; food, 277; lateral line organs, 419; light organ, 373; taste-bud complex, 410
caribbaeus, distribution, 491
coelorhynchus, swimbladder, 334
and *fasciatus*, egg sizes, 456
occa, length of retia mirabilia, 332; size, 282
Colossendeis (sea-spider, Pycnogonida), size, 203; swimming, 205
colossea, 204, luminescence, 375
Conchoecia (ostracods, Crustacea), luminescent, 349; vertical migrations, 80
spinirostris, sinking, 322
Conocara (alepocephalid fish), large eyes, 392
Corethron criophilum (diatom), 55
Corolla (Pteropoda, Mollusca), swimming, 100
Corydoras aeneus (catfish), DNA content, 264
Coryphaenoides (rat-tailed fishes, Macrouridae), relative size of eyes, 392; taste-bud complex, 410
acrolepis, oxygen consumption, 278
armatus, biomass on continental rise, 281; pelagic food, 274
colon, 136
mexicanus, distribution, 491
pectoralis, size, 137
rupestris, distribution, 492; eggs, 456; fecundity, 456; feeding habits, 274; lateral line canals, 419; retia mirabilia, 332

Coscinodiscus nodulifer (diatom), 56; distribution, 474
  radiatus, 55
Crambionella orsini (jellyfish, Rhizostomeae), in Indian Ocean, 91
Cranchia (cranchiid squid), structure, 117
  scabra, ontogenetic migrations, 439; size, 117
Creseis acicula (Pteropoda, Mollusca), 99
Cribrostomellus apertus (Foraminifera, Protozoa), 147
Crossota brunnea (Trachymedusae), 87; vertical distribution, 87
Cryptasterias personatus (sea-star, Asteroidea, Echinodermata), luminous regions, 376
Cryptopsaras (ceratioid angler-fish), 44; luminous secretion, 350–1; parasitic males, 134; stalked neuromasts, 417
  couesi, aphakic space in eye of male, 383; binocular field of eyes, 390; larval distribution, 489; males, 452
Crystallophrissa (Aplacophora, Amphineura, Mollusca), deposit feeder, 149
  indicum, 151
Culeolus (sea-squirt, Ascidiacea), form and food, 296
  murrayi, 171
  suhmi, food, 173
Cyanea capillata (jellyfish, Scyphozoa), size, 75
Cyclosalpa (pelagic tunicate), luminous bodies, 346
Cyclothone (stomiatoid fish; Gonostomatidae), 42, 44, 132; distribution, 487–8; food, 254–5; lateral line system, 416; non–migrators, 327; otic capsules, 423; reflexion of light, 362; regression of swimbladder, 327, 388; sexual dimorphism, 261, 404–8, 459; similarity of light organ system in both sexes, 367; ubiquity and abundance, 122–3; vertical ranges, 246–7
  braueri, larval stages, 448; no sex reversal, 261, 449; transparency, 123
    and *alba*, *pseudopallida*, small gas-filled swimbladder, 335
    and *microdon*, fecundity, depth distribution and spawning cycle, 447–9
    and *obscura*, relative size of ventral light organs, 362
    and *signata*, *alba* and *pseudopallida*, structure of eyes, 390
  livida, distributed off West Africa, 254
  microdon, sex reversal, 261, 449; structure of olfactory organs and brain in male and female, 406
  obscura, distribution, 489
  pygmaea, a Mediterranean species, 254

Cyclothone signata, transparency, 123
Cyema atrum (bob-tailed snipe-eel, Cyemidae), jaws and fins, 131; leptocephalus, 450, 451; lateral line organs, 413, 416, 415
Cymbulia (Pteropoda, Mollusca), movement of wings, 100
Cynomacrurus (rat-tailed fish; Macrouridae), absence of chin barbel, 410; regressions, 339
Cystisoma (Amphipoda, Crustacea), 116

Danaphryne nigrifilis (ceratioid angler-fish, female), 132
Deltocyathus (solitary coral, Madreporaria, Cnidaria), depth distribution, 184
Dentalium (scaphopod mollusc), relative egg size, 462
  megathyrus, depth distribution, 149
Diacria trispinosa (Pteropoda, Mollusca), 99
Diaphus (lantern-fish, Myctophidae), aphakic aperture in eye, 383; small gape in certain species, 127
  diadematus, sexual dimorphism in head lights, 368
  dumerili, distribution, 476
  rafinesquei, blue mirrors in light organs, 356; lipid content, 335; swimbladder, 328
  taaningi, food and daily vertical migrations in Cariaco Basin, 244–5
Diaptomus pallidus freshwater copepod, Crustacea), chemoreceptors on mouthparts, 412
Dicarpa simplex (sea-squirt, Ascidiacea), depth distribution, 170
Diestothyris frontalis (lamp-shell, Brachiopoda), depth distribution, 170
Dimophyes arctica (calycophoran siphonophore), distribution, 90
Dinophysis acerta (dinoflagellate), 53
Diogenichthys (lantern-fish, Myctophidae), annual fish, 455; jaw size, 244
  atlanticus, distribution, 486
Diphyes (calycophoran siphonophore), 89; buoyancy, 319
Diretmus (beryciform fish, Diretmidae), density of retinal pigment, 382
Discolabe (physonect siphonophore), structure, 88
Discosphaera thomsoni (coccolithophore), 51
Distephanus speculum (silicoflagellate), 51
Ditropichthys storeri (whale fish, Cetunculi), adult fish and regressed eye, 391
Dolichopteryx (argentinoid fish), body form, 127
  longipes, structure of eye, 387

Doliolum nationalis (pelagic tunicate), gonozooid, 97
Dolopichthys (ceratioid angler-fish), stalked neuromasts, 417
  longicornis, fecundity, 449
Dosidicus gigas (ommastrephid squid), size, 117

Echinocucumis hispida (sea-cucumber, Holothuroidea, Echinodermata), 198
Echinomacrurus mollis (rat-tailed fish, Macrouridae), distribution, 491
Echinus affinis (sea-urchin, Echinoidea, Echinodermata), 196
Echiostoma (stomiatoid fish, Melanostomiatidae), body form, 126; small olfactory organs, 403
  tanneri, sensitivity of barbel, 365
Edgertonia argillispherula (Foraminifera, Komokiacea, Protozoa), 161
Edriolychnus (ceratioid angler-fish), female and parasitic males, 134, 454
  schmidti, fecundity, 449
Edwardsia (sea-anemone, Actiniaria, Cnidaria), 183
Electrona (lantern-fish, Myctophidae), sexual dimorphism in tail lights, 368
  antarctica, 123, 455; distribution, 477; lateral line organs on head, 418
  rissoi, size of mature eggs, 443
Elpidia (sea-cucumber, Holothuroidea, Echinodermata), abundant in certain trenches, 308, 309; low content of organic matter, 311
  glacialis, distribution, 199, 200
Enypiastes (pelagic sea-cucumber, Holothuroidea, Echinodermata), 201
Ephyrina bifida (caridean prawn, Crustacea), large eggs, 438
Epizoanthus (Zoanthidea, Cnidaria), attachments, 183
Ericmaster (sea-star, Asteroidea, Echinodermata), distribution, 498
Eryoneichus, larval stage of eryonid decapod (Crustacea), 466
Ethmodiscus rex (diatom), 59, 60
Ethmolaimus (nematode worm), 144
Ethusa (crab, Brachyura, Crustacea), 208
Ethusina (crab, Brachyura, Crustacea), 208; depth distribution, 304
Etmopterus (squaloid shark), live-bearing, 458
  hillianus, size, 140
  princeps, capture at midwater level, 274
Euchaeta (Copepoda, Crustacea), egg sacs, 429
Euchirella (Copepoda, Crustacea), feathery setae, 322; large eggs, 430
Euchlora rubra (comb-jelly, Ctenophora), presence of nematocysts, 92

Eucopia (mysid shrimp), colour, 115; low fecundity, 438
Eugaptilus laticeps (Copepoda, Crustacea), 78
Eukrohnia (arrow-worms, Chaetognatha), mesopelagic genus, 103; large eggs, 439
  bathyantarctica, 102
  hamata, brood chamber, 439; life history, 439; seasonal migrations in Antarctic, 252
Euphausia (Euphausiacea, Crustacea), 108; migrant forms, 236, 238; distribution, 239; structure of eye, 399; spherical eyes, 400, and vertical migrations, 401
  gibboides, structure of light organs and luminescence, 357-8, 359
  pacifica, coiling of rhabdomes in eyes, 401; relative buoyancy, 323
  superba (krill), 108; food, 109, 113; herbivorous nature, 229; life history and growth, 434; luminescent swarms, 357; size at maturity, 437
Euphronides (sea-cucumber, Holothuroidea, Echinodermata), tracks, 201
Euplectella (hexactinellid sponge), 162, 163, 165
Euplectellum aspergillum (hexactinellid sponge), 164
Eurycope (Isopoda, Crustacea), body form, 151, 152
  californiensis, breed throughout year, 461-2
Eurypharynx (gulper-eel, Eurypharyngidae), jaws and size, 131; sea-urchin remains in gut, 135
  pelecanoides, fecundity and lateral line organs, 415; lymph spaces, 268; egg size, 449
Eustomias (stomiatoid fish, Melanostomiatidae), aphakic space and retinal organization, 383; body form, 126
Eutheleptus atlanticus (polychaete worm), 293
Eutonina indicans (Hydroida, Cnidaria), vibration receptors of medusa, 422
Evermannella (alepisauroid fish, Evermannellidae), olfactory organs, 403

Fauveliopsis (polychaete worm), distribution, 498
  brevis, 293
  glabra, reproductive cycle, 462
Flabellum apertum (solitary coral, Madreporaria, Cnidaria), distribution, 184
  goodei, 184, 493
Folia (Venus' girdle, comb-jelly, Ctenophora), 93
Fragilariopsis kerguelensis (diatom), distribution, 474

Freyella mortenseni (sea-star, Asteroidea, Echinodermata), depth record, 192

Frieleia halli (lamp-shell, Brachiopoda), reproductive cycle, 462, 463

Funchalia, deep-sea genus of penaeid prawns (Decapoda), 111; colour, 113

Fungicyathus (solitary coral, Madreporaria, Cnidaria), depth distribution, 184

Funiculina (sea-pen, Pennatulacea, Cnidaria), form, 183
  quadrangularis, distribution, 183

Gadella (deep-sea cod, Moridae), light organ, 373

Gadomus spp. (rat-tailed fishes, Macrouridae), rather small eyes, 392

Gadus morhua (cod, Gadidae), finding a sound source, 427

Galiteuthis phylura (deep-sea squid, Mollusca), ontogenetic migrations, 439

Gasterascidia lyra (sea-squirt, Ascidiacea), food, 296

Gaussia princeps (Copepoda, Crustacea), escape response, 78; luminescence and photogenic organs, 347, 348; relative buoyancy, 322

Gempylus (snake-mackerel, trichiuroid fish), 133

Gennadas, deep-sea genus of penaeid prawns (Decapoda), 111
  elegans, egg and larval stages, 436

Geryon (crab, Brachyura, Crustacea), depth distribution, 304
  quinquidens, 208, 209; larval stages, 464

Gigantactis (ceratioid angler-fish), form of larvae, 451; progenetic males, 453

Gigantocypris (ostracod, Crustacea), 79; eyes, 80; reported luminescent, 349; size and movement, 79
  agassizii, neutrally buoyant, 322; respiratory rate, 268
  mulleri, ammoniacal uplift, 322

Gigantura (deep-sea fish, Giganturidae), structure of tubular eyes, 388

Glandiceps (abyssal acorn-worm, Enteropneusta), 466

Glaucus (nudibranch mollusc), 42

Gleba (Pteropoda, Mollusca), swimming, 100
  cordata, 99, mucus web, 100

Globigerina bulloides (Foraminifera, Protozoa), 71

Globorotalia crassiformis (Foraminifera, Protozoa), 70
  scitula, 70

Glyphocrangon (crangonoid shrimp, Crustacea), armour, 205

Gnathophausia (mysid shrimp, Crustacea), colour, 115; luminous secretion, 349

gracilis, respiratory rate, 268

ingens, 115; oxygen consumption, 484-5; sensitivity to amino-acids, 411

Golfingia (sipunculid worm), 201
  flagrifera, 155

Gonichthys (lantern-fish, Myctophidae), annual fishes, 455; gape, 127, 244
  barnesi, distribution, 477
  coccoi, 43

Gonionemus (trachyline medusa, Cnidaria), structure, 85

Gonostoma (stomiatoid fish, Gonostomatidae), life history, 446; optics of aperture around eye lens, 383
  bathyphilum, large lymph spaces, 268
  denudatum, buoyancy balance sheet, 337; and *elongatum*, *bathyphilum*, contrasts in organization, 262-4, 263, 266
  elongatum, buoyancy balance sheet, 337; density of retinal pigment, 382; high water content, 268
  gracile, length of life history, 446; sex reversal, 449

Gonyaulax polyedra (dinoflagellate), 60

Gorgonocephalus (basket star, Ophiuroidea, Echinodermata), size, habit and food, 190

Grammatostomias flagellibarba (stomiatoid fish, Melanostomiatidae), 125

Gymnoscopelus (lantern-fish, Myctophidae), gape, 127
  braueri, 455; distribution, 477

Gyrosigma sp. (diatom), 55

Hadalothuria wolfii (sea-cucumber, Holothuroidea, Echinodermata), 198, 199

Halargyreus (deep-sea cod, Moridea), eaten by gulper-eel, *Saccopharynx*, 135

Halicreas minimum (Trachymedusae), colour, 87
  rotundatum, 87

Halisiphonia galatheae (Hydroida, Cnidaria), depth record, 178

Haloptilus acutifrons (Copepoda, Crustacea), 77

Harpiniopsis excavata (amphipod, Crustacea), breed throughout year, 461-2

Hastigerina murrayi (Foraminifera, Protozoa), 70

Hastigerinella digitata (Foraminifera, Protozoa), 70

Hathrometra tenella (feather-star, Crinoidea, Echinodermata), 188

Heliococranchia pfefferi (mesopelagic squid), non-migratory, 121

Heliometra glacialis (feather-star, Crinoidea, Echinodermata), feeding posture, 188

Heterocarpus (caridean prawn, Crustacea), luminescent decoy, 374
Heteroteuthis (sepioid squid), luminous clouds, 119, 212, 350
  hawaiiensis, luminescence and background light, 360
Hexacrobylus indicus (sea-squirt, Ascidiacea), 172; structure and food, 173
Himantolophus (ceratioid angler-fish), diet of females, 255; progenetic males, 453
  groenlandicus, females, 132; female olfactory organs, 405; larvae, 451
Himantozoum (moss animal, Bryozoa), form, 169
Hippopodius hippopus (calycophoran siphonophore), buoyant swimming bells, 319, 321
Histioteuthis (deep-sea squid, Mollusca), ammoniacal uplift, 324; light organs, 119, 360
  dofleini, photosensitive vesicles, 396
  meleagroteuthis, ontogenetic migrations, 439; yellow lens and unequal size of eyes, 394–5
Holascus (hexactinellid sponge), form, 165, 167
Homarus vulgaris (lobster, Crustacea), vibration receptors, 422
Hoplostethus (beryciform fish, Trachichthyidae), light organ, 373
Hyalinoecia artifex (polychaete worm), depth range, 493
  tubicola, 203
Hyalocylis striata (Pteropoda, Mollusca), 99
Hyalonema (hexactinellid sponge), 164, 165, 167
Hydractinia (hydroid, Cnidaria), and copepod food, 84
Hydrasterias ophidion (sea-star, Asteroidea, Echinodermata), luminous regions, 376
Hygophum (lantern-fish, Myctophidae), capacious swimbladder and low lipid content, 335
  benoiti, fecundity, 444; length of retia mirabilia in swimbladder, 332
Hygrosoma petersii (sea-urchin, Echinoidea, Echinodermata), 196
Hymenaster sp. (sea-star, Asteroidea, Echinodermata), luminous regions, 376
  blegvadi, depth record, 192
Hymenocephalus (rat-tailed fish, Macrouridae), food, 276; light organ and ventral colour pattern, 373
  aterrimus, 136
  italicus, egg size, 456; inner ear, 424; light organ, 372, 373; young stages, 491
Hymenodora, deep-sea genus of caridean prawns (Decapoda), 111; luminous organs, 114
  glacialis, a bathypelagic species, 114 and *frontalis*, fecundity, 438
Hyperia (hyperiid amphipod, Crustacea), 116

Ianthina (purple snail, Mollusca), 42
Ichthyococcus (stomiatoid fish, Gonostomatidae), ventral light organs, 353
Idiacanthus (stomiatoid fish, Idiacanthidae), sexual dimorphism and luminescence; small olfactory organs, 403
  antrostomus, luminescence, 370
Ilyarachna profunda (isopod, Crustacea), breeds throughout year, 461–2
Ipnops (chlorophthalmid fish), fin pattern and sense organs, 213, 215; eyes, 292; olfactory organs, 409; in Bahamas basins, 305
  murrayi, fecundity, 456; egg size, 458; lateral line system, 420
Isistius (midwater squaloid shark), 441

Japetella (deep-sea octopod, Mollusca), subdivisions of retina, 395
  diaphana, ionic uplift, 325, 341; life history stages, 440, 443; ontogenetic migrations, 439–40; size and transparency, 119

Kamptosoma abyssale (sea-urchin, Echinoidea, Echinodermata), depth record, 195
Kesun (polychaete worm), distribution, 498
Kinetoskias (moss animal, Bryozoa), 168, 169
Kolga hyalina (sea-cucumber, Holothuroidea, Echinodermata), luminous regions, 377
Kophobelemnon (sea-pen, Pennatulacea, Cnidaria), form, 182

Laetmogone violacea (sea-cucumber, Holothuroidea, Echinodermata), luminescence, 375, 377
Lamellibranchia barhami (beard-worm, Pogonophora), size, 177
Lampadena (lantern-fish, Myctophidae), 44; aphakic space in eye, 383; fecundity, 447; observed near bottom, 272; non-migratory species, 244; size of jaws, 368; tail light organs 368; ventral light organs, 362
  chavesi and *speculigera*, length of retia mirabilia in swimbladder, 332
  speculigera, swimbladder, 331
Lampanyctodes hectoris (lantern-fish, Myctophidae), eggs, 443

Lampanyctus (lantern-fish, Myctophidae), food of ceratioids, 253; gape, 127; tail light organs, 368; vertical range, 246
  alatus, olfactory organs, 404
  cuprarius, daily vertical migrations, diet and time of feeding, 245
  macdonaldi, observed near bottom, 272
  pusillus, length of retia mirabilia in swimbladder, 332
  steinbecki, size of mature eggs, 443
Lanceola (hyperiid amphipod, Crustacea), colour, 116
Lauderia annulata (diatom), 56
Lensia subtilis, *multicrista* and *achilles* (calycophoran siphonophores), vertical distribution off the Canary Islands, 90
Lepidophanes gaussi (lantern-fish, Myctophidae), distribution, 476
  guentheri, length of retia mirabilia in swimbladder, 332
Lepidopleurus (Polyplacophora, Mollusca), 209
Lepidorhynchus (rat-tailed fish, Macrouridae), light organ and ventral colour pattern, 373
Lepidosiren (South American lungfish), lateral line system, 416
Leptacanthichthys (ceratioid angler-fish), facultative male parasitism, 453
Leptocylindrus mediterraneus (diatom), 56
Leptoderma (fish, Alepocephalidae), body and fin form, 128; large eyes, 392
Leptopenus (solitary coral, Madreporaria, Cnidaria), depth distribution, 184
Leptostomias (stomiatoid fish, Melanostomiatidae), moderately developed olfactory organs, 404
Lestidiops (alepisauroid fish, Paralepididae), light organ, 357
Lestidium (alepisauroid fish, Paralepididae), light organ, 357
  atlanticum, 130
Leucochlamys (ophidioid fish), egg clutches, 458
Leucothoe (comb-jelly, Ctenophora), hair cells, 422
Levinsenella (moss animal, Bryozoa), form, 169
  magna, 169
Limacina (Pteropoda, Mollusca), structure, 98
Limulus (horse-shoe crab, Chelicerata), genetic variability, 500
Linophryne (ceratioid angler-fish), eye structure in male, 391; tubular eyes in free-living males, 127, and depth range, 388
  arborifera, female, 391; light organ system of female, 364–5

macrorhinus, olfactory organs in free-living male, 405
  quinqueramosus, diet of female, 256
Liocranchia (cranchiid squid genus), structure, 117
  reinhardti, 118; ontogenetic migrations, 439; size, 117
Lionurus (rat-tailed fish, Macrouridae), depth distribution, 137; no light organ, 37, or drumming muscles, 426; small eyes, 392
  carapinus, fecundity, 456; population density, 427; size, 282; swimbladder, 334
Lipogenys gillii (notacanthiform fish), 139; ooze eater, 274–5
Lobianchia (lantern-fish, Myctophidae), vertical range, 246
  dofleini, daily vertical migrations and diet, 245; life history in Mediterranean and off Bermuda, 443–6; life history stages, 445; smallness, 480
Loligo (squid, Mollusca), photosensitive vesicles, 395
Lolliguncula brevis (coastal squid, Mollusca), olfactory organs, 412
Lophius (angler-fish, Lophiidae), 133; lateral line organs, 416; small olfactory organs, 409
Lophogaster (mysid shrimp), 115
Lophohelia (ivory-coral, Madreporia, Cnidaria), associated fauna, 303; form and distribution, 184
Lota lota (gadiform fish), structure of neuromast, 414
Lotella (deep-sea cod, Moridae), may have light organ, 373
Lucicutia bicornuta (Copepoda, Crustacea), 77
Lycenchelys (eel-pout, Zoarcidae), depth distribution, 213
Lycoteuthis (deep-sea squid, Mollusca), negative buoyancy, 324
Lyonsiella abyssicola (septibranch bivalve, Mollusca), 150

Macandrevia crania (lamp-shell, Brachiopoda), depth distribution, 170
Macrocypridina castanea (Ostracoda, Crustacea), 79
Macropinna (argentinoid fish), tubular eyes, 386–7
Macrostylis (Isopoda, Crustacea), body form, 151
  galatheae, 152
Macrourus (rat-tailed fishes, Macrouridae), large eyes, 392
  berglax, egg size, 456; distribution, 492; fecundity, 456

Malacocephalus laevis (rat-tailed fish, Macrouridae), distribution, 491; sound-producing mechanism, 426; structure of light organ, 371–2, 372

Malacosteus (rat-tailed fish, Malacosteidae, Stomiatoidea), 363; red light from cheek photophores, 363; olfactory organs, 403–4

Mallicephala (polychaete worm), distribution, 498
hadalis, 202
mirabilis, 201

Mastigoteuthis (deep-sea squid, Mollusca), ammoniacal uplift, 324; aphakic space in eye?, 394; light organs, photosensitive vesticles, 395
schmidti, 119; non-migratory, 121

Maurolicus (stomiatoid fish, Gonostomatidae), length of rete mirabile in swimbladder, 332; size of blood erythrocytes, 332; swimbladder, 335

Meator rubatra (hydroid, Medusae), 86

Megacalanus princeps (Copepoda, Crustacea), 77

Meganyctiphanes norvegica (Euphausiacea, Crustacea), coiling of rhabdomes in eyes, 401; movements of light organs, 358

Melamphaes (midwater genus of fishes, Melamphaidae), 133
pumilus, 480; distribution, 483
simus, distribution, 483
suborbitalis, 480

Melanocetus (ceratioid angler-fish), 44; feeding of males, 255; progenetic males, 453
johnsoni, diet of females, 256; free-living males, 452; larval distribution, 489
murrayi, buoyancy of female, 338; olfactory organs and brain structure of female, 405

Melanostomias (stomiatoid fish, Melanostomiatidae), high water content, 268
spilorhynchus, 125

Metapenaeus sp. (penaeid prawn, Crustacea), image formation in eyes, 398

Meteoria (ophidoid fish), egg clutches, 458

Metridia lucens (Copepoda, Crustacea), luminescence, 349

Michaelsarsia splendens (coccolithophore), 51

Mola (ocean sunfish, Molidae), food, 83

Molgula immunda (sea-squirt, Ascidiacea), 172

Molpadia danialsenni (sea-cucumber, Holothuroidea, Echinodermata), distribution, 199
violacea, distribution, 199

Monomitopus metriostoma (brotulid fish), sound-producing mechanism, 426

Mora moro (deep-sea cod, Moridae), optical properties of eye lens, 381

Muggiaea (calycophoran siphonophore), more neritic than oceanic distribution, 90

Munidopsis (squat lobster, Galatheidae), depth distribution, 304; near hydrothermal vent, 306
subsquamosa latimana, 207

Myctophum (lantern-fish, Myctophidae), luminescent sexual dimorphism, 368; size of jaws, 244; swimbladder and lipid content, 335
affine, luminescent response to luminous dial of watch, 369
nitidulum, ventral light organs, 353
punctatum, 43; fecundity, 444; length of retia mirabilia, 332; retinal pigment, 382; spectral composition of luminescence, 356; swimbladder, 328

Myriotrochus bruuni (sea-cucumber, Holothuroidea, Echinodermata), 198

Myxine glutinosa (hagfish, Myxinidae), 157

Nanomia (physonect siphonophore), 89, 320; structure and movements, 88, 90

Nansenia (argentinoid fish), aphakic space and retinal organization, 383
groenlandica, retinal structure, 382

Nausithoe rubra (jellyfish, Coronatae), colour, 91

Nautilus (Nautiloidea, Mollusca), buoyancy chamber, 325

Nealotus (trichiuroid fish), 133

Nearchaster aciculosus (sea-star, Asteroidea, Echinodermata), genetic variability, 501

Nematobrachion (Euphausiacea, Crustacea), 108; bilobed eyes, 110, 400; embryos protected, 432; fecundity, 432
boopsis, depth ranges of furcilia and adults, 433
sexspinosus, relative size of eye lobes, 402

Nematonurus (rat-tailed fish, Macrouridae), no drumming muscles on swimbladder, 426; no light organ, 371
armatus, distribution, 491; population density, 427; organization compared with gulper-eel *Saccopharynx*, 270–1

Nematoscelis (Euphausiacea, Crustacea), bilobed eyes, 109–10, 400; embryo protection and fecundity, 432, 433; sexual dimorphism, 369
difficilis, eye structure and spectral sensitivity, 401
mantis, 109
tenella, distribution, 433; luminescent camouflage, 358; size of second calyptopis, 433

Nemichthys (snipe-eel, Nemichthyidae), olfactory organs, 404

Neoceratias (ceratioid angler-fish), female and parasitic male, *454*; lateral line system, 318; no rod and light lure, 255, 364; stalked neuromasts, 417

Neopilina galatheae (Monoplacophora, Mollusca), 209, *210*

Nessorhamphus (deep-sea eel, Nessorhamphidae), 449; life history stages, *450*

Nezumia spp. (rat-tailed fishes, Macrouridae), distribution, 491; food, 455, 457; light organ, 373; relative size of eyes, 392; taste-bud system, 410

sclerorhynchus, egg size, 456; food, 273

Nitzschia bicapitator (diatom), 56

curta, 57
cylindrus, 56
frigida, 55
grunowii, 57
kerguelensis, 57
marina, 56
seriata, 57

Noctiluca (dinoflagellate), 52

Normanina tylota (Foraminifera, Komokiacea, Protozoa), *161*

Protozoa), *161*

Notacanthus (spiny-eels, Notacanthidae), dentition and food, 275–6; food, 138

chemnitzi, size, 138

Notolepis coatsi (alepisauroid fish, Paralepididae), 455; size, 129

rissoi, retinal organization, 382

Notolychnus (lantern-fish, Myctophidae), annual fish, 455

valdiviae, smallness, 480

Notoscopelus kroyeri (lantern-fish, Myctophidae), low lipid content, 335

Notostomus (caridean prawn, Decapoda), 111; ammoniacal means of buoyancy, 119, 323; colour, 113; size, 113; small eggs, 438

Notosudis (notosudid fish, Notosudidae), eye, 384

Nuculana pontonia and *darella* (bivalve molluscs), reproductive cycle, 462

Nybelinia erikssoni (ophidioid fish), embryo and adult, 457, 458

Nymphon femorale (sea-spider, Pycnogonida), 204

gracile, luminescence, 375
longitarse, depth record, 203
tripectinatum, depth record, 203

Octacnemus bythius (sea-squirt, Ascidiacea), 172; structure and food, 173; capture of food, 296

Octopoteuthis (deep-sea squid, Mollusca), 120; ammoniacal uplift, 324

nielseni, luminescence and background light, 361, 362

Octopus vulgaris (coastal octopus, Mollusca), chemosensors on suckers, 412

Ocyropsis maculata (comb-jelly, Ctenophora), swimming by oral lobes, 94

Odontomacrurus murrayi (rat-tailed fish, Macrouridae), adaptations to midwater life, 271; no chin barbel, 410; regression of tissues, 339

Oikopleura (pelagic tunicate, Larvacea), 95; luminescence, 96, 346; population growth, 224

albicans, 97

Oithona (Copepoda, Crustacea), feathery setae, 322

plumifera, vertical migrations, 243

Ommastrephes bartrami (flying squid), 117

pteropus (oceanic squid, Mollusca), heavier than sea, 324

Omosudis (alepisauroid fish, Omosudidae), dentition, 129; ventral aphakic space in eye, 385

Oncaea media (Copepoda, Crustacea), vertical migrations, 243

Oneirodes (ceratioid angler-fish), olfactory organs and brain in free-living male, 405

bulbosus, distribution, 490
eschrichti, distribution, 488; esca structure, 257
thompsoni, distribution, 490

Onychoteuthis banksi (flying squid), 117; density, 324

Ophiacantha aculeata (brittle-star, Ophiuroidea, Echinodermata), luminescence, 375

bidentata, luminescence, 192
normani, reproductive cycle, 462

Ophiaster formosus (coccolithophore), *51*

Ophiocoma scolopendrina (brittle-star, Ophiuroidea, Echinodermata), method of feeding, 190

Ophiomusium lymani (brittle-star, Ophiuroidea, Echinodermata), *191*, 192; dominance, 494; genetic variability, 500–1; reproductive cycle, 462

Ophiura lovenii (brittle-star, Ophiuroidea, Echinodermata), depth distribution, 190

Opisthoproctus (argentinoid fish), body form, 127; light organ, 127; luminescence, 357, tubular eyes, 248, 386–7

Opisthoteuthis (octopod, Mollusca), form and colour, 211

Oplophorus (caridean prawn, Crustacea), luminous secretion, 349; relatively large eyes, 397

gracilirostris and *spinosus*, fecundity, 438

Oplophorus spinosus, image formation in eyes, 398; luminescent field, 360

Ornithocercus steinii (dinoflagellate), 53

Oryzias (rice-fish, Cyprinodontidae), lateral line organs of larva, 421

Oxynotus (squaloid shark), live-bearing, 458

Oxytoxum tessalutum (dinoflagellate), 53

Pachystomias (stomiatoid fish, Melanostomiatidae), red light from cheek photophores, 126, 363

Pagetopsis macropterus (nototheniiform fish), perching on sponges, 305

Pandea rubra (hydroid, medusae), 86

Parabrotula (zoarcid fish), viviparity, 458

Paradiplospinus (midwater trichiuroid fish), form, 133

Paragissa galatheae (Gammaridea, Crustacea), 153

Paralepis atlantica (paralepidid fish, Alepisauroidea), size, 129

Paramphiacella (harpacticoid copepod, Crustacea), 144

Parapagurus (hermit-crab, Paguridea, Crustacea), 207; depth distribution, 304

Parapandulus (caridean prawn, Crustacea), benthic forms, 134, 205; life history on bottom, 460

Parathemisto (hyperiid amphipod, Crustacea), 116

Paratrachichthys prosthemius (beryciform fish, Trachichthyidae), light organ and luminescence, 373–4

Paraugaptilus (Copepoda, Crustacea), large eggs, 430; non-migratory, 430

Parazoanthus (Zoantharia, Cnidaria), attachments, 183

Pareuchaeta (Copepoda, Crustacea), egg sacs, 429; non-migratory, 430

Paroriza sp. (sea-cucumber, Holothuroidea, Echinodermata), luminous regions, 377

Pasiphaea (caridean prawn, Crustacea), benthic forms, 205

hoplocerca, large eggs, 438

Pavonaria (sea-pen, Pennatulacea, Cnidaria), form, 183

Pectinaster forcipulatus (sea-star, Asteroidea, Echinodermata), luminous regions, 376

Pegea confoederata (salp, pelagic tunicate), mucous net and food, 98

Pelagia noctiluca (jellyfish, Semaeostomae), ionic uplift, 319; luminescence, 91, 343

Pelagodiscus (lamp-shell, Brachiopoda), 169; off east coast of US, 295; suspension feeder, 298

atlanticus, 169; distribution, 170, 295; larval stage, 466, 477

Pelagonemertes (pelagic nemertean worm), colour, 105

Pelagosphaera (larva of sipunculid worm), 155

Pelagothuria (pelagic sea-cucumber, Holothuroidea, Echinodermata), 201

Pelopatides (sea-cucumber, Holothuroidea, Echinodermata), tracks, 201

Peniagone (sea-cucumber, Holothuroidea, Echinodermata), in southwest Pacific trenches, 309

theeli, luminous regions, 377

Perca (freshwater perch, Percidae), size of gas-resorbing area, 336

Peridinium (dinoflagellate), 53

Periphylla (jellyfish, Coronatae), and captured lantern-fish, 81

hyacintha, colour and predatory powers, 91

Persparsia taningi searsiid fish), 128

Pheronema (hexactinellid sponge), 163

Photobacterium phosphoreum, luminous bacterium, 342, 371

Photostomias (rat-trap fish, Malacosteidae), large cheek lights in males, 367; water content, 268

Phronima (Diogenes shrimp, Amphipoda, Crustacea), 116

Physalia physalis (cystonect siphonophore), 42; member of neuston, 88

Physetocaris microphthalma (caridean prawn, Decapoda), a bathypelagic species, 114

Physiculus (genus of deep-sea cod, Moridae), light organ, 138

japonicus (deep-sea cod, Moridae), light organ and luminescence, 373

Physophora (physonect siphonophore), structure, 88

Planktosphaera pelagica, larva of acornworm (Enteropneusta), 466

Platytroctagen (fish, Searsiidae), body form, 128

Platytroctes (fish, Searsiidae), body form, 128

Pleurobrachia pileus (sea-gooseberry, comb-jelly, Ctenophora), 29

Pleuromamma (Copepoda, Crustacea), egg laying, 429

xiphias, *abdominalis*, *pisekei*, *gracilis*, and *borealis*, size and vertical migrations off Canary Islands, 242, 243

Podolampas bipes (dinoflagellate), 53

Polyacanthonotus (spiny-eel, Notacanthidae), feeding, 276

rissoanus, 139

Polycheles (eryonid decapod, Crustacea), distribution, 208; larval stages, 466

Polyipnus (hatchet-fish, Sternoptychidae, Stomiatoidea), distribution, 245, 484

Polyipnus laternatus, 124

Porcellanaster (sea-star, Asteroidea, Echinodermata), distribution, 498
  coeruleus, 193

Porpita (Chondrophora, Cnidaria), 42

Pourtalesia aurorae (sea-urchin, Echinoidea, Echinodermata), depth record, 195
  miranda, 195

Priapulus abyssorum (priapulid worm), 155
  atlanticus, larval stages, 464

Primnoella krampi (horny coral, Gorgonacea, Cnidaria), depth record, 179, 181

Protomyctophum (lantern-fish, Myctophidae), sexual dimorphism in tail lights, 368; swimbladder and lipid content, 335
  anderssoni, 455
  tenisoni, distribution, 477

Pseudoeunotia doliolus (diatom), distribution, 474

Pseudomma affine (mysid shrimp, Crustacea), structure of eyes, 400

Pseudostichopus (sea-cucumber, Holothuroidea, Echinodermata), depth distribution, 199

Psychropotes longicauda (sea-cucumber, Holothuroidea, Echinodermata), 200, 201

Pterosagitta draco (arrow-worm, Chaetognatha), cosmopolitan at epipelagic levels, 103; eggs, 438–9

Pterotrachea coronata (Heteropoda, Mollusca), ionic uplift, 319; predation, 101

Pterygioteuthis (deep-sea squid, Mollusca), 120; buoyancy, 324; light organs, 360; photosensitive vesicles, 395, 396
  microlampas, luminescence and background light, 360

Pyrosoma (pyrosomid, pelagic tunicate), luminescence and photogenic bodies, 346
  spinosum, 97

Pyroteuthis addolux (deep-sea squid, Mollusca), luminescence and background light, 360
  margaritifera, light organs, 119

Raja bathyphilus (skate, Rajidae), depth distribution, 212

Renilla (sea-pen, Pennatulacea, Cnidaria), luminescence, 375

Rhincalanus (Copepoda, Crustacea), egg laying, 429
  gigas, ontogenetic migrations, 252, 430
  nasatus and *cornutus*, size and vertical distribution off Canary Islands, 242

Rhizammina alta (Foraminifera, Protozoa), 147

Rhizocrinus (sea-lily, Crinoidea, Echinodermata), 189; in Palau Trench, 295

Rhizophysa spp. (cystonect siphonophores), structure, 88

Rhizosolenia hebetata forma *semispina* (diatom), 55

Rhynchohyalus (argentinoid fish), tubular eyes and feeding, 248

Rosacea (calycophoran siphonophore, Cnidaria), luminescence, 345

Rossia (sepioid squid), form and colour, 211;
  on continental slope, 141; symbiotic luminous bacteria, 350

Ruvettus (oil-fish, Trichiuroidea), size, 133

Saccopharynx (gulper-eel, Saccopharyngidae), food, 131, 135; organization compared with that of *Nematonurus*, 270–1; size, 131
  harrisoni, 132

Sagitta (arrow-worm, Chaetognatha), planktonic genus, 101
  bipunctata, enflata and *hexaptera*, cosmopolitan at epipelagic levels, 103
  gazellae, endemic in Southern Ocean, ontogenetic migrations, 103
  lyra, decipiens and *macrocephala*, mesopelagic species, 103
  maxima and *gazellae*, rather similar life histories, 439
  setosa, density and swimming speed, 321

Salpa (pelagic tunicate), luminous bodies, 346; oxygen consumption, 83
  (Thalia) democratica, 97
  maxima, ionic uplift, 319

Sandalops (deep-sea squid, Mollusca), photosensitive vesicles, 395

Sardinops caerulea (Californian sardine, Clupeidae), energy budget, 244

Scalpellum (barnacle, Cirripedia, Crustacea), form, habit and distribution, 174
  regium, 175
  vitreum, distribution, 175

Sciadonus (ophidioid fish), egg clutches, 458

Sclerocrangon (crangonoid shrimp, Crustacea), armour, 205; egg brooding, 460

Scopelarchus analis (alepisauroid fish, Scopelarchidae), eyes with yellow lens, 248, 388

Scotoplanes (sea-cucumber, Holothuroidea, Echinodermata), 200

Scypholanceola (hyperiid amphipod, Crustacea), colour, 116

Searsia (fish, Searsiidae), lateral line canals, 414; light organs, 127; luminous secretions, 350; retinal pigment, 382

Sepia (cuttlefish, Mollusca), buoyancy chamber, 325; photosensitive vesicles, 395

Sepiola (sepioid squid), form and colour, 211; on slope, 141; symbiotic luminous bacteria, 350

Sergestes (sergestid prawn, Penaeidea, Crustacea), 111; colour, 113; light organs, 114, 360; spiny larvae, 114; transparency, 239; vertical distribution and food, 239–40

and *Sergia*, food of species off Bermuda, 240

challengeri, dermal photophores with lenses, 114

corniculum, 113

japonicus, watery tissues, 268; protein content compared with *splendens*, 268

phorcus and *similis*, relative buoyancy, 323

robustus, structure of eyes and depth range, 398, 400

and *japonicus*, lipid reserves and muscle reduction, 323

similis, egg size and spawning time, 435

Sergia (sergestid prawn, Penaeidea, Crustacea), colour and light organs (if present), 239, 114; photophore lens, 360; spiny larvae, 114; vertical distribution, 239–40

japonicus, absence of photophores, 114

lucens, life history and food, 434–5; larval stage, 436; growth, 437, 443

Serolis (Isopoda, Crustacea), eye structure and depth ranges, 397

bromleyana, size, 153

Serrivomer (mesopelagic eel, Serrivomeridae), jaws and teeth, 131; olfactory organs, 404

Siboglinum caulleryi (beard-worm, Pogonophora), 176

fjordicum, early life history stages, 465

Skeletonema costatum (diatom), 55

Solmissus albaescens (Narcomedusae), daily vertical migrations in Adriatic Sea, 87

Somniosus microcephalus (sleeper shark, Squaloidea), size, 141

Spadella (arrow-worm, Chaetognatha), benthic genus, 101; 'lateral line' organ, 421

Spatangus purpureus (sea-urchin, Echinoidea, Echinodermata), feeding method, 197

Sphagemacrurus grenadae (rat-tailed fish, Macrouridae), egg size, 456

Spheroides maculatus (puffer-fish), DNA content, 264

Spinocalanus magnus and *spinosus* (Copepoda, Crustacea), size and vertical distribution off Canary Islands, 242

Spiratella helicina (Pteropoda, Mollusca), and deep scattering layer in Beaufort

Sea region of arctic, 100; food of baleen whales, 100

Spirula (deep-sea sepioid, Mollusca), 326; buoyancy control, 325–6; light organ, 119; photosensitive vesicles, 395; size of eggs and larvae, 439; structure of eyes, 393, 394

Squalogadus modificatus (rat-tailed fish, Macrouridae), distribution, 491

Stannoma coralloides (Xenophyophoria, Protozoa), 160

Stannophyllum jenosum (Xenophyophoria, Protozoa), 160

Steindachneria argentea (gadiform fish, Steindachneriidae), light organ and luminescence, 373, 374

Stenobrachius (lantern-fish, Myctophidae), size of jaws, 244

leucopsarus, food off California, 245

Stephalia coronata (physonect siphonophore), near bottom, 142

Stephanoscyphus (scyphoid polyp, Scyphozoa), form and attachment, 178–9, 180

mirabilis, medusae belong to *Nausithoë*, 179

simplex, medusae may belong to *Nausithoë punctata*, 179

Stereomastis (erynoid decapod, Crustacea), larval stages, 466

Sternoptyx diaphana (hatchet-fish, Sternoptychidae), inner ear, 424; oral light organs, 365, 367, 366

Stomias (stomiatoid fish, Stomiatidae), light organs, 126; migratory, 327; olfactory organs, 404

affinis, distribution, 486

atriventer, luminescence, 370

Storthyngura benti (abyssal isopod, Crustacea), 461

novae-zelandiae, embryos, 461

Stylephorus (deep-sea fish, Stylephoridae), tubular eyes and small jaws, 387

Stylocheiron (Euphausiacea, Crustacea), 108; bilobed eyes, 109–10, 400; distribution, 239; embryo protection and fecundity, 432–3; feeding time, 238; non-migrant form, 236; structure, 238

affine, *suhmii* and *maximum*, comparison of eye structure, 401, 402

elongatum, fecundity, 433

maximum, distribution, 433

suhmii, fecundity, 433; structure of eye, 399

Sulculeolaria quadrivalvis (calycophoran siphonophore), buoyant swimming bells, 319, 321

Symbolophorus (lantern-fish, Myctophidae), capacious swimbladder and low lipid content, 335

boops, distribution, 477

californiense, size of mature eggs, 443
veranyi, 43; fecundity, 444
Synaphobranchus (deep-sea eel, Synaphobranchidae), leptocephalus stage, *450*; olfactory organs, 404
kaupi, life history, 455
Syracosphaera subsalsa (coccolithophore), *51*
Systellaspis (oplophorid prawn, Crustacea), eyes, 397; lens of photophores, 360
cristata and *braueri*, large eggs, 438
debilis, *111*; angular distribution of luminescent field compared with that of hatchet-fish, 360; fecundity, 438

Taaningichthys (lantern-fish, Myctophidae), aphakic aperture in eye, 383; fecundity, 447; light organs, 368; non-migratory, 244; retia mirabilia, 337; ventral light organs, 362
bathyphilus, *123*; length of retia mirabilia, 332
Talismania (alepocephalid fish), large eyes, 392
Tarletonbeania crenularis (lantern-fish, Myctophidae), low lipid content, 335; luminescence, 357
Tatjanella grandis (echiurid worm), *156*
Terebratalia tisimani (lamp-shell, Brachiopoda), depth distribution, 170
Terebratulina retusa (lamp-shell, Brachiopoda), depth distribution, 170
Tetraodon fluviatilis (puffer-fish), DNA content, 264
Thalassicola pelagica (Radiolaria, Protozoa), flotation, 71, 317
Thalassiosira antarctica (diatom), 56
constricta, 57
diporocyclus, 56
eccentrica, 56
gracilis, 57
gravida, *55*; distribution, 474
Thalassophya pelagica (Radiolaria, Protozoa), 317
Thalia democratica (salp, Tunicata), growth of populations, 224; ionic uplift, 319
Tharyx nigrorostrum (polychaete worm), 293
Thaumatocrinus jungerseni (feather-star, Crinoidea, Echinodermata), luminescence, 375
Themisto (hyperiid amphipod, Crustacea), 116
Thorosphaera elegans (coccolithophore), *51*
Thysanoessa gregaria (Euphausiacea, Crustacea), and *parva*, depth ranges off Canaries, 238
Thysanopoda (Euphausiacea, Crustacea),

eyes and vertical migrations, 400–1; eyes in 'giant' species, 397; giant species, likely to take copepods, 254; light organs in 'giant' species, 362; mesopelagic species, 108; migrant species, 236, 238; structure, 238
acutifrons, *109*; larval stages, 432; daily growth increment, 437
cornuta and *egregia*, ovarian eggs, 433; size of larvae, 433
egregia, spinicaudata, distribution, 487; size and economy of life, 261 and *spinicaudata*, size, 108
orientalis, monacantha, pectinata, depth of furcilia and adults, 433
Tindaria callistiformis (deep-sea bivalve, Mollusca), slow rate of growth, 310
Tintinnopsis ventricosa (Tintinnoida, Protozoa), 69
Todarodes pacificus (oceanic squid, Mollusca), egg mass, 439
Tomopteris kefersteinii (pelagic polychaete worm, Tomopteridae), *104*
septentrionalis, vertical distribution, 105
Torrea candida (pelagic polychaete worm, Alciopidae), structure and sensitivity of eyes, 397
Trachonurus sulcatus (rat-tailed fish, Macrouridae), distribution, 491
Travisia (polychaete worm), distribution, 498
Trichodesmium (blue-green alga), *51*, 56; colour of blooms, 221
erythraeum, 54
Tridacna (giant clam, Mollusca), genetic variability, 500
Triphoturus mexicanus (lantern-fish, Myctophidae), food off California, 245; size of mature eggs, 443
Tripterophycis (deep-sea cod, Moridae), light organ, 373
Tuscarora nationalis (Radiolaria, Protozoa), 74
Tylaspis (hermit-crab, Paguridea), depth distribution, 304
Typhlichthys (amblyopsid fish), lateral line system, 416

Umbellula (sea-pen, Pennatulacea, Cnidaria), *182*, *182*, 295
guntheri, depth record in Atlantic, 182
Undeuchaeta major and *plumosa* (Copepoda, Crustacea), size and vertical migrations off Canary Islands, 242, 243

Valdiviella (Copepoda, Crustacea), egg sacs, 429
insignis, size of eggs, 430

Valenciennellus (stomiatoid fish, Gonostomatidae), 43; annual form, 455; silvery, 123
   tripunctulatus, 124; eye, 385, 386; life history stages, 452; selective feeding, 245; swimbladder, 331, and rete mirabile, 332
Vampyroteuthis (vampire squid, Mollusca), 117; eyes, 393; luminescence, 360; retina, 395
   infernalis, distribution and structure, 121; ontogenetic migrations, 440
Vanadis formosa (pelagic polychaete worm, Alciopidae), structure of eye, 394, 397
Velella (purple-sail, Chondrophora, Cnidaria), 42
Ventrifossa (rat-tailed fishes, Macrouridae), eyes, 392; food, 277; light organ, 373
Verruca (barnacle, Cirripedia, Crustacea), 174
Vinciguerria (stomiatoid fish, Gonostomatidae), 43; annual fishes, 455; fecundity, 447; rete mirabile (size), 332; size of erythrocytes, 332; ventral light organs, 353
   attenuata, *nimbaria* and *poweriae*, depth ranges, 446–7
   lucetia, life history, 441; life history stages, 442, 447; retinal structure, 382
   poweriae, structure of eye, 385, 386

Vitreledonella richardi (mesopelagic octopod), size and transparency, 119; non-migratory, 121
Vityazema (echiurid worm), depth distribution, 156
Vogtia (calycophoran siphonophore, Cnidaria), luminescence, 345

Willemoësia (eryonid decapod, Crustacea), 206; depth range, 304; distribution, 208
Winteria telescopa (argentinoid fish), body form, 127; eye, 384; jaws, 387

Xanthocalanus (Copepoda, Crustacea), bottom-dwelling species, 141; eggs, 429, 455; non-migratory, 430
Xenodermichthys copei (alepocephalid fish), body form, 128; buoyancy balance sheet, 337; light organs, 128; reflexion of light, 362
Xylophaga (wood-boring bivalve mollusc), 210, 211; larval stages, 464

Yarrella blackfordi (stomiatoid fish, Gonostomatidae), caught near bottom, 135

Zenkevitchiana longissima (beard-worm, Pogonophora), size, 177
Zoanthus (Zoantharia, Cnidaria), attachments, 183

General index

abyssal fauna, 492
abyssal plains, 13
Acantharia (Protozoa), structure and mode of life, 72, 73
acorn-worms (Enteropneusta), 154–5; on rich sediments, 311
active organismal transport of organic matter, 234–5
adaptations of bathypelagic fishes, 258
adenosine triphosphate (ATP), as indicator of relative metabolism in the water column, 232
advection, 24
Alcyonaria (octocorals), 179, 181–3, *181*, *182*
alepisauroid fishes (Alepisauroidaea), 129, *130*; diversity, 128; structure, dentition and food, 129, *131*; vertical migrators rare, 129, 131
alepocephalid fishes (Alepocephalidae), 135; food, 272; size of eyes, 392
Aleutian Trench, 308–9; high productivity, 308; high density of minifauna, 308; sediment instability, 308–9
amblyopsid cave fishes, lateral line system, 416, *417*
ampharetids (tube-dwelling polychaete worms), 173–4
Amphipoda (Peracarida, Crustacea), 153; pelagic forms, 116; benthic deposit-feeding forms, 151, *152*
anchor dredge, 148
angler-fishes (monkfishes, *Lophius* spp.), 133
anomuran decapod crustaceans, 208
Antarctic Bottom Water Mass, 27, 30, 31
Antarctic Circumpolar Current, 30
Antarctic Convergence, 30
Antarctic Intermediate Water Mass, 27, 28, 30, 31
Antarctic Surface Water Mass, 28
Anthozoa (Cnidaria), classification, 179; (octocorals, sea-anemones, stony corals, black corals, etc., Cnidaria), 85

Antipatharia (black corals, Zoantharia), 179, *185*; form and depth distribution, 185–6
Aphroditidae (family of polychaete worms), 202; structure and depth distribution, 201–2
Aphyonidae (viviparous ophidioid fishes), 138; life history stages, 457–8, *457*
aplacophoran molluscs, *151*; deposit feeders, 149, 151
Apoda (order of Holothuroidea), *198*; depth distribution, 198
appendicularians (Larvacea, Tunicata), filtering houses as substrates for microorganisms, 224
archaeogastropods (snails), deposit-feeding forms, 209
archibenthal fauna, 492
Arcidae (Noah's Ark shells), in the deep sea, 211
argentinoid fishes, form, eye structure and light organs, 127; tubular eyes and retinal diverticula, 386–7, *387*
arrow-worms (Chaetognatha), negative buoyancy, 321; *see also* Chaetognatha
Ascidiacea (sea-squirts, class of Tunicata), *171*, *172*; diversity, structure, mode of feeding, 170, 172; carnivorous forms, *173*; trophic types, 296
Aspidochirota (order of Holothuroidea), 199
Asteroidea (sea-stars or starfishes, Echinodermata), *193*; structure, diversity, reproduction and modes of feeding, 192–4
Astronesthidae (stomiatoid fishes), 126
Astropectinidae (Asteroidea), 194

bacteria, 56; chemosynthetic, 61; luminescence, 342; metabolic rate, 288–9; standing crop, 289
bacterioplankton, 40

557

baited automatic camera, records of benthopelagic animals, 273
bathyal fauna, 492
Bathydraconidae (dragon-fishes, Nototheniiformes), depth distribution, 212
bathylagid fishes, aphakic space in eye, 383
bathypelagic fish fauna, 44, 131–4
bathypelagic food webs, 253–8
bathypelagic zone, 43
Bathysauridae (myctophiform fishes), 213, 215
Benguela Current, 37
benthic animals, number of species, 40
benthic fauna, 44–5
benthic invertebrates, biomass off New England, 281
benthic organisms, biomass in relation to depth and productivity, 297
Benthopectinidae (Asteroidea), 193; structure and distribution, 192, 194
benthopelagic fauna, 44, 134–42, 272–80
benthopelagic fishes, 135–41, 136, 137, 139, 140; buoyancy, 135–6; length of swimbladder retia mirabilia, 332–4, 334
billfishes, 43
bioluminescence, 6; as evolutionary vestige, 342; and organisms responsible, 343–4; intensity of, 345
bivalve molluscs, eulamellibranchs, 148, 151; protobranchs, 148–9, 151; septibranchs, 149, 151; infaunal, 148–9, 150; larval types, 464; in minifauna, 292; trophic types and depth ranges, 303
black corals (Antipatharia), distribution, 498
black scabbard fishes (*Aphanopus*), 135
Black Sea, anoxic conditions, 32
blue-green algae (Cyanophyta), 54, 56, 51 distribution, 57; as nitrogen fixers, 221
bob-tailed snipe-eels (Cyemidae), lateral line system, 413, 415, 416
bottom-dwelling fishes, biomass and density off New England, 281–2
Brachiopoda (lamp-shells), 169; structure and depth distribution, 169–70; larval stages, 466, 467
Brazil Current, 37
Brisingidae (family of Asteroidea), 192, 193
brittle-stars, *see* Ophiuroidea
brotulid fishes (Ophidiiformes), 135, 140; sexual dimorphism in sonic mechanisms, 425–7, 426
Bryozoa (moss animals), 168; diversity, structure and depth distribution, 167–9
bullheads (Cottidae and Cottunculidae, Pisces), 212
buoyancy balance sheet, of midwater fishes without a gas-filled swimbladder, 337, 337, 338

buoyancy chamber (gaseous), in *Spirula*, 325–6
buoyancy of: benthopelagic animals, 338–41; micronekton, 323–4; midwater fishes, 326–35; nekton, 324–6; oceanic organisms, 5–6; siphonophores, 319; squalid sharks and chimaeroid fishes, 338; zooplankton, 316–23
buoyancy relations of benthic animals, 339–41

calcium carbonate compensation depth, 146
California Current, 38
Calycophorae (division of siphonophores), structure, movements and distribution, 90
calyptopis stages, of euphausiids, 432, 433
camouflage by colour, 362–4; red coloration and its unmasking by certain stomiatoid fishes, 363–4, 363
Cariaco Trench, anoxic conditions, 32
Caridea (prawns, group of Natantia, Decapoda), structure, 111, 111, 112; life history, 435–8; luminous clouds, 349–50
carnivores, distribution, 498
Central Water Masses, 27, 30, 31; and dwarf species of mesopelagic fishes, 480–3; 479, 480, 482
Cephalopoda (pelagic forms, Mollusca), 116–22, 118, 120; diversity, 117; epipelagic species, 117; mesopelagic species, 117–21; vertical migrations, 121; ontogenetic migrations, 121; squids as whale food, 121–2; light organs, 119
Ceratiidae (family of ceratioid anglerfishes), feeding of parasitic males, 255
ceratioid angler-fishes (Lophiiformes), 132, 133–4; angling system, 255, 257; distribution, 488–9; diversity, 255; economies in organization, 261; food and feeding, 255–6; eye structure in males, 383, 390, 391; lateral line system, 413, 415, 417; luminous secretions, 350–1; life history, 451–4, 452, 454; olfactory system, 404–8, 405
Ceriantharia (tube anemones, Zoantharia), 186
Chaetognatha (arrow-worms), 75, 102, 191–3; life history, 438–9
Chauliodontidae (stomiatoid fishes), 126
chemical sense, invertebrates, 411–13; in *Gnathophausia*, 411; in sergestids, 411–12; aesthetascs of crustaceans, 412; in cephalopods, 412–13; fishes, 402
chemosynthetic bacteria, as food of invertebrates near hydrothermal vents, 284
chimaeroid fishes, 135
chlorophthalmoid fishes, lateral line system, 420

Chondrophorae (Hydrazoa, Cnidaria), structure, 90–1
cirrate octopods (Mollusca), photographed near deep-sea floor, 141–2
Cirripedia (barnacles), *175*; diversity, habit and distribution, 174–5
Cnidaria, pelagic forms, 75, 83–92; nematocysts, 84; organization of tissues, 84–5, *85*; radial symmetry, 84–5; photographed near bottom, 272; benthic forms as passive suspension feeders, 178–86
coccolithophorids (calcareous flagellates), 50–2, *51*; distribution, 57
Colossendeidae (family of Pycnogonida), 204; size and locomotion, 203, 205
compensation depth, 58
complexity and evolution, 260
conservation of salt in ocean, 21–2
conservation of volume in ocean, 20–1
continental rise, 12
continental shelf, 11–12
continental slope (and sediments), 12–13
convergences, and oxygen in ocean, 32
convergent evolution of cephalopods and fishes, 117
Copepoda (Crustacea), 77, *78*, 76–9; bathypelagic forms, 44; in Indian Ocean, 253; bottom-dwelling species, 141; daily vertical migrations, 243, and size differences, 242; food supplies, 254; life histories, 429–31; shedding and care of eggs, 429; ontogenetic migrations, 430; luminous forms, 347, *348*, and luminous secretions, 347–8; relative numbers in water column, 253, 259, and decline of diversity with depth, 253; sinking rate and relative buoyancy, 321–2; size and depth range, 260
copepodite stages, of copepods, 431
Coriolis Force, 27, 33
Coronatae (order of Scyphozoa, Cnidaria), size, structure and colour, 91; luminescence, 345–6
cosmopolitan species, 499
Cranchiidae (oceanic squid), structure, 117
crangonoid shrimps (Decapoda, Crustacea), 205
Crinoidea (feather-stars and sea-lilies, Echinodermata), *187*; diversity, structure and habit, 186; depth range and response to currents, 294–5
croppers, 304
crustaceans, small forms of plankton, 75
Ctenophora (comb-jelly fishes), 75, 92–4, 93; luminescence, 346
Cubomedusae (box jellies, order of Scyphozoa, Cnidaria), 91
Cuvieran points of view, 2

Cystonectae (division of siphonophores), 88

Darwinian points of view, 2
deep scattering layers (sonic scattering layers), in the North Atlantic, 249–50; communities in, 250
deep-sea angler-fishes (Ceratioidea), 133–4, *132*
deep-sea animals (newly discovered), 3
deep-sea cod)Moridae), *137*, 135; diversity, structure and light organs, 137–8, 373; swimbladder as hydrophone, 425; swimming speed (*Antimora*), 279–80
deep-sea crabs, food and depth range, 304
deep-sea floor, main regions, 11–20, *12*, *14*, *16*
deep-sea prawns (Natantia, Crustacea), pelagic species, 110–15, *111–13*; biomass, 110; bottom-dwelling forms, 134; colour, 113; eyes, 398–400; light organs, 114; life histories, 114–15; number of species, 113; vertical migrations, 113–14
deep-sea sediments, *18*, 13, 15, *19*; rates of accumulation, 297
deep-sea trenches, 20, *14*, *16–17*
deimatic displays of luminescence, in stomiatoid fishes, 370
Demospongia, most diverse group of sponges, *166*, 162; diversity, habit and depth distribution, 165, *167*
Dendrochirota (order of Holothuroidea), 197–8
density in the ocean, 25–6; and buoyancy of organisms, 25–6
deposit-feeders, dominant in soft sediments, 296; megafaunal groups, 296; dominant in oligotrophic sediments, 298; echinoderms prefer eutrophic sediments, 300, *301*; distribution, 497–8
detritus, particulate composition, 231; biochemical nature, 232
diatoms, *55*; deposits, 49, 55; distribution, 56–7, 473–4; reproduction, 54; structure, 54
dinoflagellates, 52–4; *53*; distribution, 57; luminescence, 52, 54, 342–3; structure, 52; vertical movements, 60
direct development, in invertebrates, 464
dissolved organic carbon (DOC), as source of particles, 231
dissolved organic matter (DOM), quantitative studies, 222; as food of aquatic invertebrates, 222–3
distribution (near-continental), of mesopelagic fishes, 484; relation to oxygen minimum layers, 484–5
distribution: of: benthopelagic fishes, 490–2; of *Cyclothone* fishes, 487–8, *489*; of

distribution—*contd.*
  ceratioid angler-fishes, 488–9; of diatoms, 473–4; of mesopelagic and bathypelagic animals, 486–90; of mesopelagic fishes, 474–8
DNA (deoxyribonucleic acid), content in fishes, 264–5
Doliolida (order of Thaliacea), 97; life history, locomotion, mode of feeding, 97–8

early life histories, 7–8
early life history and ecology, 463–7
ears of fishes, 422–3, 424
Echinidae (regular sea-urchins, Echinoidea), 196; depth distribution, 195
echinoderms, luminescent forms, 375, 376, 377
Echinoidea (sea-urchins, Echinodermata), 196; structure, mode of feeding, and depth distribution, 194–5, 197
Echinothuriidae (flexible sea-urchins, Echinoidea), 196; size and depth distribution, 195
echiurid worms (Echiurida), 156; depth distribution and method of feeding, 156
ecological biogeography, 8–9
economical organization of bathypelagic fishes, 262, 264–9
ecosystems, mesopelagic and bathypelagic, 259
ecotrophic coefficient, 227
eel-pouts (Zoarcidae), eggs, 459; distribution, 213
eels (Anguilliformes), 131
eggs: of: arrow-worms, 439; caridean prawns, 435, 438; copepods, 429–31, 434; euphausiids, 431–2, 434; midwater fishes, 443; rat-tailed fishes, 456; sergestid prawns, 434; teleost fishes (buoyant), 441
elaphocaris stages, of sergestid prawns, 434–5, 436
elasipod sea-cucumbers (Holothuroidea, Echinodermata), buoyancy, 339
Elasipoda (order of Holothuroidea), 200, 197; structure, depth distribution and locomotion, 199–201
epibenthic sled, 148
epifauna, 45, 158–215; arthropods, 204, 206, 207, 208, 203–9; crustaceans, 206, 207, 208, 205–9; 'worms', 201–3, 202
epipelagic food webs, 221–30
epipelagic zone, 43
Equatorial Counter Currents, 34
Equatorial Undercurrents, 38
Equatorial Water Masses, 30–1
errant epifauna, 189–212
Eryonidea (Reptantia, Decapoda, Crustacea), 206; structure and depth distribution, 205, 208

erythrocytes, size of, and swimbladder function, 332
eunicids (tube-dwelling polychaete worms), 173
Euphausia superba (krill), life history and growth, 434
euphausiid shrimps (Euphausiacea, Crustacea), 109, 107–10; abundance, 108; bathypelagic species, size and food, 254–5, size and fecundity, 261; daily vertical migrations, 236–8, 237, and competition, 236; depth ranges of migrators and non-migrators, 236–8; eyes, 397, 399, structure and function of bilobed eyes, 401–2; distribution, 108, of mesopelagic and bathypelagic species, 487; life histories, 110, 431–3; light organs, 109, 357–9, 359; negatively buoyant, 323; size, 108; species diversity, 108; structure, 108, 238–9; trophic types, 108, 109–10
euphotic zone, 57–61
eurythermal marine organisms, 24–5
eutrophic distributions, in benthos, 497–8
eutrophic regions of deep-sea floor; 297–8, 300; and carnivorous echinoderms, 300, 302
eutrophic sediments, 496
evermannellid fishes, lateral line system, 418
eyes, aphakic space, 383–4, 384; fovea, 384; retinal structure and sensitivity, 381–3, 382; size and dioptric proportions, 380–1, of: bathypelagic fishes, 390, 391; bottom-dwelling fishes, 390–2, cephalopods, 393–5, crustaceans, 397–402, euphausiids, 399, 397, 400–2, pelagic polychaete worms, 397; tubular eyes, 385–8, 384, 386, 387

faecal remains, as main energy source of benthic invertebrates, 231, 287–8
feather-stars (comatulids, Crinoidea), depth distribution and mode of feeding, 188
fecundity: of ceratioid angler fishes, 449, euphausiids, 432–4; gulper-eel, 449, lantern-fishes, 444, 447, oplophorid prawns, 435, rat-tailed fishes, 456
feeding, (selective and unselective) of migrating fishes, 245
fishes, benthic species, 212–15, 214, 215; benthopelagic species, 135–41, 136, 137, 139, 140; midwater species, 122–34, 123, 124, 125, 128, 129, 130, 132
flapjack devilfish (Opisthoteuthis, Octopoda), colour and form, 211
flatworms (Turbellaria), 201
Florida Current, 37
flotation: of: pelagic radiolarians, 317, acantharians, 317, 319
flying-fishes, 43

flying squid, 43
food, of deep-sea copepods, 233, 232–3; lar-
	valinvertebrates, 463; zooplankton, 232
food chain, number of links and oceanic
	productivity, 228–30
food pyramid, 228, 227–8
food resources of benthopelagic fishes, 274–
	7; food taken away from bottom, 274
food webs, 3–4
Foraminifera (Protozoa), benthic species,
	147; diversity and depth, 146; mode of
	life, shell formation and structure, 145–
	6; prevalence and depth, 291; pelagic
	species, *70, 71*; structure and mode of
	life, 69–70
founder principle, 306–7, 501
frog fishes (Antennariidae), 133
furcilia stages, of euphausiids, 433, *432*

gammarid amphipods (Peracarida, Crusta-
	cea), *153*, 152–3; giant forms, 153
gastropods (snails, Mollusca), trophic types
	and faunal zonation, 303
genetic variability, 499–501; as shown by
	electrophoresis, 499; in euphausiids,
	500; in megafaunal benthic species,
	500–1
giant swallowers (fishes, Chiasmodonti-
	dae), 133; olfactory organs (*Chiasmo-
	don*), *403*
giganturoid fishes, 131; tubular eyes, 387
Globigerina ooze, 70
Goniasteridae (Asteroidea), 194
Gorgonacea (horny corals, Alcyonaria),
	181; structure, depth distribution, and
	rheophile nature, 178, 181; lumine-
	scence, 344
gravitational transport of organic matter,
	231–4
grey sharks, 43
growth coefficient, 227
Gulbenkian reflex of female ceratioid
	angler-fishes, 258
Gulf Stream, 37, 38
gulper-eels (Saccopharyngiformes), 44, *132*;
	eyes, 390; food, 135; lateral line system,
	413, 415; structure, food and larval
	stages, 131, 133
gustatory system of fishes, 410; in rat-tails
	(Macrouridae), 410

hadal fauna, 20, 492
hagfishes (Myxinidae), *157*; as bonanza
	strategists, 157–8; lateral line system,
	423; olfactory system, 409
halosaur fishes (Halosauridae, Notacanthi-
	formes), *139*, 135, 138; food and fin
	pattern, 277; lateral line system, 418,
	419; olfactory system, 409
harpacticoid copepods (Crustacea), 143–5,

144; abundance, 144; species diversity,
	145
hatchet-fishes (*Argyropelecus*, tubular eyes,
	386
heart-urchins (Echinoidea), 197
heat balance in ocean, 22–3
hermaphroditism	(synchronous),	in
	chlorophthalmid fishes, 458–9; in mid-
	water fishes, 460; adaptive significance,
	460
hermit-crabs (Paguridea, Anomura), 207,
	208; depth distribution, 208; success in
	deep sea, 304–5
heteropods (Gastropoda), structure, move-
	ments and carnivorous habit, 100–1
Hexactinellida (glass-sponges), *164*, 167;
	structure, diversity and distribution,
	162–3, 165
Himantolophidae (family of ceratioid
	angler-fishes), growth and food of
	males, 255
Holothuroidea (sea-cucumbers, Echinoder-
	mata), *198, 200*; structure, mode of
	feeding and distribution, 197–201
homeothermal animals in ocean, 24
horny corals, see Gorgonacea
Hydrozoa (Cnidaria), pelagic forms, 85–91
hydroids (Hydrazoa, Cnidaria), depth
	records, 178
hydromedusae, 85–6; structure and move-
	ments, 86
hydrothermal 'oases', 306–7; chemosyn-
	thetic bacteria as food of suspension-
	feeders, 306; large forms of inverte-
	brates, 306; founder principle, 306–7

Idiacanthidae (stomiatoid fishes), 126
infauna, 45, 143–58
irregular sea-urchins (Echinoidea), 195, 197
isopods	(Peracarida,	Crustacea),	*152*;
	benthic deposit-feeding forms, 151;
	structure and distribution, 151–2; giant
	forms, 153

Komokiacea (Foraminifera; Protozoa), 3,
	161; structure and distribution, 159,
	162; structure, depth range and habi-
	tat, 291
krill (*Euphausia superba*), food, 109
Kurile–Kamchatka Trench, high produc-
	tivity, 307–8; abundance of Pogono-
	phora, 308
Kuroshio, 37

Labrador Current, 37
lancet-fish (*Alepisaurus*), *129*
lantern-fishes (Myctophidae), 44, 122–3,
	123, 126–7, 253; caudal glands, 369;
	daily vertical migrations, 244–5; distri-
	bution, 476–8; eyes, size, 380, with
	aphakic space, 383, 385; form, diet and

lantern-fishes (Myctophidae)—*contd.*
light organs, 127; food, 244–5; lateral
line system, *418*; life history, 443–6,
observed near bottom, 135, 272–3;
olfactory organs, 404
Larvacea (class of Tunicata), diversity, size,
structure, mode of feeding and distribution, 94–6, *95*
larval stages (long-lived), 465–7, dispersion
of, and wide distribution of adults, 465,
467
lateral-line sense, in fishes, 413–20, *414*, *415*,
417, *418*, *419*; species with free-ending
organs, 413–17; species with mixed system of neuromasts, 418–20; in invertebrates, arrow-worms, 420–2, *421*; in
ctenophore, 422; in hydromedusa, 422;
in prawns, 422
lecithotrophic larvae, 464–5
leptocephalus larvae, 449–51, *450*
Liebig's law of the minimum, 63
life history, annual fishes, 446; arrow-
worms, 438–9; benthic animals, 458–
67; benthopelagic animals, 455–8;
cephalopods, 439–40; ceratioid angler-
fishes, 451–4, 452, 454; copepods, 429–
31; euphausiids, 431–4; fishes, 441–55;
lantern-fishes, 443–6; midwater animals, 429–55; ophidioid fishes, 457–8,
457; oplophorid prawns, 435–8; rat-
tailed fishes, 455–6; sergestid prawns,
434–5
light in the ocean, 39–40; penetration of
spectral components, 39, 43; twilight
zone, 39
light lures, 364–7; in ceratioid angler-fishes
and modes of luminescence, 364–5;
luminous barbels of stomiatoid fishes,
365; in mouth of hatchet-fish, 365–7,
366; of viper-fishes (*Chauliodus*), 367
Limidae (file-shells, Mollusca), in the deep
sea, 211
Linophrynidae (family of ceratioid angler-
fishes), feeding of parasitic males, 255
living space, as *n*-dimensional hypervolume,
471–2
luminescence, and inter-individual spacing,
370–1; in the benthopelagic fauna,
371–4; in the benthic fauna, 374–8; of
echinoderms, 375, 376, 377; of pycnogonids, 375; of sessile cnidarians, 378
luminescent camouflage, 351–62, 353–5; in
mesopelagic fishes, 351–7; in euphausiids, 357–9; in sergestid prawns, 360;
in midwater squid, 360–2
luminescent clouds, as decoys, 347, 349–50
luminescent flashes, recording of, and relative frequency in water column, 345–6
luminous gland (perianal), in green-eyed
fish (*Chlorophthalmus*), 374

luminous secretions, of sepiolid and loli-
ginid squid, 350

mako-shark family (Isuridae), 24, 43
Malacosteidae (stomiatoid fishes), 126
maldanids (tube-dwelling polychaete
worms), 173–4
Mariana Trench, low standing crop of benthos, 307–8
Mauthnerian system, 256
Mediterranean Water Mass, 28
megafauna (large invertebrates), feeding
types, 294; attached forms, 294
meiofauna (meiobenthos), 45, 143–6; biomass and depth, 290; key members,
290; population density, 144; Russian
criteria, 148
Melamphaidae (fishes, Xenoberyces), *480*
structure, 133; lateral line system, 418
Melanocetidae (family of ceratioid angler-
fish), growth and food of males, 255
Melanostomiatidae (stomiatoid fishes), 126
mesopelagic eels, size of olfactory organs,
404
mesopelagic fishes, 44, 122–30; feeding
compared with that of cephalopods,
122; length of retia mirabilia in swimbladder, 332
mesopelagic food webs, 230–52
mesopelagic zone, 43
metabolism, relative amounts in water
column (Sargasso Sea), 230
metamorphosis, of fishes, 441–3
metanauplius larva, of copepods, 430
microbiota, in deep-sea sediments, 290
micro-flagellates, 52
micronekton, 42; size, activity, form and
organization, 106–7
micro-zooplankton, 68–74; as herbivores,
223
mid-ocean ridge system, 15–20; attached
organisms, 19–20; and primary productivity, 19
miniature infauna (minifauna), dominant
forms, 148
minifauna (= American 'macrofauna'), percentage composition of dominant
forms, 292
mixed layer, 23–4
Molluscs, benthic forms, 209–12
Molpadonia (order of Holothuridea), *198*;
form, habitat and depth distribution,
198–9
Monoplacophora (class of Mollusca), *210*;
structure, 209
Monsoons, in the Indian Ocean and
seasonal variation in primary productivity, 223–4
Myctophidae, *see* lantern-fishes
mysid shrimps (Mysidacea, Peracarida),

pelagic forms, 115–16; modified eyes, 400; luminous secretions (*Gnathophausia*), 349

Mytilidae (mussels, Mollusca), in the deep sea, 211

Nasselaria (order of Radiolaria), 72

Natantia (swimming decapods, Crustacea), ratio of pelagic to benthic species, 205; structure and movements, 110–11

nauplius larva, of copepods, 429–31; of euphausiids, 432

nekton, 42, 106–7; epipelagic forms, 107; and micronekton, 105–34; methods of sampling, 105–6

nematode worms (Nematoda), *144*, 143; abundance, and distribution, 144–5; diversity and depth distribution, 290–1

nemertean worms (pelagic forms, Nemertea), colour and movements, 105

neogastropods (snails), predatory forms, 209

neuston, 42–3

neutral buoyancy, of benthopelagic fishes, 338–9; biological significance of, 340–1

non-migratory midwater fishes, 246–8; capacity to take large meals, 248; differences in jaw structure, 247; eye structure (in argentinoids), 248; vertical ranges, 246–7

North Atlantic Deep Water Mass, 30

North Equatorial Currents, 34, 37

North Indian Deep Water Mass, 30–1

North Pacific Current, 37

notacanth fishes (Notacanthidae, Notacanthiformes), *139*, 135, 138; food, 275–6

notosudid fishes, aphakic space in eye, 383

nototheniiform fishes, 212

number of marine organisms, 10

nutrient salts, supply of, in relation to primary productivity in: Antarctic, 65; equatorial waters, 64–5; subtropical and tropical regions, 63–4, temperate and subpolar waters, 65

Nymphonidae (family of Pycnogonida), 204; structure, 203

ocean, constant in volume, 10; origin of, 11

ocean basins, 13–15, *16*, *17*

oceanic gyres, 37

oceanic herbivores, main groups, 223

oceanic plants (phytoplankton), 2

octopods (Cephalopoda, Mollusca), benthic, 211; midwater, 440, 119–21; finned forms, 121

olfactory detection of bait, 379–80

olfactory system, of fishes, 402–9, *403*, *405*, *406*; in bottom-dwelling species, 408–9; in midwater species, 403–8; sexual dimorphism, 404–8; structure, 402–3

oligotrophic distribution, in benthos, 497–8

oligotrophic regions of deep-sea floor, 297–8

oligotrophic sediments, 496

ommastrephid squid (mollusca), 117

ontogenetic migrations, 252; of deep-sea cephalopods, 439–40

ooze-eating fish (*Lipogenys*), 274–5

Ophiuroidea (brittle-stars, Echinodermata), *191*; diversity, structure, mode of feeding and distribution, 189–92

oplophorid prawns (Caridea), neutral buoyancy, 323–4

organic food sources, of benthic animals, 283–4

organic particles, settling rate in ocean, 285

organization, of mesopelagic and bathypelagic fishes, 262–8, *263*, 267; of midwater life, 259–69; of rat-tailed fish and gulper-eel, 270–1; levels of organization, 259

Ostracoda (Crustacea), 79–80, *79*; buoyancy, 322–3; feeding types, 80; luminescence, 349; movements, 79–80; size, 79; vertical migrations, 80

oxygen, dissolved content in ocean, 31–2

oxygen consumption, of midwater crustaceans and fishes, 234; of rat-tailed fish, 278–9

oxygen minimum layers, 33; and distribution of mesopelagic fishes, 484–6

Oyashio, 37

Paralepididae (alepisauroid fishes), 128, *130*

particulate organic carbon (POC), derivation of, 231

Pectinidae (scallops, Mollusca), in the deep sea, 211

pelagic animals, number of species, 40

pelagic cnidarians, as r-strategists, 225

pelagic faunal provinces, 474–7, *475*

pelagic snails (pteropods and heteropods, Gastropoda), 98–101

pelagic tunicates, rates of population growth, 224; as *r*-strategists, 225

Penaeidea (prawns, group of Natantia, Decapoda), structure, 111, *113*

Pennatulacea (sea-pens, Alcyonaria), 179, *182*; structure, depth distribution and luminescence, 182–3

peracarid crustaceans, benthic forms, 151–4, *152*, *153*; egg-brooding and embryos, 460–1, *461*; reproductive cycles, 461–2, in minifauna, 292; plant food, 294; pelagic forms, 115–16, *115*

Peru Current, 38

Peru–Chile Trench, high productivity, 308

Phaeodaria (order of Radiolaria), 72

photosensitive vesicles, of cephalopods, 395–7, *396*
photosynthesis, 57–8
physical factors (constant), and distribution, 472
Physonectae (division of siphonophores), structure, movements, 88, 90
phytoplankton, 2, 40; buoyancy, 59–60, 316; dispersion and the food chain, 223; relative amounts of main constituents in eastern tropical Pacific, 222; sinking rate, 60
plankton, 40
planktotrophic larvae, 464
plant pigments, 57
pleuston, 42
Pogonophora (beard-worms), 3, *176*; abundance in certain trenches, 308; eggs and early stages, 464, *465*; eutrophic distribution, 177–8, 496; faunal barrier, 496–7; structure, relationships, nutrition, diversity and depth distribution, 175, *177*
polychaete worms, abundance, 154; benthic deposit-feeding forms, 151; pelagic forms, 75, *104*, 104–5; tube-dwelling forms, 173–4; structure and feeding types, 292, 294, *293*
Polyplacophora (chitons or mail shells, Mollusca), 209
Porcellanasteridae (Asteroidea), *193*; structure, depth distribution and mode of feeding, 194
pourtalesiids (irregular sea-urchins, Echinoidea), distribution, 195–6
prawns (oplophorids and pandalids), 'lateral line' system, 422
prawns (pelagic), luminous clouds, 349
priapulid worms (Priapulida), *155*, 154
primary productivity, 61–6, 62; measurements of, 61; factors influencing, 61 *et seq.*; patterns of oceanic productivity, 61–6
propulsion and hydrodynamic relations of aquatic animals, 75–6
Protozoa, epifaunal, 159–62, *160*, *161*
psychrosphere, 22
Pterasteridae (cushion stars, Asteroidea), structure, reproduction and distribution, 192
pteropods (naked forms, Gymnosomata), carnivorous habits and movements, 100
pteropods (shelled forms, Thecosomata), 99; structure, movements and mode of feeding, 98
Pycnogonida (sea-spiders), *204*; structure, mode of feeding, movement and depth distribution, 203, 205
Pyrosomida (order of Thaliacea), structure,

locomotion, luminescence and distribution, 96, 97

radiocarbon (C^{14}), as tracer of oceanic circulation, 32–3
Radiolaria (Protozoa), structure and mode of life, 72–4, *71*, *74*
rat-tailed fishes (grenadiers, Macrouridae), 135–7, *136*; diversity, distribution and structure, 136–7; distribution, 490–2; food and organization, 276–7; gustatory organization, 410; lateral line system, 418–20, *419*; life history, 455–6; olfactory system, 409; organization of midwater species, 271; number of species and depth ranges, 271; response to currents, 305; sexual dimorphism in sonic mechanisms, 425–7, *426*; size of eyes, 392; structure of light organ and luminescence, 371–3, *372*; swallow and sieve deposits, 275
rat-trap fishes (Malacosteidae), aphakic space in eye, 383
recolonization, of deep-sea benthos, 310
red clay, 15
Red Sea Water Mass, 30–1
reproductive cycles, of benthic invertebrates, 461–3
Reptantia (crawling decapod crustaceans), 205
residence time in water masses, 32–3
respiration, of inshore and deep-sea benthos, 310; of phytoplankton, 58
retia mirabilia (of fish swimbladder), countercurrent means of gas secretion and retention, 327–30; length in relation to depth ranges, 331–4
retina, structure and function in fishes, 381–3, *382*, 385
Rhizostomeae (order of Scyphozoa, Cnidaria), 91

sabellids (tube-dwelling polychaete worms), 173–4
salinity, 21
Salpida (order of Thaliacea), 97; life history and distribution, 98
Sargasso Sea, 37
sauries (fish), 43
scyphoid polyps (*Stephanoscyphus*), *180*; habit and depth distribution, 178–9
Scyphozoa (jellyfishes, Cnidaria), 85, 91
sea-anemones (Actiniaria, Zoantharia), distribution, 183
sea-cucumbers (Holothuroidea, Echinodermata), mobility and cropping activities, 311
sea-lilies (Crinoidea), *187*; mode of feeding, 188–9
sea-pens, luminescence, 344

sea-snails (fishes, Liparidae), diversity and depth record, 213; eggs and mouth-brooding, 459

sea-stars (Asteroidea, Echinodermata), trophic types and depth distribution in northeastern Pacific, 304

sea water, constancy of ionic composition, 21

Searsiidae (alepocephaliform fishes), *128*; aphakic space in eye, 383; lateral line system, 418, 414; light organ system and form, 127–8; production of luminous clouds, 350

'Seas' of the ocean, 30

seasonal changes in productivity of the equatorial Atlantic, 224

seasonal migrations (of zooplankton) in antarctic waters, 252

sedimentation, rate of, 15

sedimentation trap, 285

selection pressures, among phytoplankton, 22–2

Semaeostomae (order of Scyphozoa, Cnidaria), colour and luminescence, 91

sense organs, 6–7

sepioid squid, bottom-dwelling forms, 141, 211–12; structure, buoyancy and luminescence, 119

sergestid prawns (Sergestidae, Crustacea), daily vertical migrations, 239–40, *241*, and food, 240; life history, 434–5; negatively buoyant, 323; organization of bathypelagic species, 268; sensory structure of antennae, 411; and scent trails, 411–12

serpulids (tube-dwelling polychaete worms), 173–4

seston, 231

sexual dimorphism, in olfactory system, in *Cyclothone*, 404–7, *406*; in ceratioid fishes, 404–8, *405*

sexually dimorphic luminescent systems, 367–70; in stomiatoid fish *Idiacanthus*, 367; in rat-trap fishes, 367; in lantern-fishes, 368–9, *368*

silicoflagellates, 54

sinking rate, of pelagic protozoans, 316

Siphonophora (Hydrozoa, Cnidaria), classification, structure, movements and distribution, 87–91, *89*

sipunculid worms (Sipunculida), *155*, *154*; distribution, 303, 496–7; size and depth range, 201

size (organismal), in relation to depth in deep-sea benthos, 311

skates (Rajidae), deep-sea species and buoyancy, 212

skippers (fishes), 43

snails (Gastropoda, Mollusca), deep-sea forms, 209

snipe-eels (Cyemidae), small eyes, 390

snipe-eels (Nemichthyidae), 131

soft coral (*Alcyonium digitatum*), as suspension feeder, 178

soft corals (Alcyonacea), 179

solitary deep-sea corals, *184*; distribution, 184–5

sound-producing mechanisms, in fishes, 423–7, 426; swimbladder and drumming muscles, 425–7; means of locating a sound source, 427

sound scattering layers, 44

South Equatorial Currents, 34

spawning season, of lantern-fishes, 446

species diversity, latitudinal gradients in, 472

specific gravity, of living substance and skeletal materials, 315

Spinocalanidae (Copepoda, Crustacea), 44

sponges (Porifera), 162–7, *163*, *164*, *166*; structure and classification, 162; stalked species adapted to silty conditions, 309

Spumellaria (order of Radiolaria), 72

squalene, in buoyant liver of squaloid sharks, 338

Squaloidea (sharks), 135, 140–1; live bearing, 458

squat-lobsters (Galatheidea, Anomura), 207; depth distribution, 208; success in deep sea, 304

squid, epipelagic, negatively buoyant, 324; midwater, ammoniacal uplift and neutral buoyancy, 324–5

Stomiatidae (stomiatoid fishes), 126

stomiatoid fishes (Stomiatoidea), 123–6, *124*, *125*; classification, 126; structure, 126, light organs, 126; size of olfactory organs, 404

Subantarctic Water Mass, 27, 28, 30

Subarctic Region (Atlantic), 477

Subarctic Water Mass (Pacific), 31

Subtropical Convergence, 30

Subtropical Regions (Atlantic), 476

surface circulation of ocean, 33–9, *34*, *35*, *36*

suspension-feeders, distribution, 497; on oligotrophic sediments, 298; as rheophiles, 295; wide distribution, 300, 299

swimbladder (of teleost fishes), absent in many predatory midwater fishes, 327; buoyancy system, 327; gas retention, 328, 330; gas secreting system, 327–8, *329*; gas regulation and vertical migrations, 336–7; presence and loss and organization of midwater fishes, 262, 268; presence of, compared in mesopelagic and benthopelagic fishes, 338; regression of, and buoyancy relations in mesopelagic fishes, 335

Synallactidae (family of Aspidochirota, Holothuroidea), 199

TS diagram, 27
tanaids (Peracarida, Crustacea), 153; benthic deposit feeding forms, 151, 152
teleost fishes, production of buoyant eggs and larvae, 428
Temperate Region (North Atlantic), 476–7
temperature profiles in open ocean, 23, 23–4
terebellids (tube-dwelling polychaete worms), 173; method of feeding, 156
terrestrial vegetation, remnants of, as food of benthic invertebrates, 284
Thaliacea (class of Tunicata), 96–8, 87
thermocline, 'seasonal', 24, 64; permanent, 42, 43
thermosphere, 22, 30, 31
thresher sharks, 43
threshold of light, 43
tintinnids (Tintinnoida, Protozoa), 68–9, 69
tomopterids (pelagic polychaete worms), luminescence, 347
Tonga Trench, low standing crop of benthos, 308
trachichthyid fishes, luminous gland and luminescence, 373–4
trachyline medusae (Trachylina, Hydrozoa, Cnidaria), structure and size, 86–7, 86
Trade Winds, 34, 37
transfer efficiency, 227
transport of organic material to depths, 230 et seq.
trench faunas, 307–9; diversity in southwest Pacific trenches, 309
trichiuroid fishes, 133
tripod-fishes (Bathypterois), 135, 214; food and depth ranges, 305; olfactory organs, 409, 403; response to currents, 305; size and distribution, 213; size (and regression) of eyes, 392
trophic dimension, and the megafauna, 495–8
trophic types of invertebrates on European continental slope, 303
Tropical Region (Atlantic), 476
true (or stony) corals (Madreporia, Zoantharia), 184; form and distribution, 184–5
true crabs (Brachyura), 205, 208; form, habit and distribution, 208–9
tubular eyes (fishes), visual powers, 385–6; with yellow lenses, 388; and lens-pad in scopelarchids, 388–9, 389; and optical fold in evermannellids, 388–9
tuna family (Thunnidae), 24, 43
Tunicata, pelagic, 75, 94–8, 95, 97; luminescence, 346

turbulence and plant productivity, 59
tusk shells (Scaphopoda, Mollusca), 149

undercurrents, 38
upwelling zones, and primary productivity, 64–6
urechinids (irregular sea-urchins, Echinoidea), 195

vampire squid (Vampyroteuthis infernalis, Cephalopoda), structure and light organs, 121
variation, in melamphaid fishes, 499
vertical migrations (daily), 234–52; and: buoyancy, 335–8, community structure, 236; escape from predators, 251; gain in energy, 251; number of sound scattering layers, 249; species packing in water column, 249–50; transfer of organic material, 234; biological advantages of, 235–52; of copepods, 240, 242–4, 243; euphausiids, 236–9, 237; lantern-fishes, 244–5, 247; sergestid prawns, 239–40, 241; influence of light, 235; in relation to niche diversity and competitive exclusion, 236 et seq.
Verticordiidae (septibranch bivalves), mode of feeding, 149

water masses of ocean, 26–31, 28, 29, 479; circulation of, 32–3; in Atlantic Ocean, 28–30, 29; in Pacific and Indian Oceans, 30–1; and distribution of mesopelagic animals, 478–83
water vapour and particles ejected from ocean, 33
whale-fishes (Cetomimiformes), 44, 131–3, 391; small or regressed eyes, 390, 391
wood-boring bivalves (Pholadidae, Mollusca), depth distribution, structure and food, 210–11

Xenophyophoria (Sarcodina, Protozoa), 3, 160; size and pseudopodial system, 291–2; structure and mode of life, 159
xylophagine bivalves, as opportunists, 310

Zoantharia (Cnidaria), 179
Zoanthidea (Zoantharia), epizoic habit, 183–4
zonation, of benthic deep-sea fauna, 473, 492–5; of megafauna, 493–4; of meiofauna, 494–5
zooplankton, 40, 74–105; communities of warm and cold regions, 225; trophic structure in Indian Ocean, 226–7; size range, 42; gelatinous forms, buoyancy mechanisms, 80–1, 319; organization of Cnidarians, 81–3, and tunicates, 81